Alain Bretto
Alain Faisant
François Hennecart

Elements of Graph Theory

From Basic Concepts to Modern Developments

EMS PRESS

Authors:

Alain Bretto
Université de Caen Normandie
GREYC CNRS-UMR 6072
6, boulevard Maréchal Juin
CS 45053
14032 Caen, France
Email: alain.bretto@unicaen.fr

François Hennecart
Université de Lyon –
Université Jean Monnet Saint-Étienne
CNRS, ICJ UMR 5208
23, rue du docteur Paul Michelon
42023 Saint-Étienne, France
Email: francois.hennecart@univ-st-etienne.fr

Alain Faisant
Université de Lyon –
Université Jean Monnet Saint-Étienne
CNRS, ICJ UMR 5208
23, rue du docteur Paul Michelon
42023 Saint-Étienne, France
Email: faisant@univ-st-etienne.fr

This book is a revised English translation of the French original published as
Alain Bretto, Alain Faisant, François Hennecat, *Éléments de théorie des graphes*, deuxième édition revue et augmentée, Collection IRIS, Lavoisier, Paris, 2018

Translated by Leila Schneps, Paris.

2020 Mathematics Subject Classification: Primary 05-01, 05Cxx; Secondary 08Axx, 20-XX, 54-XX

Keywords: graph theory, hypergraph theory

ISBN 978-3-98547-017-4, eISBN 978-3-98547-517-9, DOI 10.4171/ETB/24

Bibliographic information published by the Deutsche Nationalbibliothek
The Deutsche Nationalbibliothek lists this publication in the Deutsche Nationalbibliografie; detailed bibliographic data are available on the Internet at http://dnb.dnb.de.

Published by EMS Press, an imprint of the

European Mathematical Society – EMS – Publishing House GmbH
Institut für Mathematik
Technische Universität Berlin
Straße des 17. Juni 136
10623 Berlin, Germany

https://ems.press

Cover illustration: The planar dodecahedron T_{12}
Typesetting: WisSat Publishing + Consulting GmbH, Fürstenwalde, Germany
Printed in Germany
♾ Printed on acid free paper

To the memory of my father
Alain Bretto

To the memory of my sister
Alain Faisant

To my wife and my three sons
François Hennecart

About the authors

Alain Bretto is a professor of computer science at the University of Caen in Normandy, France. He has been visiting professor in the USA and in Italy. His research is devoted to graph theory and hypergraphs, particularly their algebraic and topological aspects. A renowned expert in these areas, he is also interested in their mathematical applications: image analysis, finance, and information technology.

Alain Faisant was an associate professor of mathematics at the University of Saint-Étienne, France, where he spent the major part of his career, and still pursues research as professor emeritus. His work concentrated mainly on number theory, and he is the author of a popular textbook on the subject: *L'équation diophantienne du second degré* (Second Degree Diophantine Equations), Hermann, 1991.

François Hennecart is a professor of mathematics at the University of Saint-Étienne. His research is devoted to number theory, especially additive and combinatorial problems, two areas in which graph theory plays a particularly important role.

Preface

Avant donc que d'écrire, apprenez à penser.
Selon que notre idée est plus ou moins obscure,
L'expression la suit, ou moins nette, ou plus pure.
Ce que l'on conçoit bien s'énonce clairement,
Et les mots pour le dire arrivent aisément.
[...]
Hâtez-vous lentement, et, sans perdre courage,
Vingt fois sur le métier remettez votre ouvrage
Polissez-le sans cesse et le repolissez ;
Ajoutez quelquefois, et souvent effacez.
[...]
Il faut que chaque chose y soit mise en son lieu ;
Que le début, la fin, répondent au milieu ;
Que d'un art délicat les pièces assorties
N'y forment qu'un seul tout de diverses parties,
Que jamais du sujet le discours s'écartant
N'aille chercher trop loin quelque mot éclatant.

Boileau, L'art poétique, chant I (1674)

Learn then to Think, e'er you pretend to Write,
As your Idea's clear, or else obscure,
Th'expression follows perfect, or impure:
What we conceive, with ease we can express;
Words to the Notions flow with readiness.
[...]
Gently make hast, of Labour not afraid;
A hundred times consider what you've said;
Polish, repolish, every Colour lay,
And sometimes add; but oft'ner take away.
[...]
Each Object must be fix'd in the due place,
And diff'ring parts have Corresponding Grace:
Till by a curious Art dispos'd we find
One perfect whole, of all the pieces joyn'd.
Keep to your Subject close in all you say;
Nor for a sounding Sentence ever stray.

Boileau, The Art of Poetry, Canto I (1674)
Made English by Soames (1680)
Revis'd by Dryden (1683)

Graph theory has been one of the most fertile fields in mathematics in the past fifty years. This undoubtedly comes from the huge field of application of the combinatorial structures it deals with, as well as from the need to discretise many results in mathematics connected to computer science and also to quantum physics.

The mathematical field of graph theory may be said to have begun with an article by the Swiss mathematician Leonhard Euler on the famous Königsberg bridge problem [13]. Nevertheless, although there have been many uses of graphs, already in the Middle Ages [25], it will be necessary to wait until the 20th century before the publication of the first book on the subject, written by Dénes König [23].

The development of graph theory is quite similar to the development of probability theory, in which many results are due to an effort to understand how games of chance work. Graphs were first considered as mathematical curiosities, mere tools to use in the study of logic puzzles, such as the classical problem of whether the knight on a chessboard is able to visit every square exactly once and return to its original position.

However, graphs have now become indispensable in many areas of human endeavour. Following Euler, an increasing number of mathematicians grew interested in graphs. Let us name a few:

- In the 19th century, Edouard Lucas invented the following problem: An even number of girls walk out each day two by two; is there any way to organise their walks in such a way that each of them is paired exactly once with each of the others? The tools used to solve this type of problems are developed in Chapter 8 of this book.

- In 1856, William R. Hamilton studied a problem that appeared just as simple: Find a path that passes exactly once through every vertex of a graph. This problem gave rise to the concept of a Hamiltonian graph (see Chapter 2).
- In the 19th century, Gustav Kirchhoff used implicit graph-theoretical tools to study currents in electrical circuits (see Chapter 4).
- In the 19th century, Arthur Cayley, James J. Sylvester, and then, in the 20th century, George Pólya developed the notion of a tree in order to enumerate chemical compounds (see Chapter 2).
- In 1840, August F. Möbius stated the following problem: A king with five sons wishes for his kingdom to be divided into five provinces at his death, in such a way that each province has at least one common border with each of the others. This problem comes down to deciding whether the complete graph with five vertices K_5 is planar; the concept of planar graphs is introduced in Chapter 5.

One of the main difficulties in graph theory is that the language and notation are not yet fully established; sometimes they come directly from the field in which some aspect of graph theory was first employed.

We have tried throughout the book to unify definitions and results and place them in the most general framework possible. However, it is not always easy to manage using the same terminology for simple graphs, multigraphs, directed graphs (or digraphs) and so on.

In order to provide a self-contained reading, we have endeavoured to give detailed proofs as well as introduce mathematical notions throughout the book. We have paid attention to the algebraic and topological aspect of graph theory, these often being treated concisely in classical textbooks. In addition, we also included a chapter on random graphs as well as another on analysis applied to graphs which allowed us to give applications to some physical theories.

This book aims to take a person with very little knowledge of graphs and bring him or her at the threshold of research. Consequently, it has a wide potential readership, ranging from students of all levels to established researchers.

The original edition of this book appeared in 2012 under the title *Eléments de théorie des graphes*. Its second edition, still written in French, appeared in 2018 under the same title and contains additional developments on the spectral theory of graphs, random graphs and weighted graphs. The present English edition is based on the enlarged second French edition.

We hope that this work will be useful to anyone wishing to study graph theory, both from the mathematical perspective and from a computer science point of view, keeping in mind that many of the chapters deal with graph algorithms.

For further reading, a short list of classical texts on graphs can be found at the end of the book.

Caen, Saint-Étienne, May 2022
Alain Bretto, Alain Faisant, François Hennecart

Contents

Chapter 1

Basic concepts

In this chapter, we give the main definitions of graph theory. Even when given in their most abstract formulation, these remain fortunately very intuitive and thus easy to understand. We also introduce some notions concerning algorithms. Indeed, numerous properties of graphs can be translated into algorithmic terms. Certain properties are not effective (see Section 1.7.2), but others, such as the problem of isomorphism of graphs, play an important role in complexity theory. For this reason, we also introduce the notion of complexity class at the end of the chapter.

1.1 Non-oriented graphs

The notion of a graph is subject to numerous variations, because a graph *encodes* two types of information: the information given by the vertices, and that represented by the edges between two vertices (see the example in Figure 1.1). Consequently, we need to label both the vertices and the edges; note that it does not suffice to label an edge simply by giving its endpoints, since it is possible to have multiple edges between two vertices (see the example in Figure 1.2). We need a specific set to label the edges as well.

Our starting point will be to now give a first, abstract definition of a graph. We will give a second definition at the end of this chapter.

If V is a set, let $P_2(V)$ denote the set of subsets of V with 1 or 2 elements ($\{x\}$ or $\{x, y\}$, which we also denote simply by x or xy).

Convention. We use the notation $[x, y] = [y, x]$ for the set $\{x, y\}$, including the case where $x = y$. We can view $[x, y]$ as two-element *multi-set* (or *unordered list*), i.e. a collection of elements which may contain repeated terms.

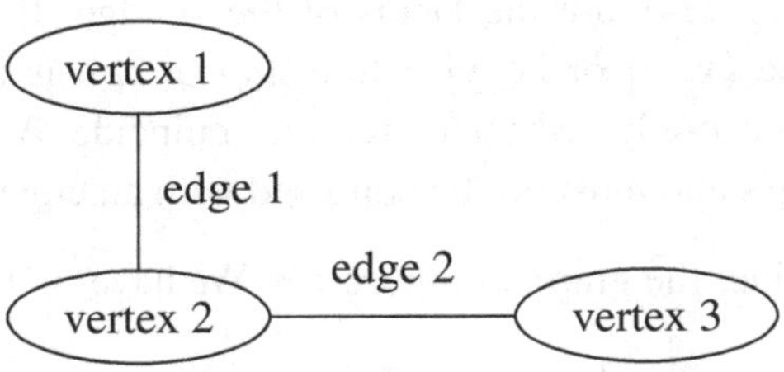

Figure 1.1. Example of graph: an edge is represented by a *line* between two vertices.

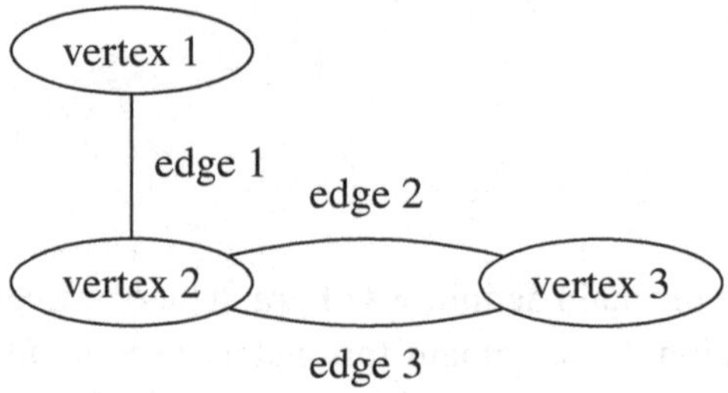

Figure 1.2. Example of graph with a *multiple edge*.

A *graph* is a triple $\Gamma = (V; E, N)$, where

- V is the set of *vertices* of the graph; it will be convenient to use the notation $V(\Gamma)$ to indicate the set of vertices of the graph Γ;
- N is a set which is used to label the edges (for example $N = \{1, 2, \dots, p\}$, $N = \{\text{blue, red, green}, \dots, \text{purple}\}$, $N = \mathbb{N} = \{0, 1, 2, 3, \dots\}, \dots$);
- $E \subset P_2(V) \times N$ is the set of *edges*, which we write as $E = E(\Gamma)$.

An edge $a \in E$ can be written $a = ([x, y], n)$ with $x, y \in V$ and $n \in N$; x and y are the *extremities* of a, and n is its label. We say that a is *incident* to x and y; x and y are said to be *adjacent*; if $x = y$, the edge is called a *loop*.

Restriction: in the graphs that we consider, we will assume that for all $x, y \in V$ the set $\{a \in E \mid x \text{ and } y \text{ incident to } a\}$ is finite. To *represent* a graph in the plane or in space, we generally *materialise* the edges by line segments or curves.

Two edges a and b are said to be *adjacent* if they have (at least) one common extremity.

The *incidence function* $\varepsilon: E \to P_2(V)$ is defined by $\varepsilon(a) = [x, y]$ if $a = ([x, y], n)$.

For fixed x, y in V, the set $\{a \in E \mid \varepsilon(a) = [x, y]\}$, of cardinality $p \geqslant 1$, is called a *p-edge*; it is the set

$$\{([x, y], n_1), ([x, y], n_2), \dots, ([x, y], n_p)\},$$

where $n_i \in N$, $i = 1, \dots, p$, are the labels of the p-edge. If $p = 1$, we speak of a *simple* edge, denoted $\{x, y\}$ or $[x, y]$, while if $p \geqslant 2$, it is a *multi-edge* or *multiple* edge. A p-loop is a p-edge whose extremities coincide. A multi-edge generally consists of several edges and must not be confused with an edge.

Example 1.1.1. Consider the graph in Figure 1.3. We have

$$V = \{x_1, x_2, x_3, \dots, x_6\}, \quad N = \{1, 2, \dots, 9\},$$

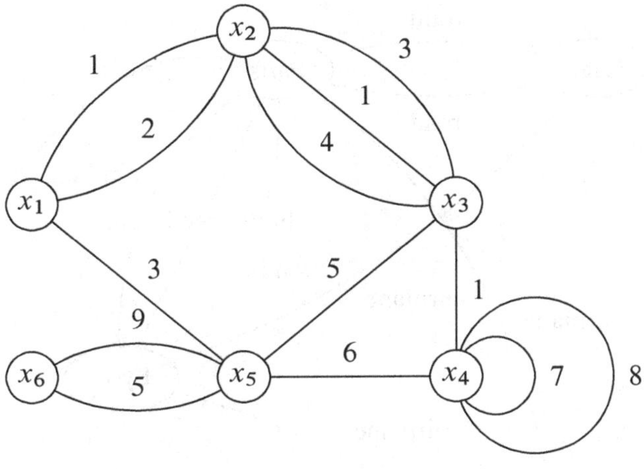

Figure 1.3. Example of a graph containing multi-edges and a 2-loop.

$$\begin{aligned}E = \{&([x_1, x_2], 1), ([x_1, x_2], 2), ([x_1, x_5], 3), ([x_2, x_3], 3), ([x_2, x_3], 1),\\ &([x_2, x_3], 4), ([x_3, x_5], 5), ([x_3, x_4], 1), ([x_4, x_5], 6), ([x_4, x_4], 7),\\ &([x_4, x_4], 8), ([x_5, x_6], 9), ([x_5, x_6], 5)\}.\end{aligned}$$

Observe that labels can be repeated; thus the edges $([x_1, x_2], 1)$ and $([x_2, x_3], 1)$ are distinguished not by their label, but by the fact that $x_3 \neq x_1$. This graph contains a 2-loop (i.e. a 2-edge whose extremities coincide), a 3-edge and two 2-edges; all the others are simple edges.

Example 1.1.2. When they are associated with a representation in the plane or in space, graphs allow us to model concrete information. This is the case of the graph in Figure 1.4, which we use to represent means of transportation between different cities represented by a *concrete* graph. We have $N = \{$airplane, road, high-speed train, train$\}$.

Other examples of graphs:

(1) The set of vertices is the set of airports of the world; the labels of N are the set of airlines: $\{$Air France, American Airlines, ...$\}$. An element $a \in E$ is an airway between two airports taken by a plane belonging to some airline, for example: $a =$ ([Paris, New York], Air France).

(2) The set of vertices is the set of French towns, the labels of N are the set of departmental roads, national roads, highways, ... An element $a \in E$ is any road between two towns, for example

$$a = ([\text{Paris, Lyon}], \text{Highway A6}).$$

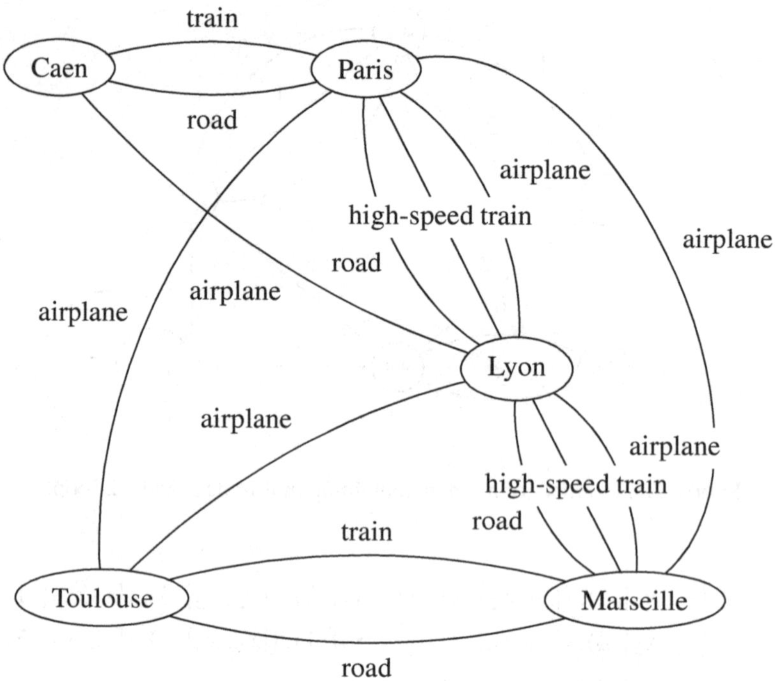

Figure 1.4. A concrete graph.

When the graph Γ has no loop and only simple edges, we say that Γ is a *simple graph* (see Section 1.4). Figure 1.5 shows a simple graph. For simple graphs, an edge having extremities x and y can be written simply as $a = \{x, y\}$.

The *order* of a graph is the number of its vertices; for a finite-order graph, the number of edges is also finite, since by hypothesis the number of edges incident to a given vertex is assumed finite.If $V = \emptyset$, the graph is called the *null* graph. *The graphs studied in this book are not null!*

In numerous problems involving graphs, it is not necessary to give explicit names to the vertices and explicit labels to the edges; the representation of the graph is often limited to a set of points (the vertices), possibly joined by segments or curves (the edges).

Example 1.1.3. The graph in Figure 1.5 represents a propane molecule. The vertices are the carbon or hydrogen atoms, the edges are the covalent bonds between these atoms. The propane molecule C_3H_8 is first represented by a *concrete* graph. In the second graph, the molecule is modelled keeping only the strictly necessary information. This graph is simple, and its order is equal to 11.

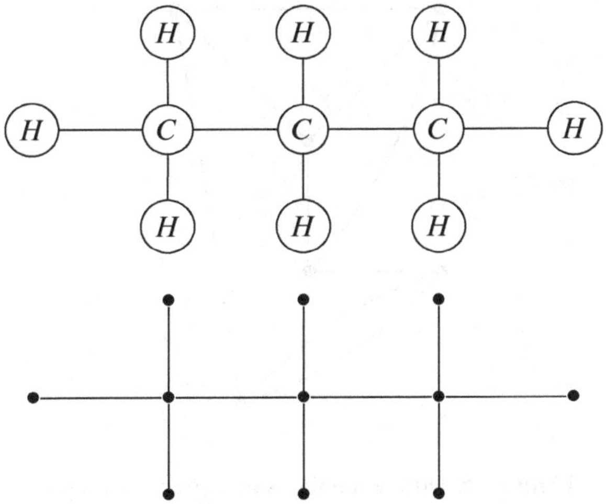

Figure 1.5. Two graphs representing the propane molecule C_3H_8.

1.1.1 Degree

The *degree* of a vertex $x \in V$ of the graph Γ is the number of edges incident to x: *by definition*, a loop incident to x contributes, *twice* in computing the degree of x. The degree of x will be written $d_\Gamma(x)$ or simply $d(x)$; it corresponds to the number of occurrences of the vertex x as an extremity of an edge $a \in E$:

$$d(x) = |\{a \in E \mid \exists y \neq x \text{ such that } \varepsilon(a) = [x, y]\}| \\ + 2|\{a \in E \mid \varepsilon(a) = [x, x]\}|,$$

where $|X|$ denotes the cardinality of X.

When Γ is a simple graph, the degree of a vertex x is equal to the number of vertices adjacent to x.

A vertex of degree 1 is called a *hanging vertex*, and an edge incident to a vertex of degree 1 is called a *hanging edge*.

The *minimal degree* of a graph Γ is $\delta(\Gamma) = \min\{d(y) \mid y \in V\}$, and its *maximal degree* is $\Delta(\Gamma) = \max\{d(y) \mid y \in V\}$. Its *average degree* is defined, if $V \neq \emptyset$, by the number

$$d(\Gamma) = \frac{1}{|V|} \sum_{x \in V} d(x).$$

Thus we have $\delta(\Gamma) \leqslant d(\Gamma) \leqslant \Delta(\Gamma)$.

A graph is *regular* if all its vertices have the same degree, i.e. $\delta(\Gamma) = \Delta(\Gamma) = k$: we then say that the graph is *regular of degree* k or k*-regular*.

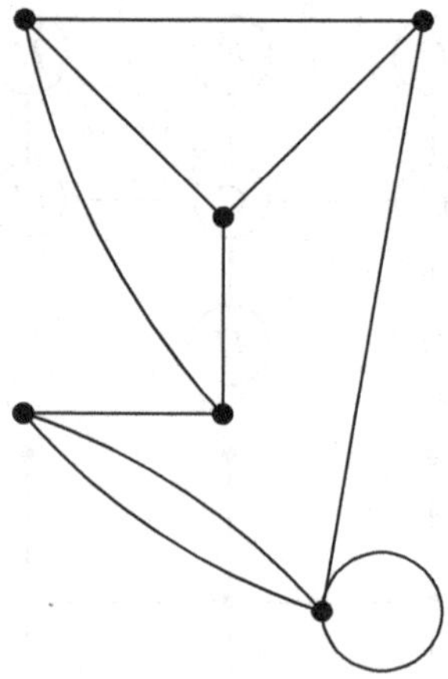

Figure 1.6. Different notions of degree in a graph.

The *distribution of degrees* of Γ is the sequence (with repetition if necessary) of degrees arranged in increasing order.

A vertex is *isolated* if its degree is zero; a p-loop with extremity x is said to be *isolated* if there exists no other edge incident to x.

Let x be a vertex of a graph Γ. We denote by $\Gamma(x)$ the set of all vertices distinct from x and adjacent to x; $\Gamma(x)$ is called the *neighbourhood* of x and its elements are the *neighbours* of x. More generally, if $X \subset V$, the neighbourhood of X is $\Gamma(X) = \bigcup_{x \in X} \Gamma(x) \backslash X$.

Example 1.1.4. The graph in Figure 1.6 is of maximal degree $\Delta(\Gamma) = 5$, of minimal degree $\delta(\Gamma) = 3$, its distribution of degrees is $(3, 3, 3, 3, 3, 5)$ and its average degree is equal to $\frac{10}{3}$.

1.1.2 Chains and cycles

Let $\Gamma = (V; E, N)$ be a graph, and let $x, y \in V$. A *chain* C from x to y is a finite sequence of edges of the type

$$([x, x_1], n_0); ([x_1, x_2], n_1); \ldots; ([x_i, x_{i+1}], n_i); \ldots; ([x_{k-1}, y], n_{k-1}).$$

Intuitively, a chain from x to y is a way of reaching y starting from x by going along various edges of Γ *without any jumps*. The vertices x and y are the extremities of the chain C. To simplify the notation, we simply write

$$C = (x, n_0, x_1, n_1, x_2, \ldots, x_{k-1}, n_{k-1}, y).$$

The number k is the *length* of the chain C.

It is clear that a chain from x to y induces a chain from y to x. As the edges of the graph are not oriented, these two chains are naturally identified. We will say that C is a chain *between* x and y.

Let C_1 be a chain between x and y and C_2 a chain between y and z; one can then imagine a *gluing* operation of these two chains at the vertex y, which creates a new chain C_3 between x and z. This operation is called *concatenation*. Naturally the concatenation of two chains can only exist when the chains have at least one common extremity. In Figure 1.4, we start from Toulouse to travel to Caen; we have a first chain which is just a single edge (Toulouse, airplane, Lyon), and a second chain (Lyon, high-speed train, Paris, train, Caen), which we can concatenate to create a chain from Toulouse to Caen.

A chain is *simple* if it does not contain the same edge twice. It is *elementary* if it does not contain the same vertex twice. Observe that every elementary chain is a fortiori simple:

$$\text{elementary} \underset{\not\Leftarrow}{\Longrightarrow} \text{simple.}$$

A chain is called *closed* if its two extremities coincide. When this is the case, every vertex of the chain can be considered to be its extremity.

A *cycle* is a *simple* closed chain; in fact a cycle has *neither head nor tail*, the numbering is a mere convenience. The length of a cycle is the number of edges it has; a loop is a cycle of length 1. A cycle is called *elementary* if it does not pass through the same vertex twice.

A *chord* in a cycle is an edge linking two non-consecutive vertices in the cycle. A graph without cycles is called *acyclic*.

A graph is said to be *connected* if for all pairs of vertices $x, y \in V$, there exists a chain between x and y: we say that the vertices x and y are *connected*. The preceding definitions are illustrated in Figure 1.7.

Example 1.1.5. In the graph shown in Figure 1.7, we have materialised the edges $([x_i, x_j], n_k)$ by curves; this graph has two connected components, and we have the chain: $(x_1, 1, x_2, 1, x_2, 1, x_4, 2, x_5, 1, x_4, 2, x_5, 2, x_6)$ between x_1 and x_6. But we also have $(x_1, 2, x_4, 1, x_3, 2, x_2, 1, x_4, 2, x_5, 3, x_6)$ which is a simple but non-elementary chain.

On the other hand, the chain $(x_1, 1, x_2, 1, x_4, 1, x_5, 1, x_6)$ is elementary and thus also simple. The chain $(x_7, 1, x_8, 3, x_9, 2, x_8)$ is simple but non-elementary. The degree of the vertex x_2 is equal to 6.

The distribution of degrees is $(2, 3, 3, 3, 4, 5, 5, 5, 6)$ and the average degree is equal to $\frac{36}{9} = 4$.

The length of the shortest chain between x_1 and x_6 is equal to 3 (we will see in the next section that this means that the distance between x_1 and x_6 is equal to 3).

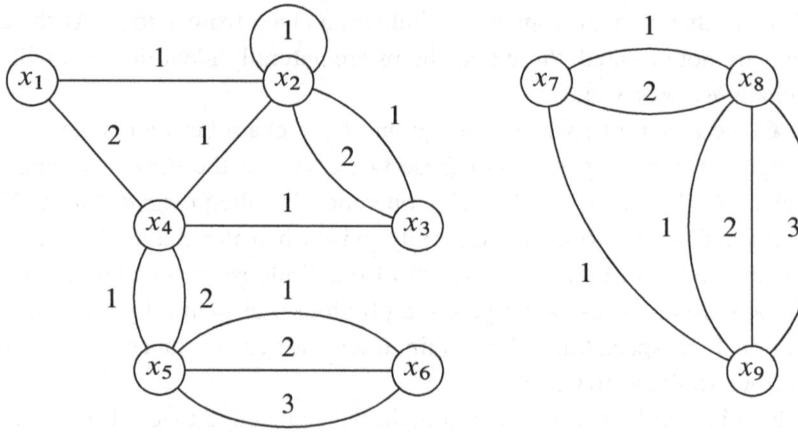

Figure 1.7. An example of a graph having two connected components.

Exercise 1.1. Let Γ be a connected graph. Show that there exists a chain between two distinct vertices if and only if there exists an elementary chain between these two vertices.

Hint: Use induction on the number of vertices of the chain, considering the penultimate vertex.

Remark 1.1.6. The relation defined by xRy if and only if $x = y$ or there exists a chain between x and y is an equivalence relation on V; the equivalence classes modulo R are called *connected components* of the graph. A connected graph thus possesses only one connected component.

1.1.3 Subgraphs

The following notions are illustrated in Figures 1.8 and 1.9.

A graph $\Gamma' = (V'; E', N')$ is called a *subgraph* of Γ if $V' \subset V$, $E' \subset E$ (so necessarily $N' \subset N$). In fact, these conditions imply that $E' \subset E \cap (P_2(V') \times N')$ since Γ' is a graph.

When $V' = V$, we say that Γ' is a *partial graph* of Γ. For example, the graph Γ minus one edge a, denoted $\Gamma \backslash a$, is a partial graph.

When $E' = \{a \in E \mid \varepsilon(a) \in P_2(V')\}$, i.e. E' keeps all the edges of Γ between the vertices in V', we say that Γ' is the *subgraph induced* by V' (or on V'), and denote it by $\Gamma' = \Gamma(V')$. The context will always make it possible to distinguish the notion of a neighbourhood $\Gamma(X)$ from that of an induced graph. For $x \in V$, $\Gamma - x$ denotes the induced subgraph on $V \backslash \{x\}$.

The partial graph generated by E' is the graph $\Gamma(E') = (V; E', N)$.

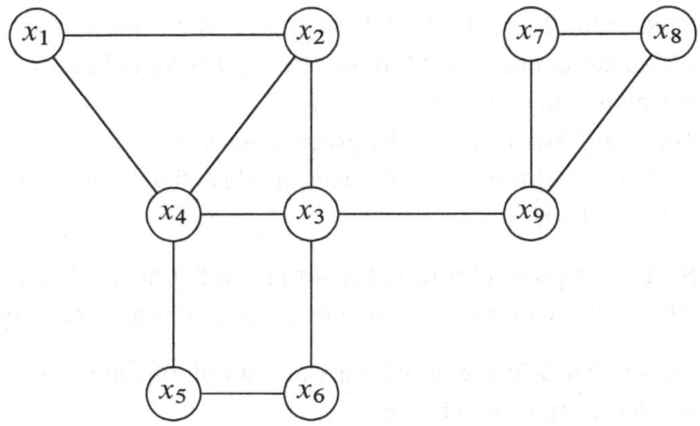

Figure 1.8. Isthmuses and cliques in graphs.

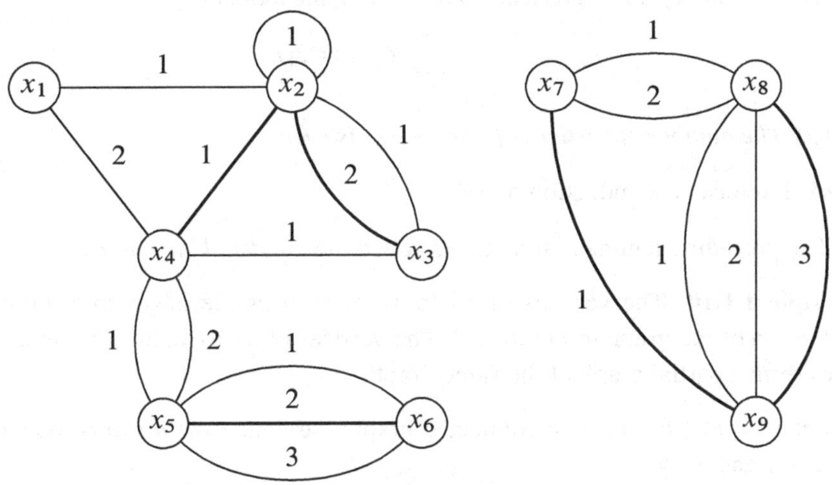

Figure 1.9. Subgraph and partial graph.

Remark 1.1.7. The data of V' and E' does not always suffice to produce a subgraph. For example, if $V = \{x, y\}$, $E = \{\{x, y\}\}$, $N = \{1\}$, $V' = \{x\}$, then the triple $(V'; E, N)$ is not a graph, so it does not correspond a fortiori to any subgraph of $\Gamma = (V; E, N)$.

A simple graph is called *complete* if it has an edge between every pair of vertices. We write K_n for the complete graph of order n.

A *clique* in a graph Γ is a complete subgraph of Γ. The maximal order of a clique in Γ is written $\omega(\Gamma)$.

In a connected graph $\Gamma = (V; E, N)$, an edge a is a *bridge* or an *isthmus* if the graph $\Gamma\backslash a$ is no longer connected. More generally, a is an isthmus of Γ if $\Gamma\backslash a$ has at least one connected component more than Γ.

The *underlying graph* of a chain is the graph whose vertices and edges are those of the chain, but without repetition of the vertices or edges that might be passed through multiple times by the chain.

Example 1.1.8. The graph in Figure 1.8 is simple and connected; the edge $\{x_3, x_9\}$ is an isthmus. The vertices x_2, x_3, x_4 form a *triangle*, that is to say a clique of order 3.

Exercise 1.2. Show that deleting an isthmus in a graph of finite order increases the number of connected components by one.

Lemma 1.1.9. *Let $\Gamma = (V; E, N)$ be a graph of finite order.*

(i) *The sum of the degrees is equal to twice the number of edges:*

$$\sum_{x \in V} d(x) = 2|E|.$$

(ii) *The number of vertices of odd degree is even.*

Proof. Exercise: use induction on $|E|$. ∎

The preceding lemma is sometimes called the *handshaking lemma*.

Example 1.1.10. The vertices of x_2 to x_9 as well as the edges in boldface form a subgraph of the graph in Figure 1.9. The vertices of x_1 to x_9 and the non-boldface edges form a partial graph of the same graph.

Let $\Gamma = (V; E, N)$ be a connected graph; we define the *distance* between two vertices x and y by

- $d(x, y)$ is the smallest chain length between x and y if $x \neq y$;
- $d(x, x) = 0$.

Exercise 1.3. Show that (V, d) is a metric space (see Section 1.10).

It can be useful to generalise this distance to graphs Γ which are not necessarily connected, by setting $d(x, y) = +\infty$ if x and y are not in the same connected component of Γ.

The *diameter* of a graph $\Gamma = (V; E, N)$ is defined by

$$\operatorname{diam}(\Gamma) = \sup\{d(x, y) \mid x, y \in V\} \in \mathbb{N} \cup \{+\infty\}.$$

The *eccentricity* of a vertex x is defined by

$$\operatorname{ecc}(x) = \sup\{d(x, y) \mid y \in V\} \in \mathbb{N} \cup \{+\infty\}.$$

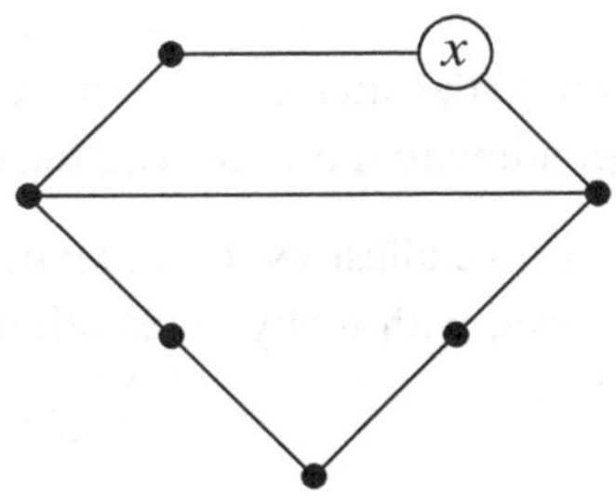

Figure 1.10. Diameter and eccentricity.

A vertex is *central* in Γ if it has minimal eccentricity: $\operatorname{ecc}(x) \leqslant \operatorname{ecc}(y)$ for all $y \in V$.

For a simple graph, we also define the *girth* $\mu(\Gamma)$ as the length of the smallest cycle (setting $\mu(\Gamma) = +\infty$ if the graph has no cycle).

Example 1.1.11. The graph in Figure 1.10 is simple and connected, its diameter is equal to 3; its girth is equal to 4; the eccentricity of the vertex x is equal to 3.

Example 1.1.12. The *Rubik's cube* is a spectacular illustration of the diameter. The problem is represented by a graph whose vertices are the $8!\,3^7 12!\,2^{10}$ ($\approx 4.3 \cdot 10^{19}$) configurations of the *Rubik's cube*; an edge between two configurations x and y means that one can pass from x to y by one *move*. There are 18 possible moves, as follows: there is a fixed cube in the centre of each of the 6 faces (giving the colour of the face), and for each of the 6, there are 3 possible rotations of the face around this central cube, of angles $\pi/2$, π and $3\pi/2$. The graph thus constructed is regular of degree 18.

By sophisticated algorithmic considerations, Rokicki, Kociemba, Davidson and Dethridge have shown (July 2010) that the diameter of this graph is equal to 20. This means that starting from any configuration, one can return to the solved cube position in a maximum of 20 moves! It remains to find a purely mathematical proof of this fact.

Proposition 1.1.13. *Let $\Gamma = (V; E, N)$ be a graph. Then its girth satisfies*

$$\mu(\Gamma) \leqslant 2\operatorname{diam}(\Gamma) + 1.$$

Proof. Let C be a cycle of Γ having length $\mu(\Gamma)$. We use the method of proof by contradiction.

If $\mu(\Gamma) \geqslant 2\operatorname{diam}(\Gamma) + 2$, then C has two vertices x, y whose distance in the subgraph, induced by the vertices of C, is at least equal to $\operatorname{diam}(\Gamma) + 1$. As C is a cycle of minimal length, the distance from x to y in Γ is the same as in C. But then $d(x, y) > \operatorname{diam}(\Gamma)$, which is a contradiction. ∎

Remark 1.1.14. We may note the analogy with the relation $C = \pi D$ connecting the circumference C (corresponding to the girth) and the diameter D of a circle.

The notions of diameter, smallest cycle, eccentricity, etc. are of major importance in the study of networks, with many applications to real-life networks such as telecommunications networks.

1.2 Connected decomposition

Let $\Gamma = (V; E, N)$ be a graph, and let $(V_\alpha)_{\alpha \in L}$ be the family of connected components of Γ (equivalence classes of R, see Section 1.1.2, Remark 1.1.6). We write $\Gamma_\alpha = (V_\alpha; E_\alpha, N_\alpha)$ for the subgraph induced by V_α.

Lemma 1.2.1. *We have:*

(i) *For all $\alpha \in L$, Γ_α is connected.*

(ii) *If $\Gamma' = (V'; E', N')$ is a connected subgraph of Γ, then there exists $\alpha \in L$ such that $V' \subset V_\alpha$.*

(iii) *If $\alpha, \beta \in L$, $\alpha \neq \beta$, then no vertex of V_α is adjacent to a vertex of V_β, so $E = \bigsqcup_{\alpha \in L} E_\alpha$ (where $\bigsqcup$ represents the disjoint union).*

Proof. (i) By definition of connectedness.

(ii) Suppose $V' \neq \emptyset$, and let $x \in V'$ and V_α be the connected components containing x. For all $y \in V'$, there exists a chain between x and y in $\Gamma(V')$ so a fortiori in Γ_α. This implies that $V' \subset V_\alpha$.

(iii) Let $x \in V_\alpha$ and $y \in V_\beta$. There is no chain between x and y, a fortiori no edge. ∎

Corollary 1.2.2. *Let $\Gamma_i = (V_i; E_i, N_i)$, $i = 1, 2$, be two connected subgraphs of a graph Γ. If Γ_1 and Γ_2 intersect, in the sense that $V_1 \cap V_2 \neq \emptyset$, then $\Gamma_1 \cup \Gamma_2 = (V_1 \cup V_2; E_1 \cup E_2, N_1 \cup N_2)$ is connected.*

Proof. Exercise. ∎

1.3 Oriented graphs

An *oriented graph* or *digraph* $\vec{\Gamma}$ is a triple $\vec{\Gamma} = (V; \vec{E}, N)$ defined as follows:

- V is the set of *vertices*; notation $V = V(\vec{\Gamma})$;
- $\vec{E} \subset V \times V \times N$ is the set of *arcs*; notation $\vec{E} = \vec{E}(\vec{\Gamma})$;
- N is a set of labels for the arcs.

An arc $a \in \vec{E}$ will be written as $a = ((x, y), n)$: the arc goes from x to y.

In what follows, we will often write Γ for $\vec{\Gamma}$.

In this context, we make use of two *incidence functions*:

$$i\colon \vec{E} \to V \quad \text{and} \quad t\colon \vec{E} \to V$$

defined for each $a = ((x, y), n))$ by

$$\begin{aligned} i(a) &= x, \quad \text{the } \textit{initial} \text{ vertex of } a, \\ t(a) &= y, \quad \text{the } \textit{terminal} \text{ vertex of } a. \end{aligned}$$

Like for non-oriented graphs, we say that the arc a is *incident* to x and y and that y is *adjacent* to x. A *loop* is an arc a such that $i(a) = t(a)$.

For fixed $x, y \in V$, the set $\{a \in \vec{E} \mid i(a) = x, t(a) = y\}$ of cardinality p is called a *p-arc*; if $p = 1$, we call it a *simple arc*; if $p \geqslant 2$, it is a *multi-arc*.

Let $x \in V$. We define

$$\begin{aligned} d^-(x) &= \operatorname{card}\{a \in \vec{E} \mid t(a) = x\}, \quad \text{the } \textit{input semi-degree} \text{ of } x, \\ d^+(x) &= \operatorname{card}\{a \in \vec{E} \mid i(a) = x\}, \quad \text{the } \textit{output semi-degree} \text{ of } x. \end{aligned}$$

The *degree* of x is defined by $d(x) := d^-(x) + d^+(x)$. If $d(x) = 0$, the vertex x is said to be *isolated*.

If $d^+(x) = 0$ and $d^-(x) > 0$, then x is a *sink*.

If $d^-(x) = 0$ and $d^+(x) > 0$, then x is a *source*.

If there exists $k \in \mathbb{N}$ such that for any vertex x, we have $d^+(x) = k$, the digraph is said to be *output semi-regular*. If we have the analogous property for the input degrees, the digraph is *input semi-regular*. The digraph is called *regular* if it is both input and output semi-regular.

Lemma 1.3.1. *Let $\vec{\Gamma} = (V; \vec{E}, N)$ be a digraph of finite order. Then*

$$\sum_{x \in V} d^+(x) = \sum_{x \in V} d^-(x) = |\vec{E}|.$$

Proof. Use induction on the number of arcs. ∎

A *path* C from x to y is a finite sequence of the form

$$C = (x, a_0, x_1, a_1, x_2, \dots, x_{k-1}, a_{k-1}, y),$$

where $k \geqslant 1$, $i(a_j) = x_j$ and $t(a_j) = x_{j+1}$, $0 \leqslant j \leqslant k-1$ (fixing $x_0 = x$ and $x_k = y$); the integer k is the *length of the path*. Like for non-oriented graphs, we can also define a path using the labels of the arcs:

$$C = (x, n_0, x_1, n_1, x_2, \dots, x_{k-1}, n_{k-1}, y),$$

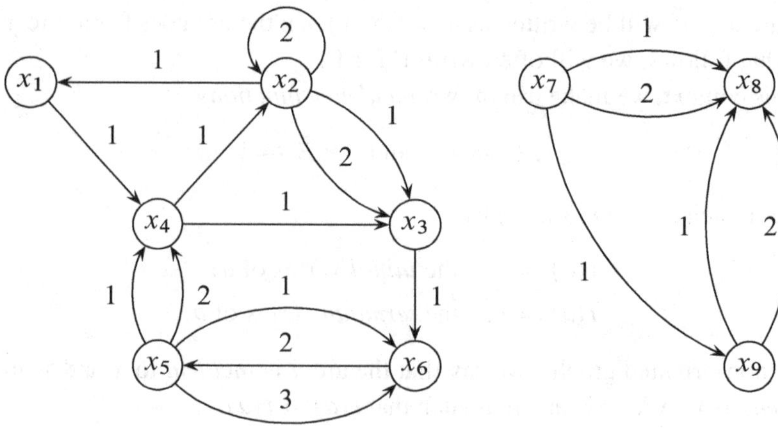

Figure 1.11. Example of a (non-connected) oriented graph.

where n_i is such that $((x_i, x_{i+1}), n_i) \in \vec{E}$. If there exists a path from x to y, we will say that y is *accessible* starting from x.

A path is *simple* if it does not contain the same arc twice; it is *elementary* if it does not contain the same vertex twice.

A *circuit* is a simple path whose extremities coincide; a digraph is said to be *circuit-free* if it does not contain any circuit.

A digraph is *strongly connected* if for all couples of distinct vertices x, y, there exists a path from x to y *and* a path from y to x. We note that a digraph containing a source or a sink cannot be strongly connected.

Example 1.3.2. In the graph shown in Figure 1.11, there exists a 2-arc between x_2 and x_3; the vertex x_7 is a source and the vertex x_8 is a sink. The left-hand part of this digraph is a strongly connected component, but the other part is not. The path $(x_1, 1, x_4, 1, x_2, 1, x_1)$ is a circuit. The input semi-degree of x_2 is equal to 2 and its output semi-degree to 4.

Exercise 1.4. Show that in a digraph $\vec{\Gamma} = (V; \vec{E}, N)$, the relation $\ll$ defined by

$$x \ll y \iff x = y \text{ or there exists a path from } x \text{ to } y$$

is a preorder relation on V (the reader will find the definition of a preorder in Section 1.10).

The equivalence classes of the binary relation R associated to the preorder $\ll$:

$$xRy \iff x \ll y \text{ and } y \ll x,$$

are the *strongly connected components* of the oriented graph $\vec{\Gamma}$.

Here is a little vocabulary table comparing terms for oriented and non-oriented graphs:

Non-oriented	Oriented
Edge	Arc
Chain	Path
Cycle	Circuit
Connected	Strongly connected

The *underlying graph* of a digraph is the graph obtained by forgetting the orientation of the arcs: each arc $((x, y), n)$ is replaced by the corresponding edge $([x, y], n)$. In particular, if $x = y$, it yields a loop in the graph.

Conversely, we can attribute an *orientation* to every non-oriented graph by choosing an orientation on each edge, i.e. by replacing the edge $([x, y], n)$ by either the arc $((x, y), n)$ or the arc $((y, x), n)$.

A connected graph $\Gamma = (V; E, N)$ is an *orientable graph* if there exists an orientation of its edges such that the resulting oriented graph is strongly connected.

Theorem 1.3.3. *A connected graph of finite order Γ is orientable if and only if each edge is contained in at least one cycle.*

Proof. The condition is clearly necessary.

Conversely, we can assume that the graph is *loopless* (i.e. has no loops) or multi-edges. We write Γ' for a maximal orientable subgraph of Γ. We can orient the edges of Γ' in such a way that it is strongly connected. If all the edges of Γ are contained in Γ', then $\Gamma = \Gamma'$ is orientable. Otherwise, as the graph is connected, we can find an edge a_0 incident to a vertex x_0 of Γ'; by hypothesis, this edge is part of a cycle in Γ, and thus there exists a chain $C = (x_0, a_0, x_1, a_1, x_2, \dots, a_{k-1}, x_k)$ starting from Γ' at x_0 and returning for the first time to Γ' at x_k. We orient this chain *linearly* (there are two ways to do this) by defining the arcs $(x_0, x_1), \dots, (x_{k-1}, x_k)$. It follows easily that $\Gamma' \cup C$ is strongly connected, which contradicts the maximality of Γ'. ■

Remark 1.3.4. Note that the presence of an isthmus is incompatible with orientability.

In analogy with graphs, we can define a *sort* of distance $d(x, y)$ on a digraph, by considering the paths of shortest length between each couple (x, y) of vertices. But the symmetry property usually attached to a distance $d(x, y) = d(y, x)$ does not necessarily hold.

Let $\Gamma = (V; E, N)$ be a graph (resp. $\vec{\Gamma} = (V; \vec{E}, N)$ a digraph). A *valuation* or *weight* on the edges (resp. on the arcs) is a map $\omega: E$ (resp. $\vec{E}) \to X$ from the set of edges (resp. the set of arcs) to another set X (for example $X = \mathbb{R}$). An oriented graph equipped with a valuation will give rise to the notion of a network (see Chapter 4).

1.4 Simple graphs

A graph (resp. a digraph) is *simple* if it contains no loops or multi-edges (resp. no multi-arcs). For a simple graph, the set N is not useful; we just write $\Gamma = (V; E)$, where $E \subset P_2(V)$, if it is non-oriented, and $\vec{\Gamma} = (V; \vec{E})$, where $\vec{E} \subset V \times V$, if it is oriented. We also note that to describe a chain (resp. a path), it suffices to give the list (resp. the ordered list) of the vertices it passes through.

Exercise 1.5. (i) Let $\Gamma = (V; E)$ be a simple graph; show that the incidence function $\varepsilon: E \to P_2(V)$ is injective.

(ii) Let $\vec{\Gamma} = (V; \vec{E})$ be a simple oriented graph; show that $(i, t): \vec{E} \to V \times V$ is injective.

It is easy to check that a simple graph of order n has at most $\frac{n(n-1)}{2} = \binom{n}{2}$ edges (where for $0 \leqslant k \leqslant n$, $\binom{n}{k} = \frac{n!}{(n-k)!k!}$ denotes the binomial coefficient), and that a simple digraph of order n has at most $n(n-1)$ arcs.

Exercise 1.6. (i) Let $\Gamma = (V; E)$ be a simple graph of finite order; show that Γ is connected if and only if for all partitions of the set of vertices into two disjoint subsets, $V = V_1 \cup V_2$, $V_1 \cap V_2 = \emptyset$, there exists an edge a such that $\varepsilon(a) \cap V_1 \neq \emptyset$ and $\varepsilon(a) \cap V_2 \neq \emptyset$.

(ii) State and prove an analogous result for oriented graphs.

For a simple (non-oriented) graph $\Gamma = (V; E)$, the *complement* of the graph Γ is the graph $\overline{\Gamma} = (\overline{V}, \overline{E})$ where $\overline{V} = V$ and $\overline{E} = P_2(V) \backslash E$.

Exercise 1.7. Let $\Gamma = (V; E)$ be a simple graph. Prove that at least one of the two graphs Γ and $\overline{\Gamma}$ is connected. Construct a connected graph Γ such that $\overline{\Gamma}$ is also connected.

Example 1.4.1. Figure 1.12 shows the representation of a graph and its complement.

Proposition 1.4.2. *Let $\Gamma = (V; E)$ (resp. $\vec{\Gamma} = (V; \vec{E})$) be a simple graph (resp. digraph) of order n. If Γ (resp. $\vec{\Gamma}$) is connected (resp. strongly connected), then it has at least $n - 1$ edges (resp. arcs). This result is a fortiori true if Γ (resp. $\vec{\Gamma}$) is not simple.*

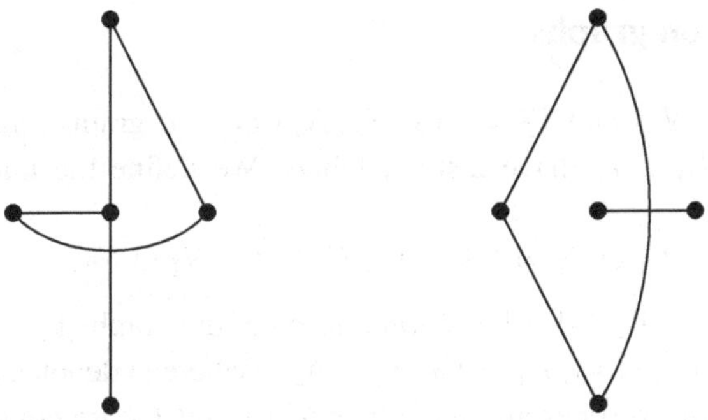

Figure 1.12. A graph and its complement.

Proof. We first consider the case of non-oriented graphs and use induction on n. If $n = 1$ or $n = 2$, the result is trivial.

Suppose that the assertion holds up to $n - 1$, and let Γ be a simple connected graph of order n and Γ' the induced subgraph obtained by deleting an arbitrary vertex x. Consider the connected components $\Gamma'_1, \dots, \Gamma'_k$ of Γ'. For each $i = 1, \dots, k$, the subgraph Γ'_i admits n_i vertices, and we have $n_1 + n_2 + \cdots + n_k = n - 1$.

By the induction hypothesis, Γ'_i has at least $n_i - 1$ edges.

Since Γ is connected, the vertex x is necessarily connected to all the connected components Γ'_i. Consequently, Γ contains at least

$$n_1 - 1 + n_2 - 1 + \cdots + n_k - 1 + k = n - 1 \text{ edges.}$$

The same proof works for digraphs. ■

Proposition 1.4.3. *Every acyclic graph $\Gamma = (V; E)$ of order n admits at most $n - 1$ edges.*

Proof. We use induction on n. If $n = 1$, we have $E = \emptyset$. If $n = 2$, there is at most one edge. Suppose the result holds up to $n - 1$, and let $\Gamma = (V; E)$ be an acyclic graph of order n. Fix an edge $a = \{x, y\}$ and consider the partial graph $\Gamma' = (V; F)$ where $F = E \backslash \{a\}$; Γ' is not connected (since otherwise we would have a chain between x and y in Γ', which would naturally yield a cycle in Γ by closing the chain with the edge a). Thus, Γ' decomposes into $k \geqslant 2$ non-empty connected subgraphs $\Gamma'_i = (V_i, F_i)$, $1 \leqslant i \leqslant k$. By the induction hypothesis, since $|V_i| \leqslant n - 1$, we have $|F_i| \leqslant |V_i| - 1$, $1 \leqslant i \leqslant k$, $(k \geqslant 2)$; hence

$$|E| = 1 + \sum_{i=1}^{k} |F_i| \leqslant 1 + \sum_{i=1}^{k} (|V_i| - 1) = 1 + n - k \leqslant n - 1.$$ ■

1.5 Operations on graphs

Let $\Gamma_1 = (V_1; E_1, N_1)$ and $\Gamma_2 = (V_2; E_2, N_2)$ be two graphs, and assume that 2 distinct edges of $E_1 \cup E_2$ have distinct labels. We define the *union* of these two graphs by

$$\Gamma_1 \cup \Gamma_2 = (V_1 \cup V_2; E_1 \cup E_2, N_1 \cup N_2).$$

We will denote by $\Gamma_1 + \Gamma_2$ the *disjoint union* of the graphs $\Gamma_i, i = 1, 2$. It corresponds to the graph $(V_1 \sqcup V_2; E_1 \sqcup E_2, N_1 \sqcup N_2)$ (where $\sqcup$ denotes a disjoint union), which means that the vertices and the edges of Γ_1 and Γ_2 are considered to be distinct. For example, when we consider all the connected components $C_i, 1 \leqslant i \leqslant k$, of a graph Γ having a finite number of connected components, then we naturally have $\Gamma = C_1 + C_2 + \cdots + C_k$.

We define the *intersection* of two graphs Γ_1 and Γ_2 by

$$\Gamma_1 \cap \Gamma_2 = (V_1 \cap V_2; E_1 \cap E_2, N_1 \cap N_2),$$

and the *difference* by

$$\Gamma_1 \backslash \Gamma_2 = (V_1 \backslash V_2; E_1 \backslash E_2, N_1 \backslash N_2).$$

We can easily extend these concepts to digraphs.

1.6 Representations of graphs

Let $\Gamma = (V; E, N)$ be a graph (resp. $\vec{\Gamma} = (V; \vec{E}, N)$ a digraph) of finite order, with $V = \{x_1, x_2, x_3, \ldots, x_n\}$.

Edge list (resp. arc list). We can describe Γ (resp. $\vec{\Gamma}$) by its number of vertices n, and its edges (resp. arcs), as follows: for each couple of vertices between which there is at least one edge (resp. at least one arc), we list the labels of these edges (resp. these arcs). A p-edge (resp. a p-arc) between x_i and x_j will be written

$$(x_i, x_j; n_{i_1}, n_{i_2}, \ldots, n_{i_p})$$

if n_{i_j} is the label of the edge (resp. arc) a_{i_j} between x_i and x_j (resp. from x_i to x_j). We form the set A_{x_i} of all the multi-edges (resp. all the multi-arcs) emerging from x_i and written as above. In this way, we obtain the list $A_{x_1}; A_{x_2}; A_{x_3}; \ldots; A_{x_n}$ (see Example 1.6.1 and Figure 1.13).

Multiplicity list. We can also characterise Γ (resp. $\vec{\Gamma}$) on the one hand by its number of vertices n and on the other hand by a list of sets $A_{x_1}; A_{x_2}; A_{x_3}; \ldots; A_{x_n}$ of couples of type (x_j, p), where the first coordinate of a couple in A_{x_i} is a vertex x_j adjacent to x_i, and the second coordinate is the multiplicity p of the multi-edge (resp. of the multi-arc) having extremities x_i and x_j.

We note that via these two representations, certain edges (resp. certain arcs) are repeated: this can be useful for some applications. We can, however, easily remove the redundancy.

Adjacency matrix. Let $\Gamma = (V; E, N)$ be a graph (resp. $\vec{\Gamma} = (V; \vec{E}, N)$ a digraph) of finite order, with $V = \{x_1, x_2, x_3, \dots, x_n\}$. We can represent this (di)graph by a square matrix of order n, $A = (\alpha_{i,j})$, with coefficients in $\mathbb{N} := \{0, 1, 2, 3, \dots\}$, where $\alpha_{i,j}$ is the number of edges (or arcs) between x_i and x_j. This matrix is called the *adjacency matrix* of the graph Γ (resp. of the digraph $\vec{\Gamma}$).

Incidence matrix. Let $\Gamma = (V; E, N)$ be a graph of finite order. If $|V| = n$, with $V = \{x_1, x_2, \dots, x_n\}$, and $|E| = m$, with $E = \{a_1, \dots, a_m\}$, the *incidence matrix* of Γ is the $n \times m$ matrix $M = (\alpha_{i,j})$ with coefficients in $\{0, 1\}$, defined by $\alpha_{i,j} = 1$ or 0 according to whether the vertex x_i is or is not incident to the edge a_j. In the case of a digraph $\vec{\Gamma} = (V; \vec{E}, N)$, the incidence matrix is the $n \times m$ matrix $M = (\alpha_{i,j})$ with coefficients in $\{-1, 0, 1\}$ defined by $\alpha_{i,j} = -1$ if x_i is the initial vertex of the arc a_j, $\alpha_{i,j} = 1$ if x_i is the terminal vertex of the arc a_j, and $\alpha_{i,j} = 0$ if x_i is not incident to a_j.

Example 1.6.1. Consider the graph Γ shown in Figure 1.13.

Representation of Γ by its edge list:

- the number of vertices is 6;
- $A_{x_1} = \{(x_1, x_2; n_1, n_2); (x_1, x_4; n_7); (x_1, x_5; n_6)\}$;
- $A_{x_2} = \{(x_2, x_3; n_3, n_4, n_5); (x_2, x_1; n_1, n_2)\}$;
- $A_{x_3} = \{(x_3, x_2; n_3, n_4, n_5); (x_3, x_5; n_9); (x_3, x_6; n_{11})\}$;
- $A_{x_4} = \{(x_4, x_1; n_7); (x_4, x_5; n_8)\}$;
- $A_{x_5} = \{(x_5, x_1; n_6); (x_5, x_3; n_9); (x_5, x_4; n_8); (x_5, x_6; n_{10})\}$;
- $A_{x_6} = \{(x_6, x_3; n_{11}); (x_6, x_5; n_{10})\}$.

Representation of Γ by its multiplicity list:

- the number of vertices is 6;
- $A_{x_1} = \{(x_2, 2); (x_4, 1); (x_5, 1)\}$;
- $A_{x_2} = \{(x_1, 2); (x_3, 3)\}$;
- $A_{x_3} = \{(x_2, 3); (x_5, 1); (x_6, 1)\}$;
- $A_{x_4} = \{(x_1, 1); (x_5, 1)\}$;
- $A_{x_5} = \{(x_1, 1); (x_3, 1); (x_4, 1); (x_6, 1)\}$;
- $A_{x_6} = \{(x_3, 1); (x_5, 1)\}$.

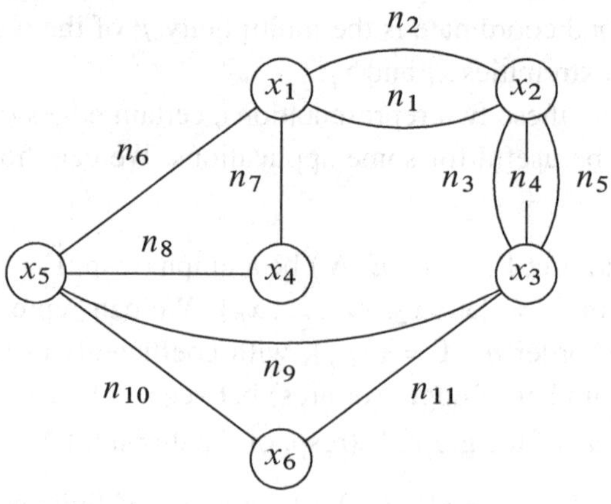

Figure 1.13. Representation of a graph by its edge list and by its multiplicity list.

Exercise 1.8. Give the adjacency matrix of the graph in Figure 1.13.

1.7 Algorithms and complexity theory

As we will see throughout this book, graphs are particularly useful in modelling concrete problems. Of course, modelling a problem is good, but solving it is even better! For this, we will need to use a special tool: algorithms. The goal of this section is to recall some basic notions of algorithms as well as the underlying mathematics.

Informally, an *algorithm* is a finite sequence of instructions. It takes data, called the *input*, and produces new data, called the *output*. An algorithm must terminate no matter what the input, and yield a correct solution. For example, an algorithm to build a building would be:

- create the foundations;
- construct the walls;
- install the windows;
- put on the roof;
- put in the electricity;
- install the plumbing;
- ...

We assume that the reader knows about the *basic building blocks* of algorithmics (assignations, declarations, conditional loops, etc.). We will now discuss the performance of algorithms.

1.7.1 Time complexity of an algorithm

The *time complexity of an algorithm* corresponds to the calculation time needed by the algorithm to solve a problem. This is a criterion of the efficiency of the algorithm, and also a way of comparing two solutions to the same problem. The cost in calculation time depends on the nature and number of the elementary operations that the algorithm performs. The notion of an *elementary operation* contains both basic operations (addition, subtraction, etc.), and other operations such as assignation, comparison, or other similar *simple* functions (or functions considered to be simple). We assume that the realisation of an elementary operation necessitates an amount of time proportional to a constant. Thus, to evaluate the complexity of an algorithm, we must evaluate the number of elementary operations it performs.

The *spatial complexity of an algorithm* is the quantity of memory it uses throughout the entire execution of the algorithm. We note that spatial complexity is always less than or equal to time complexity. Indeed, an operation of writing in memory can be considered as taking a constant time c, so writing an object of size L in memory takes an amount of time equal to $L \times c$. Furthermore, time and spatial complexity are correlated: to minimise the calculation time, one needs to use more memory. The cost of storing information, called *memory*, has an overall tendency to decrease with technological progress, so that it is legitimate to concentrate our attention mainly on time complexity. This depends mainly on the size of the data, for example the number of vertices and edges of a graph. It can also depend on the length of the coding of the data: thus for example when the datum is an integer, the cost will depend on the length of the coding of that integer in bits.

There exist several types of complexity:

- worst case complexity;
- average complexity;
- best case complexity.

Let $\mathcal{A}$ be an algorithm. For each input e, let $|e|$ denote its size and $T_{\mathcal{A}}(e)$ the calculation time of the result yielded by the algorithm starting with the input e. The *worst case calculation time* is given by

$$T_{\sup,\mathcal{A}}(n) = \sup_e\{T_{\mathcal{A}}(e) : |e| = n\}.$$

Conversely, the *best case calculation time* is

$$T_{\inf,\mathcal{A}}(n) = \inf_e\{T_{\mathcal{A}}(e) : |e| = n\}.$$

The average calculation time depends on a probability distribution p on the input set of data e of size n.

Let $\{p(n)\}$ be a probability distribution on the universe of input data of size n. We define the *average calculation time* of the algorithm $\mathcal{A}$ by

$$T_{\text{moy},\mathcal{A}}(n) = \sum_{e \mid |e|=n} p(e)T_{\mathcal{A}}(e).$$

In practice, the probability distribution is usually very difficult to estimate, so that one needs to resort to the hypothesis that the data are equiprobable, i.e. that the probability distribution is uniform. In this case, we have $p(e) = \frac{1}{N(n)}$ where $N(n)$ is the number of inputs of size n, and thus

$$T_{\text{moy},\mathcal{A}}(n) = \sum_{e \mid |e|=n} \frac{1}{N(n)} T_{\mathcal{A}}(e).$$

This quantity is the expectation of the random variable *calculation time*.

The simplest and most meaningful version of calculation time is the worst case time. In studying complexity, it is generally the only one that is used.

In short, the complexity of an algorithm is measured as a function of the size n of the input data: this function will be compared to *simple* functions $f(n)$ such as $\log n$, n^k, e^{kn} and all combinations of these, and we will simply write the complexity as $O(f(n))$. The complexity is *linear* if $f(n) = kn$, *polynomial* if $f(n) = n^k$, *exponential* if $f(n) = e^{kn}$. The choice of how to measure the size n of the data used to measure the complexity is important: for an integer, it is its length, and for a graph it can be its order or the number of its edges; for a rectangular matrix it is both the size of the matrix and that of its coefficients, etc. An algorithm is said to be linear, polynomial or exponential, referring to its complexity.

Iterative processes. These are algorithms which proceed sequentially with no line of the program ever being read twice (for example Algorithm 1.1).

What is the cost of this algorithm? It first performs an assignation, then the body is a loop which consists in an assignation followed by an elementary operation. Since the algorithm passes through the loop $n-1$ times, we obtain

$$T(n) = c + (n-1) \times (c+o),$$

where c is the constant cost of a single assignation, and o is the constant cost of one elementary operation.

We can now give a more precise definition of the time complexity of an algorithm $\mathcal{A}$: it is a function $n \mapsto f(n)$ where $f(n)$ is the maximal number of calculation steps (to each instruction or elementary operation we associate a cost called a *calculation step*) that $\mathcal{A}$ must use to solve a problem having an input of size n.

```
Function Factorial(n: integer):
    begin
        Fact := 1
        for i from 2 to n do
            Fact := Fact × i
        end
        return Fact
    end
EndFunction
```

Algorithm 1.1. Iterative factorial function.

In order to correctly formulate complexity, we need the following definition. Let $f: \mathbb{N} \to \mathbb{R}$, $g: \mathbb{N} \to \mathbb{R}^+$ be two functions: the function f is said to be *bounded above* by g if

$$\exists k > 0,\ n_0 \in \mathbb{N} \text{ such that if } n > n_0, \text{ then } |f(n)| \leqslant kg(n).$$

This means that for all sufficiently large n, the value $|f(n)|$ is always smaller than or equal to a constant times the value $g(n)$. In this case we say that the function f belongs to the class $O(g)$. By an abuse of language, we write $f = O(g)$, when in fact we should really write $f \in O(g)$. For example, we have $n^3 + 4n^2 + 13n + 2017 = O(n^3)$.

We can now give a precise formulation of average complexity. The function f is said to be *bounded above and below* by g if

$$\exists k_1, k_2 > 0,\ n_0 \in \mathbb{N},\ \forall n > n_0, \quad k_1 g(n) \leqslant |f(n)| \leqslant k_2 g(n).$$

We write $f(n) \in \Theta(g(n))$ or, again by an abuse of language, $f(n) = \Theta(g(n))$.

This means that asymptotically, $|f|$ is equal to g up to a factor.

Exercise 1.9. Let $c > 0$. Show that

- $g = O(g)$ and $c\,O(g) = O(g)$;
- $O(g) + O(g) = O(g)$ and $O(g_1) + O(g_2) = O(\max(g_1, g_2))$;
- $O(O(g)) = O(g)$;
- if $f(n) = \Theta(c \cdot g(n))$, then $f(n) = \Theta(g(n))$;
- if $f, g: \mathbb{N} \to \mathbb{R}^+$, then the relation $f(n) = \Theta(g(n))$ is reflexive and symmetric.

To compute the complexity of an algorithm, we must evaluate the complexity of each instruction. In most cases, the complexity of assignations, comparisons and elementary operations is constant. For blocks of instructions, we compute the maximal number of passages in the body of the block.

For Algorithm 1.1, we obtain a complexity of $O(c) + O((n-1) \times (c+o)) = O(\max(c, (n-1) \times (c+o)) = O(n)$.

Another example is the Fibonacci sequence, which is the sequence of integers $(u_n)_{n \in \mathbb{N}}$ defined recursively by

$$\begin{cases} u_0 = 0, \\ u_1 = 1, \\ u_n = u_{n-1} + u_{n-2} \quad \forall n \geqslant 2. \end{cases}$$

Input: Integer n;
Output: *Fib*[n];
begin
  Declare a table *Fib* with $n+1$ entries;
  Fib[0] := 0;
  Fib[1] := 1;
  for i *from* 2 *to* $n-1$ **do**
    Fib[i] := *Fib*[$i-1$] + *Fib*[$i-2$];
  end
  return *Fib*[n]
end

Algorithm 1.2. Iterative Fibonacci procedure.

The two first lines are assignations, which are performed in constant time: the complexity of this algorithm is thus reduced to that of the loop which is repeated $n-3$ times, so it is $O(n)$.

Another example is the bubble sorting algorithm (cf. Algorithm 1.3).

Computing the complexity of this algorithm is a bit more complicated. We pass $n-1$ times through the first loop, and for each passage in this first loop we pass $i-1$ times in the second, so we have a total of $\sum_1^{n-1} i$ passages. In the worst case, we will have a permutation at every passage, so the complexity is given by

$$c_{\text{perm}} \times \left(\sum_{i=1}^{n-1} i\right) = c_{\text{perm}} \times \frac{n(n-1)}{2}$$

(where c_{perm} is the cost of a permutation), which has complexity $O(n^2)$.

We will deal with the complexity of recursive algorithms in Chapter 3.

```
Input: Table T; Integer n;
Output: T sorted;
begin
    for i from n to 2 do
        for j from 1 to i − 1 do
            if T[j] > T[j + 1] then
                permute T[j] and T[j + 1] in the table T;
            end
        end
    end
    return T sorted;
end
```

Algorithm 1.3. Bubble sorting function.

1.7.2 Complexity classes

Complexity theory is without a doubt one of the most important parts of theoretical computer science. This theory is strongly related to the notion of *effective mathematics*: informally, we say that a mathematical problem is effective if it can be solved by an algorithm. For example, calculating the roots of second-degree equation or computing the eigenvalues of a matrix are effective problems.

We say that an algorithm is *efficient* for solving a problem if the number of calculation steps needed to give a solution to the problem is a function (of the size of the data) which grows at most like a polynomial function. In this case, we say that the problem lies in the class $\mathcal{P}$ (for "polynomial"). There are many problems for which we simply do not know if they lie in this class, so we need to define other classes.

A decision problem (i.e. a problem with yes/no solution) lies in the class $\mathcal{NP}$ (for "non-deterministic polynomial") if one can check a solution of this problem in polynomial time. We will see some examples of this below. A question springs immediately to mind: if one can check a solution in polynomial time, is the solution itself then polynomial? In other words, do we always have $\mathcal{P} = \mathcal{NP}$?

This question, which seems quite simple, is one of the main challenges for mathematicians - and for computer scientists worthy of the name! It is the first of the eight problems on the famous list by Clay entitled *Problems for the 21st century*.

Clearly we have $\mathcal{P} \subseteq \mathcal{NP}$. It is the other inclusion which is difficult. Let P_1 and P_2 be two problems. We say that P_1 is *polynomially reducible* to P_2 if there exists an algorithm to solve P_2 which can be transformed in polynomial time into an algorithm to solve P_1.

We also define another important complexity class: the class called $\mathcal{NP}$-complete, or $\mathcal{NPC}$. It is the subset of $\mathcal{NP}$ defined as follows: a decision problem P lies in the class $\mathcal{NPC}$ if $P \in \mathcal{NP}$ and if every problem in the class $\mathcal{NP}$ is polynomially reducible to P. Naturally, if one could give a solution in polynomial time to any $\mathcal{NPC}$ problem, then we would automatically conclude that $\mathcal{P} = \mathcal{NP}$.

1.8 Definition of a graph starting from the incidence function

The incidence function ε allows us to give an equivalent definition of the notion of graph, as a triple $\Gamma = (V; A, \varepsilon)$ where $\varepsilon: A \to P_2(V)$.

Indeed, starting from a graph $\Gamma = (V; E, N)$, we set $A = E$ and take ε as in Section 1.1.

Conversely, let $\Gamma = (V; A, \varepsilon)$. For fixed $x, y \in V$, set $T_{x,y} = \{a \in A \mid \varepsilon(a) = [x, y]\}$, which is a finite set. For all x, y such that $T_{x,y} \neq \emptyset$, a p-edge between x and y is defined by

$$E_{x,y} = \{([x, y], 1), ([x, y], 2), \ldots, ([x, y], p)\}$$

with $p = |T_{x,y}|$, the cardinality of $T_{x,y}$.

In this case $N_{x,y} = \{1, 2, 3, \ldots, |T_{x,y}|\}$, and thus

$$N = \bigcup_{x,y \in V \mid T_{x,y} \neq \emptyset} N_{x,y} \text{ and } E = \bigsqcup_{x,y \in V \mid T_{x,y} \neq \emptyset} E_{x,y}.$$

It follows that $\Gamma = (V; E, N)$ is a graph in the sense of Section 1.1.

Why did we introduce the *auxiliary* set A? The description of the graph Γ by the triple $(V; A, \varepsilon)$ with $A = E$ has a big advantage when A is the concrete representation of actual objects: it is then a concrete realisation of the graph. We must not confuse an edge that is a triple with the object that it represents. For example, the set of vertices is a set of points in the plane, the labels (elements of N) are non-negative integers. An element $a \in A$ is a curve between two points in the plane that we label by a non-negative integer. We can now identify A with E: A is a realisation of E.

Analogously, a digraph can also be described by the incidence functions $i, t : \Gamma = (V; \vec{E}, i, t)$, where $i, t: \vec{E} \to V$ (see Section 1.3).

1.9 Isomorphisms of graphs. Automorphism groups

As with every mathematical structure, it is useful to be able to compare two (di)graphs with each other. The notion of an isomorphism formalises this idea.

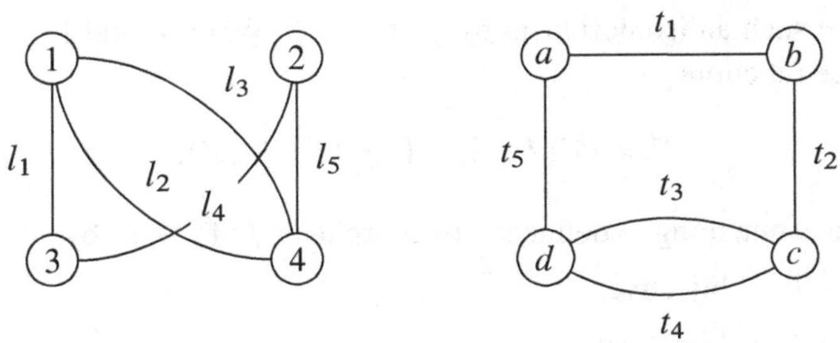

Figure 1.14. Two isomorphic graphs.

Let $\Gamma = (V; E, N)$, $\Gamma' = (V', E', N')$ be two graphs. An *isomorphism* between these two graphs is a bijection f:

$$f : V \sqcup E \to V' \sqcup E'$$

where $V \sqcup E$ denotes the disjoint union, satisfying

- $f|_V : V \to V'$ is bijective;
- $f|_E : E \to E'$ is bijective;
- for every edge $a = ([x, y], n) \in E$, there exists $n' \in N'$ such that $f|_E(a) = ([f|_V(x), f|_V(y)], n')$.

Example 1.9.1. Consider the two graphs in Figure 1.14: 1 is sent to c, 4 to d, 3 to b and 2 to a. The edge $([1, 4], l_3)$ is sent to the edge $([d, c], t_4)$; the edge $([1, 4], l_2)$ is sent to the edge $([d, c], t_3)$; the edge $([1, 3], l_1)$ is sent to the edge $([b, c], t_2)$; the edge $([2, 3], l_4)$ is sent to the edge $([a, b], t_1)$; and the edge $([2, 4], l_5)$ is sent to the edge $([a, d], t_5)$.

We then have the *incidence property*: let $\varepsilon, \varepsilon'$ be the incidence functions of graphs Γ and Γ' resp.:

$$\forall a \in E : f(\varepsilon(a)) = \varepsilon'(f(a)). \qquad \text{[I]}$$

This incidence property is expressed by the following *commutative diagram*, which expresses the equality $f \circ \varepsilon = \varepsilon' \circ f$:

$$\begin{array}{ccc} E & \xrightarrow{\varepsilon} & P_2(V) \\ \downarrow{\scriptstyle f} & & \downarrow{\scriptstyle f} \\ E' & \xrightarrow{\varepsilon'} & P_2(V') \end{array}$$

where it is understood that if $A \in P_2(V)$, $f(A) = \{f(x) \mid x \in A\} \in P_2(V')$.

We denote such an isomorphism by $f:\Gamma \to \Gamma'$. When Γ and Γ' are defined by their incidence functions

$$\Gamma = (V; E, \varepsilon), \quad \Gamma = (V'; E', \varepsilon'),$$

it comes to the same thing to define the isomorphism $f:\Gamma \to \Gamma'$ by

- $f|_V: V \to V'$ is bijective;
- $f|_E: E \to E'$ is bijective;
- $f \circ \varepsilon = \varepsilon' \circ f$.

Remark 1.9.2. When the graphs are simple, the data of $f|_V$ determines $f|_E$, thanks to the incidence property [I]. We will return to this point in Section 7.3.1.

For an *isomorphism of digraphs*, the definition is the same: the arcs are written $a = ((x, y), n)$, and the incidence property becomes

$$\forall a \in \vec{E} : f(i(a)) = i'(f(a)) \text{ and } f(t(a)) = t'(f(a)), \qquad [\vec{\mathrm{I}}]$$

where i, i', t, t' are the incidence functions of graphs $\vec{\Gamma} = (V; \vec{E}, N)$ and $\vec{\Gamma}' = (V'; \vec{E}', N')$.

Exercise 1.10. (i) Let $\Gamma = (V; E, N)$ and $\Gamma' = (V'; E', N')$ be two isomorphic graphs of finite order. Show that these two graphs must have

- the same number of vertices;
- the same number of edges;
- the same distribution of degrees;
- the same number of connected components.

(ii) Adapt the statement for oriented graphs.

As shown by Figure 1.15 and the example following it, the conditions in the above exercise are necessary but not sufficient.

Example 1.9.3. The two graphs in Figure 1.15 have the same number of edges and the same number of vertices. They also have the same distribution of degrees. However, they are not isomorphic. Indeed, under any isomorphism, in the first graph the vertex x_3 is of degree 3, so it would have to be sent to the vertex y_2 which is the only one of degree 3, but x_3 has two neighbours of degree 2 and one neighbour of degree 1, whereas y_2 has one neighbour of degree 2 and two neighbours of degree 1.

An *automorphism* is an isomorphism of a graph to itself.

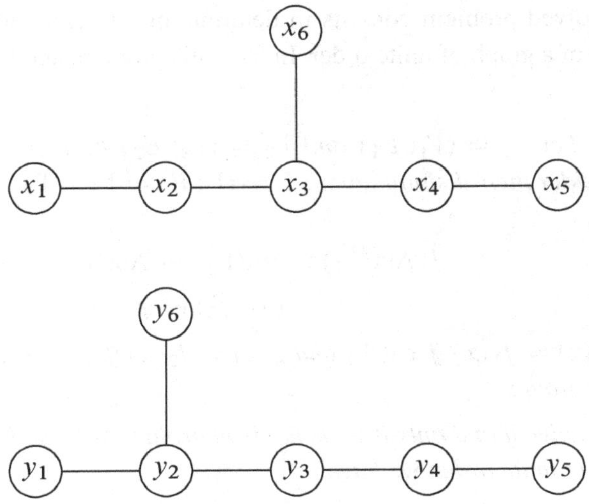

Figure 1.15. Two non-isomorphic graphs.

Exercise 1.11. Let $\Gamma = (V; E, N)$ be a graph. Show that the set of automorphisms of Γ equipped with the composition law for maps forms a group (see Section 1.10): the identity element Id_Γ is the identity map of $V \sqcup E$, and the inverse of f is f^{-1}. We denote this group by $\mathrm{Aut}(\Gamma)$.

The notion of a group is certainly one of the most simple and fruitful notions of all of mathematics. Group theory arises in mathematics, of course, but also in physics, chemistry, engineering, etc. Simply put, the automorphism group of an object allows us to *measure* the symmetries of that object. In our case, if the cardinality of $\mathrm{Aut}(\Gamma)$ is large, it means that the graph has a lot of *symmetries*. More explicitly, it has many vertices and/or edges that play an identical role.

The problem of knowing when two graphs are isomorphic, known as GI, is one of the most important problems in complexity theory. It is a relatively old problem and is still not solved; more precisely, we do not know in which complexity class the problem lies. Does there exist an algorithm of polynomial complexity to decide whether two given graphs are isomorphic? It is easy to see that the problem lies in the class $\mathcal{NP}$. Indeed, if we have a (potential) isomorphism f between two graphs, then checking whether f is a bijection that preserves adjacency can be done in polynomial time. But we do not know if the problem is in the class $\mathcal{P}$, or in the class $\mathcal{NPC}$. However, it seems [22] that the graph isomorphism problem may not lie in either of those two classes. If so, it would mean that GI would somehow lie between the two. If this were the case, we would then necessarily have $\mathcal{P} \neq \mathcal{NP}$.

Another unsolved problem consists in determining the cardinality of the automorphism group of a graph of finite order. In fact, this problem and the preceding one are related.

Theorem 1.9.4. *Let $\Gamma_1 = (V_1; E_1)$ and $\Gamma_2 = (V_2; E_2)$ be two simple connected graphs, and consider their disjoint union: $\Gamma = (V; E) = \Gamma_1 + \Gamma_2$.*

(i) *The map*

$$j\colon \mathrm{Aut}(\Gamma_1) \times \mathrm{Aut}(\Gamma_2) \to \mathrm{Aut}(\Gamma),$$
$$(f_1, f_2) \mapsto g,$$

where $g(x) = f_1(x)$ if $x \in V_1$ and $g(x) = f_2(x)$ if $x \in V_2$, is a morphism of injective groups.

(ii) *j is surjective if and only if Γ_1 is not isomorphic to Γ_2; so in this case, if the graphs are finite order, we have:*

$$|\mathrm{Aut}(\Gamma)| = |\mathrm{Aut}(\Gamma_1)| \times |\mathrm{Aut}(\Gamma_2)|.$$

Proof. (i) is immediate since f_1 and f_2 act separately on V_1 and V_2.

(ii) Suppose that Γ_1 is not isomorphic to Γ_2. By construction, $\Gamma = \Gamma_1 + \Gamma_2$ is the decomposition of Γ into connected components. Let $f \in \mathrm{Aut}(\Gamma)$; $f(\Gamma_1)$ is clearly a subconnected graph of Γ, so by Lemma 1.2.1, we have $f(V_1) \subset V_2$ or $f(V_1) \subset V_1$ (idem for $f(V_2)$).

- Either $f(V_1) \subset V_2$; then we have $f(V_1) = V_2$: indeed otherwise $V_2 \setminus f(V_1) \neq \emptyset$ and $V_2 \setminus f(V_1) \subset f(V_2)$; $f(V_2)$ is connected and by Lemma 1.2.1, $f(V_2) \subset V_2$. Thus $f(V_1) \cup f(V_2) = f(V_1 \cup V_2) \subset V_2$, but $f(V_1 \cup V_2) = V_1 \cup V_2$, which is a contradiction. As the graphs are simple, we deduce from $f(V_1) = V_2$ that $f(E_1) = E_2$; it follows that $f|_{\Gamma_1}\colon \Gamma_1 \to \Gamma_2$ is an isomorphism, which is a contradiction.
- Or $f(V_1) \subset V_1$; hence $f(V_1) = V_1$, $f(V_2) = V_2$ (for the same reasons as above), and since the graphs are simple, $f(E_1) = E_1$, $f(E_2) = E_2$, so $f = j(f_1, f_2)$, where $f_1 = f|_{\Gamma_1}$, $f_2 = f|_{\Gamma_2}$. Consequently, for all $f \in \mathrm{Aut}(\Gamma)$ there exists $(f_1, f_2) \in \mathrm{Aut}(\Gamma_1) \times \mathrm{Aut}(\Gamma_2)$ such that $j(f_1, f_2) = f$.

Suppose that $\varphi\colon \Gamma_1 \to \Gamma_2$ is an isomorphism. We construct $\psi \in \mathrm{Aut}(\Gamma)$ by $\psi(x) = \varphi(x)$ if $x \in V_1$, $\psi(x) = \varphi^{-1}(x)$ if $x \in V_2$ (since the inverse of an isomorphism is an isomorphism). It is clear that $\psi \notin \mathrm{Im}(j)$. ∎

When the graphs Γ_1 and Γ_2 are isomorphic, the group $\mathrm{Aut}(\Gamma_1 + \Gamma_2)$ is more complicated. We study this question in Section 9.2.1.

In this way, the problem of graph isomorphism can be reduced to the problem of computing the order (cardinality) of the group of automorphisms of a graph: if there exists an algorithm capable of computing the order of the automorphism group of a graph in polynomial time, then GI lies in the class $\mathcal{P}$. If we could show that such an algorithm does not exist, we would win $ 1,000,000 (the amount of the Clay foundation prize for solving this problem), because it would imply the strict inclusion $\mathcal{P} \subsetneq \mathcal{NP}$!

1.10 Complements: some basic structures

A *distance* on a set X is a map $d: X \times X \to \mathbb{R}$ satisfying the following properties: for all $x, y, z \in X$,

(i) $d(x, y) \geqslant 0$ and $d(x, y) = 0$ if and only if $x = y$,

(ii) $d(x, y) = d(y, x)$,

(iii) $d(x, z) \leqslant d(x, y) + d(y, z)$.

Then (X, d) is called a *metric space*.

Metric spaces are special cases of topological spaces, for which we can naturally define certain important notions such as limits or *open balls*

$$\mathcal{B}_d(x, r) = \{y \in E \mid d(x, y) < r\} \quad (r > 0),$$

which play a fundamental role. We note that if d is the distance attached to a graph (defined in Section 1.1.3), then for $1 < r < 2$, we have $\mathcal{B}_d(x, r) = \{x\} \cup \Gamma(x)$.

An *equivalence relation* on a set E is a binary relation R satisfying:

(i) xRx for all $x \in E$ (reflexivity),

(ii) $\forall x, y \in E, xRy \Rightarrow yRx$ (symmetry),

(iii) $\forall x, y, z \in E, xRy$ and $yRz \Rightarrow x \leqslant z$ (transitivity).

For $x \in E$, $R(x) = \{y \in X \mid xRy\}$ is called the equivalence class of x modulo R and there is a decomposition of E into subsets $E = \bigsqcup_{x \in \overline{E}} R(x)$, where $\overline{E}$ is a set of representatives of the equivalence classes modulo R. Conversely, it is easy to show that every decomposition of E into subsets yields an equivalence relation.

A *preordering* on a set E is a binary relation $\leqslant$ on E that satisfies:

(i) $x \leqslant x$ for all $x \in E$ (reflexivity),

(ii) $\forall x, y, z \in E, x \leqslant y$ and $y \leqslant z \Rightarrow x \leqslant z$ (transitivity).

We say that $(E, \leqslant)$ is a *preordered set*. To the preordering $\leqslant$, we associate a binary relation R defined by

$$xRy \quad \text{if and only if} \quad x \leqslant y \text{ and } y \leqslant x.$$

Then R is an equivalence relation called *equivalence associated to the preorder* $\leqslant$.

When moreover

(iii) $x \leqslant y$ and $y \leqslant x \Rightarrow x = y$ (antisymmetry),

we say that $\leqslant$ is an order relation or just an *ordering* (or even an *order*) on E.

If the ordering $\leqslant$ satisfies

(iv) $\forall x, y \in E : x \leqslant y$ or $y \leqslant x$ (all elements are comparable),

we say that $\leqslant$ is a *total order*; otherwise the ordering is called a *partial order*.

Let $(E, \leqslant)$ be an ordered set, and let $A \subset E$. We say that $x \in E$ is an *upper bound* of A if for all $a \in A$, we have $a \leqslant x$ (if $A = \emptyset$, then every $x \in E$ is an upper bound of A). If x is an upper bound of A and $x \in A$, we say that x is the *largest element* of A; this element is obviously unique and we denote it by $\max(A)$. Similarly, we say that $x \in E$ is a *lower bound* of A if for all $a \in A$, we have $x \leqslant a$ (if $A = \emptyset$, then every $x \in E$ is lower bound of A). If x is a lower bound for A and $x \in A$, we say that x is the *smallest element* of A; this element is obviously unique and we denote it by $\min(A)$.

When the set of upper bounds of A admits a smallest element s, we say that s is the *least upper bound* or *supremum* of A and denote it by $s = \sup(A)$. Similarly, the *greatest lower bound* or *infimum* $\inf(A)$ of A is the largest lower bound of A, if it exists.

An element $m \in A$ is called a *maximal element* of A if

$$x \in A \text{ and } m \leqslant x \implies x = m.$$

Similarly, an element $m \in A$ is said to be a *minimal element* of A if

$$x \in A \text{ and } x \leqslant m \implies x = m.$$

In a finite ordered set, every non-empty subset admits (at least) one maximal element and one minimal element. It can also have several minimal or maximal elements.

A *group* G is a set equipped with an internal composition law

$$G \times G \to G, \quad (a, b) \mapsto a \cdot b$$

such that

- for all $a, b, c \in G$, $a \cdot (b \cdot c) = (a \cdot b) \cdot c$; a law satisfying this property is said to be *associative*;
- there exists $e \in G$ such that for every element a in G, we have $e \cdot a = a \cdot e = a$; this element is necessarily unique and is called the *identity* element of G;
- for every $a \in G$ there exists $x \in G$ such that $a \cdot x = x \cdot a = e$; this element is necessarily unique and is called the *inverse* of a and written a^{-1}.

If for all $a, b \in G$ we have $a \cdot b = b \cdot a$, we say that the group is *commutative or abelian*; in this case the composition law is often written additively as $a + b$.

Chapter 2

Some remarkable graphs

In this chapter, we will select from the grand flora of graphs a few special families that arise naturally in various domains such as telecommunications, operations research and parallel processing.

The first type of graph we will consider are the bipartite graphs; the second part is devoted to trees and arborescences. The third part is devoted to graphs with no circuits, and finally, the fourth gives an introduction to Eulerian and Hamiltonian (di)graphs. To illustrate the effective nature of the results, we will display algorithms throughout. Planar graphs have been purposely omitted from this chapter, as they will be the focus of study in Chapter 5.

2.1 Bipartite graphs

Let $k \geqslant 1$ be an integer. A *simple* graph $\Gamma = (V; E)$ is said to be *k-partite* if there exists a partition of the set of its vertices into disjoint sets $V = V_1 \sqcup \cdots \sqcup V_k$ such that none of the induced subgraphs $\Gamma(V_i)$, $i = 1, \dots, k$, contains an edge. Thus in particular, a 1-partite graph is nothing more than a graph with no edges. The subsets V_i, $i = 1, \dots, k$, form the *k-partition* associated to the k-partite graph Γ.

A k-partite graph is said to be *complete* if

$$\forall i, j \in \{1, \dots, k\},\ i \neq j,\ \forall x \in V_i,\ \forall y \in V_j, \text{ we have } \{x, y\} \in E.$$

If $|V_i| = n_i$, $1 \leqslant i \leqslant k$, we write $K_{n_1,n_2,\dots,n_k}$ for a *model* of the corresponding complete k-partite graph. When $k = 1$, the graph $K_{1,n}$ is called a *star*.

For $k = 2$, we obtain an important class of graphs, the *bipartite graphs*.

Example 2.1.1. The graphs in Figure 2.1 are 3-partite; the subsets of the 3-partition are showed enclosed in dotted lines. The second one is the complete 3-partite graph $K_{2,2,3}$.

Bipartite graphs are characterised by the following theorem.

Theorem 2.1.2. *Let $\Gamma = (V; E)$ be a simple graph of finite order. Then Γ is bipartite if and only if it contains no cycle of odd length.*

Proof. A graph is bipartite if and only if all of its connected components are bipartite. Thus we can assume that Γ is connected.

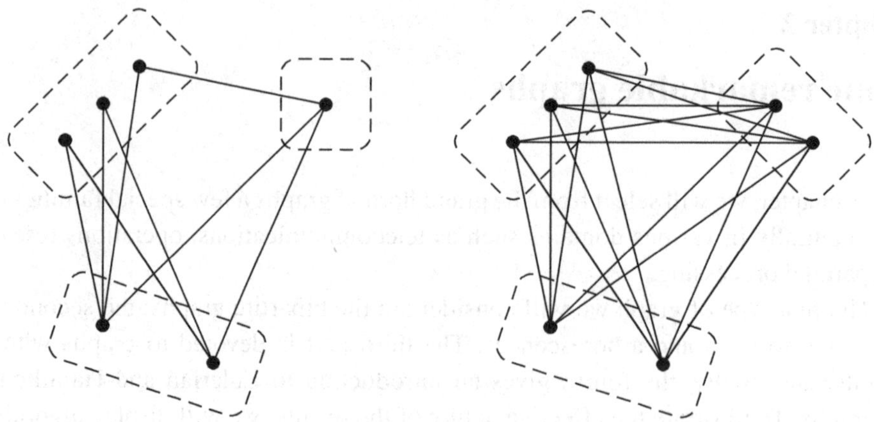

Figure 2.1. Two 3-partite graphs.

Only if direction: suppose that $\Gamma = (V; E)$ is bipartite, and write $V = V_1 \sqcup V_2$. Let $C = (x_0, a_1, x_1, a_2, \ldots, a_{k-1}, x_k)$ be a cycle (a simple closed chain) with $x_k = x_0$. We may assume that $x_0 \in V_1$. Since V_1 has no edge, we must then have $x_1 \in V_2$, but similarly V_2 has no edge, so we must have $x_2 \in V_1$. By induction, we then find that $x_i \in V_1$ if i is even and $x_i \in V_2$ if i is odd. But since $x_k = x_0$ lies V_1, it follows that k must be even.

If direction: suppose that Γ does not contain any odd cycle. Let x be an arbitrary vertex of Γ, and let d be the distance of the graph Γ considered as a metric space (see Section 1.1.3). Set $V_1 = \{y \in V \mid d(x, y) \text{ is even}\}$ and $V_2 = \{y \in V \mid d(x, y) \text{ is odd}\}$; then we have $V = V_1 \sqcup V_2$ and $x \in V_1$. Let us show that there is no edge between any pair of vertices of V_1, or any pair of vertices of V_2. For this, we reason by contradiction.

If there exists an edge $a = \{u, v\}$ with $u, v \in V_1$, then since the graph is connected, there exist two elementary chains C_u and C_v of minimal length connecting x to u and x to v. By the definition of V_1, $d(x, u)$ and $d(x, v)$ are both even, so C_u and C_v are even length chains. Note that the chains C_u and C_v may have common vertices. Starting from x, let z denote the final vertex common to both C_u and C_v. By minimality, the lengths of the subchains of C_u and C_v from x to z are both equal to $d(x, z)$, and in particular they have the same parity. It follows that the lengths of the subchains of C_u and C_v connecting z to u and z to v, which are of respective lengths equal to $d(z, u)$ and $d(z, v)$ (again by minimality), also have the same parity. Thus, the chain obtained by going from z to u along C_u, then from u to v along the edge $a = \{u, v\}$, then from v to z along C_v is a cycle of odd length $d(z, u) + 1 + d(v, z)$, contradicting the hypothesis.

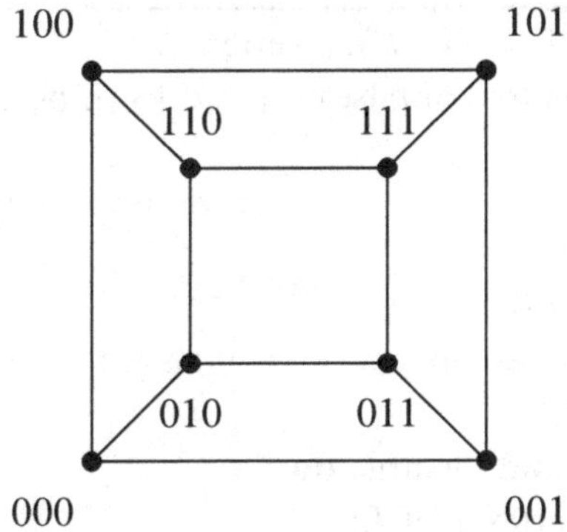

Figure 2.2. A model of the 3-cube.

If there exists an edge $a = \{u, v\}$ with $u, v \in V_2$, then since the graph is connected, there exist two elementary chains C_u and C_v of minimal length connecting x to u and x to v. By the definition of V_2, $d(x, u)$ and $d(x, v)$ are both odd, so C_u and C_v have odd length. Exactly like in the preceding case, there exists a final vertex z common to both C_u and C_v, and the chain obtained by going from z to u along C_u, then from u to v along $\{u, v\}$, then from v to z along C_v is a cycle of odd length, again contradicting the hypothesis. ■

In the general case, we perform the algorithm on each of the connected components successively.

One important example of bipartite graphs is given by the *k-cubes* Q_k, $k \geqslant 1$, also called hypercubes. A k-cube is a simple graph whose set of vertices is $V = \{0, 1\}^k$ and whose set E of edges is defined as follows: if $x = (x_1, \ldots, x_k)$ and $y = (y_1, \ldots, y_k)$ lie in V, then $\{x, y\} \in E$ if and only if $x_i = y_i$ for all except exactly one index i_0, $1 \leqslant i_0 \leqslant k$. Thus Q_1 consists of 2 vertices joined by one edge and Q_2 is a square. The 3-cube is shown in Figure 2.2.

Exercise 2.1. Prove that a cube Q_k is a regular bipartite graph of degree k with 2^k vertices.

Let us now consider Algorithm 2.1, where the graph is identified by the lists $\Gamma(y)$ of vertices adjacent to y for each $y \in V$.

Theorem 2.1.3. *Algorithm* 2.1 *determines whether or not a simple connected graph of finite order is bipartite. If it is bipartite, the algorithm yields a* 2*-partition with complexity* $O(|E|)$.

Proof. The proof is left as an exercise. ■

```
Input: a connected graph Γ = (V; E), a vertex x;
Output: bipartite (Boolean true or false), V1 and V2 (if possible);
begin
    L := ∅;                      /* intermediate list initialised to the empty set */
    d(x) := 0;                   /* the separation from x to x fixed to 0 */
    bipartite := true, V1 := ∅;
    /* a boolean variable and first subset of the partition of the vertices */
    Add x to L;
    while L is non-empty and bipartite do
        Remove the first vertex y of L;
        foreach u ∈ Γ(y) do
            if d(u) is not initialised then
                d(u) := d(y) + 1;
                Add u to L;
            else if d(u) == d(y) then
                bipartite := false;
            end
        end
    end
    if bipartite then
        foreach u ∈ V do
            if d(u) is even then V1 := V1 ∪ {u};
        end
    end
    V2 := V\V1;
end
```

Algorithm 2.1. Bipartite graph algorithm.

Proposition 2.1.4. *Let $\Gamma = (V; E)$ be a connected bipartite graph of finite or infinite order. Then the 2-partition $V = V_1 \sqcup V_2$ is unique.*

Proof. Fix an arbitrary vertex x, and as in Theorem 2.1.2, set

$$V_1 = \{y \in V \mid d(x, y) \text{ is even}\} \quad \text{and} \quad V_2 = \{y \in V \mid d(x, y) \text{ is odd}\}$$

so that $V = V_1 \sqcup V_2$ and $x \in V_1$.

Let $V = W_1 \sqcup W_2$ be an arbitrary bipartition of V, and suppose that $x \in W_1$. For all $y \in V$, let $C_y = (x, a_1, x_1, \ldots, x_{n-1}, a_n, y)$ be an elementary chain of minimal length $n = n_y = d(x, y)$ between x and y (with the convention that $x_n = y$). By the minimality of C_y, we have $d(x, x_k) = k$, $k \geqslant 1$, so $x_k \in V_1$ or V_2 according to whether k is even or odd. Similarly, since $V = W_1 \sqcup W_2$ is a bipartition, we also have $x_k \in W_1$ or W_2 according to whether k is even or odd. We then have

- $W_1 \subset V_1$: indeed let $y \in W_1$; as $x \in W_1$, the length of C_y is even, hence $y \in V_1$;
- $W_2 \subset V_2$: let $y \in W_2$; since $x \in W_1$, n_y is odd so $y \in V_2$.

Since $V = V_1 \sqcup V_2 = W_1 \sqcup W_2$, it follows that $W_1 = V_1$ and $W_2 = V_2$. ∎

Observe that without the hypothesis of connectedness, the proposition does not hold.

2.2 Trees and arborescences

2.2.1 Trees

Genealogical family trees are a typical example of tree graphs. The Darwinian theory of evolution also uses a tree to represent the species, although in fact the mechanism of evolutive fusion (endosymbiosis, hybridation and gene transfer) actually indicates a model closer to a network.

A *tree* is a connected graph with no cycle; thus, a tree is necessarily simple. There exist several useful characterisations of trees.

Theorem 2.2.1. *Let $\Gamma = (V; E)$ be a simple graph. Then the following assertions are equivalent:*

(i) Γ *is a tree.*

(ii) Γ *is connected and each edge is an isthmus.*

(iii) *Any two distinct vertices are joined by exactly one elementary chain.*

(iv) Γ *has no cycle, and adding a new edge creates exactly one elementary cycle.*

If Γ is of finite order, these assertions are also equivalent to each of the following assertions:

(v) Γ *has no cycle and* $|E| = |V| - 1$.

(vi) Γ *is connected and* $|E| = |V| - 1$.

Proof. (i) $\Rightarrow$ (ii): Delete the edge $a = \{x, y\}$; if the subgraph thus obtained is connected, then there is a chain C and thus an elementary chain (see Exercise 1.1) between x and y which does not contain the edge a. Thus, concatenating C and a, we obtain a cycle in Γ which contains the edge a. This is a contradiction.

(ii) $\Rightarrow$ (iii): Since the graph is connected, each pair of vertices is joined by at least one elementary chain. If there exists a pair of vertices $\{x, y\}$ joined by two distinct elementary chains C_1 and C_2, then there is an edge in one of these chains, say $a \in C_2$, such that $a \notin C_1$. But such an edge must give rise to a cycle. Indeed, starting from x, consider the first vertex z at which the elementary chain C_2 is different from C_1. Let z' be the first vertex *after* z where the chain C_2 rejoins C_1. We cannot have $z' = z$ since C_2 is elementary, so this induces two disjoint subchains from z to z', which thus naturally form an elementary cycle by concatenation. The edges of this cycle are obviously not isthmuses.

(iii) $\Rightarrow$ (iv): If we add an edge $a = \{x, y\}$, then since by (iii) we know that x and y are already joined by exactly one elementary chain, we see that concatenation with a produces an elementary cycle. If adjoining a also created a second elementary cycle, we would then have a second elementary chain between x and y, which is a contradiction.

(iv) $\Rightarrow$ (i): If Γ_1 and Γ_2 were two connected components of Γ, then by adding an edge between Γ_1 and Γ_2, we would not create any cycle.

(i) $\Rightarrow$ (v): Follows from Propositions 1.4.2 and 1.4.3.

(v) $\Rightarrow$ (vi): Let $\Gamma = \Gamma_1 + \cdots + \Gamma_k$ be the decomposition of Γ into connected components (see Section 1.5); each $\Gamma_i = (V_i; E_i)$ is connected, so by Proposition 1.4.2 (i), it has $|E_i| \geqslant |V_i| - 1$ edges and no cycles, so by Proposition 1.4.3, we have $|E_i| \leqslant |V_i| - 1$; hence $|E_i| = |V_i| - 1$. Then

$$|E| = \sum_{1 \leqslant i \leqslant k} |E_i| = \sum_{1 \leqslant i \leqslant k} (|V_i| - 1) = |V| - k$$

since $|E| = |V| - 1$ by hypothesis. It follows that $k = 1$, i.e. Γ is connected.

(vi) $\Rightarrow$ (ii): If we delete an edge, we have $|E| = |V| - 2$, but then by Proposition 1.4.2, the graph obtained is no longer connected. ■

Example 2.2.2. The degree of relationship in a family tree is the distance between the different vertices in the corresponding tree; to compute it, one must go back up as far as the first common ancestor of the two vertices.

A *forest* is a graph with no cycle (so necessarily a simple graph). Thus, the connected components of a forest are trees. A *leaf* in a forest is a vertex of degree 1. Trees and forests are shown in Figure 2.3.

Corollary 2.2.3. *Let Γ be a forest having n vertices and k connected components. Then Γ has $n - k$ edges.*

Proof. It suffices to apply (vi) of Theorem 2.2.1 to each of the k connected components of Γ. ■

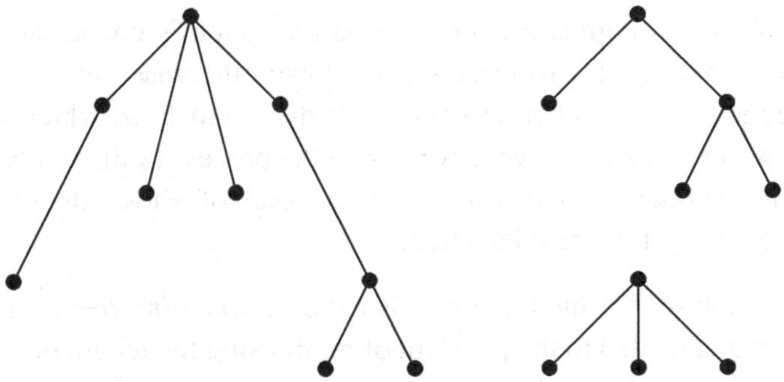

Figure 2.3. A forest with three trees and eleven leaves.

Corollary 2.2.4. *Let $\Gamma = (V; E)$ be a simple graph of finite order having k connected components. Then the graph has no cycle if and only if $|E| = |V| - k$.*

Proof. If Γ has no cycle, it is a forest, and we have $|E| = |V| - k$ by Corollary 2.2.3.

Conversely, if $|E| = |V| - k$, let $(\Gamma_i)_{1 \leqslant i \leqslant k}$, $\Gamma_i = (V_i; E_i)$, denote the connected components of Γ. By Proposition 1.4.2, $|E_i| \geqslant |V_i| - 1$ for all i; it follows that $|E| \geqslant |V| - k$, but as by hypothesis we have $|E| = |V| - k$, we must have $|E_i| = |V_i| - 1$ for each i. But Γ_i is connected, so by (vi) of Theorem 2.2.1, Γ_i is a tree, so it has no cycle; hence Γ has no cycle. ∎

Proposition 2.2.5. *A tree $\Gamma = (V; E)$ having $|V| \geqslant 2$ vertices admits at least two leaves.*

Proof. The sum of degrees in a graph is equal to twice the number of edges (see Lemma 1.1.9). We have $|E| = n - 1$ and

$$2n - 2 = \sum_{x \in V} d(x).$$

If t denotes the number of vertices of degree 1, then there are $n - t$ vertices of degree $\geqslant 2$, and we have

$$2n - 2 = \sum_{x \in V} d(x) = t + \sum_{x \in V,\, d(x) \geqslant 2} d(x) \geqslant t + 2(n - t),$$

so $t \geqslant 2$. ∎

Proposition 2.2.6. *A graph $\Gamma = (V; E, N)$ of finite order admits a tree $\Gamma' = (V; E', N')$ as partial graph if and only if it is connected.*

Proof. A graph which admits a connected partial graph is connected. Conversely, suppose Γ is connected: if it is a tree we are done; otherwise, by Theorem 2.2.1, it contains an edge a which is not an isthmus. If the graph $\Gamma\backslash a$, which is connected, is a tree then we are done; otherwise we iterate the process (a finite number of times since $|E|$ is finite) until we obtain a connected graph all of whose edges are isthmuses. By Theorem 2.2.1 (ii), this must be a tree. ■

A partial graph of Γ which is a tree is called a *spanning tree* of Γ. Here is an important example, related to the problem of minimising the length of networks connecting a set of points in the plane. Let $n \geqslant 1$ be an integer. A *Steiner tree* is a tree whose vertices are n points $x_1, x_2, \ldots, x_n$ in the Euclidean plane $\mathbb{R}^2$ with integral coordinates, called *Steiner points*, joined in such a way that the sum of distances $\|x_i - x_j\|$, $1 \leqslant i < j \leqslant n$, is minimal. The problem of determining a Steiner tree of minimal length (for the usual distance in $\mathbb{R}^2$) is called the *Euclidean Steiner* problem. It has been studied intensively by numerous authors [17]. We can show that this problem lies in the class $\mathcal{NPC}$ (see [15]).

Let $\Gamma = (V; E)$ be a connected graph, equipped with a weight (or valuation) on the edges: $\omega\colon E \to \mathbb{R}_+$. The very simple Algorithm 2.2 is due to Kruskal (1956). It yields a spanning tree $\mathcal{A}$ of minimal weight $\sum_{a\in E(\mathcal{A})} \omega(a)$.

Assume that the edges are sorted in order of increasing weight:

$$E = \{a_1, a_2, \ldots, a_m\} \quad \text{with } \omega(a_1) \leqslant \omega(a_2) \leqslant \cdots \leqslant \omega(a_m).$$

```
Input: A graph Γ = (V; E), where the vertices are sorted in order of increasing
       weight, a weight function ω;
Output: A spanning tree T;
begin
   T := ∅ ;                                              /* initialised tree */
   for i from 1 to m do
      if a_i does not form a cycle with edges of T then
         Add a_i to T;
      end
   end
end
```

Algorithm 2.2. Kruskal's algorithm.

Theorem 2.2.7. *Let $\Gamma = (V; E)$ be a simple connected graph with $m \geqslant 1$ edges. Then Algorithm 2.2 constructs a spanning tree T of minimal weight with complexity $O(m \log m)$.*

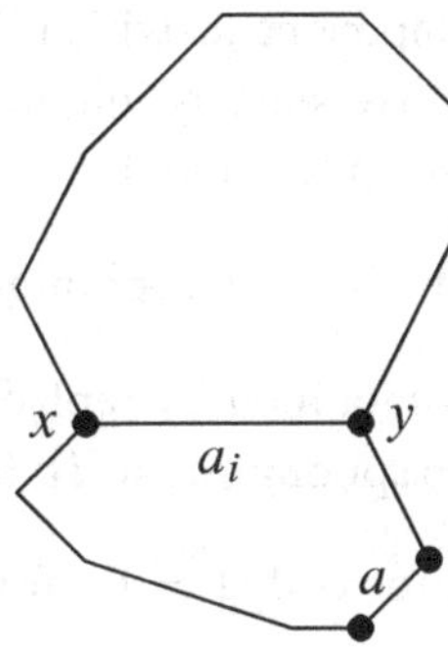

Figure 2.4. Illustration in the proof of Kruskal's algorithm.

Proof. The algorithm produces a tree T by successive addition of edges $a'_i = a_{j_i}$, $i = 1, \dots, k$. We set $E' = \{a'_1, \dots, a'_k\}$; T is the subgraph induced by all extremities of edges belonging to E'.

The tree T is a spanning tree: if there existed a vertex x of Γ incident to T but not in $V(T)$, there would be an edge $a = \{x, y\}$, $y \in V(T)$, such that $a \notin E(T) = E'$. By construction, $E' \cup \{a\}$ would contain a cycle, so x would be an element of $V(T)$.

The tree T is of minimal weight: let T_1 be an arbitrary spanning tree, and let a'_i be the first edge produced by the algorithm which is not in T_1, so $a'_1, a'_2, \dots, a'_{i-1} \in T \cap T_1$. As T_1 is a spanning tree, if we add a'_i to T_1, we obtain a cycle C in T_1. This cycle contains an edge $a \in T_1$ which is not in T (otherwise T would contain a cycle). Let us replace the edge a in T_1 by the edge a'_i: we obtain a new spanning tree T_2. Indeed, if there were an (elementary) cycle C' in T_2, then on the one hand this cycle should contain a'_i (otherwise it would be a cycle of T_1), but on the other hand $C' \backslash a'_i$ would be a chain of T_1 between x and y, where $a'_i = \{x, y\}$. Since a is not on C', we would then have a cycle of T_1 by following C from x to y passing through a, then following C' from y to x (Figure 2.4). Note moreover that T_2 has one more edge in common with T than T_1.

Furthermore, $\omega(a'_i) \leqslant \omega(a)$, otherwise $\omega(a) < \omega(a'_i)$. Since a was not chosen at stage i by the algorithm, this means that it forms a cycle C' with the edges of $\{a'_1, \dots, a'_{i-1}\}$; but a is an edge of T_1, so that $a'_1, \dots, a'_{i-1}$, so C' would be a cycle of T_1, which is a contradiction. This implies that $\omega(T_2) \leqslant \omega(T_1)$.

We can repeat this construction starting from T_2, and so forth until the procedure stops, yielding the tree $T_r = T$. We have thus transformed T_1 into T by replacing the edges one by one in such a way that the total weight at each stage is not increased. Consequently, $\omega(T) \leqslant \omega(T_1)$.

To compute the complexity, we need a table compconnected which will give us the label of the connected component to which the vertex under consideration belongs. Let $V = \{1, 2, 3, \ldots, n\}$. We start by initialising: compconnected[i] $\leftarrow i$. Let $e = \{i, j\}$ be the edge under consideration. If

$$\texttt{compconnected}[i] < \texttt{compconnected}[j],$$

then the table compconnected is modified by replacing all the entries equal to the value of compconnected[j] by compconnected[i]; in the end we have

$$\texttt{compconnected}[i] = 1 \quad \text{for all } i,$$

since by assumption every vertex must belong to the same connected component.

It is well known that sorting a list of size m can be done in $O(m \log m)$ elementary operations. Thus E can be sorted in increasing order of weight in $O(|E| \log(|E|))$.

In the worst case, we obtain a tree when we adjoin the m-th edge. For each edge $\{i, j\}$ that is added, we first compare compconnected[i] with compconnected[j]. If they are equal, this means that the edge $\{i, j\}$ forms a cycle with the other edges already considered. Otherwise and assuming that

$$\texttt{compconnected}[i] < \texttt{compconnected}[j],$$

we change all entries

$$\texttt{compconnected}[k] = \texttt{compconnected}[j]$$

into compconnected[i]. This gives a complexity of $O(|E| \times |V|)$. Note that there exist improvements in the treatment of the connected components through the algorithm which yields a complexity of $O(|E| \log(|E|)$. Thus, at worst we have $|V|$ comparisons to make, and the updating relative to all the connected components requires $O(|V| \log |E|)$ operations throughout the algorithm. Since Γ is connected, we have $|V| = O(|E|)$, and we obtain a complexity of $O(|E| \log(|E|))$. ∎

Remark 2.2.8. When we modify the order of the edges having the same weight, Kruskal's algorithm can yield several spanning trees of minimal weight. However all these spanning trees will have the same (minimum) weight.

2.2.2 Arborescences

An *arborescence* is a simple digraph $\vec{\Gamma} = (V; \vec{E})$ satisfying the following conditions:

- the underlying graph is a tree;
- there exists one vertex of input degree 0 and every other vertex has an input degree equal to 1.

The vertex which has input degree equal to 0 is called the *root* of the arborescence.

Let $\vec{\Gamma} = (V; \vec{E})$ be a digraph whose underlying graph has $k \geqslant 1$ connected components. The digraph $\vec{\Gamma} = (V; \vec{E})$ is an *oriented forest* if the connected components of the underlying graph are arborescences. A partial digraph of a digraph $\vec{\Gamma} = (V; \vec{E})$ which is an arborescence is called a *spanning arborescence*.

Proposition 2.2.9. *Let $\Gamma = (V; E)$ be a tree of finite order. Then for every vertex r of Γ, there exists an orientation such that Γ is an arborescence with root r.*

Proof. We use induction on the number of edges.

If $|E| = 0$, the property is trivial. If $|E| = 1$, the tree has an edge $a = \{r, y\}$: we can orient this edge so that r is the root.

Suppose now that the property holds for all trees with $m - 1$ edges. Let $\Gamma = (V; E)$ be a tree with m edges. Since the vertex r is not isolated, there exists an edge $a = \{r, y\}$ incident to r. If we delete a, we obtain two connected components (see Theorem 2.2.1 (ii)) both of which are trees: Γ_r and Γ_y.

We can apply the induction hypothesis to each of these two trees. Orient Γ_r and Γ_y so that their respective roots are r and y. Next, orient the edge a of r towards y. We obtain an arborescence with root r, since all the vertices of Γ have input degree equal to 1 except for r, which has input degree equal to 0. ∎

We have also the following property.

Proposition 2.2.10. *Let $\vec{\Gamma} = (V; \vec{E})$ be an arborescence, and let x be a vertex. The subdigraph $\vec{\Gamma}_x$ induced by*

$$V_x = \{y \in V \mid \textit{there exists a path from } x \textit{ to } y\}$$

is an arborescence.

Proof. It is clear that the underlying graph Γ_x is a tree. By definition, all the vertices except for x are of input degree equal to 1; $d^-(x) = 0$. The result follows. ∎

The arborescence $\vec{\Gamma}_x$ is the *arborescence attached to* x.

Let $\vec{\Gamma} = (V; \vec{E})$ be an arborescence of root r. If $a = (x, y) \in \vec{E}$, we say that x is the *father* of y and y is the *son* of x. Two vertices having the same father are *brothers*. A vertex having at least one son is called an *internal vertex*; a *branch* of $x \in V$ is a path starting from x and going all the way to a leaf.

The *height* of a vertex x is the length of the longest path that starts from x and goes to a leaf; the *height of an arborescence* is the height of its root; the *depth* of a vertex x is the length of the unique path from the root to x. By Theorem 2.1.2, there exists a unique chain between r and $x_n = x$:

$$C = (r, a_0, x_1, a_1, x_2, \ldots, a_{n-1}, x_n).$$

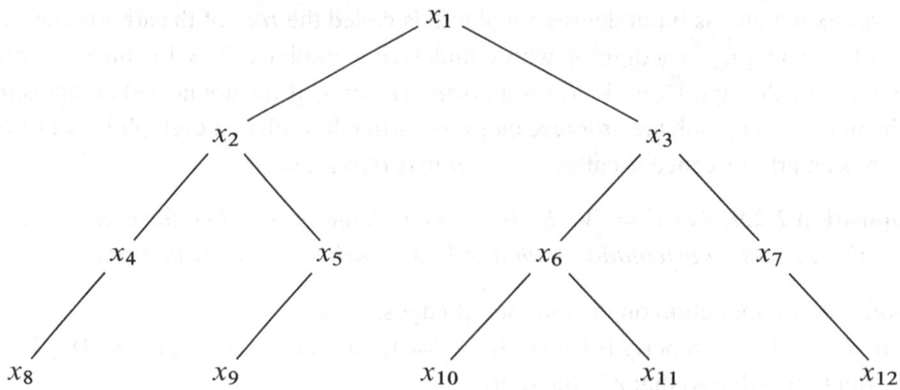

Figure 2.5. Ordered tree.

As $\vec{\Gamma}$ is an arborescence, the input degree of x_1 is equal to 1, so the oriented edge corresponding to a_0 goes from r to x_1, and the same holds for the other a_i; thus C actually gives a path from r to x.

A *level* in an arborescence is the set of all vertices of the same depth.

An *ordered arborescence* is an arborescence in which there is a total order on the set of sons of each vertex. An ordered arborescence can be drawn in the plane simply by ordering the sons of each vertex from left to right (Figure 2.5). An ordered arborescence is a *binary complete arborescence* if every vertex has either no sons or two sons.

Exercise 2.2. Show that the number of vertices in a complete binary arborescence is necessarily odd.

A complete binary arborescence is *perfect* if all the leaves are of the same depth and if every vertex which is not a leaf has exactly two sons.

Proposition 2.2.11. *A perfect binary arborescence of height h possesses* $2^{h+1} - 1$ *vertices.*

Proof. We use induction on the height. If $h = 0$, we have $n = 1$.

Suppose the property holds for all perfect binary arborescences of height h. Let $\mathcal{T}$ be a perfect binary arborescence of height $h + 1$. By deleting the root, we obtain two arborescences (by Proposition 2.2.10). Furthermore, these are both perfect. They both have height equal to h, and consequently they each possess exactly $2^{h+1} - 1$ vertices. Thus, there are exactly $2(2^{h+1} - 1) + 1 = 2^{h+2} - 1$ vertices in $\mathcal{T}$. ■

2.3 Digraphs with no circuits

Digraphs with no circuits are very important in applications to sequencing. One can use them, for example, in work planning.

Let $\vec{\Gamma} = (V; \vec{E})$ be a simple digraph with no circuit. A *topological sort* is an injective map f from V to $\mathbb{R}$ such that if $a = (x, y)$ is an arc, then $f(x) < f(y)$. When Γ is of finite order, this comes down to giving a total order on the vertices $x_1 < x_2 < \cdots < x_n$ that satisfies

$$(x_i, x_j) \in \vec{E} \implies i < j.$$

Lemma 2.3.1. *Let $\vec{\Gamma} = (V; \vec{E})$ be a simple digraph of finite order with no circuits. Then Γ contains at least one vertex x of input degree equal to* 0, *i.e.* $d^-(x) = 0$.

Proof. Let x_1 be an element of V. If $d^-(x_1) = 0$, it is finite. Otherwise there exists a vertex x_2 such that (x_2, x_1) is an arc. If $d^-(x_2) = 0$, it is finite. Otherwise there exists x_3 such that (x_3, x_2) is an arc, and so forth. Since the digraph is finite and has no circuits, we must eventually obtain a vertex x_k such that $d^-(x_k) = 0$. ∎

Proposition 2.3.2. *Let $\vec{\Gamma} = (V; \vec{E})$ be a simple digraph of finite order. Then it has no circuits if and only if the set of its vertices admits a topological sort.*

Proof. Only if: Let Γ be a simple digraph with no circuits; we use induction on the number of vertices. If $|V| = 1$, it is trivial. Assume the result holds up to rank n. Let Γ be such that $|V(\Gamma)| = n + 1$. By Lemma 2.3.1, the digraph admits a vertex u such that $d^-(u) = 0$. Delete this vertex: the induced digraph Γ_u on $V \backslash \{u\}$ has no circuits and is of order n, so by the induction hypothesis it admits a topological sort $x_1 < x_2 < \cdots < x_n$. As the vertex u has an input degree equal to 0, we obtain a topological sort on V by setting $u < x_1 < x_2 < \cdots < x_n$ (the only arcs containing u are of the type (u, x_i)).

If: Conversely, assume that Γ admits a topological sort f, and that the digraph possesses a circuit

$$(x_0, a_0, x_1, a_1, x_2, \ldots, x_t, a_t, x_0).$$

Then we have $f(x_0) < f(x_1) < \cdots < f(x_t) < f(x_0)$, which is a contradiction. ∎

Let $\vec{\Gamma} = (V; \vec{E})$ be a simple digraph with $|V| = n$. Algorithm 2.3 checks whether the digraph has a circuit or not, and determines a topological sort on its vertices if it does not.

Let A_x be the set of vertices y such that (x, y) is an arc of $\vec{\Gamma}$.

```
Input: Γ⃗; A_{x_1}, A_{x_2}, ..., A_{x_n};      /* a digraph and a list of sets of vertices */
Output: s, entrd, listopo;
/* Boolean variable indicating that Γ⃗ is without circuit, table of input degrees of all vertices,
   table of vertex numbers in the topological sort */
begin
  v := 1 ;                                     /* number of vertices considered */
  L := ∅ ;                                     /* list of vertices having d⁻(x) = 0 */
  foreach i from 1 to n do
    entrd(x_i) := 0;
  end
  foreach i from 1 to n do
    foreach x_j ∈ A_{x_i} do
      entrd(x_j) := entrd(x_j) + 1;
    end
  end
  foreach i from 1 to n do
    if entrd(x_i) == 0 then
      Add x_i to L;
    end
  end
  while L is non-empty do
    Delete the first vertex y of L;
    listopo(y) := v;
    v := v + 1;
    foreach u ∈ A_y do
      entrd(u) := entrd(u) − 1;
      if entrd(u) == 0 then
        Add u to L;
      end
    end
  end
  if v == n + 1 then
    s := false
  else
    s := true
  end
end
```

Algorithm 2.3. Topological sort.

Theorem 2.3.3. *Algorithm 2.3 determines if a simple digraph $\vec{\Gamma} = (V; \vec{E})$ has circuits or not. If not, and if the underlying graph of Γ is connected, it constructs a topological sort on the vertices with complexity $O(|V| + |\vec{E}|)$.*

Proof. Clearly, the algorithm tests whether or not the digraph Γ has a circuit by giving the list of vertices sorted *topologically*.

The first initialising **for** loop is done in $O(|V|)$, and the second in $O(|\vec{E}|)$, since the digraph is represented by a list of arcs which is run over exactly once. The construction of L is also in $O(|V|)$. Consequently, these initialisations are in $O(|V| + |\vec{E}|)$. In the **while** loop, every arc is run over at most once. Thus, this algorithm has complexity $O(|V| + |\vec{E}|)$. ■

2.4 Euler graphs and Hamiltonian graphs

2.4.1 Euler graphs

There is at least one problem which cannot be escaped by anyone undertaking the study of graph theory – and we will not break that tradition in this book! If one day you find yourself taking a walk in Kaliningrad (formerly Königsberg in Prussia), you may note that the town has no less than seven bridges. You might even ask yourself the following question: Can I start walking in one part of the town and get back to my starting point after having crossed every bridge exactly once? If you do, then you should know that a famous mathematician of the 18th century, Leonhard Euler (1707–1783), asked it already long ago. The answer is explained below.

Let $\Gamma = (V; E, N)$ be a connected graph of finite order. A cycle of Γ is called an *Euler cycle* if it passes through every edge of Γ exactly once. The graph Γ is an *Euler graph* if it contains an Euler cycle. An example of an Euler graph is shown in Figure 2.6.

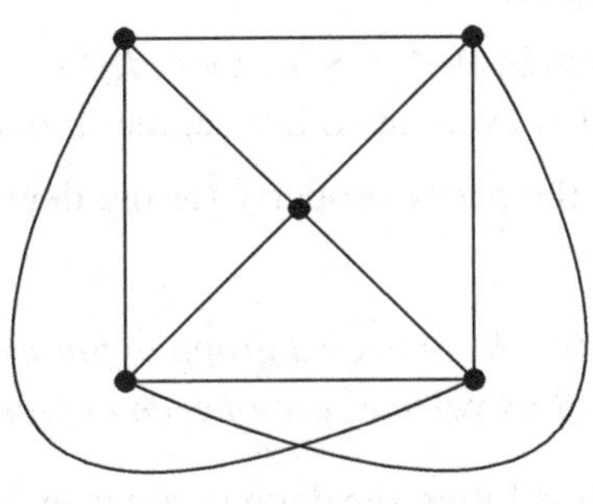

Figure 2.6. An Euler graph.

An *Euler chain* is a simple chain that contains all the edges of Γ. A *semi-Euler graph* is one that contains an Euler chain.

There exists a very simple characterisation of Euler and semi-Euler graphs based on the parity properties of the degrees of vertices. Recall (see Section 1.1.3) that the *underlying graph of a chain* C of the graph Γ is the subgraph whose vertices and edges are those of C.

Lemma 2.4.1. *Let $\Gamma = (V; E, N)$ be a graph of finite order and C a simple chain of Γ. Then in the underlying graph of C, all the vertices are of even degree except for the extremities of C (when they are distinct), which are of odd degree.*

Proof. This result seems intuitively obvious, but the proof is actually a bit complicated. We use induction on the number n of edges of the chain. For $n = 1$, we have either a loop, so one vertex of degree 2, or a simple edge, so two vertices of degree 1.

Suppose the result holds up to n, and let C be a chain with $n + 1$ edges:

$$C = (x_0, a_0, x_1, \ldots, x_n, a_n, x_{n+1}).$$

Remove a_n and x_{n+1} from C so as to obtain a chain C' with n edges. We distinguish two cases and use the induction hypothesis.

(i) If $x_n \neq x_0$, then $d_{C'}(x_0)$ and $d_{C'}(x_n)$ are odd and all the $d_{C'}(x_i)$ for $1 \leqslant i \leqslant n-1$ are even (where $d_{C'}$ denotes the degree in the subgraph C'):

- if $x_{n+1} \notin \{x_0, \ldots, x_n\}$, then $d_C(x_n)$ becomes even, $d_C(x_{n+1})$ is equal to 1 and $d_C(x_i) = d_{C'}(x_i)$ for the other vertices;
- if $x_{n+1} = x_j$ for some $j \neq 0$, then $d_C(x_j)$ becomes odd, $d_C(x_n)$ becomes even and the other degrees are unchanged;
- if $x_{n+1} = x_0$, then all vertices of the chain C are of even degree.

(ii) If $x_n = x_0$, then all vertices of the chain C' are of even degree:

- if $x_{n+1} \notin \{x_0, \ldots, x_n\}$, then $d_C(x_0)$ becomes odd, $d_C(x_{n+1}) = 1$, the other degrees remain unchanged;
- if $x_{n+1} = x_j$, for a certain $0 \leqslant j \leqslant n$, then $d_C(x_{n+1}) = d_{C'}(x_i) + 1$ is odd, $d_C(x_n) = d_{C'}(x_n) + 1$ is even, the other degrees are unchanged.

In each of these cases, the parity property for the degrees of the vertices of C is satisfied. ■

Theorem 2.4.2 (Euler, 1736). *A connected graph of finite order $\Gamma = (V; E, N)$ is an Euler graph if and only if all its vertices are of even degree.*

Proof. Only if: If the graph is Euler, the degrees are even by Lemma 2.4.1.

If: Suppose the degrees of vertices of the graph are even, and consider the set $\mathcal{S}$ of simple chains

$$C = (x_0, a_0, x_1, a_1, x_2, \dots, x_{k-1}, a_{k-1}, x_k)$$

ordered by the relation $C \leqslant C'$ if C is a subgraph of C'. Since $\mathcal{S}$ is finite, it possesses at least one maximal element (see Section 1.10)

$$C'' = (x_0, a_0, x_1, a_1, x_2, \dots, x_n, a_n, x_{n+1}).$$

The degree of x_{n+1} is even by hypothesis, and non-zero since a_n is incident to x_{n+1}, so there exists another edge a incident to x_{n+1}; thus we must have $a \in \{a_0, a_1, \dots, a_{n-1}\}$, since otherwise adding a to the chain would contradict the maximality of C''. For the same reason, *all* the edges incident to x_{n+1} lie in $\{a_0, a_1, \dots, a_n\}$, so the degree of x_{n+1} in the underlying graph of C'' is the same as in the graph Γ, and this degree is even. Consequently, by Lemma 2.4.1, we have $x_{n+1} = x_0$, so C'' is a cycle. Furthermore, it is an Euler cycle since otherwise there would exist another edge $a \notin \{a_0, \dots, a_n\}$ incident to a vertex of the cycle (since the graph is connected), $a = ([y, x_i], k)$, but then the chain

$$(y, a, x_i, a_i, \dots, x_n, a_n, x_{n+1}, x_0, a_0, x_1, a_1, \dots, x_{i-1}, a_{i-1}, x_i)$$

would be a simple chain and yet longer than C'', again a contradiction. ■

It follows that in particular, the bipartite graph $K_{n,n}$ is Euler if and only if n is even.

Corollary 2.4.3. *The graph Γ is semi-Euler if and only if it has either no vertex of odd degree or exactly two vertices of odd degree.*

Proof. Only if: If the graph is semi-Euler, the result follows from Lemma 2.4.1.

If: If it has no vertices of odd degree, then the graph is Euler by Theorem 2.4.2. If it has 2 vertices x and y of odd degree, we add an edge a between x and y, in which case the degree of x and the degree of y both become even. The graph Γ' thus obtained has only even degrees, so it is Euler by Theorem 2.4.2. Thus it has an Euler cycle of Γ', that we write as $(x, a_0, x_1, \dots, x_{n-1}, a_{n-1}, y, a, x)$. Consequently, C passes exactly once through every edge of the graph Γ'. We now delete a from this cycle, and thus obtain a simple chain $(x, a_0, x_1, \dots, x_{n-1}, a_{n-1}, y)$ which passes exactly once through every edge of Γ. This chain between x and y is thus an Euler chain in Γ, so Γ is a semi-Euler graph. ■

We will also need the following lemma.

Lemma 2.4.4. *Every cycle C in a graph $\Gamma = (V; E, N)$ of finite order can be decomposed into edge-disjoint elementary cycles $C_1, \dots, C_k$:*

$$E(C) = \bigsqcup_{i=1}^{k} E(C_i).$$

Proof. We proceed by induction on the length of the cycle. If the cycle is a loop, it is an elementary cycle. Suppose that every cycle of length less than or equal to $n-1$ decomposes into edge-disjoint elementary cycles. Let C be a cycle of length n. If no vertex of C is repeated, then C is elementary and the result is proven. Otherwise, consider the first vertex x_i of C that is passed through at least twice by C. Set $j = \min\{k > i \mid x_k = x_i\}$. The two cycles C_1 and C_2 contained in C, with the first going from x_i to x_j (the second occurrence of x_i as a vertex of C) and the second going from x_j to x_i, are both of length strictly less than n. Thus by the induction hypothesis, we can decompose these two cycles into edge-disjoint elementary cycles. Then C_1 and C_2 have no common edge since C is simple. The union of the decompositions of C_1 and C_2 into edge-disjoint elementary cycles thus forms a decomposition of C into edge-disjoint elementary cycles. ■

There can be multi-edges in an Euler cycle; however, we can prove the following corollary.

Corollary 2.4.5. *A connected graph $\Gamma = (V; E, N)$ of finite order is Euler if and only if the set of its edges can be partitioned into edge-disjoint elementary cycles.*

Proof. If the graph is Euler, it has an elementary cycle C with $E(C) = E$. Then Lemma 2.4.4 yields the result.

Conversely, let C be an elementary cycle of the partition. If C contains all the edges, we are done. Otherwise, there exists another cycle C' having a vertex x in common with C, since the graph is connected, but then, since C and C' are edge-disjoint, the concatenation of C and C' forms a cycle C''. If C'' contains all the edges, we are done; otherwise, we do the same thing as many times as necessary until we have exhausted all the elementary cycles. ■

An *Euler digraph* is a digraph containing an Euler circuit, i.e. a circuit that passes through every arc exactly once. A *semi-Euler digraph* is a digraph containing a semi-Euler path, i.e. a path that passes through each arc exactly once. A digraph $\vec{\Gamma} = (V; \vec{E})$ is *pseudo-symmetric* if for every vertex $x \in V$ we have $d^-(x) = d^+(x)$. We now state the following theorem.

Theorem 2.4.6. *Let $\vec{\Gamma} = (V; \vec{E})$ be a digraph whose underlying graph is connected. Then $\vec{\Gamma}$ is an Euler graph if and only if it is pseudo-symmetric.*

Proof. Left as an exercise for the reader. The idea is to adapt Lemma 2.4.1 by giving the input and output degrees in a path, after which the proof is similar to that of Theorem 2.4.2. ∎

2.4.2 Hamiltonian graphs

We have already seen graphs that contain an Euler cycle. The *dual* problem is that of knowing whether a graph of finite order contains an elementary cycle that passes through all its vertices (though not necessarily all its edges). Such a cycle is called a *Hamiltonian* cycle. In 1859, Sir William Hamilton proposed the following game.

Consider a dodecahedron, and identify its vertices with cities of the world. The game is called *Around the world*, and its goal is to start from a given city, visit all the others exactly once by travelling along the edges of the dodecahedron, and end by returning to the starting point. This simply comes down to finding a Hamiltonian cycle in a graph. A graph is said to be *Hamiltonian* if it possesses a Hamiltonian cycle, for example K_n and C_n.

Similarly, a graph (of finite order) is said to be *semi-Hamiltonian* if there exists an elementary chain that passes through all the vertices of the graph. Such a chain is called a *Hamiltonian* chain.

Surprisingly, while the problem of finding a Hamiltonian cycle or chain is analogous to the problem of finding an Euler cycle or chain, it turns out to be much harder. As of today there exists no simple characterisation of Hamiltonian graphs. In fact, this is an $\mathcal{NPC}$ problem.

The problem of the knight's tour on a chessboard, first posed by Euler, consists in finding a Hamiltonian cycle on the graph $\Gamma = (V; E)$, where V is the set of $8 \times 8 = 64$ squares of the chessboard, and the neighbours of a vertex $v \in V$ are the 8 squares which can be reached from any given square by a single knight's move (less than 8 if the vertex is near the edge of the chessboard). Euler himself indicated a strategy to find such a cycle: starting from a corner of the chessboard, one covers the largest possible number of squares starting near the edge and moving bit by bit towards the centre; when it is no longer possible to make a move, if there remains some squares that have not been reached, Euler then describes an inverse method for modifying the trajectory in order to reach one of them, and in this way, by trial and error, the method ends up by producing a Hamiltonian cycle.

It is natural to think that the greater the number of edges compared to the number of vertices, the greater the likelihood that a given a graph is Hamiltonian. This following statement makes this idea precise. Let $\Gamma = (V; E, N)$ be a graph of finite order n. The *closure* of Γ, written $\mathrm{Cl}(\Gamma)$ is a graph having the same vertices as Γ, i.e. $V(\mathrm{Cl}(\Gamma)) = V(\Gamma)$, but whose edges satisfy the following property:

$$\text{if } x, y \in V(\mathrm{Cl}(\Gamma)) \text{ are not adjacent, then } d(x) + d(y) < n.$$

The closure is constructed as follows. Consider all the pairs $\{x, y\}$ of non-adjacent vertices for which $d(x) + d(y) \geqslant n$ in Γ, and add an edge $a = (\{x, y\}, 1)$. To distinguish the added edges, we assume that $1 \notin N$. The construction of the closure consists in making the pairs of non-adjacent vertices x, y such that $d(x) + d(y) \geqslant n$ into neighbours.

Proposition 2.4.7. *Let $\Gamma = (V; E, N)$ be a connected loopless graph such that $|V| = n \geqslant 3$. Then the graph Γ is Hamiltonian if and only if its closure $\mathrm{Cl}(\Gamma)$ is Hamiltonian.*

Proof. Clearly if Γ is Hamiltonian then $\mathrm{Cl}(\Gamma)$ is also Hamiltonian (because we add only edges, but no vertices). Suppose now that $\mathrm{Cl}(\Gamma)$ is Hamiltonian. Then $\mathrm{Cl}(\Gamma)$ is obtained by adding m' simple edges, which we denote by $e_k = (\{x_k, y_k\}, 1)$, $1 \leqslant k \leqslant m'$, yielding successive graphs

$$\Gamma_k = (V; E_k, N \cup \{1\}), \quad E_k = E_{k-1} \cup \{e_k\}, \quad 1 \leqslant k \leqslant m',$$
$$E = E_0 \subset E_1 \cdots \subset E_k \subset \cdots \subset E_{m'} \text{ and } \Gamma_{m'} = \mathrm{Cl}(\Gamma).$$

Let C be a Hamiltonian cycle of $\mathrm{Cl}(\Gamma)$; it contains the edge $e_{m'}$, so we can write

$$C = (t_0, a_0, t_1, a_1, \ldots, a_{n-2}, t_{n-1}, a_{n-1}, t_n) \text{ with } t_n = t_0$$

where $V = \{t_0, t_1, \ldots, t_{n-1}\}$, $x_{m'} = t_0$, $y_{m'} = t_{n-1}$, $a_{n-1} = e_{m'} = \{x_{m'}, y_{m'}\}$. By construction, in $\Gamma_{m'-1}$, the vertices $x_{m'}$, $y_{m'}$ are not adjacent, so $d(x_{m'}) + d(y_{m'}) \geqslant n$.

Consider the sets of indices

$$I = \{i \mid t_i \text{ adjacent to } x_{m'} \text{ in } \Gamma_{m'-1}\},$$
$$J = \{j \mid t_{j-1} \text{ adjacent to } y_{m'} \text{ in } \Gamma_{m'-1}\}.$$

Since the graph is loopless, we have $0 \notin I$. Since $x_{m'}$ is not adjacent to $y_{m'} = t_{n-1}$ in $\Gamma_{m'-1}$, we have $I \subset \{1, \ldots, n-2\}$ and $J \subset \{2, \ldots, n-1\}$ (recall that $n \geqslant 3$), so that we have

- $I \cup J \subset \{1, \ldots, n-1\}$, so $|I \cup J| \leqslant n-1$,
- $|I| + |J| = d(x_{m'}) + d(y_{m'}) \geqslant n$.

It follows that $I \cap J \neq \emptyset$. Let $l \in I \cap J$; t_{l-1} is adjacent to $y_{m'}$, t_l is adjacent to $x_{m'}$ (in Γ_{m-1}) and $\{t_{l-1}, t_l\}$ is an edge of C. Then the cycle

$$C' = (t_0, a_0, t_1, \ldots, a_{l-2}, t_{l-1}, \{t_{l-1}, y_{m'}\}, y_{m'}, a_{n-2}, t_{n-2}, \ldots, a_{l+1}, t_{l+1}, a_l, t_l, \{t_l, x_{m'}\})$$

with $x_{m'} = t_0$, is Hamiltonian in $\Gamma_{m'-1}$ (Figure 2.7).

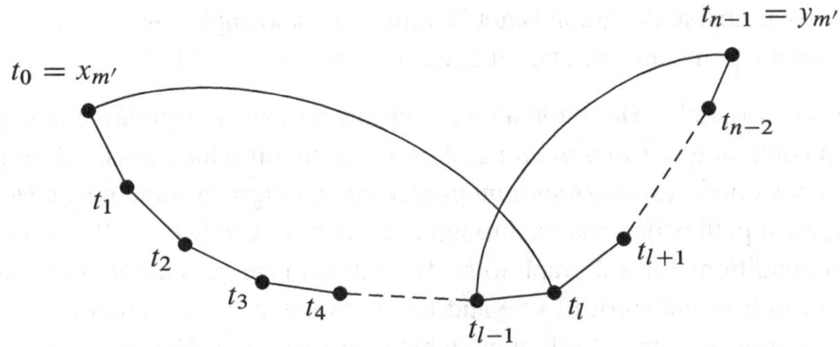

Figure 2.7. Representation of the cycle obtained in the proof of Proposition 2.4.7 with $a_0 = \{t_0, t_1\}, \ldots, a_{l-1} = \{t_{l-1}, t_l\}, \ldots, a_{n-2} = \{t_{n-2}, t_{n-1}\}$.

We apply the preceding construction starting from $\Gamma_{m'-1}$, and show analogously that $\Gamma_{m'-2}$ contains a Hamiltonian cycle. Thus, proceeding step by step, we finally obtain a Hamiltonian cycle in Γ. ■

We deduce from this the following theorem, first proved by Ø. Ore.

Theorem 2.4.8 (Ore, 1961). *Let $\Gamma = (V; E)$ be a simple connected graph with $|V| = n \geqslant 3$. If $d(x) + d(y) \geqslant n$, for every pair of non-adjacent vertices x, y, then the graph is Hamiltonian.*

Proof. By hypothesis, the closure $\mathrm{Cl}(\Gamma)$ of Γ is a complete graph, so it is Hamiltonian. The result then follows from Proposition 2.4.7. ■

We note that the connectedness follows from the condition on the sum of the degrees (exercise). This theorem generalises the earlier result that we now present, originally due to G. A. Dirac.

Corollary 2.4.9 (Dirac, 1952). *Let $\Gamma = (V; E)$ be a simple connected graph with $|V| = n \geqslant 3$. If $d(x) \geqslant n/2$ for every vertex x, then the graph is Hamiltonian.*

Proof. Indeed, if $d(x) \geqslant n/2$ for all $x \in V$, then for every pair of vertices (adjacent or not), we have $d(x) + d(y) \geqslant n$. But by Theorem 2.4.8, this implies that the graph is Hamiltonian. ■

It is unfortunately very difficult in general to determine whether the closure of a graph is Hamiltonian.

Exercise 2.3. Show that a simple graph $\Gamma = (V; E)$ having n vertices and m edges where $m \geqslant \frac{1}{2}(n-1)(n-2) + 2$ is Hamiltonian.

Hint: Assume that the graph is not Hamiltonian, and apply Theorem 2.4.8. Recall that in a simple graph, the number of edges is at most $n(n-1)/2$.

One can also define Hamiltonian digraphs, as follows. A *Hamiltonian digraph* is a digraph containing a *Hamiltonian circuit*, i.e. a circuit which passes through each vertex exactly once. A *semi-Hamiltonian digraph* is a digraph containing a *Hamiltonian path*, i.e. a path which passes through each vertex exactly once. There is a set of sufficient conditions for a digraph to be Hamiltonian (resp. semi-Hamiltonian). Let us state a result in the spirit of Ore and Dirac. We start by formulating an equivalence that is simple to prove, following which we will prove Meyniel's theorem, and complete it with several corollaries.

Proposition 2.4.10. *Let $\vec{\Gamma} = (V; \vec{E})$ be a simple digraph whose underlying graph is connected. Let $\vec{\Gamma}' = (V'; \vec{E}')$ be the digraph obtained by adding a vertex x' in such a way that for all $y \in V$, all the couples (x', y) and (y, x') are arcs. Then $\vec{\Gamma}$ admits a Hamiltonian path if and only if the digraph $\vec{\Gamma}'$ admits a Hamiltonian circuit.*

Proof. Left as an exercise. ■

Theorem 2.4.11 (Meyniel, 1973). *Let $\vec{\Gamma} = (V; \vec{E})$ be a simple, strongly connected digraph of order $n \geqslant 2$. If $d(x) + d(y) \geqslant 2n - 1$ for every non-adjacent pair of vertices $x, y \in V$, then the digraph is Hamiltonian.*

Meyniel's theorem follows from the following proposition.

Proposition 2.4.12. *Let $\vec{\Gamma} = (V; \vec{E})$ be a simple, strongly connected digraph, and assume that $\vec{\Gamma}$ contains no Hamiltonian circuit. Let C be an elementary circuit of maximum length k in $\vec{\Gamma}$. Then there exists $v \in W = V \setminus V(C)$ and s contiguous vertices $x_0, x_1, \dots, x_{s-1}$ on C, with $2 \leqslant s \leqslant k$, such that*

- $(x_0, v) \in \vec{E}$,
- $(x_j, v), (v, x_j) \notin \vec{E}$, $j = 1, \dots, s-1$,
- $d(v) + d(x_{s-1}) \leqslant 2|V| - s$.

Proof. Note first that $k \leqslant |V| - 1$, since otherwise C would be a Hamiltonian circuit of $\vec{\Gamma}$. Hence $W = V \setminus V(C) \neq \emptyset$ and $|W| = |V| - k$. We separate two cases.

(1) Assume that $\vec{\Gamma}$ contains no elementary path

$$(x, v_1, v_2, \dots, v_t, y),$$

where $x \neq y \in V(C)$, $t \geqslant 1$ and the v_j for $j = 1, \dots, t$ lie in W.

Since $\vec{\Gamma}$ is strongly connected and $W \neq \emptyset$, there exists a vertex x_0 on C and $v \in W$ such that $(x_0, v) \in \vec{E}$. There also exists an elementary path connecting v to x_0. If this path meets C a first time at $y \neq x_0$, there exists an elementary path

$(x_0, v_1 = v, v_2, \ldots, v_t, y)$ with $v_j \in W$, $j = 1, \ldots, t$, which is a contradiction. Thus we have $y = x_0$, and consequently there is an elementary circuit

$$C' = (x_0, v_1 = v, v_2, \ldots, v_t, x_0) \quad \text{with } v_j \in W \text{ for } j = 1, \ldots, t.$$

Let x_1 denote the vertex following x_0 along C.

Firstly, we see that x_0 is the only vertex of C which is a neighbour of v. Hence, writing $\mathbb{1}_P \in \{0, 1\}$ for the indicator of the property P (i.e. the function which takes value 1 if P holds and 0 otherwise), we obtain

$$d(v) + d(x_1) \leqslant 2 + 2(k-1) + \sum_{w \in W \setminus \{v\}} (\mathbb{1}_{(v,w) \in \vec{E}} + \mathbb{1}_{(w,x_1) \in \vec{E}} + \mathbb{1}_{(w,v) \in \vec{E}} + \mathbb{1}_{(x_1,w) \in \vec{E}}).$$

Secondly, if there existed a vertex $w \in W \setminus \{v\}$ such that (v, w, x_1) were a path in $\vec{\Gamma}$, then (x_0, v, w, x_1) would be an elementary path of $\vec{\Gamma}$, which contradicts the hypothesis. If there existed a vertex $w \in W \setminus \{v\}$ such that (x_1, w, v) were a path in $\vec{\Gamma}$, then we would have the path $P = (x_1, w, v = v_1, v_2, \ldots, v_t, x_0)$ in $\vec{\Gamma}$, which would contain the elementary path $(x_1, w, v_i, v_{i+1}, \ldots, v_t, x_0)$, where v_i is the *last successor* of w in P: this is a contradiction. Hence for $\{u, u'\} = \{v, x_1\}$ and for all $w \in W \setminus \{v\}$,

$$\mathbb{1}_{(u,w) \in \vec{E}} + \mathbb{1}_{(w,u') \in \vec{E}} \leqslant 1,$$

which gives

$$d(v) + d(x_1) \leqslant 2 + 2(k-1) + 2(|W| - 1) = 2|V| - 2.$$

This proves the result for $s = 2$.

(2) Assume that $\vec{\Gamma}$ contains an elementary path

$$C' = (x, v_1, v_2, \ldots, v_t, y)$$

where $x \neq y \in V(C)$, $t \geqslant 1$ and the v_j, $j = 1, \ldots, t$, all lie in W. Consider one such path C' for which the *oriented distance* $r \geqslant 1$ on C between any two vertices of C is minimal. Set $x_0 = x$ and $x_r = y$, and let $C = (x_0, x_1, \ldots, x_r, \ldots, x_{k-1}, x_k = x_0)$. Note that if we had $r = 1$, we would obtain the elementary circuit

$$(x_0, v_1, \ldots, v_t, x_1, \ldots, x_{k-1}, x_0) \quad \text{with } |V(C)| + t > |V(C)| \text{ vertices},$$

which contradicts the hypothesis that C is maximal. Hence $r \geqslant 2$.

We will need the following combinatorial lemma.

Lemma 2.4.13. *Let $K = (x_0, x_1, x_2, \ldots, x_{k-1})$ be a (not necessarily elementary) path on a simple digraph $\vec{\Gamma} = (V; \vec{E})$, and let $x \in V$ be a vertex not equal to any x_i.*

Assume that for all $i = 0, \ldots, k-2$, at least one of the two couples (x_i, x), (x, x_{i+1}) does not lie in $\vec{E}$.

Then the number $|\vec{E}(x \leftrightarrow K)|$ of arcs in either direction between the vertex x and the vertices of K is less than or equal to $|V(K)| + 1$.

Proof. Set

$$V_1 = \{v \in V(K)\backslash\{x_{k-1}\} \mid \exists i \text{ such that } v = x_i \text{ and } (x_i, x) \in \vec{E}\},$$
$$V_2 = \{v \in V(K)\backslash\{x_{k-1}\} \mid \exists i \text{ such that } v = x_i \text{ and } (x, x_{i+1}) \in \vec{E}\}.$$

Then $|\vec{E}(x \leftrightarrow K)| \leqslant |V_1| + |V_2| + 2$. But we have $V_1 \cup V_2 \subseteq V(K)\backslash\{x_{k-1}\}$ and $V_1 \cap V_2 = \emptyset$. Hence

$$\begin{aligned}|\vec{E}(x \leftrightarrow K)| &\leqslant |V_1| + |V_2| + 2 = |V_1 \cup V_2| + |V_1 \cap V_2| + 2\\ &\leqslant |V(K)| + 1.\end{aligned}$$

■

Set $v = v_1$. We see that

- $(x_0, v) \in \vec{E}$;
- v is not a neighbour of any vertex among $x_1, \ldots, x_{r-1}$, since otherwise we would have the elementary paths (x_0, v, x_j) or $(x_j, v, v_2, \ldots, v_t, x_r)$ contradicting the minimality of r.

It remains to show that there exists s such that $2 \leqslant s \leqslant r$ and $d(v) + d(x_{s-1}) \leqslant 2|V| - s$. We make the following observations.

Firstly, if the number of arcs $|\vec{E}(v \leftrightarrow K)|$ between v and the vertices of the path $K = (x_r, x_{r+1}, \ldots, x_{k-1}, x_k = x_0)$ were greater than or equal to $k - r + 3$, then by the above lemma, we would have a path (x_j, v, x_{j+1}), which would contradict the maximality of the length of C.

Secondly, there can be no path (x_j, w, v) or (v, w, x_j) in $\vec{\Gamma}$ with $w \in W\backslash\{v\}$ and $1 \leqslant j \leqslant r-1$, since this would again contradict the minimality of r by yielding the elementary paths $(x_j, w, v, v_2, \ldots, v_t, x_r)$ or (x_0, v, w, x_j).

Thirdly, let $s \geqslant 2$ be the greatest integer such that $\vec{\Gamma}$ contains an elementary path C'' from x_r to x_0 for which

$$V(C'') = \{x_r, \ldots, x_{k-1}, x_0, \ldots, x_{s-2}\}.$$

The concatenation of elementary paths C'' and C' is an elementary circuit of $\vec{\Gamma}$ passing through

$$|V(C')| + |V(C'')| - 2 = (t+2) + (k - r + s - 1) - 2 = k + t - r + s - 1$$

vertices. Thus, since C is assumed maximal, we find that $s \leqslant r - t + 1 \leqslant r$. Let $z \in V(C)\backslash V(C'')$. Lemma 2.4.13 implies that $|\vec{E}(z \leftrightarrow C'')|$ is less than or equal to

$|V(C'')| + 1 = k - r + s$. Moreover, there are plainly at most $2(r - s)$ arcs between x_{s-1} and $x_s, \dots, x_{r-1}$. Hence, given the facts established above, we have

$$d(v) + d(x_{s-1}) \leqslant (k - r + 2) + (k - r + s) + 2(r - s) + \sum_{w \in W \setminus \{v\}} \left(\mathbb{1}_{(w,v) \in \vec{E}} + \mathbb{1}_{(v,w) \in \vec{E}} + \mathbb{1}_{(w,x_{s-1}) \in \vec{E}} + \mathbb{1}_{(x_{s-1},w) \in \vec{E}}\right).$$

But for $\{u, u'\} = \{v, x_{s-1}\}$ and all $w \in W \setminus \{v\}$, we have

$$\mathbb{1}_{(u,w) \in \vec{E}} + \mathbb{1}_{(w,u') \in \vec{E}} \leqslant 1,$$

and hence

$$\begin{aligned} d(v) + d(x_{s-1}) &\leqslant k - r + 2 + k - r + s + 2(r - s) + 2(|W| - 1) \\ &\leqslant k - r + 2 + k - r + s + 2(r - s) + 2(|V| - k - 1) \\ &= 2|V| - s. \end{aligned}$$

This concludes the proof of Proposition 2.4.12. ■

We now give several corollaries of Meyniel's theorem.

Corollary 2.4.14. *Every strongly connected complete simple digraph admits a Hamiltonian circuit.*

Corollary 2.4.15 (Woodall's theorem, 1972). *Every simple strongly connected digraph $\vec{\Gamma}$ of order n satisfying $d(x) \geqslant n$ for all $x \in V(\vec{\Gamma})$ admits a Hamiltonian circuit.*

Corollary 2.4.16 (Ghouila-Houri's theorem, 1960). *Let $n \geqslant 2$ and let $\vec{\Gamma} = (V; \vec{E})$ be a simple strongly connected digraph of finite order n. If $d^-(x) \geqslant n/2$, $d^+(x) \geqslant n/2$ for all $x \in V$, then the digraph is Hamiltonian.*

Let $\Gamma = (V; E)$ be a simple graph. The *line-graph* of Γ is the simple graph $L(\Gamma) = (V'; E')$ defined as follows:

- $V' = E$: the vertices of $L(\Gamma)$ are the edges of Γ;
- if $a, b \in V'$, then $\{a, b\} \in E'$ if and only if the edges a and b are incident.

Proposition 2.4.17. *Let $\Gamma = (V; E)$ be a simple graph and Γ' a subgraph of Γ. Then $L(\Gamma')$ is an induced subgraph of $L(\Gamma)$.*

Proof. Let $u, v, w \in V(\Gamma')$ be such that $\{u, v\}$ and $\{v, w\}$ are two vertices of $L(\Gamma')$. They are adjacent in $L(\Gamma)$. As v is a vertex of Γ', we conclude that $\{u, v\}$ and $\{v, w\}$ are adjacent in $L(\Gamma')$. ■

Exercise 2.4. Show that if a simple graph $\Gamma = (V; E)$ is Euler, then its line-graph is Euler and Hamiltonian.

Let us end this chapter by showing how the famous *travelling salesman problem* is actually a problem on Hamiltonian graphs. The problem is as follows: a travelling salesman needs to visit several cities and then return to his starting point, while travelling the shortest possible distance. We may reformulate this problem in terms of networks as follows. Let $\Gamma = (V; E)$ be a graph whose vertices are the cities, each edge being a road between two cities, and define a valuation ω on the edges of Γ which gives the distance between two cities. The problem is then precisely that of finding a Hamiltonian cycle of minimal weight.

Chapter 3

(Di)graphs and data structures

The goal of algorithmics is the conception, the evaluation and the optimisation of computational methods to find solutions to given problems. Since manipulating data is the major task of computer science, it is absolutely indispensable to organise the work efficiently. The data structures arising from (di)graphs allow for simple and elegant algorithms.

Lists and stacks are one-dimensional dynamic structures which can be viewed as a particular type of graph, in which each element has a single (possibly empty) son. By generalising these dynamic structures to the multidimensional case, we obtain trees and arborescences that are very practical for structuring data, for example:

- a book is divided into chapters, each of which is itself divided into sections, themselves divided into paragraphs;
- an arithmetic expression can be viewed as a tree: the leaves are the variables, and the other vertices are the operators;
- the flowchart of a business is usually an arborescence;
- the organisation of files and directories in Unix or Linux systems is an arborescence.

This chapter is devoted to the type of data structures that correspond to trees and arborescences. We will only work with *finite simple graphs*. Most definitions and algorithms based on arborescence representations can be written efficiently and naturally using recursion, which is the subject of the first section below.

3.1 Recursive procedures

We saw some iterative algorithms in Chapters 1 and 2, but one can also conceive of functions which actually call upon themselves. These are called *recursive functions*, and they are based on recurrent processes. The advantage of this approach is the readability of algorithms, because the iterative form is often more complex and less intuitive. It is also an elegant way of solving problems, thanks to its simplicity and rapidity.

To properly understand the difference between these two approaches, we need to introduce the notion of *program stacking*. A program stack is a memory zone which is reserved to each program. Its function is to stock the local variables and parameters of the routines that constitute the program. We explain various stacking structures at the end of this chapter.

Suppose that we have a program using two procedures, of which `Procedure_1` calls `Procedure_2`. When `Procedure_2` is called, the processor executes the corresponding lines of code while keeping the values of its local variables, then emerges from this procedure and continues with `Procedure_1`. Thus, we need to stock the variables and the parameters of the procedure at each call. In a recursive program, the program stack manages the function calls and the local parameters, which are stacked as many times as there are calls. When the stopping condition is reached, the variables are popped (removed from the stack starting with the last one placed there). The stopping condition is especially important since it is the signal to start popping, and thus avoids stack overflow and saturation of the machine's memory capacity. An example is given in Algorithm 3.1.

```
Function Fact_rec(n: integer): integer
    begin
        if n == 0 then
            return 1;
        else
            return n × Fact_rec(n − 1);
        end
    end
EndFunction
```

Algorithm 3.1. Recursive factorial function.

How does Algorithm 3.1 work? Consider the case $n = 3$.

- *Stacking*
  - First call: `Fact_rec(3)`, $n = 3 \neq 0$, the program executes the alternative "Otherwise: 3 × `Fact_rec(2)` = ?";
  - Second call `Fact_rec(2)`, $n = 2 \neq 0$, the program executes the alternative "Otherwise: 2 × `Fact_rec(1)` = ?";
  - Third call `Fact_rec(1)`, $n = 1 \neq 0$, the program executes the alternative "Otherwise: 1 × `Fact_rec(0)` = ?";
  - Fourth call `Fact_rec(0)`, $n = 0$, the program returns 1.
- *Popping*
  - Return the value 1;
  - Compute 1×1;
  - Return the value 1;

- Compute 2×1;
- Return the value 2;
- Compute 3×2;
- Return of the value 6;

Algorithm 3.1 uses a *simple recursion*, i.e. it contains only a single recursive call.

Algorithm 3.2 uses *multiple recursion*, i.e. it uses several recursive calls. There is another type, called *mutual* or *crossed recursion*, consisting of several recursive functions which call each other.

We also have *nested recursive* functions, which are of the type in Algorithm 3.3. This algorithm makes a recursive call within a recursive call. Note that there are two essential steps to a recursive algorithm:

(i) The first consists in giving one or more stopping conditions.

(ii) The second is the recursive call.

```
Function Fibo_rec(n: integer): integer
    begin
        if n ≤ 1 then
            return 1;
        else
            return Fibo_rec(n − 2) + Fibo_rec(n − 1);
        end
    end
EndFunction
```

Algorithm 3.2. Recursive Fibonacci function.

```
Function Fn_rec_imb(m: integer, n: integer): integer
    begin
        /* Initialisation of the function in (m, 0) */
        if n == 0 then
            return v_m;
        else
            return Fn_rec_imb(Fn_rec_imb(m, n − 1), n − 1);
        end
    end
EndFunction
```

Algorithm 3.3. Example of a nested recursive function.

3.1.1 Complexity of a recursive algorithm

To compute the complexity of a recursive algorithm, it is not surprising that we need to solve recursive equations. Therefore, we begin this paragraph by introducing some ideas about how to solve this type of equation.

Linear recursive equation.

(1) A linear recursive equation of order 1 is an equality of the form

$$T(n) + g(n)T(n-1) = f(n);$$

(2) A linear recursive equation of order 2 is an equality of the form

$$T(n) + g(n)T(n-1) + h(n)T(n-2) = f(n).$$

We can define linear equations of any order, but we will only consider the cases of order 1 and 2 in this book.

Solving equations of order 1. We state the result without proof (exercise) in the case where g is constant.

Proposition 3.1.1. *Let $f\colon \mathbb{N} \to \mathbb{R}$ and $a, c \in \mathbb{R}$. Then the equation*

$$T(0) = a, \qquad T(n) = cT(n-1) + f(n), \quad n \geqslant 1,$$

has a unique solution given by

$$T(n) = c^n a + \sum_{i=1}^{n} c^{n-i} f(i), \quad n \geqslant 0.$$

Consider Algorithm 3.1, which recursively computes the factorial of an integer. We compute its complexity $T(n)$.

If $n = 0$, the cost is a comparison of integers, denoted c_e. If $n \neq 0$, the cost of Fact_rec(n) is c_e plus the cost of computing Fact_rec($n-1$) plus the cost of a multiplication of integers m_e. Thus, the complexity must satisfy the equation:

$$T(n) = c_e + T(n-1) + m_e.$$

Without loss of generality, we may assume that c_e and m_e are constants, so we obtain

$$T(n) = T(n-1) + k, \quad (k = c_e + m_e).$$

If we now apply Proposition 3.1.1 with $c = 1$ and $f(n) = 1$, we obtain $T(n) = 1 + nk$. We conclude that Algorithm 3.1 of the function *recursive factorial* Fact_rec has complexity $O(n)$.

Solving equations of order 2. Consider the linear recursive equation of order 2 given by

$$T(n) + g(n)T(n-1) + h(n)T(n-2) = f(n). \tag{3.1}$$

The associated *homogeneous equation* is

$$T(n) + g(n)T(n-1) + h(n)T(n-2) = 0. \tag{3.2}$$

Because the difference between any two solutions of the complete equation must satisfy the homogeneous equation, the general solution of the complete equation (3.1) is the sum of the general solution of the linear homogeneous equation (3.2) and of a particular solution of the equation with right-hand side (3.1).

In what follows, we will assume that $g(n)$ and $h(n)$ are positive constants. When g and h are constant, we have the following result.

Proposition 3.1.2. *Let $a, b \in \mathbb{R}$ be two numbers not simultaneously zero. Consider the following homogeneous recursive equation:*

$$T(n) = aT(n-1) + bT(n-2). \tag{3.3}$$

Let $\mathbf{S}$ *denote the set of all sequences $(T(n))_{n\geq 0}$ satisfying the induction* (3.3) *and $P = X^2 - aX - b \in \mathbb{R}[X]$ the associated characteristic polynomial. Then*

(i) $\mathbf{S}$ *is an $\mathbb{R}$-vector space isomorphic to $\mathbb{R}^2$ via the isomorphism*

$$\begin{cases} \mathbb{R}^2 \to \mathbf{S}, \\ (T(0), T(1)) \mapsto (T(n))_{n\geq 0}. \end{cases}$$

Let $\text{Vect}\{\mathcal{F}\}$ denote the vector subspace generated by the family $\mathcal{F}$ of $\mathbb{R}^{\mathbb{N}}$.

(ii) *If $\Delta = a^2 + 4b > 0$, then $\mathbf{S} = \text{Vect}\{(s^n)_{n\geq 0}; (t^n)_{n\geq 0}\}$, where s and t are the two real roots of P.*

(iii) *If $\Delta = 0$, then $\mathbf{S} = \text{Vect}\{(s^n)_{n\geq 0}; (ns^{n-1})_{n\geq 0}\}$, where s is the double root of P.*

(iv) *If $\Delta < 0$, then $\mathbf{S} = \text{Vect}\{(s^n \cos(n\theta))_{n\geq 0}; (s^n \sin(n\theta))_{n\geq 0}\}$, where $se^{i\theta}$ and $se^{-i\theta}$ are the two complex conjugate roots of P $(0 < \theta < \pi)$.*

Proof. We content ourselves with giving the main points of the argument.

(i) Observe that $\mathbf{S}$ is a vector subspace of $\mathbb{R}^{\mathbb{N}}$ and that $T(0)$ and $T(1)$ determine the sequence completely.

(ii) Find solutions that are geometric sequences $(r^n)_{n\geq 0}$; this comes down to showing that $P(r) = 0$. The sequences $(s^n)_{n\geq 0}$ and $(t^n)_{n\geq 0}$ are linearly independent, since the couples $(1, s)$ and $(1, t)$ are independent in $\mathbb{R}^2$.

(iii) Then check that $(ns^{n-1})_{n\geqslant 0}$ is indeed a solution of (3.3), and that it is linearly independent of $(s^n)_{n\geqslant 0}$ since the couples $(1, s)$ and $(0, 1)$ are independent in $\mathbb{R}^2$.

(iv) Note that $(s^n e^{in\theta})_{n\geqslant 0}$ is a *complex* solution of (3.3), so its real and imaginary parts lie in $\mathbf{S}$, and they are linearly independent since $\sin\theta \neq 0$. ■

Finding the general solution of the equation

$$T(n) - aT(n-1) - bT(n-2) = f(n)$$

is thus essentially based on finding a particular solution, which is quite a delicate operation in general.

As an illustration, let us compute the complexity of Algorithm 3.2 for the *recursive Fibonacci* function `Fibo_rec`, i.e. the number of operations $T(n)$ in the worst case, where

$$\begin{aligned} &T(0) = 1, \quad T(1) = 1, \\ &T(n) = T(n-2) + T(n-1) + 1 \quad \text{if } n \geqslant 2. \end{aligned}$$

The homogeneous equation relative to the sequence $(T(n))_{n\geqslant 0}$ is

$$s_n = s_{n-2} + s_{n-1}. \tag{3.4}$$

Its characteristic equation is thus $x^2 - x - 1 = 0$.

We obtain two distinct real roots

$$\phi = \frac{1+\sqrt{5}}{2} \quad \text{and} \quad -\frac{1}{\phi} = \frac{1-\sqrt{5}}{2}.$$

Thus the general solution of the homogeneous equation (3.4) is given by

$$s_n = \lambda\Big(\frac{1+\sqrt{5}}{2}\Big)^n + \mu\Big(\frac{1-\sqrt{5}}{2}\Big)^n, \quad n \geqslant 0.$$

The *right-hand side* of our recursive equation is constant with $f(n) = 1$. Thus, it is easy to see that there is a constant sequence which is a particular solution, because 1 is not a root of the characteristic equation. The sequence in question is $u_n = -1$, $n \geqslant 0$. Finally, once we have identified the coefficients of the linear combination that gives the general solution, the complexity $T(n)$ is

$$T(n) = \Big(1 + \frac{1}{\sqrt{5}}\Big)\Big(\frac{1+\sqrt{5}}{2}\Big)^n + \Big(1 - \frac{1}{\sqrt{5}}\Big)\Big(\frac{1-\sqrt{5}}{2}\Big)^n - 1, \quad n \geqslant 0.$$

Thus the complexity is $\Theta(\phi^n)$, which implies that the recursive Algorithm 3.2 is less efficient from this point of view than the iterative algorithm (Algorithm 1.2) from Chapter 1, whose complexity was $O(n)$.

3.2 The return of trees and arborescences

Using trees and arborescences makes it possible to perform dynamic operations such as searching for a predecessor or successor, or for a minimum or a maximum. They make it possible to perform deletions and insertions in a time proportional to the height. Thus, they are very efficient representations of data in computer science.

In this chapter, we will concentrate on arborescences (Section 2.2.2). A tree can be oriented by fixing a root r (Proposition 2.2.9), which induces an arborescence. Precisely, the unique orientation is obtained as follows: for every neighbour x_i of r, we define the arc (r, x_i), then we do the same with the neighbours of x_i that are different from r, and so forth.

We assume that *all parameters are pass-by-reference*.

Most programming languages, such as *C*, *C++* or *Java*, are *typed languages*, which means that the type of each of variable is declared beforehand, for example **integer:** n; **real number:** x; **character:** a; **character string:** u; **table** (resp. **chain, list**): $T\,[size\_max]$. The names of the declared variables make it possible to store their values in different parts of the machine's memory as a function of their type.

A *data structure* is a way of defining a new type of variable that is in fact a group of several items of data (called fields), which may be of different types, under a single name. The declaration of a structure is given as follows.

Struc *name_of_the_structure*{
    type of field_1: field_1
    type of field_2: field_2
    ⋮
    type of field_k: field_k
}

Example 3.2.1. We define the structure *student* consisting of 4 different fields:

Struc *student*{
    character string: lastname
    character string: firstname
    integer: age
    character string: student_id_number
}

Here is how to declare a variable of type **student:**

student: E

Here is an example of how to initialise the fields:

E.lastname := "martin"
E.firstname := "marcel"
E.age := 20
E.student_id_number := "1234"

and how to display the contents of different fields of the variable E:

display E.firstname, E.lastname, E.age, E.student_id_number.

We now give a new definition of arborescences. An *arborescence* is defined recursively as follows:

- either it is empty,
- or it consists of a vertex called the *root* together with a (possibly empty) sequence of non-empty arborescences.

3.2.1 Representation of an arborescence in the son-brother format

For each vertex y that is not a leaf of the arborescence, we put an arbitrary order $x_1, \ldots, x_k$ on its sons. We say that x_1 is *the son* of y, that x_{i+1} is *the brother* of x_i, $i = 1, \ldots, k-1$, and that *the brother* of x_k is empty. If y is a leaf, we say that *its son* is empty. Thus each vertex is joined to *its* son and to *its* brother.

To represent an oriented forest (i.e. a family of arborescences), we add a new root ω, joined by an outgoing arc to each of the roots of the arborescences in the forest. This gives a single arborescence.

Struc *vertex*{
  integer or **real number** or **character string:** value
  vertex: son
  vertex: brother
}

Struc *arborescence*{
  vertex: root
}

We saw that an arborescence is characterised recursively by its root. When we apply the command **return** to an arborescence, the result is the arborescence constructed starting from its root.

Recall that a *leaf* is a vertex of degree 1 different from the root. The *height* of a vertex x in an arborescence (viewed as an oriented graph) is the length of the longest path starting from x; the *height of an arborescence* is the height of its root; the *depth* of a vertex x is the length of the unique path from the root to x.

Height of an arborescence. To illustrate the notion of an arborescence, we first give some functions used to compute the height of a vertex and of an arborescence (Algorithms 3.4, 3.5 and 3.6).

```
Function Height_arborescence(A: arborescence): integer
    begin
        return Height_vertex(A.root);
    end
EndFunction
```

Algorithm 3.4. Function to compute the height of an arborescence.

```
Function Height_vertex(current: vertex): integer
    begin
        if current == null then
            return −1;
        else
            return Height_siblings(current.son) + 1;
        end
    end
EndFunction
```

Algorithm 3.5. Function to compute the height of a vertex.

```
Function Height_siblings(current: vertex): integer
    integer: height_son, height_brother;
    begin
        if current == null then
            return −1;
        else
            height_son := Height_siblings(current.son);
            height_brother := Height_siblings(current.brother);
            return max(1 + height_son, height_brother);
        end
    end
EndFunction
```

Algorithm 3.6. Recursive function to compute the maximal height of a sibling.

Next, we give the procedures to display the value of the vertices in an arborescence (Algorithms 3.7 and 3.8).

Algorithm 3.8 follows a *preorder traversal*: each vertex is treated at the first visit. A *postorder traversal* is when each vertex is treated at the final visit, and an *inorder traversal* is when each vertex is treated between the two recursive calls.

The notation <> means "different from".

```
Procedure Display_arborescence(A: arborescence)
    begin
        Display_vertex(A.root);
    end
EndProcedure
```

Algorithm 3.7. Procedure to display an arborescence.

```
Procedure Display_vertex(current: vertex)
    begin
        if current <> null then
            display current.value;
            Display_vertex(current.son);
            Display_vertex(current.brother);
        end
    end
EndProcedure
```

Algorithm 3.8. Procedure to display a vertex.

Theorem 3.2.2. *Algorithm 3.7 has time complexity $\Theta(n)$, where n is the number of vertices of the arborescence.*

Proof. Since Algorithm 3.7 is a simple call which takes a constant time t_0, its complexity can be evaluated starting from that of the recursive Algorithm 3.8. Let $T(n)$ denote the complexity for an arborescence with n vertices. If the arborescence is empty, then $T(0) = t_1$ where $t_1 > 0$ is simply the time needed for the test *current* <> *null*. Write $t_2 > 0$ for the time (assumed constant) outside of the recursive all, during the first execution of Algorithm 3.8, when $n > 0$. Suppose $n > 0$, and make the induction hypothesis that for $k < n$, we have $T(k) = \alpha k + t_1$. Let x be a vertex of the arborescence. The left subarborescence has $k \geqslant 0$ vertices, so the right subarborescence has $n - k - 1$. Consequently, the execution time is

$$T(n) = T(k) + T(n - k - 1) + t_2.$$

Thus

$$\begin{aligned}T(n) &= T(k) + T(n-k-1) + t_2\\ &= (\alpha k + t_1) + (\alpha(n-k-1) + t_1) + t_2\\ &= \alpha n + (2t_1 + t_2 - \alpha).\end{aligned}$$

The choice $\alpha := t_1 + t_2$ leads to $T(n) = (t_1 + t_2)n + t_1$ for all integers n, which gives the complexity $\Theta(n)$ of the statement. ∎

We will give a more complete study of the complexity of algorithms on trees in Section 3.2.6.

Exercise 3.1. Write a procedure `Display_vertex_post` to display vertices according to the postorder traversal.

Leaves of an arborescence. We can count the leaves as in Algorithm 3.9. We determine if a vertex is a leaf based on the auxiliary function in Algorithm 3.10 and also on the function in Algorithm 3.11, which counts the number of leaves situated under a given vertex.

```
Function Count_leaves_tree(A: arborescence): integer
    begin
        return Count_leaves_under(A.root);
    end
EndFunction
```

Algorithm 3.9. Function to count the leaves of an arborescence.

```
Function Is_leaf(current: vertex): Boolean
    begin
        if current.son == null then
            return true;
        else
            return false;
        end
    end
EndFunction
```

Algorithm 3.10. Function to determine if a vertex is a leaf.

```
Function Count_leaves_under(current: vertex): integer
    begin
        if current == null then
            return 0;
        else
            if Is_leaf(current) then
                return 1;
            else
                return Count_leaves_under(current.son) +
                 Count_leaves_under(current.brother);
            end
        end
    end
EndFunction
```

Algorithm 3.11. Function to count the leaves under a vertex.

Copying an arborescence. The function in Algorithm 3.12, which copies an arborescence, is very useful. It needs Algorithm 3.13.

```
Function Copy_tree(A: arborescence): arborescence
    arborescence: tmp;
    begin
        tmp := nouvarbo();                    /* creation of a new arborescence */
        tmp.root := Copy_vertex(A.root);
        return tmp;
    end
EndFunction
```

Algorithm 3.12. Function to duplicate an arborescence.

3.2.2 Binary arborescences

Binary arborescences are particularly useful for representing decision problems, or for storing variables efficiently. A binary arborescence is

- either empty,
- or consists of a root and two binary subarborescences: the left one and the right one.

```
Function Copy_vertex(current: vertex): vertex
    vertex: ver;
    begin
        if current == null then
            return null;
        else
            ver := newvertex();                       /* creation of a new vertex */
            ver.value := current.value;
            ver.son := Copy_vertex(current.son);
            ver.brother := Copy_vertex(current.brother);
            /* initialisation of the new vertex */
            return ver;
        end
    end
EndFunction
```

Algorithm 3.13. Function to copy a vertex.

In practice, we use the following structure:

```
Struc vertex{
    integer or real number or character string: value
    vertex: left                                      /* the left son */
    vertex: right                                     /* the right son */
}

Struc arbor_binary{
    vertex: root
}
```

A level in a binary arborescence is *filled* when every vertex in the level has exactly two sons. Two binary arborescences A_1 and A_2 are *isomorphic* if

- either A_1 and A_2 are empty,
- or the left subarborescence of A_1 is isomorphic to the left subarborescence of A_2 and the right subtree arborescence of A_1 is isomorphic to the right subarborescence of A_2.

This definition allows us to differentiate between the two expressions $a \times b - c$ and $c - a \times b$ represented in Figure 3.1 by two non-isomorphic binary arborescences.

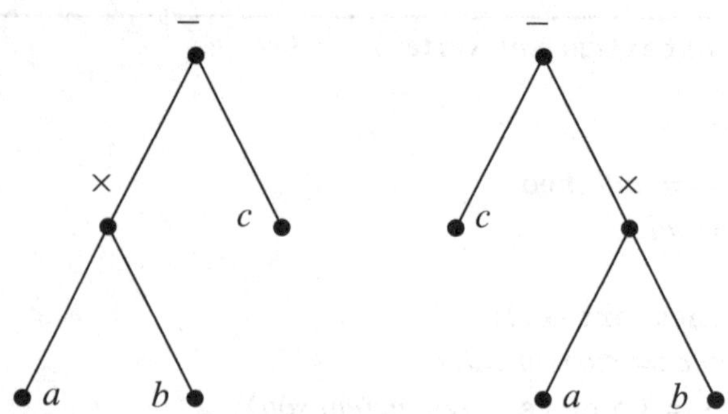

Figure 3.1. Two non-isomorphic binary arborescences, which are isomorphic as digraphs.

Isomorphism. Algorithms 3.14 and 3.15 check if two arborescences are isomorphic as arborescences.

```
Function Is_isomorphic(A1: arbor_binary, A2: arbor_binary): Boolean
    begin
        return Iso_under(A1.root, A2.root);
    end
EndFunction
```

Algorithm 3.14. Function to decide whether two binary arborescences are isomorphic.

```
Function Iso_under(v1: vertex, v2: vertex): Boolean
    begin
        if v1 == null or v2 == null then
            return v1 == v2 ;        /* returns the Boolean value of the comparison */
        if v1 <> null and v2 <> null then
            return Iso_under(v1.left, v2.left) and
              Iso_under(v1.right, v2.right);
        else
            return false;
        end
    end
EndFunction
```

Algorithm 3.15. Function to compare the binary arborescences coming out of two vertices.

Exercise 3.2. Adapt the procedures in Algorithms 3.14 and 3.15 from the case of general arborescences to binary arborescences.

3.2.3 Binary search arborescences

Binary arborescences are efficient structures for representing data. However, certain operations such as deletion or insertion of a vertex are not easy for arbitrary binary arborescences. To better perform these operations, we introduce some particular binary arborescences. For this, we first add an order on the vertices.

A *binary search arborescence* is a binary arborescence the values of whose vertices are ordered according to a (not necessarily total) order satisfying the following property: for each vertex x, we assume that

- all the vertices of the subarborescence to the left of x have value strictly less than the value of x;
- all the vertices of the subarborescence to the right of x have value strictly greater than the value of x.

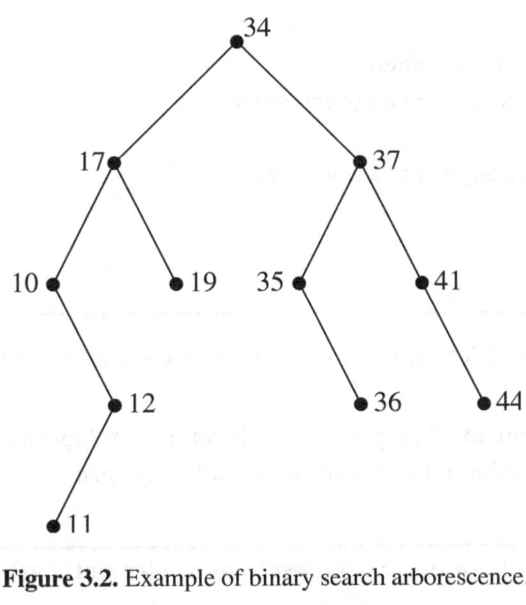

Figure 3.2. Example of binary search arborescence.

There are three particularly relevant operations on binary search arborescences: search, insertion, and deletion of an element. An *element* here means the value that a vertex can take, generally an **integer**, a **real number** or a **character string.** In the following algorithms, we will use the phrase *arborescence* to signify a binary search arborescence.

Searching for an element. Both Algorithms 3.16 and 3.17 allow to search for an element in a tree.

```
Function Find_vertex(A: tree, x: element): vertex
    /* returns a vertex of A of value x */
    begin
        return Seek_vertex(A.root, x);
    end
EndFunction
```

Algorithm 3.16. Function to find a vertex of given value.

```
Function Seek_vertex(current: vertex, x: element): vertex
    /* returns the first vertex coming out of current having the value x */
    begin
        if current == null or current.value == x then
            return current;
        end
        if current.value > x then
            return Seek_vertex(current.left,x);
        else
            return Seek_vertex(current.right, x);
        end
    end
EndFunction
```

Algorithm 3.17. Function to find a vertex of given value under a vertex.

Insertion of an element. The goal of the functions in Algorithms 3.18, 3.19 is to insert a new leaf of value x in a binary search arborescence.

```
Function Insert_element(A: arborescence, x: element): arborescence
    begin
        /* returns the new arborescence after insertion of a leaf of value x */
        Insert_leaf(A.root, x);
        return A;
    end
EndFunction
```

Algorithm 3.18. Function to insert an element in a binary search arborescence.

```
Procedure Insert_leaf(current: vertex, x: element)
  begin
    if current.value > x then
      if current.left == null then
        current.left := newvertex();          /* creation of a new vertex */
        current.left.value := x;
        current.left.left := null;
        current.left.right := null;
      else
        Insert_leaf(current.left, x);
      end
    else
      if current.right == null then
        current.right := newvertex();         /* creation of a new vertex */
        current.right.value := x;
        current.right.left := null;
        current.right.right := null;
      else
        Insert_leaf(current.right, x);
      end
    end
  end
EndProcedure
```

Algorithm 3.19. Recursive procedure to insert a new leaf of value x coming from the vertex *current* into the subarborescence.

We can also insert an element x at the root of a binary search arborescence A with root r.

Method. To do this, we first cut the arborescence A into two subarborescences A_1 and A_2, and then construct a new arborescence A' where x is the root of A', A_1 is the left subarborescence of A', and A_2 is the right subarborescence of A'. The cut is made as follows.

(1) *If x is strictly less than the value of the root of A*, then we cut the left subarborescence of A into two subarborescences A_{11} and A_{12} in such a way that the values of all the vertices of A_{11} are less than x and the values of all the vertices of A_{12} are greater than x.

- For the left subarborescence of A', we take $A_1 = A_{11}$; thus, it is the left cut of the left sons of A.
- For the right subarborescence of A' we take the arborescence whose root is r, whose left subarborescence is A_{12}, and whose right subarborescence is the right subarborescence of A.

(2) *If x is strictly greater than the value of the root of A*, then we cut the right subarborescence of A into two subarborescences A_{21} and A_{22} in such a way that the values of all the vertices of A_{21} are less than x and the values of all the vertices of A_{22} are greater than x.

- For the left subarborescence of A', we take the arborescence whose root is r, whose left subarborescence is the left subarborescence of A, and whose right subarborescence is the subarborescence A_{21}.
- For the right subarborescence of A', we take $A_2 = A_{22}$; thus, it is the right cut of the right sons of A.

Exercise 3.3. Show that the above method generates a binary search arborescence.

Here follows the procedure to cut an arborescence.

```
Procedure Cut_tree(v: vertex, v1: vertex, v2: vertex, x: element)
  begin
    if v == null then
      v1 := null;
      v2 := null;
    else
      if x < v.value then
        v2 := v;
        Cut_tree(v.left, v1, v2.left, x);
      else
        v1 := v;
        Cut_tree(v.right, v1.right, v2, x);
      end
    end
  end
EndProcedure
```

Algorithm 3.20. Procedure to cut an arborescence.

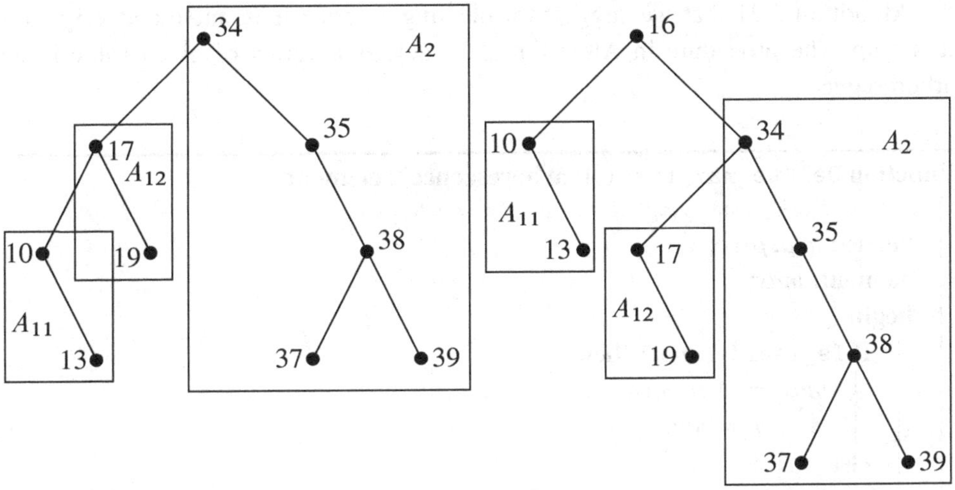

Figure 3.3. Example of cutting: insertion of the value 16 at the root of the arborescence. As the value 16 is less than the value of the root (34) of the initial arborescence, we cut the left subarborescence of the initial arborescence into two subarborescences: the subarborescence whose values are less than 16 and the subarborescence whose values are greater than 16. We insert 16 at the root of a new arborescence whose left subarborescence and right subarborescence are the subarborescences obtained according to the method described above.

Example 3.2.3. Figure 3.3 shows an example of cutting.

Deletion. The deletion of an element is also an important operation. Suppose the vertex v of value x exists. Then

- either v is a leaf and its deletion does not involve any reorganisation of the arborescence;
- or v has only one son, in which case we replace v by this son;
- or v has two sons, in which case we delete the vertex v' of maximum value of the left subarborescence of root v, and put it in the place of v, and we replace the vertex v' by the root of its left subarborescence.

Exercise 3.4. Show that the above method yields a binary search arborescence.

Remark 3.2.4. To apply this method, we need to find and delete the vertex of maximum value in a binary search arborescence (we assume that the value to be deleted does lie in the arborescence).

Algorithm 3.21 that we suggest for deleting a vertex is iterative and very easy to set up. The procedure in Algorithm 3.22 deletes a vertex of given value in an arborescence.

```
Function Delete_max_tree(A: arborescence): element
    /* returns the maximum of arborescence A and deletes it */
    vertex: tmp, prec;
    element: max;
    begin
        if Is_leaf(A.root) then
            max := A.root.value;
            A.root := null;
        else
            if A.root.right == null then
                max := A.root.value;
                A.root := A.root.left;
            else
                tmp = A.root;
                while tmp.right <> null do
                    prec := tmp;
                    tmp := tmp.right;
                end
                max := tmp.value;
                prec.right := tmp.left;
            end
        end
        return max;
    end
EndFunction
```

Algorithm 3.21. Deletion of the vertex of maximum value in a binary search arborescence, returning this value.

```
Procedure Delete_vertex(A: arborescence, x: element)
    vertex: current;
    element: max;
    arborescence: tmp;
    begin
        if A.root <> null then current := A.root;
        else
            while current.value <> x do
                if x < current.value then current := current.left;
                else current := current.right;
            end
            if Is_leaf(current) then current := null;
            else
                if current.left == null then
                    /* there is only one son, the right */
                    current.value := current.right.value;
                    current.left := current.right.left;
                    current.right := current.right.right;
                    /* we can also simplify these three assignations by using */
                    current := newvertex(current.right);
                else
                    if current.right == null then
                        /* there is only one son, on the left */
                        current := newvertex(current.left);
                    else
                        /* the vertex to be deleted has 2 sons */
                        tmp.root := current.left;
                        max := Delete_max_tree(tmp);
                        current.value := max;
                        /* replace the value to be deleted by the maximum of the left
                           subarborescence */
                    end
                end
            end
        end
    end
EndProcedure
```

Algorithm 3.22. Deletion of a vertex of given value x.

3.2.4 Priority search trees, heaps

A priority search arborescence is a particular binary arborescence used to implement a type of data called a *priority queue*. These structures are particularly interesting in the case of a queue for a set of processes which are associated with priorities.

In a binary arborescence, the *depth* of a vertex x is the length of the unique path from the root to x; a *level* is the set of all the vertices of the same depth.

A level of height h is *full* if it contains 2^h vertices. A binary arborescence of height h is said to be *left quasiperfect* if it satisfies the following properties: all the levels except possibly the final one are full, and in this case the leaves of the final level are *packed* on the left.

A *priority arborescence* is a binary arborescence that is left quasiperfect and whose vertices have a valuation that is a partial order called the *priority*, which satisfies the following property: no vertex has a value greater than that its father. Figure 3.4 gives an example of a priority arborescence.

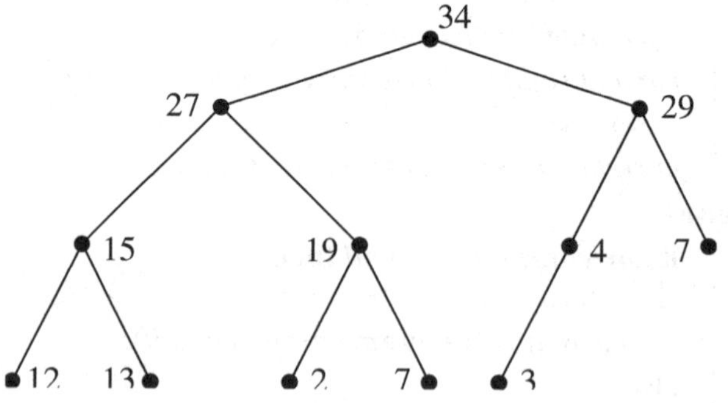

Figure 3.4. Priority search arborescence. The same value can appear several times in a priority search arborescence: here 7.

The coding of a left quasi perfect arborescence is done via an indexed table starting from 0. The size of the table *size_max* is an integer fixed beforehand, sufficiently large to allow all the manipulations that may possibly occur to be entered into the table.

Struc *arborescence_quasiperfect_left*{
  integer: *size* /* number of vertices of the arborescence */
  table of elements: *tab*[*size_max*]
}

We write $\lfloor x \rfloor$ for the *floor of* x, i.e. the largest integer less than or equal to the real number x.

The indices of the table are distributed as follows:

- the root is at the index 0;
- if a vertex is at the index i, then
  - its left son is at the index $2i + 1$,
  - its right son is at the index $2i + 2$,
  - its father is at the index $\lfloor \frac{i-1}{2} \rfloor$ (*the root has no father*);
- if n is the number of vertices of the arborescence and if $i > \frac{n-1}{2}$, then i is a leaf.

A priority search arborescence whose vertices are indexed as above is called a *heap*. The declaration of the structure relative to a heap is identical to that a left quasi perfect arborescence.

Example 3.2.5. The table $T = [35, 27, 29, 1, 7, 6]$ corresponds to the arborescence in Figure 3.5.

Insertion and deletion. Naturally, we need to be able to add and delete elements of a heap in such a way that the resulting structure is still a heap. The procedure in Algorithm 3.23 adds the value x to the heap T in such a way as to conserve the heap structure.

```
Function Add_element(T: heap, x: element): heap
    integer: i, j;
    begin
        T.size := T.size + 1;
        i := T.size − 1;
        j := ⌊(i−1)/2⌋;
        while i ⩾ 1 and x > T.tab[j] do
            T.tab[i] := T.tab[j];
            i := j;
            j := ⌊(i−1)/2⌋;
        end
        T.tab[i] := x;
        return T;
    end
EndFunction
```

Algorithm 3.23. Function adding an element to a heap.

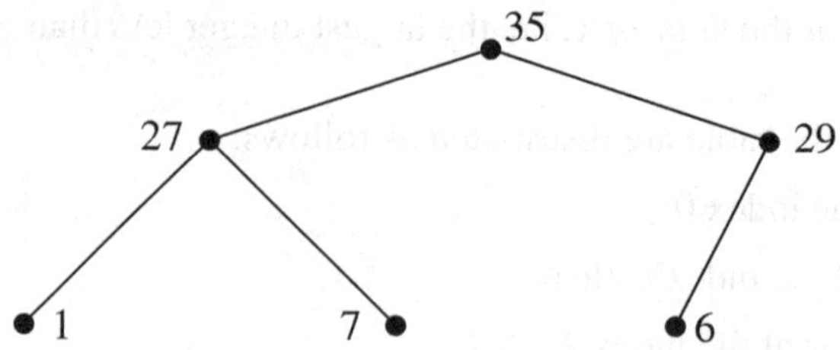

Figure 3.5. Coding of an arborescence via a table.

Add element to a heap; remove maximum from a heap. Algorithm 3.24 allows removing the maximum element from a heap.

- We remove $T.tab[0]$
- We remove $x := T.tab[T.size - 1]$ (*value of the rightmost leaf*)
- $size := size - 1$
- $i := 0$
- As long as $T.tab[i]$ has one or two sons whose value is greater than x:
  - *following* := index of the son of $T.tab[i]$ which has the greatest value
  - $T.tab[i] := T.tab[following]$
  - $i := following$
  - $T.tab[i] := x$

```
Function Delete_max_heap(T: heap): element
    integer: next, i, n;
    element: x, max;
    Boolean: done;
    begin
        max := T.tab[0];
        T.size := T.size − 1;
        n := T.size;
        x := T.tab[n];
        next := 0;
        i := 0;
        done := false;
        while done == false do
            if i > ⌊(n−1)/2⌋ then done := true;
            else
                if i == ⌊(n−1)/2⌋ or T.tab[2i + 1] > T.tab[2i + 2] then
                    next := 2i + 1;
                else
                    next := 2i + 2;
                end
            end
            if T.tab[next] > x then
                T.tab[i] := T.tab[next];
                i := next;
            else
                done := true;
            end
        end
        T.tab[i] := x;
        return max;
    end
EndFunction
```

Algorithm 3.24. Function to delete the maximum in a heap and return its value.

Heap sorting. We can sort a table T thanks to the heap structure in Algorithm 3.25. We can now sort T by using the function `Delete_max_heap` which deletes the maximum in a heap (see Algorithm 3.26).

```
Function Create_heap(T: table of elements): heap
    integer: n := T.size;
    /* extract the size of the table T */
    heap: A := newtas();
    /* initialisation of an empty heap */
    begin
        for i from 0 to n − 1 do
            A := Add_element(A,T[i]);
        end
        return A;
    end
EndFunction
```

Algorithm 3.25. Function creating a heap starting from a table.

```
Function Sort_table(T: table of elements): table
    integer: n := T.size;
    element: max;
    table: tab[n];
    heap: A := Create_heap(T);
    begin
        for i from 0 to n − 1 do
            max := Delete_max_heap(A);
            tab[i] := max;
        end
        return tab;
    end
EndFunction
```

Algorithm 3.26. Function to heap sort a table.

3.2.5 AVL arborescences

A binary search arborescence is said to be an *AVL* (for Adelson–Velskii–Landis) if it satisfies the following "balance" property: at every vertex, the heights of the left subarborescence and of the right subarborescence differ by at most 1.

We represent an AVL arborescence as a binary search arborescence by adding to the structure *vertex* an additional field containing the height of the subarborescence having this vertex as root:

Struc *vertex*{
  integer or **real** or **character string:** value
  vertex: left
  vertex: right
  integer: height
}

Struc *AVL*{
  vertex: root
}

An example of an AVL arborescence is given in Figure 3.6.

Exercise 3.5. Give a recursive algorithm which tests whether a binary search arborescence is an AVL.

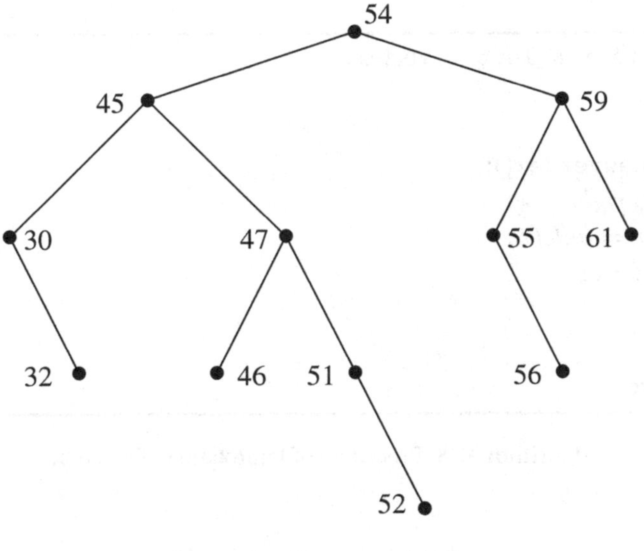

Figure 3.6. Example of AVL.

The advantage of the AVL structure lies in the fact that it avoids the degenerate cases (in which the left and right subarborescences have a large difference of depth). But adding or deleting elements in an AVL can create imbalance in the arborescence, making it necessary to rebalance it. To do this, we apply *rotations* to non-empty arborescences.

Exercise 3.6. What happens if the vertex C in the left-hand arborescence of Figure 3.7 lies under the tree B_d?

Rotations. In this paragraph, we describe the rotations that can be applied to arborescences. In the following paragraph, we will give algorithms to balance an arborescence using rotations. Algorithms 3.27 and 3.28 implement left rotation corresponding to Figure 3.8. Similar algorithms can be easily written for a right rotation (see Figure 3.7 for an understanding of how it works). Figures 3.9 and 3.10 illustrate the composition of right and left rotations.

```
Procedure Rotation_left_tree(A: AVL)
    begin
        Rotation_left(A.root);
    end
EndProcedure
```

Algorithm 3.27. Procedure of left rotation of an arborescence.

```
Procedure Rotation_left(v: vertex)
    vertex: w;
    begin
        w := newvertex();
        w := v.right;
        v.right := w.left;
        w.left := v;
        v := w;
    end
EndProcedure
```

Algorithm 3.28. Procedure of left rotation of a vertex.

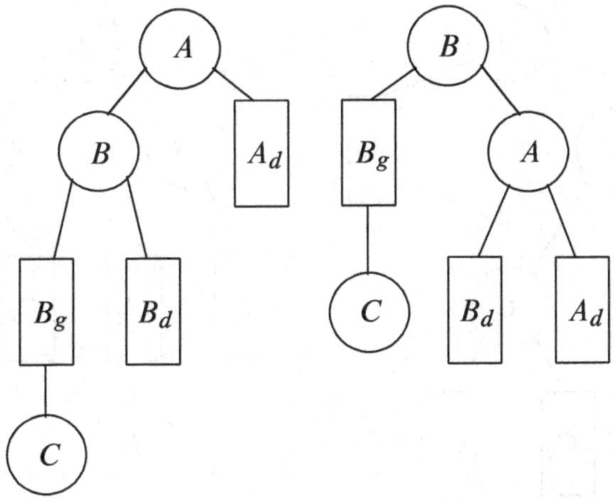

Figure 3.7. Right rotation. In the left-hand tree, the subtrees A_d, B_d and B_g have height difference at most 1 when we do not take into account C. Thus, the binary tree without vertex C is an AVL. The addition of C creates an imbalance in the heights. We rebalance them by performing a rotation to the right.

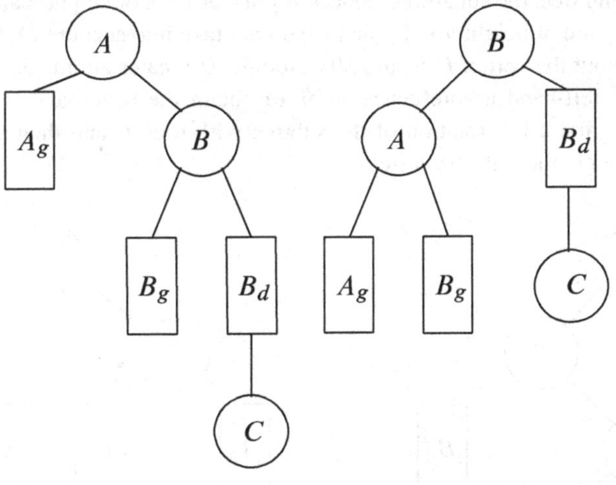

Figure 3.8. Left rotation. It is the symmetric case in comparison with Figure 3.7. In order to rebalance the heights of the tree structure, we perform instead a rotation to the left.

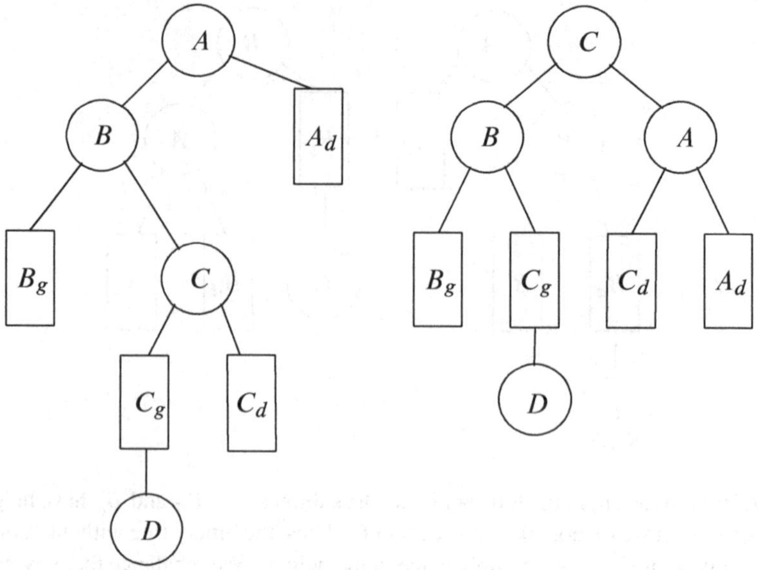

Figure 3.9. Left rotation of the left subarborescence followed by a right rotation of the whole tree. In the left-hand tree, the subarborescences B_g and A_d are of height h and the subarborescences C_g and C_d are of height $h-1$ when we do not take into account D. Consequently, the arborescence without the vertex D is an AVL. Adding D creates an imbalance in the heights. We rebalance the left-hand arborescence so as to obtain the right-hand tree by performing a double rotation: first a left rotation of the subtree with root B and then a right rotation of the whole tree. We get the right-hand tree.

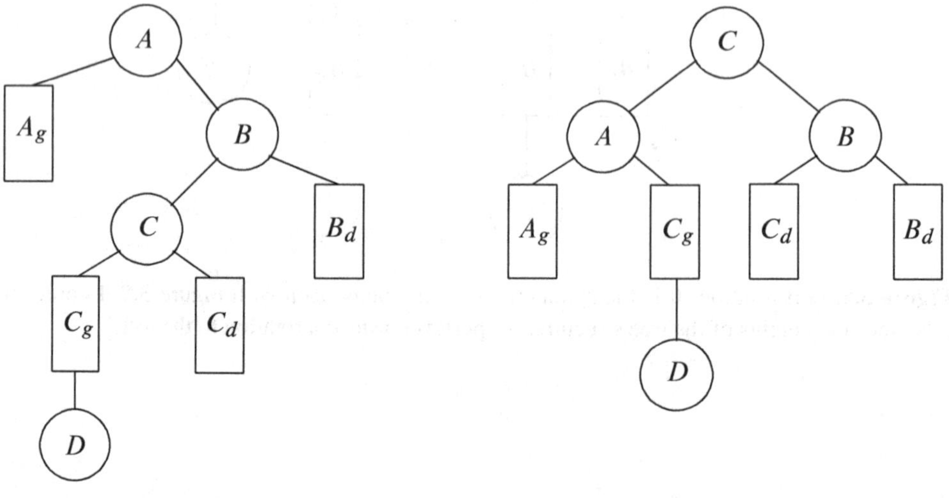

Figure 3.10. Right rotation of the right-hand subarborescence followed by a left rotation of the whole tree. We proceed as in Figure 3.9 except that we first perform in the left-hand tree the right rotation of the subtree with root B and then a left rotation. We get the right-hand tree.

Balancing. Let A be an arborescence, and let A_g and A_d be its left and right subarborescences respectively. Suppose that there is an imbalance between A_g and A_d, i.e. that

$$|A.root.left.height - A.root.right.height| = 2.$$

We have Algorithms 3.29 and 3.30 to rebalance the arborescence.

```
Procedure Balance_tree(A: AVL)
  begin
    Balance_subtree_struc(A.root);
  end
EndProcedure
```

Algorithm 3.29. Balancing an arborescence.

```
Procedure Balance_subtree_struc(v: vertex)
  begin
    if v.left.height − v.right.height == 2 then
      if v.left.left.height < v.left.right.height then
        Rotation_left(v.left);
        Rotation_right(v);
      else
        Rotation_right(v);
      end
    else
      if v.left.height − v.right.height == −2 then
        if v.right.right.height < v.right.left.height then
          Rotation_right(v.right);
          Rotation_left(v);
        else
          Rotation_left(v);
        end
      end
    end
  end
EndProcedure
```

Algorithm 3.30. Balancing a subarborescence.

Insertion. We need to be able to insert a vertex into an AVL in such a way that whole structure remains an AVL (Algorithms 3.31 and 3.32).

```
Function Insert_vertex(v: vertex, A: AVL): AVL
  begin
    Insert_vertex(v, A.root);
    return Balance_tree(A);
  end
EndFunction
```

Algorithm 3.31. Procedure to insert a vertex into an AVL.

```
Procedure Insert_vertex_root(v: vertex, w: vertex)
  begin
    if w == null then
      w := newvertex();
      w.value := v.value;
      w.height := 0;
      w.left := v.left;
      w.right := v.right;
    else
      if v.value < w.value then Insert_vertex_root(v, w.left);
      else Insert_vertex_root(v, w.right);
    end
  end
EndProcedure
```

Algorithm 3.32. Procedure to insert a vertex into an AVL.

Exercise 3.7. Write an algorithm to delete a vertex of given value from an AVL.

3.2.6 Properties of binary arborescences

In what follows, all arborescences will be assumed to have at least one vertex. We have already seen that a perfect binary arborescence of height h possesses $2^{h+1} - 1$ vertices. Thus, in every binary arborescence, the number of vertices must satisfy $n \leqslant 2^{h+1} - 1$.

Proposition 3.2.6. *Let A be a binary arborescence of height h with n vertices. Then* $\log_2(n+1) - 1 \leqslant h \leqslant n - 1$.

Proof. We know that for an arborescence, the number m of edges is equal to $n - 1$. Thus $h \leqslant n - 1$. Furthermore, we have $n \leqslant 2^{h+1} - 1$, so $\log_2(n+1) \leqslant h + 1$, i.e. $\log_2(n+1) - 1 \leqslant h$. ■

Corollary 3.2.7. *Let A be a binary arborescence of height h with n vertices. Then $h \leqslant n \leqslant 2^{h+1} - 1$.*

Proposition 3.2.8. *Let A be a binary arborescence with n vertices and f leaves. Then $f \leqslant \frac{n+1}{2}$.*

Proof. We use induction on the number n of vertices. The statement holds for $n = 1$. Suppose the property holds for a binary arborescence having at most $n - 1$ vertices. Let A be a binary arborescence having n vertices. Then by definition, A can be decomposed as $A = \langle r, A_g, A_d \rangle$, where r is the root of A and A_g (resp. A_d) is the left (resp. right) subarborescence of A. Thus the number of vertices of A_g is $n_1 < n$, and the number of vertices of A_d is $n_2 < n$. By the induction hypothesis, the number of leaves of A_g is $f_1 \leqslant \frac{n_1+1}{2}$ and the number of leaves of A_d is $f_2 \leqslant \frac{n_2+1}{2}$. But $f = f_1 + f_2 \leqslant \frac{n_1+1}{2} + \frac{n_2+1}{2}$. Since $n = n_1 + n_2 + 1$, we have $f \leqslant \frac{n_1+n_2+1}{2} = \frac{n+1}{2}$. ■

Proposition 3.2.9. *Let A be a binary arborescence of height h with n vertices and f leaves. Then $\log_2(f) \leqslant h$.*

Exercise 3.8. Prove this proposition.

Theorem 3.2.10. *Let A be an AVL of height h with n vertices. Then*

$$\log_2(n+1) \leqslant h + 1 \leqslant \alpha \log_2(n+2),$$

where $\alpha = (\log_2(\phi))^{-1}$ and $\phi = \frac{1+\sqrt{5}}{2}$, the golden ratio.

Proof. We always have $n \leqslant 2^{h+1} - 1$, so $\log_2(n+1) \leqslant h + 1$. Let $f(h)$ denote the minimal number of vertices of an AVL. Then $f(h) = 1 + f(h-1) + f(h-2)$. Indeed, the AVL has a minimal number of vertices, so the difference between the left and right arborescences is 1 (Figure 3.11).

The sequence defined by $F(h) = f(h) + 1$ is a shifted Fibonacci sequence: $F(0) = 2$, $F(1) = 3$, $F(2) = 5, \ldots$, and so we get the following equality: $F(h) = \frac{1}{\sqrt{5}}(\phi^{h+3} - \phi^{-(h+3)})$, $h \geqslant 0$. Hence $n + 1 \geqslant F(h) > \frac{1}{\sqrt{5}}(\phi^{h+3} - 1)$ and thus

$$h + 3 < \log_\phi(\sqrt{5}(n+2)) < \frac{\log_2(n+2)}{\log_2(\phi)} + 2.$$ ■

Corollary 3.2.11. *For an AVL, we have $h = O(\log_2(n))$.*

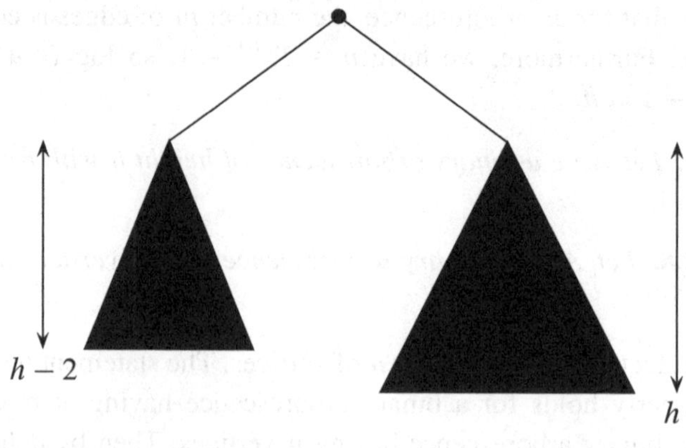

Figure 3.11. Structure of an AVL having a minimal number of vertices.

3.3 Time complexity of algorithms on binary arborescences

Most of the classical operations on binary search arborescences concern the operations of searching for, inserting or deleting an element. The algorithms described in the preceding sections descend within the arborescence by moving from fathers to right or left sons until the desired result is reached (finding, insertion or deletion of an element). Consequently, the vertices encountered during the various recursive calls form a descending chain coming out of the root of the arborescence. Thus, the execution is in $O(h)$ time.

Two different cases can occur:

- each vertex possesses only a single left or right son; the arborescence is then reduced to a chain, and the complexity of classical operations is $O(n)$;
- the arborescence is perfect; then we have $n = 2^{h+1} - 1$ and $h + 1 = \log_2(n + 1)$; the height h is of order $O(\log_2(n))$, so the classical operations are of complexity $O(\log_2(n))$.

One can prove (but it is not easy) that the average height of a binary search arborescence is $O(\log_2(n))$.

For an AVL, insertion and deletion are also $O(\log_2(n))$. Simple or double rotations take a constant time $O(1)$. Furthermore, we have the following result (which we state without proof).

Theorem 3.3.1. • Insertion: *to rebalance an AVL arborescence after an insertion, a single or double rotation suffices.*

- Deletion: *to rebalance an AVL arborescence after a deletion, one may need up to $O(\log_2(n))$ single or double rotations.*

The basic structure of a heap is a quasi-perfect left arborescence. From this we deduce the following proposition.

Proposition 3.3.2. *Let T be a heap with height $h \geqslant 1$ and number of vertices n. Then $h = \lfloor \log_2(n) \rfloor$. Thus, the height h of a heap has complexity $O(\log_2(n))$.*

Proof. As the basic structure of a heap is a quasi-perfect left arborescence, all the levels up to $h - 1$ are full. Thus we have the following bounds:

$$2^h + 1 \leqslant n \leqslant 2^{h+1} - 1, \text{ so } 2^h \leqslant n \leqslant 2^{h+1} - 1;$$

thus in particular $h \leqslant \log_2(n) < h + 1$. Hence $h = \lfloor \log_2(n) \rfloor$. ∎

By Proposition 3.3.2, the classical operations on heaps are $O(\log_2(n))$. Observe that the search for the maximal element takes constant time $O(1)$.

Exercise 3.9. Show that the complexity of the heap sort (see Algorithm 3.26) is $O(n \log_2(n))$.

3.4 Back to graphs

In Chapter 1 we saw how to code a (di)graph. In this section, we will give more detail on representations of (di)graphs in computer science. In what follows, we assume that all graphs are simple; note however that all the results can easily be adapted to multigraphs.

3.4.1 Representation by the adjacency matrix

A (di)graph $\Gamma = (V; E)$ with n vertices can be represented by its adjacency matrix $M(\Gamma)$, which is the $n \times n$ matrix given by

$$M[x; y] = \begin{cases} 1 & \text{if } \{x, y\} \in E \quad \text{(or for digraphs } (x, y) \in E\text{)}, \\ 0 & \text{otherwise.} \end{cases}$$

For a non-oriented graph, the adjacency matrix is symmetric. Let us give an algorithmic procedure for the adjacency matrix.

```
Struc graph_mat{
    integer: n                                  /* number of vertices of the (di)graph */
    table of integers: tab[size_max][size_max]
}
```

where *size_max* is fixed and chosen large enough to cover all needs.

Exercise 3.10. Adapt the representation of a (di)graph by its adjacency matrix to (di)graphs equipped with values associated to the edges, to multigraphs, and to multigraphs equipped with values associated to the edges.

3.4.2 Representation by a table of adjacency lists

We can also represent a (di)graph $\Gamma = (V; E)$ with n vertices by a table of n linked lists: the i-th list is the list of neighbours of the vertex indexed by i in the table.

```
Struc vertex{
    integer or real or character string: value
    vertex: following
}

Struc list{
    vertex: first
    vertex: last
}

Struc graph_mat{
    integer: n                          /* number of vertices of the (di)graph */
    table of lists: tab[size_max]
}
```

In what follows, the word graphs will generally also indicate digraphs unless specifically indicated otherwise.

3.4.3 Remarks on the complexity of these representations

For an algorithm which necessitates a complete reading of a (di)graph with n vertices and m edges (or arcs), the cost will be at least $O(n^2)$ in the case of a adjacency matrix, under the assumption that reading or writing a single coefficient in the matrix takes constant time $O(1)$.

For the representation of a (di)graph by a table of adjacency lists, the cost will be at least that of reading the structure *graph_mat*, so again under the assumption that reading or writing a vertex takes constant time, it has complexity $O(n + m)$.

Determining if an edge (or an arc) exists then costs

- $O(1)$ for the adjacency matrix;
- $O(n)$ for the table of adjacency lists.

We can also represent a (di)graph by lists of successors and lists of predecessors. When the (di)graph is not very dense, i.e. when the number of edges (or of arcs) is small compared to the square of the number of vertices (i.e. $m = O(n^2)$), the table of adjacency lists is preferable.

3.4.4 (Di)graph traversal

Breadth-first. *Idea*: starting from a vertex v, we visit the vertices situated at a distance of 1 from v which have not already been visited, then the vertices at distance 2 from v which have not already been visited, etc.

We will use the queue data structure (first in, first out). We also need a Boolean table *mark*[ ], declared as a general variable, which will mark the vertices that have already been visited so as not to visit them twice. The **queue** structure and the procedures of enqueuing and dequeuing are defined at the end of this chapter.

```
Procedure Breadth_first_search(Γ: graph, v: vertex)
    vertex: w, current;
    queue: F;
    table: mark[size_max];
    /* if it was not already declared in the second algorithm below */
    begin
        for every vertex w of Γ do
            mark[w] := 0;
        end
        mark[v] := 1;
        Enqueue(F, v);
        repeat
            current := Dequeue(F);
            for every successor w of current do
                if mark[w] == 0 then
                    Enqueue(F, w);
                    mark[w] := 1;
                end
            end
        until F == empty queue;
    end
EndProcedure
```

Algorithm 3.33. Breadth-first search starting from a vertex.

Remark 3.4.1. Algorithm 3.33 only allows us to visit vertices that can be accessed through the vertex v, which is insufficient when the (di)graph is not connected. In that case, to visit the entire graph, we will need to use a vertex marking that allows us to know which vertices have not already been visited in the course of the preceding traversal (cf. Algorithm 3.34).

However, we need to pay careful attention to the declaration of our variables so as to be sure they are not redundant between the two algorithms, or that one algorithm does not erase the variables of the other. Here, we have declared and initialised the table *mark*[ ] in both algorithms.

```
Procedure Breadth_first_search_tot(Γ: graph)
    vertex: v, w;
    table: mark[size_max];
    begin
        for every vertex v of Γ do
            mark[v] := 0;
            for every successor w of v do
                if mark[w] == 0 then Breadth_traversal(Γ, w);
            end
        end
    end
EndProcedure
```

Algorithm 3.34. Breadth-first traversal of a graph with several connected components.

Complexity of breadth-first search. The complexity of Algorithm 3.33 comes from the loop "**for** every successor w of *current* **do**".

If the (di)graph is represented by its adjacency matrix, then

- either all the vertices are accessible from v, in which case it is necessary to read the matrix entirely, which can be done in $O(n^2)$ time. Consequently, there is only a single call to the loop "**for** every vertex v of Γ **do**" in Algorithm 3.34. Thus we see that Algorithm 3.34 has complexity $O(n^2)$.
- Or the vertices are not all accessible from v, and consequently the loop "**for** every successor w of *current* **do**" has complexity $O(kn^2)$, where k is the quotient of the number of vertices accessible from v by the total number of vertices. In the loop "**for** every vertex v of Γ **do**" of Algorithm 3.34, there will be multiple calls (equal to the number of connected components of the graph, or of strongly

connected components for a digraph). In the end this loop, which calls Algorithm 3.33 several times, will have read the entire matrix. Thus, Algorithm 3.34 has complexity $O(n^2)$.

If the (di)graph is represented by a table of adjacency lists, we reason as above, and this loop will have complexity $O(n + m)$ (edges are traversed only once).

Depth-first search. *Idea*: go as far as possible along a chain (resp. a path) starting from a vertex v. For this process, we start from v and then call the algorithm recursively for each vertex neighbouring v. As for breadth-first search, we use a table *mark*[ ], declared as a general variable, to mark the vertices of the graph so as not to visit them twice.

```
Procedure Depth_first_search(Γ: graph, v: vertex)
    vertex: w;
    begin
        mark[v] := 1;
        for every successor w of v do
            if mark[w] == 0 then Depth_first_search(Γ, w);
        end
    end
EndProcedure
```

Algorithm 3.35. Depth-first search of a graph starting from a vertex.

```
Procedure Depth_first_search_tot(Γ: graph)
    vertex: v;
    begin
        for every vertex v of Γ do
            mark[v] := 0;
        end
        for every vertex v of Γ do
            if mark[v] == 0 then Depth_traversal(Γ,v);
        end
    end
EndProcedure
```

Algorithm 3.36. Depth-first search of a graph with one or more connected components.

Remark 3.4.2. Since these algorithms pass variables by reference, one could put the table *mark*[ ] as a parameter in Algorithms 3.35 and 3.36 if needed, especially in Algorithm 3.35:

Procedure `Depth_first_search`(Γ: **graph**, v: **vertex**, *mark*: **table**)

Exercise 3.11. Study the complexity of the two procedures above for depth-first search (Algorithms 3.35 and 3.36).

Differences between the algorithm of breadth-first search and the algorithm of depth-first search. We saw that the algorithm of breadth-first traversal uses a **queue** type data structure. We then ave a recursive version of the depth-first traversal procedure; however, there also exists an iterative version. Let us briefly analyse the analogies between these two approaches. We need Algorithms 3.37 and 3.38.

```
Procedure Visit_vertex(v: vertex)
    vertex: w;
    begin
        mark v;
        for every vertex w adjacent to v do
            add w to the end of the list L;
        end
    end
EndProcedure
```

Algorithm 3.37. Visiting vertices adjacent to a vertex v.

Observe that this outline corresponds to both breadth-first search and depth-first search. In general, every type of search of the vertices of a graph is based on the following description:

(1) place the current vertex in the structure;

(2) remove the first available vertex of the structure;

(3) place all the yet-unexplored neighbours in the structure;

(4) as long as the structure is non-empty, return to step (2).

The only difference is in the actual choice of data structure: **queue** or **stack**.

```
Procedure Traversal_outline(Γ: graph)
    list: L := null;
    tree: A := null;
    vertex: v, w;
    begin
        choose a vertex v;
        Visit_vertex(v);
        while L is non-empty do
            remove the vertex w from the beginning of L;
            if w is not marked then
                add the vertex w to V(A);
                add the edge {v, w} to E(A);
                Visit_vertex(w);
            end
        end
    end
EndProcedure
```

Algorithm 3.38. Outline of graph traversal.

3.5 Complements

3.5.1 Simple data types

In computer science, most languages are *typed*, i.e. every variable has an associated

- name;
- data type.

The name is used to identify and recognise the variable, while the data type determines its nature. Defining the type of a variable tells the memory how many octets need to be allocated to store the variable. Recall that a *bit* is the binary unit of the quantity of information; it generally takes only the values 0 or 1. A block of 8 consecutive bits forms a *byte* or *octet*. Thus, whenever we declare a variable, it is necessary to tell the compiler what type of data the variable will contain. The exact manner of declaration depends on the language.

Basic types. Several basic types of variable can be declared: **integer**, **real number**, **character**, **chain**, **Boolean**. The declaration is done as follows:

integer: x;
$x := 10$;

or more concisely:

integer: $x := 10$;

or for example:

chain: $s :=$ "character string";

A table is a more elaborate basic type of variable. A table is a set of elements, all of the same type, which are identified by indices. A table has a maximal size; it can have one, two or several dimensions, each of which must have a maximal size.

The declaration

table of elements: $T[N]$;

defines a one-dimensional table with N *cells*, namely $T[0], T[1], \dots, T[N-1]$, each configured so as to contain one *element* (an **integer**, a **real number** or a **chain**, etc.); the parameter N is the maximal size of the table, i.e. the maximum number of cells which can be called by the algorithm.

The indices of a table start at 0. An element of the table is accessed by giving the name of the table and the position of the element. For example $T[3]$ will give the 4-th element of the one-dimensional table T. To store a 3×2 matrix, for example, we can use a table of the type $Mat[3][2]$. To display all the elements of this matrix, we will need a nested double "**for**... **do**" loop. The six elements of the matrix are identified by their positions:

$$Mat[0][0], Mat[0][1], Mat[1][0], Mat[1][1], Mat[2][0], Mat[2][1].$$

3.5.2 Functions

A function or procedure is a set of instructions that are grouped in order to perform a specific set of operations. A function can take zero, one or several parameters, and it returns a certain number of values (possibly none). A function is declared as in Algorithm 3.39.

```
Function Name_of_the_function(parameter_1: type of the parameter_1,
  parameter_2: type of the parameter_2, ..., parameter_k: type of the
  parameter_k): type of the returned value
    Declaration of local variables;
    begin
        Instruction_1;
        Instruction_2;
        ⋮
        Instruction_q;
    end
EndFunction
```

Algorithm 3.39. Structure of a function.

In general, the function returns one or more values using the command **return**. To execute a function, one calls it from the interior of a program, providing it with the necessary parameter values if needed:

Function_name($x_1, x_2, \ldots, x_k$).

Example 3.5.1. Consider the example of Algorithm 3.40. The function takes two integers as arguments, computes their sum and returns the answer as an integer. We call it as follows in a command window:

> Sum(7, 6)	*type the command line then* ↵
> 13	*display the result*

```
Function Sum(x: integer, y: integer): integer
    integer: z;
    begin
        z := x + y;
        return z;
    end
EndFunction
```

Algorithm 3.40. Compute the sum of two integers.

3.5.3 Passing parameters to functions

There exist two ways of passing parameters to a function.

Passing by value. A variable x is declared and initialised in a function `Function_1`, then passed as a parameter to another function `Function_2` by a call of the form `Function_2`(x). This variable is then copied to a stack (called the call stack). It is this copy that will be used by `Function_2`.

This means that `Function_2` does not use the variable initialised in `Function_1`. Consequently, on coming out of `Function_2`, if one does not return its value nor affect it to another variable, the changes made by `Function_2` to x will not be stored and will be lost;

Passing by reference. The variable is passed in `Function_2` using its address in memory. The function `Function_2` will make changes to the base object and not to a copy, as when passing by value.

Example 3.5.2. Consider the function in Algorithm 3.41, which exchanges two objects (here integers), and the following lines of commands:

```
x := 3;
y := 10;
Exchange(x, y);          /* gives the result 10, 3 */
display x;               /* gives 3 */
display y;               /* gives 10 */
```

Above, we used copies of x and a copy of y as local variables of the function `Exchange`, so that the actual values of x and y themselves were not modified.

However, when variables are passed by reference, then they may be modified during a procedure. That is the case for some of the algorithms given in this chapter.

```
Function Exchange(x: integer, y: integer)
    integer: z;
    begin
        z := x;
        x := y;
        y := z;
        display x, y;
    end
EndFunction
```

Algorithm 3.41. Function to exchange the order of two integers.

Remark 3.5.3. As a general rule, when a structure does not return anything, we will prefer to use a procedure.

3.5.4 Linear structures

Stacks. A stack is a data structure in which adding and removing elements can be done only at the top of the pile. The term used is LIFO (for *last in, first out*). We can implement a stack in the form of a table or a list.

In table form, we have

```
Struc stack{
    table of elements: tab[size_max]
    integer: size
}
```

By convention, a stack will be empty when $size = 0$. The operations needed to manipulate stacks are

- push: adding an element to the top on the pile;
- pop: removing an element from the top of the stack and returning it;
- stack empty?: returns true if the stack is empty and false otherwise;
- number of elements in the stack: returns the number of elements in the stack.

Pushing. Algorithm 3.42 allows to push an element x onto the stack P.

```
Function Push(P: stack, x: element): stack
    begin
        if P.size < size_max then
            P.tab[P.size] := x;
            P.size := P.size + 1;
            return P;
        else
            display "Stack full";
            return P;
        end
    end
EndFunction
```

Algorithm 3.42. Function to push x onto P.

Popping. To pop elements off the stack, we need two functions: the first returns the last element that was pushed (Algorithms 3.44), and the second modifies the stack (Algorithm 3.45). We start with Algorithm 3.43 to test if a stack is empty.

```
Function Non_empty(P: stack): Boolean
    begin
        if P.size ⩾ 1 then return true;
        else return false;
    end
EndFunction
```

Algorithm 3.43. Function to test if a stack is empty.

```
Function Pop_element(P: stack): element
    begin
        if Non_empty(P) then
            return P.tab[P.size − 1];
        else
            display "Stack empty";
            return k0;
        end
    end
EndFunction
```

Algorithm 3.44. Function that pops and returns the last element of a stack.

```
Function Dequeue(P: stack): stack
    begin
        if Non_empty(P) then
            P.size := P.size − 1;
            return P;
        else
            display "Stack empty";
            return P;
        end
    end
EndFunction
```

Algorithm 3.45. Function to pop the last element of P and return the resulting stack.

The element k_0 in Algorithm 3.44 is an element chosen previously to indicate that the stack is empty, which is generally necessary if one needs to continue with subsequent operations.

The implementation by linked list is done using a dynamic allocation of memory:

Struc *cell*{
  integer or **real number** or **character string:** value
  cell: following
}

Struc *linked_list*{
  cell: first
}

We also have the elementary functions in Algorithms 3.46, 3.47 and 3.48.

```
Function Non_empty(P: linked_list): Boolean
  begin
    if P.first == null then return false;
    else return true;
  end
EndFunction
```

Algorithm 3.46. Function to test if a linked_list is empty.

```
Function Push(P: linked_list, x: element): linked_list
  cell: C;
  begin
    C.value := x;
    C.next := P.first;
    P.first := C;
    return P;
  end
EndFunction
```

Algorithm 3.47. Function to push an element onto a linked_list.

```
Function Dequeue(P: linked_list): linked_list
    cell: C;
    begin
        if Non_empty(P) then
            C := P.first;
            P.first := C.next;
            return P;
        else
            display "Linked list empty";
            return P;
        end
    end
EndFunction
```

Algorithm 3.48. Function to pop the last element of a linked_list.

Queues. Queues are used to deal with objects that are waiting on hold. It works exactly like a waiting list according to the FIFO principle: *first in, first out*. The typical example would be people waiting in line in an administrative office. Every person who arrives gets a number which allows them to be called up to the counter in the order of arrival. Doing this means we need to act with respect to both the first arrival and the latest arrival.

Struc *cell_in_queue*{
  integer or **real number** or **character string:** value
  cell_in_queue: following
}

Struc *file*{
  cell_in_queue: first
  cell_in_queue: last
}

Algorithms 3.49, 3.50 and 3.51 illustrate the operations of *enqueuing* (adding an element to the end of a queue) and *dequeuing* (removing the first element of the queue). Observe that in Algorithm 3.51, the cell *F.last* keeps all the listed elements in the inverse order even after dequeuing.

```
Function Non_empty(F: file): Boolean
    begin
        if F.first == null then
            return false;
        else
            return true;
        end
    end
EndFunction
```

Algorithm 3.49. Function to test if a queue is empty.

```
Function Push(F: file, x: element): file
    cell_in_queue: C;
    begin
        C.value := x;
        C.next := null;
        if Non_empty(F) then
            F.last.next := C;
            C.next := null;
            F.last := C;
        else
            F.first := C;
            F.last := C;
        end
        return F;
    end
EndFunction
```

Algorithm 3.50. Function to add a last element to a queue.

```
Function Dequeue(F: file): file
    cell_in_queue: C;
    begin
        if Non_empty(F) then
            if F.first.following == null then
                F.first := null;
                return F;
            else
                C := F.first;
                F.first := C.following;
                return F;
            end
        else
            display "error";
            return F;
        end
    end
EndFunction
```

Algorithm 3.51. Function to remove the first element in a queue.

Chapter 4

Connectedness and flows in networks

It is very important to be able to determine *to what extent* a graph is connected, or more precisely, how many vertices or edges need to be removed in order to disconnect it. This notion of connectedness is particularly relevant for analysing the *vulnerability* of networks. In Chapter 1, we defined the notion of a network as a (di)graph with a valuation. Examples of applications could be an individual who needs to drive from her home to her job using the network of roads, or a business that needs to send its manufactured products to the markets of various countries using the full travel network of routes by road, sea and air.

Networks are indispensable tools in our globalised society. Their practical use in areas such as transport, communication, and distribution is fundamental. This was the motivating reason for the development of the mathematical analysis of (di)graphs, which gave rise to the theory of flows. In this chapter, we will introduce the elementary properties of networks. Throughout the entire chapter, the word (di)graph refers only to *simple (di)graphs of finite order*.

4.1 Vertex-connectedness and edge-connectedness

Let $\Gamma = (V; E)$ be a simple graph of finite order with at least one vertex, and let k be a non-negative integer. Then Γ is said to be *k-vertex-connected* (or simply *k-connected*) if

(i) $|V| > k$ and

(ii) for every subset $X \subset V$ such that $|X| < k$, the induced subgraph $\Gamma(V \setminus X)$ is connected.

We note that a k-connected graph is also h-connected for every non-negative integer $h \leqslant k$. Any graph with at least one vertex is automatically 0-connected.

When the graph $\Gamma = (V; E)$ possesses at least one vertex, we can also define its level of *vertex-connectedness*. This number, which we denote by $\kappa(\Gamma)$, is the largest integer $k \geqslant 0$ such that Γ is k-vertex-connected. Thus, $\kappa(\Gamma)$ is the minimal number of vertices that must be removed to disconnect the graph, i.e. to make it non-connected:

$$\kappa(\Gamma) = \min\{h \geqslant 0 \mid \exists X \subset V \text{ such that } |X| = h \text{ and } \Gamma(V \setminus X) \text{ non-connected}\}.$$

If $|V| = 1$, then $\kappa(\Gamma) = 0$ and Γ is connected. When $|V| \geqslant 2$, $\kappa(\Gamma) = 0$ means that Γ is not connected. For the complete graph K_n of order $n \geqslant 2$, it is easy to see that $\kappa(K_n) = n - 1$. For the cycle C_n of order $n \geqslant 4$, we have $\kappa(C_n) = 2$.

At the end of this chapter, we will give a characterisation of k-connectedness expressed in terms of disjoint chains (Whitney's theorem, see Section 4.5).

For $k \geqslant 0$, there is a dual notion to k-vertex-connectedness, namely k-edge-connectedness. We say that Γ is *k-edge-connected* if

(i) $|E| > k$ and

(ii) for every subset $F \subset E$ such that $|F| < k$, the subpartial graph $(V, E \backslash F)$ obtained by removing the edges of F is connected.

A graph satisfying this definition must have at least one edge and two vertices. More precisely, in a k-edge-connected graph, every vertex is adjacent to at least k other vertices. This means that the degree of each vertex is greater than or equal to k, so that the number of edges satisfies $|E| \geqslant k|V|/2$ by Lemma 1.1.9 (i).

When $|E| \geqslant 2$, we define the *edge-connectedness* of Γ, denoted $\lambda(\Gamma)$, to be the largest integer k such that Γ is k-edge-connected; thus it is the minimal number of edges that must be deleted to disconnect the graph:

$$\lambda(\Gamma) = \min\{h \geqslant 0 \mid \exists F \subset E \text{ such that } |F| = h \text{ and } (V; E \backslash F) \text{ non-connected}\}.$$

When $|E| \geqslant 2$, then $\lambda(\Gamma) = 0$ means that Γ is not connected. These notions of connectedness are illustrated in Figure 4.1.

The following results give the relations between vertex-connectedness and the edge-connectedness.

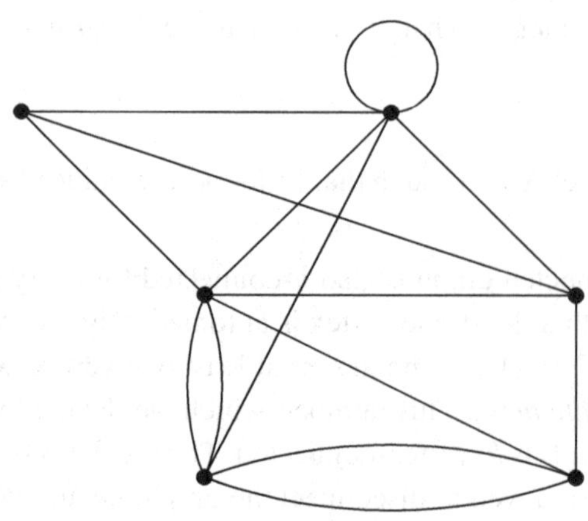

Figure 4.1. A 3-vertex-connected and 3-edge-connected graph. Observe that this graph satisfies $\kappa(\Gamma) = \lambda(\Gamma) = 3$.

Lemma 4.1.1. *Let k and t be two natural integers satisfying $1 \leqslant t \leqslant k-1$, and let $\Gamma = (V; E)$ a k-vertex-connected graph. Then every partial graph Γ' obtained by deleting t edges is $(k-t)$-vertex-connected.*

Proof. We proceed by induction on t. Suppose $t = 1$, so $k \geqslant 2$. Let $a = \{x, y\} \in E$ and $\Gamma' = \Gamma \backslash a$; we must show that Γ' is $(k-1)$-vertex-connected. Let $X \subset V(\Gamma') = V$ with $|X| = k-2$; we need to ensure that Γ'', the graph induced by Γ' on $V \backslash X$, is connected. First choose $z \in V \backslash (X \cup \{x, y\})$ (which is possible since $|V| > k \geqslant |X \cup \{x, y\}|$). It suffices to show that any $u \in V \backslash X$ can be joined to z in Γ''. But

- if $u \neq x$, then $z, u \in V \backslash (X \cup \{x\})$, and since Γ is k-vertex-connected, we can join them even after removing $|X \cup \{x\}| = k-1$ vertices. Thus there exists a chain C in Γ between z and u that does not pass through any vertex of $X \cup \{x\}$. In particular, this chain cannot pass through the edge a, so it is a chain of Γ''.
- If $u = x$, then $z, u \in V \backslash X \cup \{y\}$ and for the same reason, there is a chain C between z and u that avoids $X \cup \{y\}$, so it also avoids a.

Suppose now that the property holds up to $t-1$, and let Γ' be the graph Γ minus t edges. For any one edge a of these t edges, we can write $\Gamma' = \Gamma'' \backslash a$ where Γ'' is the graph Γ with the other $t-1$ edges removed. By the induction hypothesis, Γ'' is k'-vertex-connected, where $k' = k - (t-1)$. Applying the case $t = 1$ to Γ'' and $\Gamma' = \Gamma'' \backslash a$, we find that Γ' is $(k'-1)$-vertex-connected, i.e. $(k-t)$-vertex-connected. ∎

The next result establishes the relation between the edge-connectedness $\lambda(\Gamma)$ and the minimal degree $\delta(\Gamma)$ of a graph Γ.

Lemma 4.1.2. *Let $\Gamma = (V; E)$ be a connected graph with at least one edge. Then $\lambda(\Gamma) \leqslant \delta(\Gamma)$.*

Proof. By hypothesis, the minimal degree $\delta(\Gamma)$ is non-zero. Let x be a vertex of degree equal to $\delta(\Gamma)$. By eliminating the $\delta(\Gamma)$ edges incident to x, we obtain at least two connected components, one of which is reduced to the vertex x. ∎

As illustrated in Figure 4.2, it may happen that $\lambda(\Gamma) < \delta(\Gamma)$.

Proposition 4.1.3. *Let $\Gamma = (V; E)$ be a graph with at least one edge. Then*

$$\kappa(\Gamma) \leqslant \lambda(\Gamma) \leqslant \delta(\Gamma).$$

Proof. If $\kappa(\Gamma) = 0$, then the result follows from Lemma 4.1.2. Otherwise, take a set Y of $\kappa(\Gamma) - 1$ edges of Γ. Then the graph Γ is $\kappa(\Gamma)$-vertex-connected. By Lemma 4.1.1, the partial graph obtained by deleting the edges of Y is 1-vertex-connected. Consequently, deleting edges in the set Y does not disconnect the graph, so $\lambda(\Gamma) > \kappa(\Gamma) - 1$. ∎

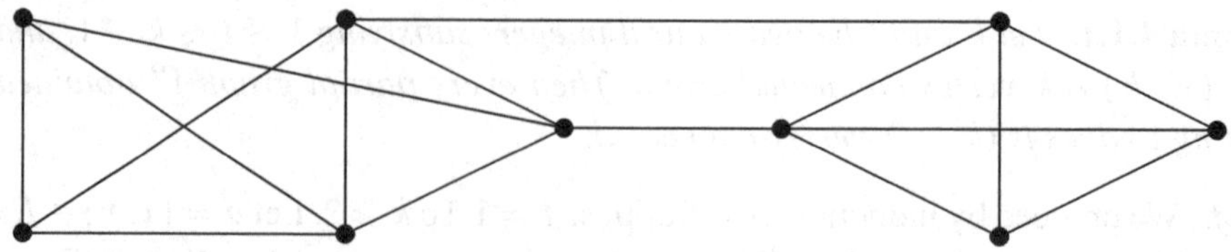

Figure 4.2. A graph with minimal degree equal to $\delta(\Gamma) = 3$ whereas $\lambda(\Gamma) = 2$.

Exercise 4.1. Let $\Gamma = (V; E)$ be a k-vertex-connected graph. First show that the number of edges of Γ is at least equal to $\lceil \frac{k|V|}{2} \rceil$. Then characterise the simple graphs $(V; E)$ such that $|E| = k|V|/2$.

Hint: use Lemma 1.1.9.

Let $\Gamma = (V; E)$ be a connected graph. A vertex $x \in V$ is said to be an *articulation vertex* of Γ if the induced graph $\Gamma(V \setminus \{x\})$, written simply $\Gamma - x$, is not connected. Graphs which are 2-vertex-connected (and thus also those which are k-vertex-connected for $k \geqslant 2$) have no articulation vertex. This notion is *dual* to that of an isthmus, which is an edge e such that $\Gamma \setminus e$ is not connected.

Articulation vertices and isthmuses are indicators of the fragility of a connected graph. In particular, 2-vertex-connected graphs and 2-edge-connected graphs are very important in the conception of telecommunication networks, since their structure guarantees that communications can continue even if one particular communication line fails. The next paragraph is devoted to this type of graph.

4.2 2-vertex-connected graphs

Let $\Gamma = (V; E)$ be a graph and $W \subset V$. The induced subgraph $B = \Gamma(W)$ is called a *block* of Γ if it is connected, has no articulation vertex and is maximal for the inclusion order in the set of connected induced subgraphs with no articulation vertex. This means that

(i) B is connected;

(ii) B has no articulation vertex;

(iii) if $W' \supset W$ and $B' = \Gamma(W')$ is connected and has no articulation vertex, then $B' = B$.

More precisely, let $\mathcal{A}$ be the set of subsets X of V such that $\Gamma(X)$ is connected and has no articulation vertex: the blocks B correspond to the maximal elements of $\mathcal{A}$, but $M \in \mathcal{A}$ is maximal for inclusion (see Section 1.10) if and only if $B = \Gamma(M)$ is a block.

Remark 4.2.1. There is actually a more precise result: if $X \in \mathcal{A}$, then there exists a maximal $M \in \mathcal{A}$ such that $X \subset M$. Thus, every connected induced graph with no articulation vertex is contained in a block. In particular, this is the case when the induced graph is an elementary cycle.

Let us give some further details about the blocks in a graph. Let $B = \Gamma(W)$ be a block.

- If $|W| = 1$, B is an isolated vertex.
- If $|W| = 2$, B is an isthmus.
- If $|W| \geqslant 3$, B is maximal 2-connected, i.e. we have

$$W' \supset W \text{ and } \Gamma(W') \text{ 2-connected} \implies W' = W.$$

Indeed, one can show (exercise) that when $|W| \geqslant 3$, then $\Gamma(W)$ is connected and has no articulation vertex if and only if $\Gamma(W)$ is 2-connected.

Remark 4.2.2. A block with at least 3 vertices cannot be a tree: trees are *thin* from the point of view of connectedness, since every vertex of a tree except for the leaves is an articulation vertex.

Lemma 4.2.3. *Let $\Gamma = (V; E)$ be a graph of finite order, and let $\mathcal{B}$ denote the (finite) set of its blocks.*

(i) *If $B_1 = (V_1; E_1)$, $B_2 = (V_2; E_2)$ are two distinct blocks (i.e. $V_1 \neq V_2$) then $|V_1 \cap V_2| \leqslant 1$.*

(ii) *If Γ is connected, its articulation vertices are exactly the intersections of distinct blocks.*

(iii) *E is the disjoint union of its blocks: $E = \bigsqcup_{B \in \mathcal{B}} E(B)$.*

Proof. (i) Suppose that $|V_1 \cap V_2| \geqslant 2$. Let $\Gamma_3 = \Gamma(V_1 \cup V_2)$. Since $V_1 \cap V_2 \neq \emptyset$, the subgraph Γ_3 is connected (using a result proved in Chapter 1 that states that for two connected subgraphs with common vertices, their union is also connected). Let $z \in V_1 \cup V_2$. By hypothesis, $(V_1 \cap V_2)\backslash\{z\} \neq \emptyset$. Thus deleting z does not disconnect the subgraph Γ_3, so this subgraph has no articulation vertex. By the maximality of the blocks B_1 and B_2, both contained in Γ_3 which is connected and has no articulation vertex, we must have $\Gamma_3 = B_1$ and $\Gamma_3 = B_2$; thus we find that $B_1 = B_2$.

(ii) Let x be an articulation vertex of Γ. By definition, $\Gamma - x = \Gamma(V\backslash\{x\})$ is not connected, so there exist two vertices y, y' of $\Gamma - x$ which are not connected. By the connectedness of Γ, there exists an elementary chain C_0 in Γ from y to y'. Clearly C_0 must pass through x, since y and y' are not connected in $\Gamma - x$. The chain C_0 decomposes as $C_0 = C \cup C'$ where C goes from y to x and C' from x to y'. The vertex z (resp. z') adjacent to x on the chain C (resp. C') is not connected to z'

(resp. z) in $\Gamma - x$. Thus z and z' lie in two distinct connected components of $\Gamma - x$. The subgraph $\Gamma(\{x, z\})$ (resp. $\Gamma(\{x, z'\})$) with two vertices is connected and has no articulation vertex, so it is contained in a block B (resp. B') of Γ (see remark above). As z and z' are not connected in $\Gamma - x$, they are also not connected in $(B \cup B') - x$, so x is an articulation vertex of the graph $B \cup B'$. Thus we cannot have $B = B'$, since otherwise $B \cup B' = B$ would not have any articulation vertex. Hence $B \neq B'$ and x lies in two different blocks.

Conversely, let x belong to two distinct blocks B and B' of Γ. Then on the one hand $B \cup B'$ is connected, and on the other hand $B \cup B'$ has an articulation vertex z by the maximality property of the blocks B and B'. This vertex z is necessarily equal to x, otherwise we can assume it is in $B - x$. But then $B - z$ is connected (since a block has no articulation vertex) and B' is connected, so $(B \cup B') - z$ is connected since x (which is different from z) is a vertex of both B and B'. But then z is not an articulation vertex of $B \cup B'$, which is a contradiction. Consequently, we have $x = z$ and it is an articulation vertex of $B \cup B'$. Let $y \in V(B)$ and $y' \in V(B')$ be vertices adjacent to x in B and B' respectively. If x were not an articulation vertex of Γ, there would exist an elementary chain C connecting y to y' but avoiding x. Closing the chain by adding the edges $\{y', x\}$ and $\{x, y\}$, we would obtain an elementary cycle C' that passes through x and traverses the blocks B and B'. But this is a contradiction given Remark 4.2.1.

(iii) Let $a = \{x, y\} \in E$ be an edge. The subgraph $\Gamma(\{x, y\})$ with two vertices is clearly connected and has no articulation vertex. By the remark above, it is then contained in a block B of which a is an edge. By (ii) above, a cannot belong to more than one block, so B is the only block of Γ containing a. ∎

Figure 4.3 gives an example of the decomposition of a graph into blocks.

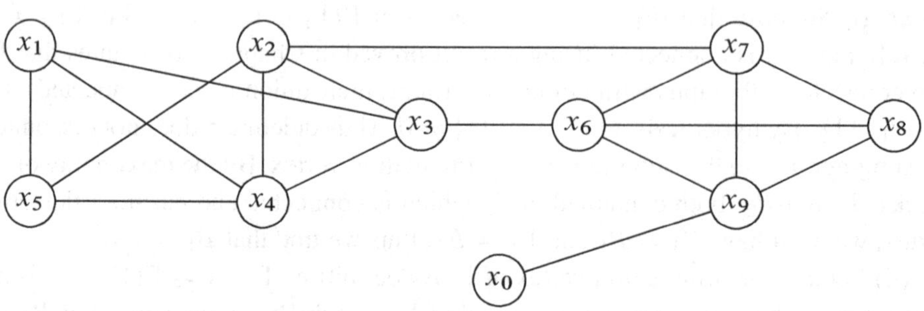

Figure 4.3. A graph $\Gamma = (V; E)$ with 4 blocks, $B_1 = \Gamma(\{x_1, x_2, x_3, x_4, x_5\})$, $B_2 = \Gamma(\{x_3, x_6\})$, $B_3 = \Gamma(\{x_6, x_7, x_8, x_9\})$ and $B_4 = \Gamma(\{x_9, x_0\})$.

Exercise 4.2. Show that all connected graphs with at least two vertices possess at least two vertices that are not articulation vertices.

Hint: use Propositions 2.2.5 and 2.2.6.

Starting from the blocks and articulation vertices of a graph $\Gamma = (V; E)$, we define an associated graph as follows:

- the vertices are the blocks and the articulation vertices of Γ (if any);
- there is an edge between an articulation vertex x and a block B if $x \in B$.

This graph is called the *block graph* of Γ and denoted by $\flat(\Gamma)$. If Γ possesses at least two blocks, then $\flat(\Gamma)$ is a bipartite graph, with blocks on the one hand and articulation vertices on the other.

A chain between two vertices of Γ can be described as a succession of chains between vertices of the same block, where any passage between two blocks must pass through the unique articulation vertex of Γ that is common to both of them (see Lemma 4.2.3). To a chain of Γ between two vertices x, x' in the same block B, there corresponds

- the elementary chain on three vertices $x, \{x, B\}, B, \{B, x'\}, x'$ in $\flat(\Gamma)$, if x and x' are articulation vertices of Γ.
- The elementary chain on two vertices $x, \{x, B\}, B$, if x is an articulation vertex and x' is not.
- The trivial chain on a single vertex reduced to B, if neither x nor x' are articulation vertices.

It follows that to an elementary chain connecting two vertices from different blocks of Γ, there corresponds in $\flat(\Gamma)$ a succession of chains on three vertices, each possibly having a chain on two vertices at each of its extremities, as described above. Conversely, an elementary chain in $\flat(\Gamma)$ comes from a (not necessarily unique) elementary chain in Γ. This property enables us to show that connectedness is preserved in the block graph.

Lemma 4.2.4. *Let Γ be a connected graph. Then* (i) $\flat(\Gamma)$ *is connected and* (ii) *every articulation vertex of Γ is also an articulation vertex of* $\flat(\Gamma)$.

Proof. (i) If Γ is a block, then $\flat(\Gamma)$ is reduced to a single vertex, so it is connected. Suppose now that Γ admits several blocks. Since Γ is connected, each block of Γ contains at least one articulation vertex of Γ. Thus it suffices to show that two articulation vertices x and x' of Γ can always be connected in $\flat(\Gamma)$. But there always exists a chain between x and x' in Γ since Γ is connected, and such a chain corresponds to a chain between x and x' in $\flat(\Gamma)$ given by a succession of elementary chains on three vertices of $\flat(\Gamma)$. Thus $\flat(\Gamma)$ is connected.

(ii) Let x be an articulation vertex of Γ. Since $\Gamma - x$ is not connected, there exist two vertices y, y' of Γ adjacent to x which are not connected in $\Gamma - x$. There exists a block B (resp. a block B') containing the connected subgraph with no articulation vertex $\Gamma(\{x, y\})$ (resp. $\Gamma(\{x, y'\})$). These two blocks B and B' intersect at x (see Lemma 4.2.3). But $B - x$ and $B' - x$ are connected, so no vertex of $B - x$ is not connected to a vertex of $B' - x$ in $\Gamma - x$. Then B and B' cannot be connected in $\flat(\Gamma - x)$: indeed, if they were, there would be a succession of chains on three vertices of the type $B_1, \{B_1, x_1\}, x_1, \{x_1, B_2\}, B_2$ which would form a chain from B to B' in $\flat(\Gamma - x)$. This chain would then lift to a chain in Γ between y and y' avoiding x, which is a contradiction. Thus we have shown that $\flat(\Gamma - x)$ is not connected, so that x is indeed an articulation vertex of $\flat(\Gamma)$. ■

This gives the following property.

Proposition 4.2.5. *Let Γ be a connected graph. Then its block graph $\flat(\Gamma)$ is a tree.*

Proof. We saw in Lemma 4.2.4 that $\flat(\Gamma)$ is connected. It remains to show that Γ has no cycles. Suppose there exists an elementary cycle C in $\flat(\Gamma)$. This cycle would necessarily be even, since $\flat(\Gamma)$ is bipartite. Thus, at least two of the vertices of the cycle C in $\flat(\Gamma)$ would have to be articulation vertices x and y of Γ. However, deleting either one of these vertices does not disconnect the cycle, so x and y would not be articulation vertices of $\flat(\Gamma)$. But this contradicts Lemma 4.2.4. ■

Corollary 4.2.6. *Let Γ be a connected graph with at least one articulation vertex (so at least three vertices). Then there exist at least two blocks each containing only one articulation vertex of Γ.*

Proof. As Γ admits at least one articulation vertex, it possesses at least two blocks, so $\flat(\Gamma)$ has at least three vertices, namely the articulation vertex and the two blocks. As $\flat(\Gamma)$ is a tree by Proposition 2.2.6, it contains at least two leaves x_1, x_2 corresponding to the hanging edges $p_i = \{B_i, x_i\}, i = 1, 2$: B_i, which corresponds to a block in Γ, is the vertex of degree 1 in $\flat(\Gamma)$ of the hanging edge $\{B_i, x_i\}$; it is attached to the rest of the graph by its unique incident edge $\{B_i, x_i\}$, where x_i is an articulation vertex of the block B_i in Γ. Note that we may have $x_1 = x_2$. Each of these leaves also constitutes a block in $\flat(\Gamma)$, since deleting x_i disconnects B_i from the rest of the graph in $\flat(\Gamma)$. This means that the vertex B_i in $\flat(\Gamma)$ is adjacent only to the articulation vertex x_i; the vertices of B_i other than x_i are not articulation vertices. Thus the two blocks B_1 and B_2 each possess only one articulation vertex of Γ. ■

Exercise 4.3. Draw the block graph of the graph shown in Figure 4.3.

We will now give several characterisations of k-vertex-connectedness and k-edge-connectedness in terms of disjoint chains. These are due to Whitney (1932), whose theorem will be proved at the end of this chapter.

The following theorem gives a characterisation of 2-vertex-connected graphs, not in term of connectedness but in terms of cycles.

Theorem 4.2.7. *Let $\Gamma = (V; E)$ be a simple connected graph with at least three vertices. Then the following assertions are equivalent:*

(a) *the graph Γ is 2-vertex-connected;*

(b) *for any two vertices, there exists an elementary cycle that passes through both of them;*

(c) *for any vertex and any edge, there is an elementary cycle that passes through both of them;*

(d) *for any two edges, there is an elementary cycle that passes through both of them.*

Proof. (a) $\Leftrightarrow$ (b): cf. Theorem 4.5.4.

(d) $\Rightarrow$ (c) $\Rightarrow$ (b): elementary.

(a) $\Rightarrow$ (c): take an edge $a = \{x, y\}$ and a vertex u distinct from x and of y. Add a vertex v on the edge a so that a is replaced by two new edges $\{x, v\}$ and $\{v, y\}$. The graph Γ' thus obtained is 2-vertex-connected: removing v from Γ' comes down to removing a from Γ, but Γ is 2-edge-connected by Proposition 4.1.3. By (b), Γ' has an elementary cycle that passes through u and v, so this cycle must also pass through x and y since these are the only two neighbours of v.

(b) $\Rightarrow$ (d): by the same reasoning as above, but considering the graph Γ'' obtained by cutting each of two edges a and b in two. ■

Exercise 4.4. Complete Theorem 4.2.7 by proving that 2-connectedness is equivalent to each of the following assertions:

(e) for any three vertices x, y and z, there exists an elementary cycle passing through all of them;

(f) for any three vertices x, y and z, there exists an elementary cycle passing through x and y but avoiding z.

Theorem 4.2.7, particularly the characterisation (b) of 2-connectedness, actually yields an algorithm that allows us to construct 2-vertex-connected graphs. This method is called the *chain method*. It works as follows:

Step 1: Start from a cycle Γ_1.

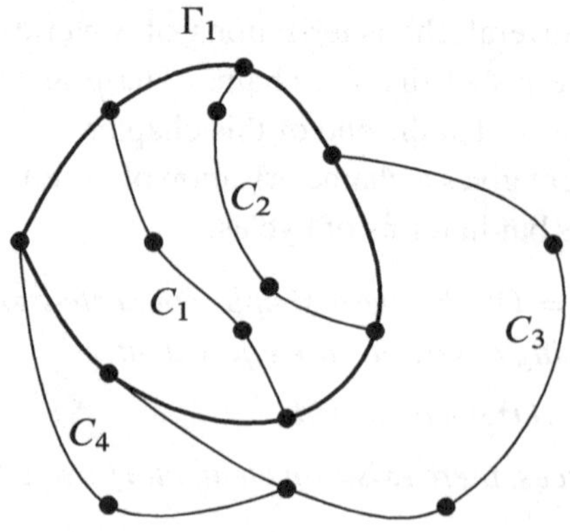

Figure 4.4. Construction of a 2-connected graph with the chain method. The cycle Γ_1 is the starting cycle: we successively add the chains C_1, C_2, C_3 and C_4.

Step 2: Add to this cycle an elementary chain of arbitrary length whose only vertices in common with the initial cycle are the (distinct) extremities of this chain. In this way, we obtain a graph Γ_2 that remains 2-connected.

Step n: To the graph Γ_{n-1}, we add an elementary chain whose only vertices in common with Γ_{n-1} are its extremities. This again yields a 2-connected graph Γ_n.

The result of this method is shown in Figure 4.4. The advantage of this method resides principally in the fact that it actually allows us to construct all finite order 2-connected graphs. This is the meaning of the following result.

Proposition 4.2.8. *Let Γ be a simple graph with at least 3 vertices. Then Γ is 2-vertex-connected if and only if it can be constructed using the chain method.*

Proof. By Theorem 4.2.7, a graph constructed by the chain method is 2-vertex-connected. We use induction on the number of steps n needed to reach the graph Γ by the chain method.

If $n = 1$, then the graph is a cycle, so it is already 2-connected. Suppose that n steps have been performed to obtain the graph Γ. Let C be the chain with extremities u and v added to the graph Γ_{n-1} at the n-th and last step. Let x and y be two vertices of Γ. If x and y were already present in the preceding step, the induction hypothesis applies. If x was present in step $n-1$ and y was created in step n (Figure 4.5), then an elementary cycle passes through the vertices u, v and x of Γ_{n-1} by the characterisation (e) of 2-connectedness. Taking the part of this cycle that connects u to v passing through x and adjoining it to the chain C yields an elementary cycle that passes through x and y. If x and y both lie on C, then it suffices to take an element-

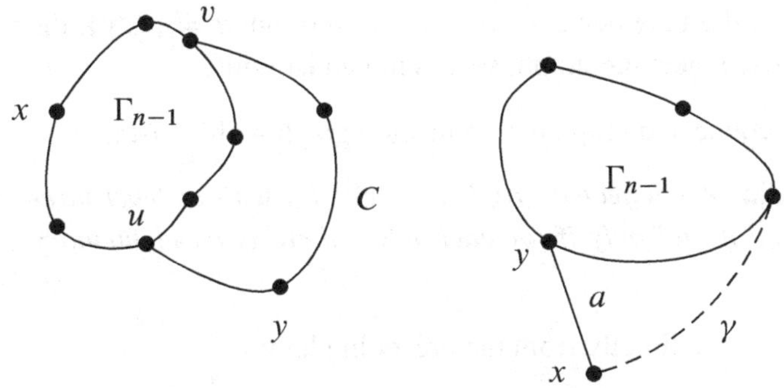

Figure 4.5. Illustration of the chain method.

ary chain connecting u to v and concatenate it with the chain C. This proves that Γ satisfies characterisation (b) of 2-connectedness of Theorem 4.2.7.

Conversely, if $\Gamma = (V; E)$ is 2-vertex-connected with $|V| \geqslant 3$, it must contains a cycle C, otherwise it would be a tree. Let $\Gamma' = (V'; E')$ be a maximal subgraph in the set of subgraphs of Γ that can be obtained by the chain method starting from C. Suppose that $V' \neq V$. Then since the graph Γ is connected, there exists an edge $a = \{x, y\} \in E$ such that $x \in V \setminus V'$ and $y \in V'$. The graph Γ is 2-vertex-connected so $\Gamma - y$ is connected; thus there exists an elementary chain γ connecting x to one of the vertices t of V'. Choosing a γ of minimal length, we may assume that t is the only vertex of γ which lies in V' (Figure 4.5).

Then $a \cup \gamma$ is a chain between y and t which does not meet V' except at its extremities, and contains some $x \notin V'$. It is not possible to adjoin a new chain to Γ' by the chain method since this would contradict the maximality of Γ', so we must have $V' = V$. Furthermore, every edge of Γ that lies in Γ' can be viewed as an elementary chain with extremities in Γ'. ■

4.3 2-edge-connected graphs

We begin by showing how to characterise isthmuses.

Lemma 4.3.1. *Let $\Gamma = (V; E)$ be a simple graph. An edge is an isthmus if and only if no elementary cycle passes through this edge.*

Proof. If $a = \{x, y\}$ is not an isthmus, then deleting it does not disconnect the graph. Thus there exists an elementary chain from x to y (see Exercise 1.1), so adjoining the edge $\{x, y\}$ creates an elementary cycle passing through a.

Conversely, if an elementary cycle passes through $a = \{x, y\}$, then deleting this edge cannot disconnect the graph, so a is not an isthmus. ■

A 2-edge-connected graph is a connected graph with no isthmus.

Corollary 4.3.2. *A connected graph $\Gamma = (V; E)$ with at least three vertices is 2-edge-connected if and only if for each edge, there is an elementary cycle passing through it.*

Proof. This follows directly from the preceding lemma. ■

As in the case of 2-vertex-connected graphs, there exists a chain method that makes it possible to generate 2-edge-connected graphs.

Step 1: Start with a cycle Γ_1 (which exists by the above).

Step 2: Add to this cycle a simple elementary chain whose only vertices in common with the cycle are its the extremities x and y. (The difference with the chain method for 2-vertex-connected graphs described above is that here we allow $x = y$.) We obtain a graph Γ_2.

Step n: Continuing as above, we add to the graph Γ_{n-1} an elementary chain whose only vertices in common with Γ_i are its extremities (which may possibly be equal). This yields a 2-edge-connected graph Γ_n.

The next result is obtained similarly to the analogous result for 2-vertex-connected graphs.

Theorem 4.3.3. *A connected graph Γ is 2-edge-connected if and only if it can be constructed by the edge-version chain method.*

Exercise 4.5. Prove Theorem 4.3.3 by adapting the proof of Theorem 4.2.8.

Recall that a connected graph $\Gamma = (V; E)$ is orientable if there exists an orientation of its edges such that the oriented graph is strongly connected.

Theorem 4.3.4. *Let $\Gamma = (V; E)$ be a simple connected graph with at least three vertices. Then the graph Γ is 2-edge-connected if and only if it is orientable.*

Proof. If the graph is orientable, then by Theorem 1.3.3, there is at least one cycle passing through each edge, so there is at least one elementary cycle (see Exercise 1.1). By Corollary 4.3.2, the graph is then 2-edge-connected.

Suppose now that the graph is 2-edge-connected. By Corollary 4.3.2, there is an elementary cycle C passing through each edge. Choose an arbitrary orientation of C. If $\Gamma = C$, then Γ is a circuit and the result is proven. Otherwise there exists a vertex $x \notin C$ adjacent to a vertex y of the circuit C. The edge $a = \{x, y\}$ is on an elementary chain which leaves C starting at y and returns to C for the first time at a vertex t

(which may be equal to y, see Theorem 4.3.3). Orient this elementary chain in such a way that the orientation is compatible with the previous chosen orientation of C, so as to obtain a cycle that passes through the edge a. The oriented subgraph thus obtained, which we denote by Γ_2, is strongly connected. At the n-th step, consider an edge $a = \{x, y\}$, where y is a vertex of Γ_{n-1} and x does not lie on Γ_{n-1}. This edge is on an elementary chain with extremities y and t on Γ_{n-1}. It can be oriented so that the edge a lies on a circuit of the new graph Γ_n.

To conclude, one iterates this process until exhaustion of all the edges. ∎

4.4 Flows in a network

Intuitively, a *flow* in a network can be viewed as the flowing of a substance along the edges of a graph. One can envision for example petrol circulating in a system of pipelines, water in sewage pipes, electricity in electric cables, vehicles on roads, telephone calls, email, or internet information; all of these can be modelled by flows in networks. Because of these applications, the theory of network flows has become a very important aspect of graph theory. As we will see, the theory of flows has a large number of applications in many different domains.

4.4.1 Definitions

A *network* $\mathcal{R}$ is a digraph $\Gamma = (V; \vec{E})$ that is simple (i.e. has neither multiarcs nor loops), of finite order, and having

(i) a *single source* s and a *single sink* p with a path from s to p;

(ii) a valuation $c\colon \vec{E} \to \mathbb{N}$ called the *capacity*.

A network is written $\mathcal{R} = (V; \vec{E}, s, p, c)$. In principle, the underlying graph is not necessarily connected and can possess isolated vertices. However, the only object of study related to the notions introduced above is the oriented subgraph generated by the vertices of the connected component containing s and p.

A *cut* of the network $\mathcal{R} = (V; \vec{E}, s, p, c)$ is a partition $V = X \sqcup \overline{X}$ of V into two disjoint subsets such that $s \in X$ and $p \in \overline{X}$. A cut is simply written $(X, \overline{X})$. The *capacity of the cut* $(X, \overline{X})$, written $c(X, \overline{X})$, is the sum of capacities of the arcs $a \in \vec{E}$ from X to $\overline{X}$:

$$c(X, \overline{X}) = \sum_{\substack{a \in \vec{E} \\ i(a) \in X,\, t(a) \in \overline{X}}} c(a),$$

where i and t are the functions defined in Section 1.3.

A cut $(X, \overline{X})$ is called a *minimum cut* if its capacity is minimum, i.e. if for every cut $(X', \overline{X'})$ of $\mathcal{R}$, we have $c(X, \overline{X}) \leqslant c(X', \overline{X'})$.

A map $f : \vec{E} \to \mathbb{R}_+$ is a *flow* in the network $\mathcal{R}$ if it satisfies the following properties:

(a) *Capacity constraint:* for every $a \in \vec{E}$, $0 \leqslant f(a) \leqslant c(a)$.

(b) *Conservation of the flow:* for every $x \in V \setminus \{s, p\}$,

$$\sum_{\substack{a \in \vec{E} \\ t(a)=x}} f(a) = \sum_{\substack{a \in \vec{E} \\ i(a)=x}} f(a).$$

This condition expresses the fact that the flows entering and leaving a given vertex x are equal. We can extend the definition of f to $V \times V$ by setting $f(x, y) = 0$ if $(x, y) \notin \vec{E}$.

One very visual example is that of a road network, whose intersections are represented by the vertices of the network, while each road between two intersections is represented by an oriented edge; for two-way roads, we place an arc in each direction. In this situation, the capacity measures the maximum number of vehicles that can circulate on a given road during, say, a one-hour time period. In this situation, one might wish to study the circulation of vehicles using different itineraries from the point s to the point p.

Let $X, Y \subseteq V$, and define the subset of $\vec{E}$:

$$\vec{E}(X, Y) = \{a \in \vec{E} \mid i(a) \in X, t(a) \in Y\}.$$

When $X = \{x\}$ and $Y = \{y\}$ are singletons, we simply write $\vec{E}(X, Y) = \vec{E}(x, y)$, which is empty if there is no arc from x to y. If there is an arc $a = (x, y)$, then we write $f(x, y)$ for $f(a)$. This allows us to extend the flow function to all pairs (X, Y) of subsets of V, by setting

$$f(X, Y) = \sum_{a \in \vec{E}(X,Y)} f(a) - \sum_{a \in \vec{E}(Y,X)} f(a)$$

We see immediately that for all $X, Y \subseteq V$, we have

$$f(X, Y) \in \mathbb{R}, \quad f(X, Y) = -f(Y, X) \quad \text{and} \quad f(X, X) = 0.$$

This function f satisfies something like a bilinearity condition:

$$f(X, Y \cup Z) = f(X, Y) + f(X, Z) \quad \text{if } Y \cap Z = \emptyset.$$

Furthermore, writing $f(x, V)$ for $f(\{x\}, V)$, we have

$$\forall x \in V, \quad f(x, V) = \sum_{a \in \vec{E} : i(a)=x} f(a) - \sum_{a \in \vec{E} : t(a)=x} f(a),$$

so by conservation of the flow, $f(x, V) = 0$ if $x \neq s, p$. Similarly, we have $f(V, x) = 0$ if $x \neq s, p$. Furthermore, we have

- $f(s, V) = \sum_{a \in \vec{E}: i(a)=s} f(a)$,
- $f(V, p) = \sum_{a \in \vec{E}: t(a)=p} f(a)$.

Proposition 4.4.1. *For every cut $(X, \overline{X})$ of the network $\mathcal{R}$, we have*

$$f(X, \overline{X}) = f(s, V) = f(V, p).$$

Proof. We have

$$\begin{aligned} f(X, \overline{X}) &= f(X, V \backslash X) = f(X, V) - f(X, X) \\ &= f(X, V) = f(s, V) + f(X \backslash s, V). \end{aligned}$$

If $x \in X \backslash s$, we have $x \neq s, p$, so $f(x, V) = 0$ and consequently $f(X \backslash s, V) = 0$, hence $f(X, \overline{X}) = f(s, V)$. The analogous reasoning gives the other equality:

$$f(X, \overline{X}) = f(V, \overline{X}) = f(V, p) + f(V, \overline{X} \backslash p) = f(V, p),$$

since $f(V, x) = 0$ if $x \neq s, p$. ■

The common value $f(X, \overline{X})$ of the cuts will be called the *total value of the flow* and denoted by $\omega(f)$. A flow f is said to be *maximum* in $\mathcal{R}$ if it is of maximum total value, i.e. if $\omega(f) \geqslant \omega(f')$ for every flow f' in $\mathcal{R}$.

Thus we obtain the following expressions for computing a flow f in a network.

Corollary 4.4.2. *Let $\mathcal{R}$ be a network equipped with a flow f. Then*

$$\omega(f) = \sum_{a \in \vec{E}: i(a)=s} f(a) = \sum_{a \in \vec{E}: t(a)=p} f(a).$$

If $(X, \overline{X})$ is a cut of $\mathcal{R}$, then

$$\omega(f) = \sum_{i(a) \in X, t(a) \in \overline{X}} f(a) - \sum_{t(a) \in X, i(a) \in \overline{X}} f(a).$$

Proof. These formulas express the identities $\omega(f) = f(s, V) = f(V, p) = f(X, \overline{X})$. ■

Example 4.4.3. Figure 4.6 shows a flow in a network. Each arc has an attached value representing the flow and another value in brackets representing the capacity. The total value of the flow is $\omega(f) = 14 + 10 + 2 + 4 - 1 + 5 = 34$. There is a cut $X = \{s, x_1, x_2, x_3, x_4, x_5, x_6\}$ and $\overline{X} = \{x_7, x_8, x_9, x_{10}, p\}$. The capacity of this cut is $c(X, \overline{X}) = 17 + 13 + 4 + 4 + 6 = 44$.

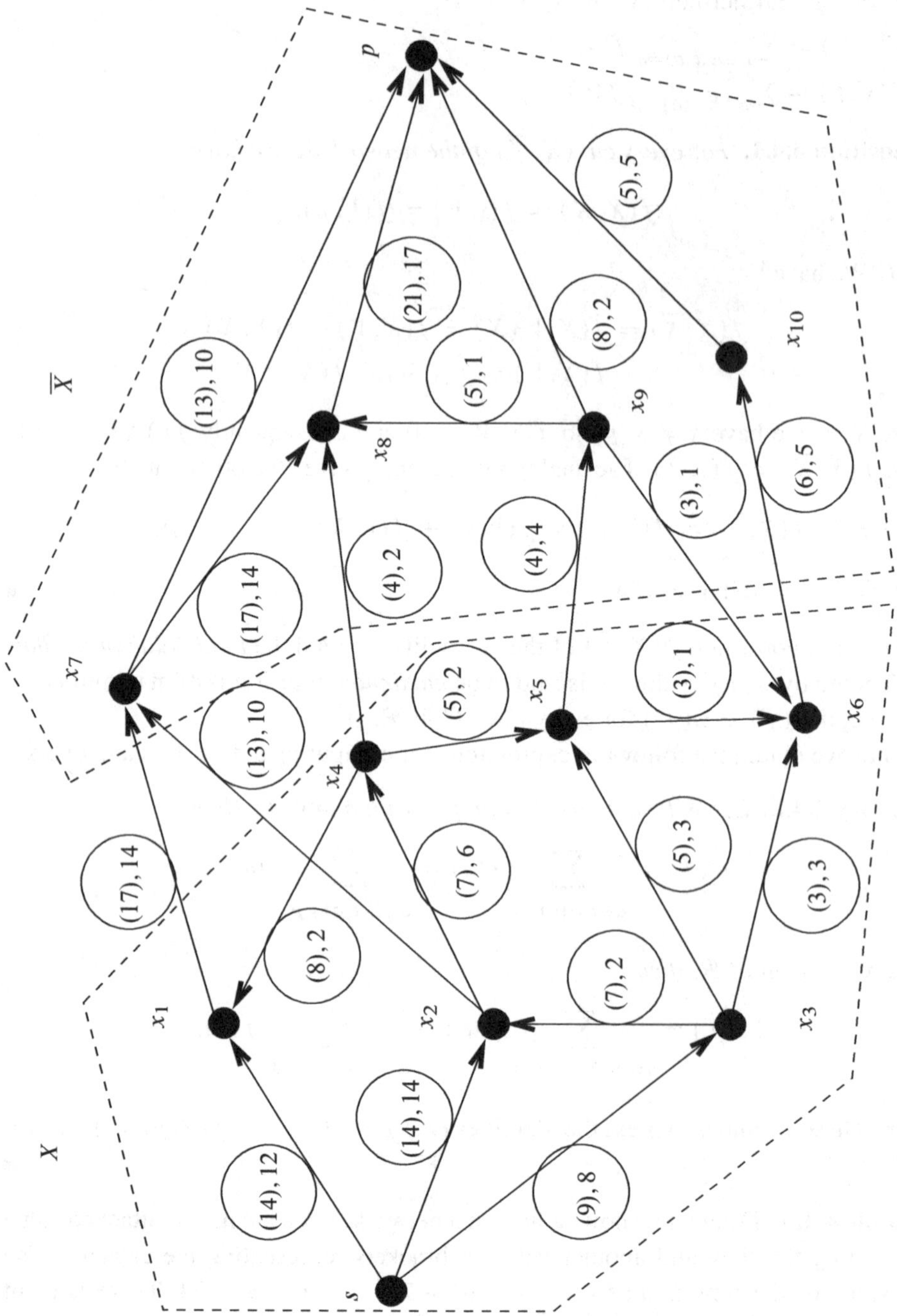

Figure 4.6. Example of a flow in a network.

Remark 4.4.4. (i) Consider a digraph with several sources $(s_i)_{i\in\{1,2,\ldots,k\}}$ and several sinks $(p_j)_{j\in\{1,2,\ldots,l\}}$. We can make it into a network as follows. First, we add a vertex s that will be considered as a *supersource*, and an arc (s, s_i) with infinite (or sufficiently large) capacity for each $i \in \{1, 2, \ldots, k\}$. Next, we do the same for the sink, by adding a *supersink* p and arcs (p_j, p), $j \in \{1, 2, \ldots, l\}$, again with appropriate capacities.

(ii) If we want to know what is passed from a vertex x to a vertex y in a digraph, we can always proceed by creating a network; we add a new vertex s, an arc from s to x, a new vertex p and an arc from y to p, so that every path from s to p induces a path from x to y.

4.4.2 The Ford–Fulkerson theorem

By the capacity constraint axiom, we have

$$\omega(f) = f(X, \overline{X}) \leqslant c(X, \overline{X})$$

for every cut $(X, \overline{X})$ of $\mathcal{R}$. Consequently, the total value of the flow is always less than or equal to the smallest possible capacity of a cut. The theorem of Ford and Fulkerson shows that there always exists a maximum flow which attains this bound. Before proving this theorem, we need to introduce the notion of an *improvable chain*.

Let f be a flow in a network $\mathcal{R} = (V; \vec{E}, s, p, c)$. If C is an elementary chain connecting s to p (which exists by definition) in the underlying graph of $\mathcal{R}$, then an oriented edge of C from s to p in $\mathcal{R}$ will be called a *positive arc* of C, while an oriented edge of C from p to s in $\mathcal{R}$ will be called a *negative arc* of C. A chain C from s to p in the underlying graph of $\mathcal{R}$ is called *improvable* if the flow of each positive arc can be increased while respecting the capacity of the arc and the flow of each negative arc can be decreased (i.e. is strictly positive). Thus, in an improvable chain C from s to p, we have

- $f(a) < c(a)$ for every positive arc a of C;
- $f(a) > 0$ for every negative arc a of C.

For example, in Figure 4.6, the following chains from s to p are improvable:

- $s, x_3, x_5, x_6, x_{10}, p$;
- $s, x_3, x_5, x_4, x_8, x_9, p$;
- s, x_1, x_7, x_8, p;
- s, x_1, x_7, p.

Let C be an improvable chain from s to p in a network $\mathcal{R}$. We define the default value on a by

$$\Delta_a = \begin{cases} c(a) - f(a) & \text{if } a \text{ is a positive arc of } C, \\ f(a) & \text{if } a \text{ is a negative arc of } C. \end{cases}$$

We write $\Delta_C = \min_{a \in C}\{\Delta_a\}$.

Proposition 4.4.5. *Let f be a flow in a network $\mathcal{R} = (V; \vec{E}, s, p, c)$. The following assertions are equivalent:*

(i) *f is a maximum flow.*

(ii) *$\mathcal{R}$ contains no improvable chain from s to p.*

(iii) *There exists a cut $(X, \overline{X})$ such that $\omega(f) = c(X, \overline{X})$.*

Proof. (i) $\Rightarrow$ (ii): Let f be a maximum flow of $\mathcal{R}$. *Suppose that there exists an improvable chain C from s to p.* Let $f' \colon \vec{E} \to \mathbb{R}$ be the map defined by

$$f'(a) = \begin{cases} f(a) + \Delta_C & \text{if } a \text{ is a positive arc of } C, \\ f(a) - \Delta_C & \text{if } a \text{ is a negative arc of } C, \\ f(a) & \text{if } a \text{ is not an arc of } C. \end{cases}$$

Then f' is a flow in $\mathcal{R}$ since $f(a) - \Delta_C \geqslant 0$. Thus the capacity constraint axiom is satisfied.

We must now show that f' satisfies the property of conservation of the flow. Let $x \in V \setminus \{s, p\}$. If x is not on C, there is nothing to prove since the flows of arcs incident to x are not modified. Now assume $x \in C$, $x \neq s, p$, and let a, b, in that order, be the arcs incident to x in C. We need to show that $f'(x, V) = 0$. There are four cases:

- $t(a) = x = t(b)$.

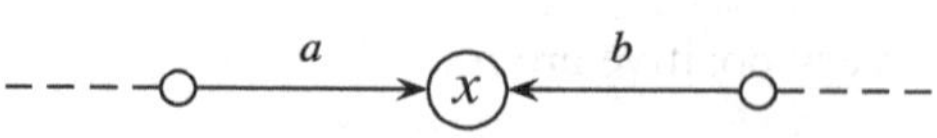

  Thus a is a positive arc and b a negative arc. Consequently, $f'(a) = f(a) + \Delta_C$ and $f'(b) = f(b) - \Delta_C$, and thus $f'(x, V) = f(x, V) + \Delta_C - \Delta_C = 0$.

- $i(a) = x = i(b)$.

  Thus a is a negative arc and b a positive arc. Consequently, $f'(a) = f(a) - \Delta_C$ and $f'(b) = f(b) + \Delta_C$, and thus $f'(x, V) = f(x, V) - \Delta_C + \Delta_C = 0$.

- $i(a) = x = t(b)$.

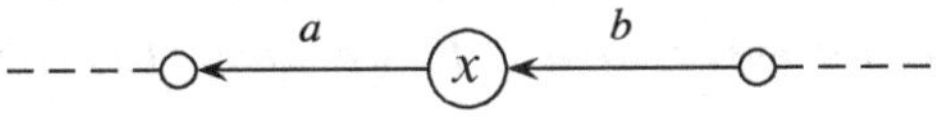

Thus a and b are negative arcs. Hence $f'(a) = f(a) - \Delta_C$ and $f'(b) = f(b) - \Delta_C$. Thus we have

$$\sum_{\substack{a' \in \vec{E} \\ t(a')=x}} f'(a') = \sum_{\substack{a' \in \vec{E} \setminus a \\ t(a')=x}} f(a') + f'(a) = \sum_{\substack{a' \in \vec{E} \setminus a \\ t(a')=x}} f(a') + f(a) - \Delta_C.$$

Similarly,

$$\sum_{\substack{a' \in \vec{E} \\ i(a')=x}} f'(a') = \sum_{\substack{a' \in \vec{E} \setminus b \\ i(a')=x}} f(a') + f(b) - \Delta_C.$$

Hence

$$f'(x, V) = \sum_{a' \in \vec{E} : t(a')=x} f'(a') - \sum_{a' \in \vec{E} : i(a')=x} f'(a') = f(x, V) = 0.$$

- $t(a) = x = i(b)$.

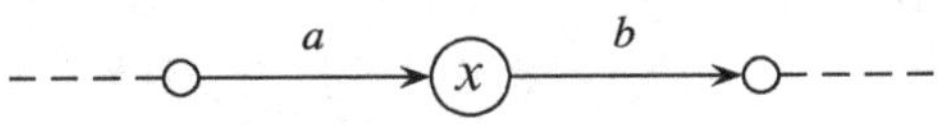

Thus a and b are positive arcs, and we conclude as above that $f'(x, V) = 0$.

We now need to show that the flow has increased by Δ_C. Every arc incident to s (resp. to p) is an arc whose initial (resp. terminal) vertex is s (resp. p). Consequently, the arc a (resp. a') of C which has s (resp. p) as initial (resp. terminal) vertex has flow value $f'(a) = f(a) + \Delta_C$ (resp. $f'(a') = f(a') + \Delta_C$). Thus $f'(s, V) = f(s, V) + \Delta_C = f'(V, p) = f'(V, p) + \Delta_C$. Therefore f is not maximum, which is a contradiction. This proves that there is no improvable chain from s to p.

(ii) ⇒ (iii): *suppose there is no improvable chain from s to p in* $\mathcal{R}$. Let

$$X = \{s\} \cup \{x \in V \mid \text{such that there is an improvable chain from } s \text{ to } x\}.$$

Then $(X, \overline{X})$ is a cut of $\mathcal{R}$, since $p \notin X$ by hypothesis. Thus for every arc a such that $i(a) \in X$ and $t(a) \in \overline{X}$, we have $f(a) = c(a)$, and for every arc a' such that $t(a') \in X$ and $i(a') \in \overline{X}$, we have $f(a) = 0$. Consequently, by Corollary 4.4.2, we have

$$\omega(f) = f(X, \overline{X}) = \sum_{\substack{i(a) \in X \\ t(a) \in \overline{X}}} f(a) = c(X, \overline{X}).$$

(iii) ⇒ (i): Since we must have $\omega(f) \leqslant c(X', \overline{X'})$ for all cuts $(X', \overline{X'})$, we conclude here that $\omega(f)$ is maximum and that $(X, \overline{X})$ is a minimum cut. ∎

Corollary 4.4.6. *Let $\mathcal{R} = (V; \vec{E}, s, p, c)$ be a network. Then there exists a maximum flow f in $\mathcal{R}$ such that for every arc $a \in \vec{E}$, we have $f(a) \in \mathbb{N}$.*

Proof. Let f_0 be the flow in $\mathcal{R}$ such that for every arc $a \in \vec{E}$, $f_0(a) = 0$. Either this flow is maximum and we are done, or there exists an improvable chain C from s to p. The value Δ_C introduced before Proposition 4.4.5 is an integer since $c\colon \vec{E} \to \mathbb{N}$. Consequently, the flow f' constructed as in the proof of Proposition 4.4.5 takes integral values. Either it is maximum and we are done, or it is not maximum, in which case we iterate the process until we obtain a maximum integer flow. ∎

Proposition 4.4.5 implies the following fundamental result.

Theorem 4.4.7 (Ford and Fulkerson, 1956). *The maximum total value of a flow on a network $\mathcal{R}$ is equal to the capacity of a minimum cut.*

```
Procedure Ford_Fulkerson(R: network)
  begin
    for a ∈ E do
      f(a) := 0;
    end
    foreach x ∈ X do
      while there exists an improvable path from s to x do
        Let C = (a_1, a_2, a_3, ..., a_k) be such a path;
        foreach a_i ∈ C do
          if a_i is a positive arc in C then
            f(a_i) := f(a_i) + Δ_C;
          else
            f(a_i) := f(a_i) − Δ_C;
          end
        end
      end
    end
  end
EndProcedure
```

Algorithm 4.1. The Ford–Fulkerson algorithm.

Algorithm 4.1 is due to Ford and Fulkerson. It was improved by Edmonds and Karp in 1972. Starting from a network, it yields a maximum flow f and a minimum cut $(X, \overline{X})$; in particular we have $\omega(f) = c(X, \overline{X})$.

The Ford–Fulkerson algorithm, in its initial version, runs as follows: take a network $\mathcal{R} = (V; \vec{E}, s, p, c)$ and initialise the flow on the arcs to 0.

(i) Let $X = \{s\} \cup \{x \in V \mid \text{there is an improvable chain from } s \text{ to } x\}$. If $p \notin X$, then f is a maximum flow and $(X, \overline{X})$ is a minimum cut.

(ii) Otherwise, choose an improvable chain C and determine the value of Δ_C on this chain. Then update the value of the flow on C by replacing $f(a)$ by $f(a) + \Delta_C$ or $f(a) - \Delta_C$, according to whether a is a positive or negative arc of C. Return to step one.

The Ford–Fulkerson algorithm can be viewed as a *greedy algorithm*: in the *while* loop, it searches for a path with positive capacity and increases the flow on this path as much as possible, which has the effect of saturating the capacity of at least one arc, then it starts again searching for another path, and continues this process until there are no more improvable paths. Suppose for simplicity that the capacities are natural numbers: then it will take at most $\omega(f)$ iterations of Algorithm 4.1 to find the maximum flow. Each iteration to find an improvable path has complexity $O(|\vec{E}|)$, so the procedure has complexity $O(\omega(f)|\vec{E}|)$. This can actually be rather slow in practice.

The improvement due to Edmonds and Karp lies in choosing the improvable paths at each step of the algorithm in such a way as to reduce the number of iterations. At each iteration, the algorithm seeks the improvable path between the source s and the vertex t that possesses the smallest number of arcs. For this, it uses a breadth (or depth) traversal, initialising the weight on each arc to 1. It can be shown that this approach, which is used in Algorithm 4.4, minimises the number of iterations. It uses two functions, namely Algorithms 4.2 and 4.3.

Remark 4.4.8. In Algorithm 4.2 the data structure used is a queue. If we use a stack instead, the breadth-first traversal becomes a depth-first traversal.

Theorem 4.4.9. *The Edmonds–Karp algorithm has complexity $O(nm^2)$, where $n = |V|$ and $m = |\vec{E}|$.*

Proof. We give a sketch of the proof. Since the algorithm only concerns the connected component of the underlying graph containing s and p, we may assume that it is connected. Thus $m \geqslant n - 1$.

Computing the complexity of Algorithm 4.4 depends on the **repeat** loop. Since the function `Breadth_traversal` has complexity $O(n + m)$ because it visits each vertex and each arc at most once, the complexity is $O(m)$.

Each time that the **repeat** loop finds an improvable chain, it increases it so as to obtain a chain with an edge $\{x, y\}$ that is saturated, i.e. an edge of minimal capacity. The distance from s to x is thus less than the distance from s to y. The saturated edge can only be used again if the flow has itself undergone an increase in the opposite direction, so that the distance from s to y has become less than the distance from s to x.

```
Function Breadth_traversal(ℛ = (V; E⃗, s, p, c): network): table
  table: pred;                        /* declaration of the table of vertices */
  file: Q;
  begin
    /* initialisation of the table of vertices */
    foreach x ∈ V do pred[x] := null;
    Q := s;
    while Q ≠ ∅ do
      x := Dequeue(Q);
      foreach y ∈ V such that (x, y) ∈ E⃗
       with c(x, y) > 0 and pred[y] = null do
        pred[y] := x;
        if y == p then return pred;
        Add y to Q;
      end
    end
    return −1;
  end
EndFunction
```

Algorithm 4.2. Breadth search adapted to the Edmonds–Karp algorithm.

```
Function Improvable_chain(ℛ = (V; E⃗, s, p, c): network, pred: table): file
  file: C;
  vertex: x;
  begin
    initialise C;
    x := p;
    while x ≠ s do
      add (pred[x], x) to C;
      x := pred[x];
    end
    return C;
  end
EndFunction
```

Algorithm 4.3. Compute improvable chains.

```
Function Edmonds_Karp(ℛ = (V; E⃗, s, p, c): network): table
    real: δ;
    table: pred;                       /* declaration of the table of vertices */
    table: flow;                       /* declaration of the table of the flow on the arcs */
    table: Δ;                          /* declaration of the table of defaults */
    file: C;                           /* declaration of the list of arcs */
    begin
        /* initialisation of the flow on the arcs and defaults */
        foreach a ∈ E⃗ do
            flow[a] := 0;
            Δ[a] := ∞;
        end
        /* call on breadth-first traversal to find an improvable path */
        repeat
            pred := Breadth_traversal(ℛ);
            if pred == −1 then return flow;
            /* Compute an improvable chain C */
            C := Improvable_chain(ℛ);
            /* Compute the minimal default δ = Δ_C on C */
            foreach a ∈ C do
                if a is a positive arc of C then Δ[a] := c(a) − flow[a];
                else Δ[a] := flow[a];
            end
            δ := min_{a∈C}{Δ[a]};
            foreach a ∈ C do
                if a is a positive arc of C then
                    flow[a] := flow[a] + δ;
                    c(a) := c(a) − δ;
                else
                    c(a) := c(a) + δ;
                end
            end
        endrepeat
    end
EndFunction
```

Algorithm 4.4. Edmonds–Karp algorithm.

The distance of s to any other vertex is bounded above by the number of vertices n. If the distance does not decrease, an edge can be saturated at most n times. Thus the only way to modify the distances is to increase the flow.

We remove the saturated edges, which does not decrease the distance between s and the other vertices, and we add the edges oriented in the opposite direction only if their capacity is zero, which also does not decrease the distance. Thus the total number of operations performed by the algorithm that actually increase the flow is $O(nm)$. Taking account of the breadth-first traversal, we conclude that the Edmonds–Karp algorithm has complexity $O(nm^2)$. ■

Exercise 4.6. Explain exactly what the algorithm produces, starting from the network in Figure 4.6.

4.5 Applications of flows in a network

We will now describe some applications of flows, in particular to questions of connectedness. A set A of arcs of a digraph Γ is said to be *arc-separating* for two vertices s and p of Γ if every elementary path from s to p passes through an arc of A. The set A is said to be *minimum arc-separating* if in addition it is of minimal cardinality, i.e. if for every arc-separating set A' we have $|A| \leqslant |A'|$. Two paths are said to be *arc-disjoint* if they have no common arc.

Similarly, a set A of edges of a graph Γ is said to be *edge-separating* for two vertices s and p of Γ if every elementary chain between s and p passes through an edge of A. The set A is said to be *minimum edge-separating* if in addition for every edge-separating set A' we have $|A| \leqslant |A'|$.

When the capacity of a digraph $\Gamma = (V; \vec{E}, c)$ satisfies $c(a) = 1$ for all $a \in \vec{E}$, the capacity of a cut $(X, \overline{X})$ of Γ is equal to the number of arcs starting in X and ending in $\overline{X}$. In this situation we refer to the *cardinality* of the cut $(X, \overline{X})$. The following lemma deals with this type of digraph.

Lemma 4.5.1. *Let $\Gamma = (V; \vec{E}, c)$ be a digraph equipped with a capacity c such that for all $a \in \vec{E}$, $c(a) = 1$. Assume that there exists a path from a vertex s to a vertex p. Then the cardinality of a minimum cut for the vertices s and p (i.e. a partition $V = X \sqcup \overline{X}$ such that $s \in X$ and $p \in \overline{X}$) is equal to the cardinality of a minimum arc-separating set for s and p.*

Proof. Let α be the minimal cardinality of any arc-separating set for s and p, and let β be the cardinality of a minimum cut for s and p.

$\alpha \leqslant \beta$: Let $(X, \overline{X})$ be a cut for s and p of cardinality β. The set $A = \{a \in \vec{E} \mid i(a) \in X, t(a) \in \overline{X}\}$ is clearly an arc-separating set of cardinality $c(X, \overline{X})$. Consequently, $\beta = c(X, \overline{X}) = |A| \geqslant \alpha$.

$\beta \leqslant \alpha$: Let A be an arc-separating set for s and p of cardinality α. Let X denote the set of vertices that can be reached starting from s (including s) but avoiding A. Since A is arc-separating for s and p, we have $p \notin X$, so $(X, \overline{X})$ is a cut for s and p. If $a \in \vec{E}$ satisfies $i(a) \in X$ and $t(a) \in \overline{X}$, let γ be a path from s to $i(a)$ avoiding A. If a does not lie in A, then the concatenation of γ and a yields a path from s to $t(a)$ that avoids A, i.e. $t(a) \in X$, which is a contradiction; thus $a \in A$. It follows that the cut $(X, \overline{X})$ satisfies $c(X, \overline{X}) \leqslant |A| = \alpha$. Hence $\beta \leqslant \alpha$. ■

Exercise 4.7. Let Γ be as in the statement of Lemma 4.5.1. Find a bijection between the minimum arc-separating sets and the cuts of minimal cardinality.

We are now in a position to give a version of Menger's theorem in terms of minimal cardinalities of arc-separating (resp. edge-separating) sets. Recall that we only consider loopless (di)graphs. Two chains are said to be edge-disjoint if they have no edge in common.

Theorem 4.5.2 (Menger, arc and edge versions). (i) *Let $\Gamma = (V; \vec{E})$ be a simple digraph, and let s, p be two non-adjacent vertices of Γ. Then the maximum number of arc-disjoint elementary paths from s to p is equal to the cardinal of a minimum arc-separating set for s and p.*

(ii) *Let $\Gamma = (V; E)$ be a simple graph and let s, p be two non-adjacent vertices of Γ. Then the maximum number of elementary edge-disjoint chains between s and p is equal to the cardinal of a minimum edge-separating set for s and p.*

Proof. (i) If there exists no path from s to p, the result is obvious. Otherwise, define a capacity c by setting $c(a) = 1$ for all $a \in \vec{E}$. Note that $(V; \vec{E}, s, p, c)$ is not necessarily a network, since s and p are not necessarily a source and a sink, but if needed, it suffices to add two vertices s_0 and p_0 and two arcs connecting s_0 to s and p to p_0 respectively, to obtain a bona fide network (see Remark 4.4.4). Let k be the maximum number of arc-disjoint elementary paths from s to p; denote these paths by $C_1, \dots, C_k$. Let l be the minimal cardinality of an arc-separating set for s and p.

(α) Let us show that $l \geqslant k$. Construct a flow f between s and p as follows:

$$f(a) = \begin{cases} 1 & \text{if } a \text{ is an arc of one of the } C_i,\ i = 1, \dots, k, \\ 0 & \text{otherwise.} \end{cases}$$

Let us show that this is really a flow. On the one hand, the capacity constraint is trivially verified, and on the other, for $x \neq s, p$, the sum $\sum_{a \in \vec{E}: t(a) = x} f(a)$ is exactly the number of arcs among the C_i ending at x. As the C_i are arc-disjoint, this number is equal to the number of arcs belonging to the different C_i and starting from x, namely to the sum $\sum_{a \in \vec{E}: i(a) = x} f(a)$. Thus conservation of the flow is satisfied.

We have $\omega(f) = \sum_{i(a)=s} f(a)$, and this counts the number of arcs a in the C_i starting at s, so since the C_i are arc-disjoint, we have $\omega(f) = k$.

Let $(X, \overline{X})$ be a minimum cut of the network. Then $c(X, \overline{X}) \geqslant \omega(f) = k$. Hence, by Lemma 4.5.1, $l \geqslant k$.

(β) Conversely, let A be a minimum arc-separating set (for s and p), and let $a \in A$. Since $A \backslash \{a\}$ is not arc-separating, there exists an elementary path C from s to p that passes through a. Let Γ' be the graph obtained by removing the arcs of C from Γ. The set $A \backslash \{a\}$ is a minimum arc-separating set in Γ' for s and p. Iterating this reasoning $l - 1$ times, we produce l arc-disjoint paths. Thus $l \leqslant k$.

(ii) The case of graphs can be reduced to the case of digraphs as follows. Replace all the edges of the graph Γ by two arcs with opposite directions. This gives a digraph, which we denote by $\vec{\Gamma}$. A system of edge-disjoint elementary chains between s and p induces a system of arc-disjoint elementary paths from s to p. Conversely, to every system of arc-disjoint elementary paths from s to p in $\vec{\Gamma}$, there corresponds a system of edge-disjoint elementary chains between s and p in Γ. Let k be the maximum number of edge-disjoint elementary chains between s and p in Γ. Then k is also the maximum number of arc-disjoint elementary paths from s to p. Thus there exists an arc-separating set A in $\vec{\Gamma}$, with cardinality equal to k. This set A is also an edge-separating set in Γ. This proves the assertion. ■

We will use these results to prove Menger's theorem. Historically, this theorem was first proved in this form. Before we give a new definition: a set S of vertices in a (di)graph is said to be vertex-separating for the vertices s and p if any path (resp. chain) from s to p passes through S. It is called a minimum vertex-separating set if its cardinal is minimum.

Theorem 4.5.3 (Menger, vertex version, 1927). *Let Γ be a digraph (resp. a graph). Let s and p be two non-adjacent vertices of Γ. Then the maximum number of vertex-disjoint elementary paths (resp. chains) between s and p is equal to the cardinality of a minimum vertex-separating set for s and p.*

Proof. The result is easy to prove in the digraph case. The case of a graph Γ then follows by applying the result to a digraph whose underlying graph is Γ.

Let $\Gamma = (V; \vec{E})$ be a digraph. We will transform this digraph so as to be able to apply Theorem 4.5.2. Define an associated digraph Γ' as follows:

- the *vertices of* Γ' are s, p and every vertex $x \neq s, p$ of Γ is doubled, becoming two vertices x' and x'';
- the *arcs of* Γ' are every arc from s to x in Γ is replaced by an arc from s to x' in Γ'. Every arc from x to p in Γ is replaced by an arc from x'' to p in Γ'. Every arc from x to y in Γ with $x, y \neq s, p$ is replaced by an arc from x'' to y'. Finally, between each pair of new vertices x' and x'', we add an arc.

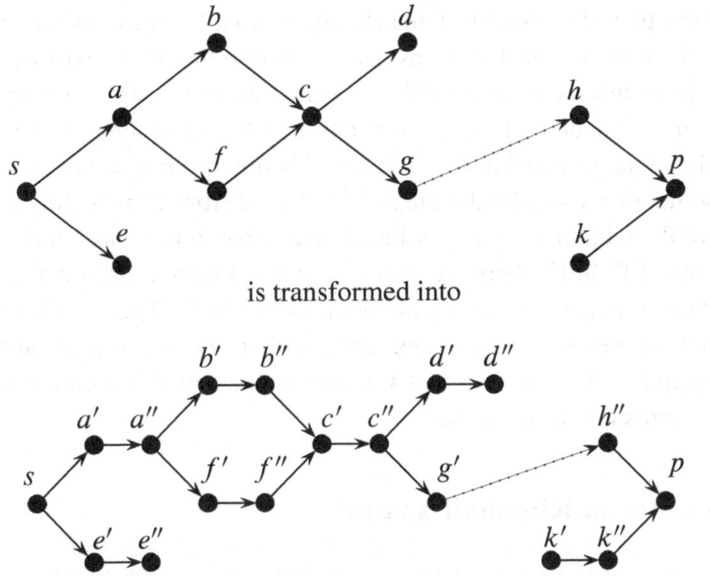

Figure 4.7. Transformation of a graph by insertion of vertices. The paths $abcd$ and $afcg\dots h$ have a common vertex c in Γ. They become $a'a''b'b''c'c''d'd''$ and $a'a''f'f''c'c''g'\dots h''$ in Γ', and have a common edge $c'c''$.

An illustration of this transformation is given in Figure 4.7.

It is clear that vertex-disjoint elementary paths of Γ become arc-disjoint elementary paths of Γ'. Conversely, two non-vertex-disjoint elementary paths of Γ induce two elementary paths of Γ' which are not arc-disjoint. Consequently, by Theorem 4.5.2, the maximum number of vertex-disjoint elementary paths between s and p in Γ is equal to the cardinality of a minimum arc-separating set A between s and p in Γ'. Replacing each arc of A from x'' to y' by an arc from x' to x'', we obtain a set A' which is still an arc-separating set between s and p and has the same cardinality as A by minimality. Thus we can restrict ourselves to minimum arc-separating sets A' of Γ' containing only arcs of the form (x', x''). These are in natural correspondence with the vertex-separating sets of Γ, by identifying the arcs from x' to x'' with the vertices x they come from. ■

Exercise 4.8. Prove an edge version of Theorem 4.5.3.

We conclude this paragraph with the following result, which is due to H. Whitney.

Theorem 4.5.4 (Whitney, 1932). *A graph Γ is k-vertex-connected if and only if every pair of vertices is connected by at least k vertex-disjoint chains.*

Proof. If every pair of vertices is connected by at least k vertex-disjoint chains, then Γ is clearly k-vertex-connected. Conversely, suppose that there exist two vertices s and p that are connected by at most $k-1$ vertex-disjoint chains. If s and p are not adjacent, then by Theorem 4.5.3 there exists a vertex-separating set for s and p of cardinality less than or equal to $k-1$. Thus Γ is not k-vertex-connected. If s and p are adjacent, then we consider the graph Γ' obtained from Γ by adding a vertex z to the middle of the edge from s to p, which is attractive in the sense that it is adjacent to every vertex of Γ. In Γ', there are still at most $k-1$ vertex-disjoint chains between the two vertices s and p. As s and p are not adjacent in Γ', Theorem 4.5.3 applied to Γ' shows that there exists a vertex-separating set between s and p of cardinality less than or equal to $k-1$. Thus Γ' is not k-vertex connected. But then it is clear that Γ cannot be k-vertex connected either. ∎

4.6 Complements: Kirchhoff's laws

These laws, which date back to 1845, complete Ohm's law (1827). They concern the circulation of energy in an electrical circuit.

- 1st law $[K_1]$: the potential drop in a closed circuit is zero;
- 2nd law $[K_2]$: the algebraic sum of currents entering one node of a closed circuit is zero.

These two laws make it possible to compute currents and potentials in any electrical network.

Example 4.6.1. In Figure 4.8, we have

$[K_2]$: $I_1 = I_2 + I_3$ (currents in O).

$[K_1]$: $z_1 I_2 - \mathcal{E} = 0$ (square $\mathcal{C}$); hence $\frac{1}{z_1} = \frac{I_2}{\mathcal{E}}$.

$[K_1]$: $z_2 I_3 - \mathcal{E} = 0$; hence $\frac{1}{z_2} = \frac{I_3}{\mathcal{E}}$.

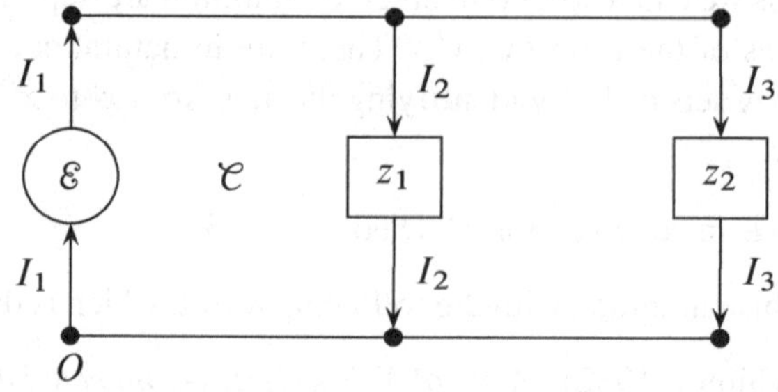

Figure 4.8. Circuit with two parallel impedances.

If we try to replace this circuit by a single impedance z with the same intensity I_1 as the generator [equivalent circuit], we must have $\frac{1}{z} = \frac{I_1}{\mathcal{E}}$ by K_1; then

$$\frac{1}{z} = \frac{I_1}{\mathcal{E}} = \frac{I_2 + I_3}{\mathcal{E}} = \frac{1}{z_1} + \frac{1}{z_2} = \frac{z_1 + z_2}{z_1 z_2};$$

thus we recover the classical result $z = \frac{z_1 z_2}{z_1 + z_2}$.

Similarly, using Kirchhoff's laws, we could show that a sequence of two impedances is equivalent to their sum.

Example 4.6.2. The circuit in Figure 4.9 can be solved step by step using the equivalent circuits. This is not always possible, as we can see from the circuit in Figure 4.10.

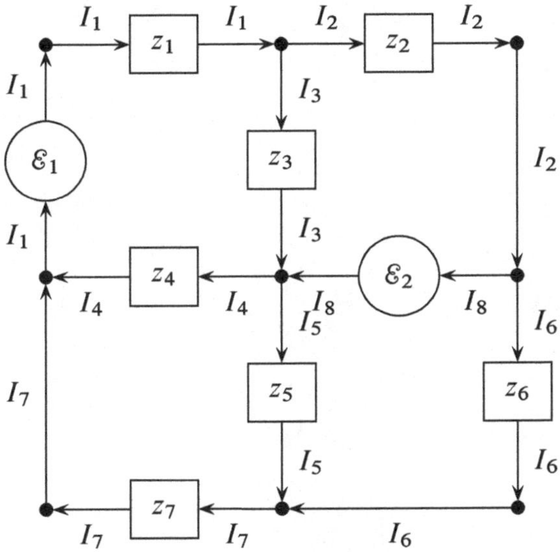

Figure 4.9. Representation of an electrical circuit.

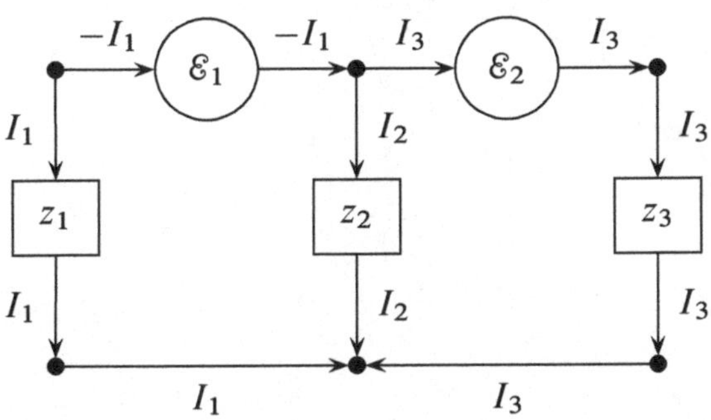

Figure 4.10. Representation of an electrical circuit.

Example 4.6.3. Concerning Figure 4.10, we have the equations

$[K_1]$: $\mathcal{E}_1 = z_2 I_2 - z_1 I_1$,

$[K_1]$: $\mathcal{E}_2 = z_3 I_3 - z_2 I_2$,

$[K_2]$: $I_1 + I_2 + I_3 = 0$.

The combinatorial interpretation of $[K_2]$ is simple: the electrical circuit is represented by an oriented graph, and the intensities then form a flow in the graph (which comes down to taking a network of infinite capacity).

Chapter 5

Planar graphs

This chapter is devoted to a particularly topological question concerning graphs, namely whether a graph Γ can be represented (or drawn) on a plane or another surface. The classical notion of a representation of a graph on a surface S is as follows:

- the vertices are represented by points of S, still called vertices;
- the edges are represented by simple curves (i.e. curves with no double points) on the surface S between the vertices, in such a way that two curves can only intersect at the vertices on S.

This operation is also called *embedding* Γ in S. Viewing graphs this way can be very important, for example in the conception of integrated circuits in electronics. We will concentrate mostly on the case where S is the Euclidean plane $\mathbb{R}^2$.

Even though it is perfectly possible to define oriented planar graphs, we only consider non-oriented graphs in this chapter. The notions we use here lie at the intersection of topology, algebraic topology and geometry, and they are often quite difficult to define rigorously. We have included some useful background for the material in this chapter in Section 5.8.

5.1 Drawings

The Euclidean space $\mathbb{R}^k$, $k \geqslant 1$, is equipped with the usual norm

$$\|(x_1, \dots, x_k)\| = \sqrt{x_1^2 + \cdots + x_k^2}.$$

Let T be a topological subspace (see the definition in Paragraph 5.8) of $\mathbb{R}^k$, and let $x, y \in T$. A *geometric arc* from x to y is a continuous injective map

$$\xi \colon [0, 1] \to T$$

satisfying $\xi(0) = x$ and $\xi(1) = y$. A *simple curve* γ between x and y is the image of an arc from x to y or from y to x: $\gamma = \mathrm{Im}(\xi) = \xi([0, 1]) \subseteq T$. We write $\mathcal{C}_T$ for the set of simple curves on T.

A *drawing* F of a graph $\Gamma = (V; E, N)$ on T is an injective map

$$\sigma \colon V \sqcup E \to T \sqcup \mathcal{C}_T$$

such that the restriction $\sigma|_V$ (resp. $\sigma|_E$) of σ to V (resp. E) is an injective map of V (resp. E) to T (resp. $\mathcal{C}_T$).

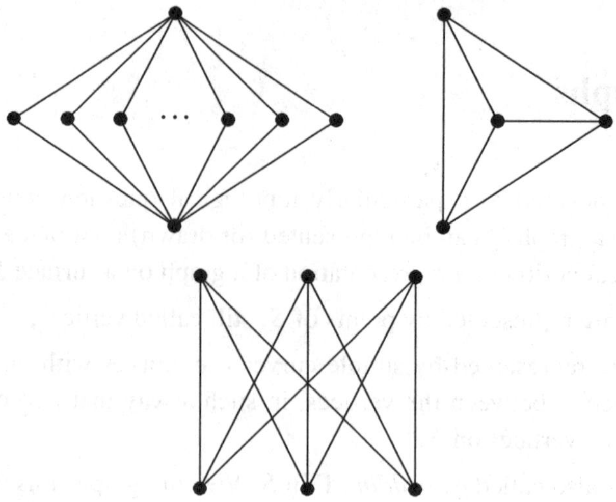

Figure 5.1. Embedding of $K_{2,n}$, K_4 and drawing of $K_{3,3}$ in $\mathbb{R}^2$.

It is understood that if e and e' have the same extremities (i.e. when multi-edges are present), we have $\sigma(e) \neq \sigma(e')$; however, the curves may possibly intersect at points other than their extremities.

By an abuse of language, the set $F = \sigma(V) \cup \bigcup_{e \in E} \sigma(e)$ is still called a drawing of Γ on T; the curves $\sigma(e)$, $e \in E$, are still called edges and the points $\sigma(v)$, $v \in V$, vertices of the drawing F.

Example 5.1.1. The spaces T in which graphs are *drawn* are usually $T = \mathbb{R}, \mathbb{R}^2$, $\mathbb{R}^3$, but also $T = \{x \in \mathbb{R}^3, \|x - s\| = r\}$, the sphere of centre s and radius r in $\mathbb{R}^3$; we can also consider the question drawing graphs on the torus (which has the shape of a tire or the surface of a doughnut) in $\mathbb{R}^3$ and other topological surfaces. However, most of the figures in this book are representations of graphs on $T = \mathbb{R}^2$.

If the curves intersect only at vertices, we say that the drawing is an *embedding* of Γ in T.

Proposition 5.1.2. *Every graph $\Gamma = (V; E, N)$ of finite order admits an embedding into $\mathbb{R}^3$.*

Proof. Let D be a line in $\mathbb{R}^3$, and fix $|V|$ distinct points x_v, $v \in V$ on D. Take $|E|$ distinct half-planes P_e, $e \in E$, all of which have the line D as boundary (like the fletching found at the back end of an arrow). For each edge $e \in E$ with extremities v, w, draw a simple curve γ_e in the half-plane P_e between x_v and x_w (for example an arc of a circle, or a circle if e is a loop). Then $F = \{x_v, v \in V\} \cup \bigcup_{e \in E} \gamma_e([0,1])$ is a drawing of Γ in $\mathbb{R}^3$ corresponding to an embedding of Γ. ∎

Observe that this construction can be adapted to the infinite case as long as V and E have at most the cardinality of the continuum, i.e. if there exists an injection of $V \sqcup E$ into $\mathbb{R}$).

Example 5.1.3. A non-closed elementary chain admits an embedding into $\mathbb{R}$; an elementary cycle $C_n, n \geqslant 3$, admits an embedding into $\mathbb{R}^2$ but not into $\mathbb{R}$; this is also the case for $K_{1,n}, K_{2,n}, n \geqslant 3$; the complete graphs K_3 and K_4 can be embedded in $\mathbb{R}^2$; K_5 and $K_{3,3}$ can be drawn on the plane, like every finite order graph, but they cannot be embedded in the plane; this is a more difficult question that we will consider below. See Figure 5.1.

The main object of the next paragraph is the question: *which graphs admit an embedding into the plane?*

5.2 Planar graphs

A graph of finite order is said to be *planar* if it admits an embedding in the plane $\mathbb{R}^2$. For example, the graphs $K_{2,n}$, K_3 and K_4 are planar.

Remark 5.2.1. The drawing F that results from embedding the graph $\Gamma = (V; E, N)$ can be viewed as an isomorphic graph (in the sense of graph isomorphisms), which makes it possible to transport the combinatorial properties of the graph to its drawing; but the embedding allows us to also make use of the topological properties of the plane $\mathbb{R}^2$.

5.2.1 Fundamentals of the topology of $\mathbb{R}^n$

For the details and the proofs of the results in this paragraph, see Section 5.8. Let A be a topological subspace of $\mathbb{R}^n$.

- A is said to be *connected* if there exists no non-trivial partition of A into two disjoint open sets: in other words, if U, V are open sets of A such that $A = U \cup V$ and $U \cap V = \emptyset$, then either $U = \emptyset$ or $V = \emptyset$.
- For $x \in A$, we write $C(x)$ for the *connected component* of x, i.e. the largest connected subspace (for the partial ordering by inclusion) of A containing x. We have the *connected decomposition* of A,

$$A = \bigsqcup_{x \in \tilde{A}} C(x),$$

where $\tilde{A}$ is a set of representatives of equivalence classes for the relation $x \sim y$ if and only if $y \in C(x)$.

- If A is an open set of $\mathbb{R}^n$, then its connected components are open sets, because $\mathbb{R}^n$ is locally connected.
- If A and B are connected and $A \cap B \neq \emptyset$, then $A \cup B$ is connected.
- The connected subspaces of $\mathbb{R}$ are exactly the intervals (closed or not, bounded or not) of $\mathbb{R}$;
- If $f: A \subset \mathbb{R}^m \to \mathbb{R}^n$ is continuous and if A is connected, then $f(A)$ is connected (Bolzano's theorem).

5.2.2 Polygonal lines

For $x, y \in \mathbb{R}^k$, $x \neq y$, the (closed) *segment* with extremities x and y is the subset

$$[\![x, y]\!] = \{(1-t)x + ty,\ 0 \leqslant t \leqslant 1\}.$$

We define the subsets $[\![x, y[\![$, $]\!]x, y]\!]$ and $]\!]x, y[\![$ analogously; the segment $[\![x, y]\!]$ is homeomorphic to the closed interval $[0, 1]$ in $\mathbb{R}$, via the parametrisation

$$\gamma: \begin{cases} [0,1] \to \mathbb{R}^k, \\ \quad t \mapsto (1-t)x + ty. \end{cases}$$

Thus, by Bolzano's theorem, it is a connected subspace of $\mathbb{R}^k$.

A *polygonal line* is the union of a finite number of segments placed *end to end* with no intersection:

$$P = [\![x_0, x_1]\!] \cup [\![x_1, x_2]\!] \cup \cdots \cup [\![x_{n-1}, x_n]\!], \quad (n \geqslant 1).$$

It is a particular type of simple curve: it suffices to define $\gamma: [0, 1] \to \mathbb{R}^2$ by $\gamma(t) = (1-t)x_i + tx_{i+1}$, if $\frac{i}{n} \leqslant t \leqslant \frac{i+1}{n}$, $0 \leqslant i \leqslant n-1$); x_0 and x_n are the extremities of P, the other x_i, $i = 1, \ldots, n-1$, are called the *angular points* of P; P is said to be *closed* if $n \geqslant 3$ and $x_0 = x_n$. We set $\mathring{P} = \gamma(]0, 1[) = P \backslash \{x_0, x_n\}$, and call it the *interior* of P, which is actually a somewhat misleading term since it is not actually an open set in the plane; rather, the term refers to the interior $]a, b[$ of intervals $[a, b]$ in $\mathbb{R}$.

We will use the intuitive notion of *point of no return*.

Lemma 5.2.2. (i) *Let $\gamma: [0, 1] \to \mathbb{R}^2$ be a geometric arc with extremities $\gamma(0) = z$, $\gamma(1) = x$, and let r be such that $0 < r < \|z - x\|$. Then there exists a unique u, $0 < u < 1$, such that*

$$\begin{aligned} t > u &\implies \|\gamma(t) - z\| > r, \\ t = u &\ \ :\ \ \|\gamma(u) - z\| = r. \end{aligned}$$

We say that $\gamma(u)$ is the exit point *of γ from the disk $\overline{D}(z, r)$.*

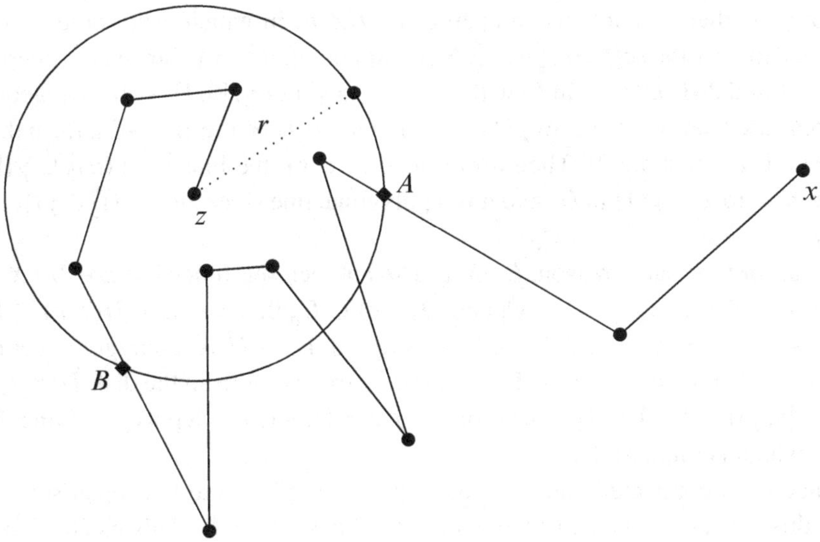

Figure 5.2. Points of no return: exit point A, entry point B.

(ii) *Let* $\gamma: [0,1] \to \mathbb{R}^2$ *be a geometric arc with* $\gamma(0) = x$, $\gamma(1) = z$, *and let* r *be such that* $0 < r < \|z - x\|$. *Then there exists a unique* v, $0 < v < 1$, *such that*

$$\begin{aligned} t > v &\implies \|\gamma(t) - z\| < r, \\ t = v &\ \ : \ \ \|\gamma(v) - z\| = r. \end{aligned}$$

We say that $\gamma(v)$ *is the entry point of* γ *into the disk* $\overline{D}(z, r)$.

Proof. (i) Let $A = \{t,\ 0 \leqslant t \leqslant 1 : \|\gamma(t) - z\| \leqslant r\} = \gamma^{-1}(\overline{D}(z, r))$. Then A is closed, and non-empty since $0 \in A$, so $u := \sup A$ belongs to A. If $t > u$, then $t \notin A$, so $\|\gamma(t) - z\| > r$. If we had $\|\gamma(u) - z\| < r$, then since $\|\gamma(1) - z\| > r$, by the *passing-customs theorem* (see Lemma 5.8.2), we would have t', $u < t' < 1$ such that $\|\gamma(t') - z\| = r$, hence $t' \in A$ with $t' > u$, which is impossible. If u' were another exit point, then we would necessarily have $u' > u$ or $u > u'$, so $\|\gamma(u') - z\| > r$ or $\|\gamma(u) - z\| > r$, which is impossible.

(ii) We proceed similarly with the set $\{t,\ 0 \leqslant t \leqslant 1 : \|\gamma(t) - z\| \geqslant r\}$. ■

Lemma 5.2.3. *Let* U *be a connected open set of* $\mathbb{R}^2$. *Then for every pair of points in* U, *there exists a polygonal line in* U *connecting them.*

Proof. This proof is a good example of reasoning with connectedness. Choose $x \in U$, and define X to be the set of $z \in U$ for which there exists a polygonal line in U (possibly reduced to $\{x\}$) between x and z.

If $z = x$, then U contains an open disk $D(x, r)$ in which every point w can be connected to x by the segment $[\![w, x]\!]$. Now suppose that $z \neq x$. Since U is open, there exists a closed disk $\overline{D}(z, r)$ in U with $r < \|z - x\|$. Let $\gamma\colon [0, 1] \to U$ be a geometric polygonal arc from $\gamma(0) = z$ to $\gamma(1) = x$, and let $\gamma(u)$ be the exit point from the disk $\overline{D}(z, r)$ (cf. Lemma 5.2.2). Then for all $w \in D(z, r)$, the line $[\![w, \gamma(u)]\!] \cup \gamma([u, 1])$ goes from w to $x = \gamma(1)$ in U, and it is a polygonal line since $[\![w, \gamma(u)]\!] \cap \gamma([u, 1]) = \{\gamma(u)\}$.

For almost the same reason, $U \backslash X$ is also an open set. Indeed, if $z \in U \backslash X$, there exists r such that $0 < r < \|z - x\|$ and $\overline{D}(z, r) \subset U$; then we have $D(z, r) \subset U \backslash X$, otherwise there exists $w \in D(z, r) \cap X$, so $\gamma\colon [0, 1] \to U$ is a polygonal geometric arc from $\gamma(0) = w$ to $\gamma(1) = x$. Let $\gamma(u)$ be its exit point from the disk $\overline{D}(z, r)$; then the line $[\![z, \gamma(u)]\!] \cup \gamma([u, 1])$ goes from z to x in U, and it is a polygonal line. Hence $z \in X$, which is impossible.

Since U is connected and we have a partition of U into two open sets X and $U \backslash X$, this partition must be trivial, so we must have $X = U$. This means that every $z \in U$ is connected to x by a polygonal line. It follows that any two points in U can always be connected by a polygonal line. ∎

Proposition 5.2.4. *Let $\Gamma = (V; E, N)$ be a planar graph. Then Γ admits an embedding such that all of its edges are polygonal lines.*

Proof. Let $\sigma\colon V \sqcup E \to \mathbb{R}^2 \sqcup \mathscr{C}_{\mathbb{R}^2}$ be an embedding of Γ into $\mathbb{R}^2$, associated to the drawing F. We will construct an embedding of Γ having the same vertices, but whose edges will be replaced by polygonal lines.

We start by erasing all the loops. We then define

$$\delta = \frac{\min\{\|\sigma(v) - \sigma(w)\| : v, w \in V,\ v \neq w\}}{2}.$$

Let e be an edge with extremities v, w, $v \neq w$, associated to the geometric arc $\xi_e\colon [0, 1] \to \mathbb{R}^2$: $\xi(0) = v$, $\xi(1) = w$. We need to *straighten* ξ_e into a polygonal line that does not intersect any other edge. For this, δ allows us to move away from the points of confluence that are the vertices; we then use Lemma 5.2.3. Let v_e be the exit point of ξ_e from the disk $\overline{D}(\sigma(v), \delta)$, and let w_e be the entry point of ξ_e into $\overline{D}(\sigma(w), \delta)$.

Define the open set

$$\Omega_e = \mathbb{R}^2 \backslash \Big(\bigcup_{f \in E \backslash \{e\}} \sigma(f) \cup \bigcup_{x \in V \backslash \{v, w\}} \overline{D}(\sigma(x), \delta/2) \Big).$$

The connected components of Ω_e are open (since $\mathbb{R}^2$ is locally connected) and both v_e, w_e lie in the same connected component U_e of Ω_e (because the arc portion of ξ_e from v_e to w_e is connected in Ω_e).

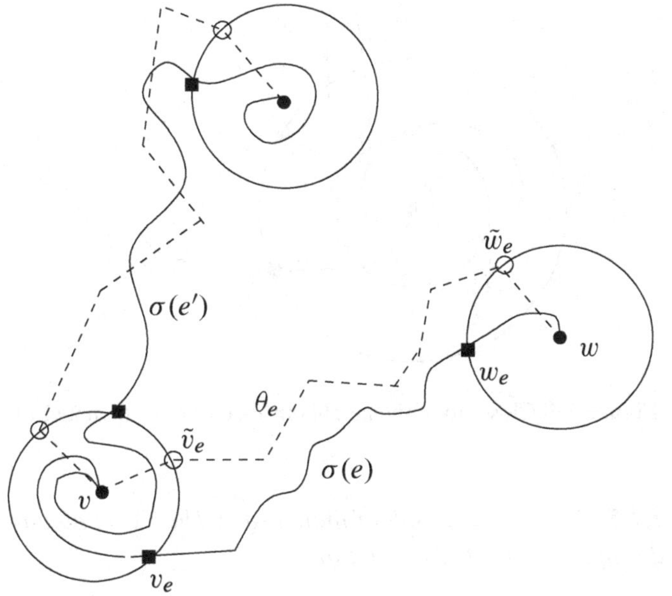

Figure 5.3. Illustration of the proof of Proposition 5.2.4 (we write v and w instead of $\sigma(v)$ and $\sigma(w)$).

By Lemma 5.2.3, there exists a polygonal line P_e in U_e connecting v_e to w_e. Let $\tilde{v}_e$ denote the point at which P_e leaves $\overline{D}(\sigma(v),\delta)$ and let $\tilde{w}_e$ denote the point at which P_e enters $\overline{D}(\sigma(w),\delta)$. Finally, let $\tilde{P}_e$ be the corresponding arc portion between these two points.

The union $\theta_e = [\![\sigma(v),\tilde{v}_e]\!] \cup \tilde{P}_e \cup [\![\tilde{w}_e,\sigma(w)]\!]$ is a polygonal line connecting $\sigma(v)$ to $\sigma(w)$ and not intersecting any other edge.

We recommence this construction with all the (non-loop) edges of Γ. It is then very easy to add loops: a loop of vertex v can be realised by a little triangle of vertex $\sigma(v)$. ∎

In order to avoid some delicate problems that can be posed by curves, we will assume that all representations of edges consist of polygonal lines, which is always possible by Proposition 5.2.4. Nevertheless, for aesthetic reasons and above all to avoid any confusion between vertices and angular points on polygonal lines, we draw all the figures with … curves.

The *simplified graph* Γ_s of $\Gamma = (V;E,N)$ is the simple graph obtained by deleting the loops from Γ and replacing each multi-edge by a simple edge. The following result once again appears obvious, yet in fact has a proof that requires a great deal of care!

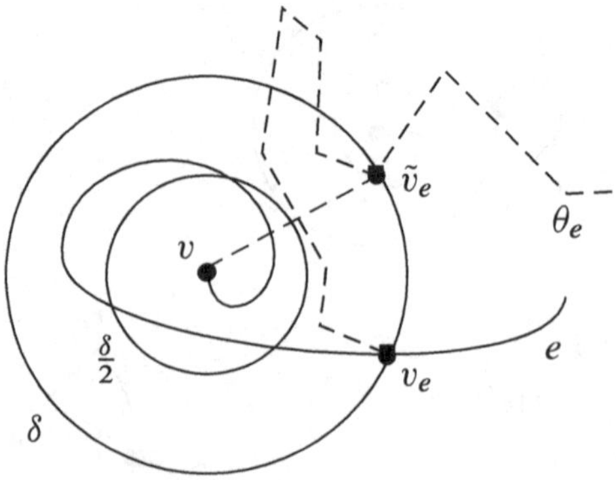

Figure 5.4. Close-up of the neighbourhood of $\sigma(v)$ (denoted v).

Proposition 5.2.5. *Let Γ be a graph of finite order. Then Γ is planar if and only if the associated simplified graph Γ_s is planar.*

Proof. If the graph is planar, then its simplification is obtained by erasing edges; therefore it remains planar. Suppose now that the simplified graph $\Gamma_s = (V; E, N)$ is planar, with an embedding by polygonal arcs

$$\sigma: V \sqcup E \to \mathbb{R}^2 \sqcup \mathscr{C}_{\mathbb{R}^2}.$$

Define $\delta > 0$ as follows: for each vertex v, there exists a sufficiently small real number $r_v > 0$ such that the disk $\overline{D}(\sigma(v), r_v)$ intersects edges e adjacent to $\sigma(v)$ only in their first segment, and does not intersect any other edge. Set $\delta = \min\{r_v : v \in V\}$.

Doubling an edge. Let $e = \{v, w\} \in E$ be an edge represented by a polygonal line $\gamma: [0,1] \to \mathbb{R}^2$, $\gamma(0) = v$, $\gamma(1) = w$ (simplified notation: we *drop* the embedding σ). Set $P = \gamma([0,1])$.

Consider the open set Ω_e and the points v_e and w_e introduced in the proof of Proposition 5.2.4 (Γ_s is loopless). The portion P_e of the polygonal line P between v_e and w_e is a connected subset of Ω_e, so it is contained in a connected component U_e of Ω_e. The open set U_e is *erased* from $\sigma(f)$, $f \in E$, $f \neq e$.

For each $z \in P_e$, there exists $r_z > 0$ such that $D_z = D(z, r_z)$ is contained in U_e. Then $\mathcal{U} = \{D_z \cap P_e : z \in P_e\}$ is an open cover of P_e, to which we apply Lemma 5.8.5. There exists a sequence of disks $D_i = D_{z_i}$, $i = 1, \ldots, m$ to which we adjoin the disks $D_0 = D(v, r_v)$ and $D_{m+1} = D(w, r_w)$, so that the segments $U_i = D_i \cap P$, $i = 0, \ldots, m+1$, connect v to w. Furthermore, v_e is in U_1 and in no other U_i; w_e is in U_m and in no other U_i, and $U_i \cap U_j \neq \emptyset$ if and only if $|i - j| \leqslant 1$. Since P can zigzag in the interior of D_i, we forget it, and consider for each $i =$

$0, \dots, m$ the extremities c_i, d_i of the closed *biconvex lens* $\overline{D_i} \cap \overline{D_{i+1}}$. The points $c_i, d_i, c_{i+1}, d_{i+1}$ are distinct ($c_i, d_i \in \overline{D_i}$ and $c_{i+1}, d_{i+1} \in \overline{D_{i+2}}$), are not aligned (they all lie on the circle $\mathcal{C}_{i+1} = \mathrm{Fr}(D_{i+1})$), and satisfy $[\![c_i, d_i]\!] \cap [\![c_{i+1}, d_{i+1}]\!] = \emptyset$.

For each i, arbitrarily choose two distinct points $x_i, y_i \in]\!]c_i, d_i[\![$. From above, the points $x_i, y_i, x_{i+1}, y_{i+1}$ are distinct and not aligned. We connect them in sequence thanks to the result stated in Exercise 5.1 above, so that the segments $[\![x_i, x_{i+1}]\!]$ and $[\![y_i, y_{i+1}]\!]$ do not intersect (it may be necessary to exchange x_{i+1} and y_{i+1}). We obtain two disjoint polygonal lines $[\![x_0, x_1]\!] \cup [\![x_1, x_2]\!] \cup \dots \cup [\![x_{m-1}, x_m]\!]$ and $[\![y_0, y_1]\!] \cup [\![y_1, y_2]\!] \cup \dots \cup [\![y_{m-1}, y_m]\!]$. We complete them with two polygonal lines P_1 and P_2 from v to w by adding the two segments $[\![v, x_0]\!]$ and $[\![x_m, w]\!]$ on the one hand, and adding the two segments $[\![v, y_0]\!]$ and $[\![y_m, w]\!]$ on the other (Figure 5.5).

Exercise 5.1. Let x, y, x', y' be four distinct points that are not aligned in the plane. Then $[\![x, x']\!] \cap [\![y, y']\!] = \emptyset$ or $[\![x, y']\!] \cap [\![y, x']\!] = \emptyset$.

Hint: assume the contrary and use a barycentric computation.

Creation of an n-loop at $\sigma(v)$ (Figure 5.6*).* After doubling edges as in the preceding step, the circle $\mathcal{C}(\sigma(v), \delta)$ possesses a finite number of distinct points v'_e, where e runs over the set of edges having v as one extremity. Take an arc A of this circle $\mathcal{C}(\sigma(v), \delta)$ between two of these points, then arbitrarily fix an ordered set of $2n$ points $x_1, y_1, \dots, x_n, y_n$ on this arc. Now we can form the n loops $[\![\sigma(v), x_i]\!] \cup [\![x_i, y_i]\!] \cup [\![y_i, \sigma(v)]\!]$, $1 \leqslant i \leqslant n$ (Figure 5.6). ■

Among the following results, some will appear obvious, but we will see that certain proofs turn out to be very difficult! This is the case in particular for Jordan's famous theorem, which asserts that a simple closed curve divides the plane into two connected components[1]. We will need a *polygonal* version of this result, which is given in Theorem 5.2.6, and another rather technical result, Proposition 5.2.7, which formulates the intuitive notion that a non-closed polygonal line does not divide a connected open set into two disjoint parts, whereas a closed polygonal line does.

Theorem 5.2.6 (Jordan). *Let P be a closed polygonal line in the plane $\mathbb{R}^2$. Then the open set $\mathbb{R}^2 \setminus P$ has exactly 2 connected components U_1, U_2, of which only one is unbounded. Furthermore,* $\mathrm{Fr}(U_1) = \mathrm{Fr}(U_2) = P$.

Proof. See Section 5.8. ■

Let U be a connected open subset of $\mathbb{R}^2$ whose boundary $\mathrm{Fr}(U)$ consists of a finite number of polygonal lines that meet pairwise at at most one point. Let P be a (closed) segment of $\mathrm{Fr}(U)$ and $z \in P$.

[1]A complete proof of this result can be found in [36].

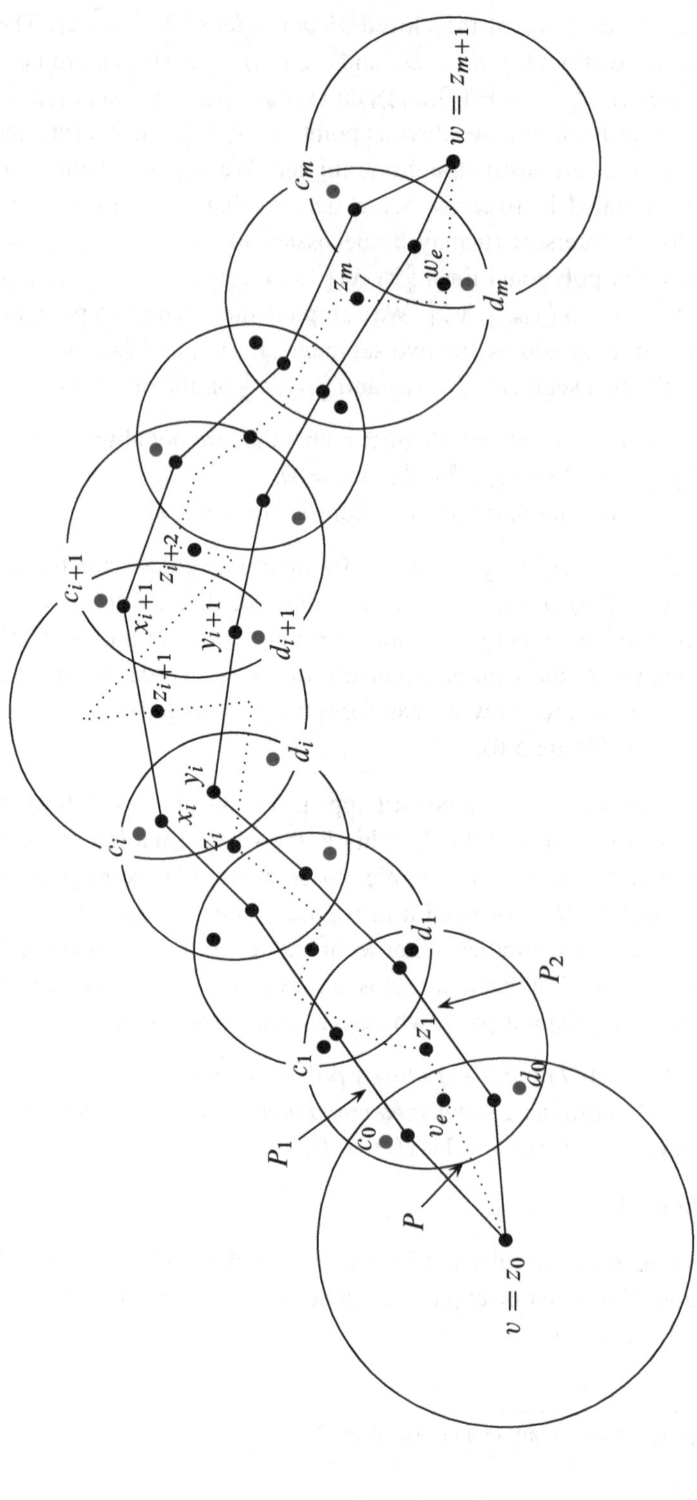

Figure 5.5. Doubling of an edge P at P_1 and P_2.

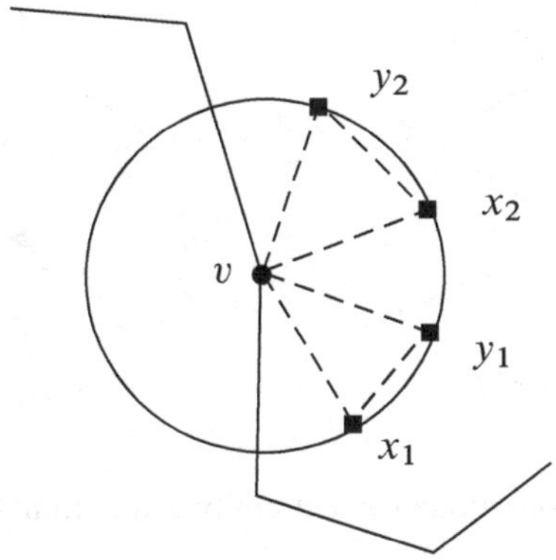

Figure 5.6. Creation of a 2-loop.

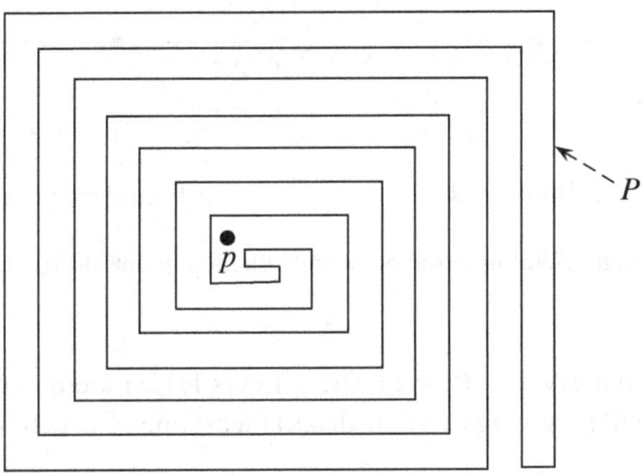

Figure 5.7. A polygonal line P and a point p. Is the point p in the interior (bounded connected component of $\mathbb{R}^2 \setminus P$) or the exterior (unbounded connected component) of the closed polygonal line P?

(α) If z is an extremity of P, we take ε sufficiently small so that the disk $D(z, \varepsilon)$ only meets the segments of $\mathrm{Fr}(U)$ that pass through z. Then the intersection $D(z, \varepsilon) \cap \mathrm{Fr}(U)$ is made up of radii of the circle $\mathscr{C}(z, \varepsilon)$, which thus define angular sections that are open sets *centred* at z, each one of which is contained in either U or $\mathbb{R}^2 \setminus \overline{U}$: indeed, they are connected and contain no point of $\mathrm{Fr}(U)$.

When there is only one radius, we will say that z is a *terminal point* of $\mathrm{Fr}(U)$; in this case the unique section is contained in U (Figure 5.8, in which $\overline{U}^c$ denotes the open set $\mathbb{R}^2 \setminus \overline{U}$).

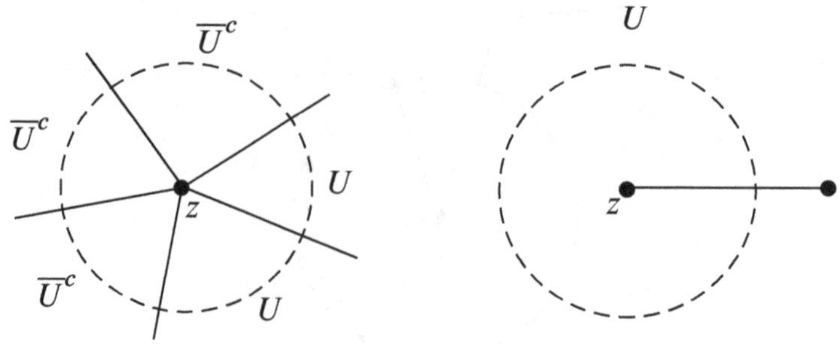

Figure 5.8. Sections around an extremity and terminal point of $\mathrm{Fr}(U)$.

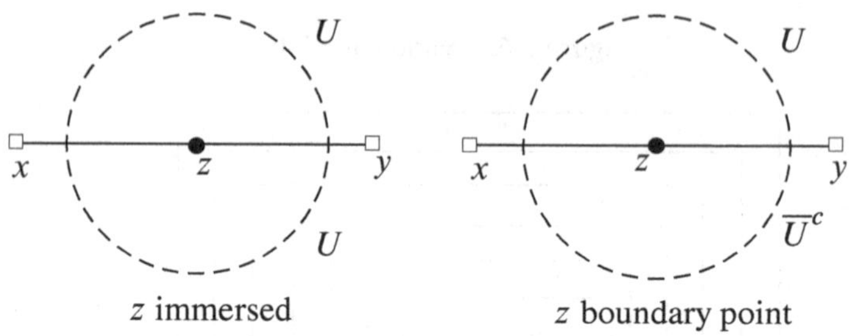

Figure 5.9. Immersed point and boundary point of $\mathrm{Fr}(U)$.

(β) If on the contrary $z \in \mathring{P}$, then $D(z, \varepsilon)$ cuts $\mathrm{Fr}(U)$ along a diameter (always for sufficiently small ε), giving two half-disks at least one of which is contained in U, since $z \in \overline{U}$.

- If the other half-disk is also contained in U, we say that z is *immersed* in U (Figure 5.9);
- Otherwise the other half-disk is contained in $\mathbb{R}^2 \backslash U$, in which case we say that z is a *boundary point* of U (Figure 5.9).

Proposition 5.2.7. *Let U be a connected open set in the plane whose boundary is made up of polygonal arcs. Take $x \in \mathrm{Fr}(U)$, $y \in \overline{U} = U \cup \mathrm{Fr}(U)$, and let P be a polygonal line connecting x and y in such a way that $\mathring{P} \subset U$.*

(i) *If $y \in U$, then $U \backslash \mathring{P}$ is connected.*

(ii) *If x and y belong to the same connected component of* $\mathrm{Fr}(U)$, *then $U \backslash \mathring{P}$ has two connected components, and P is contained in the boundary of each of them.*

(iii) *If x and y lie in two distinct connected components of* $\mathrm{Fr}(U)$, *then $U \backslash P$ is connected.*

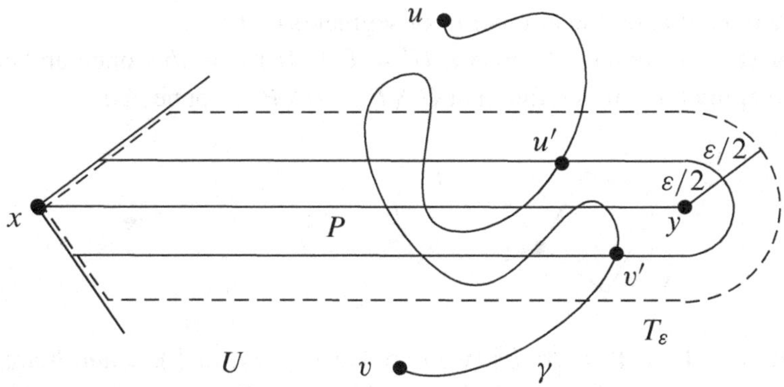

Figure 5.10. Tube truncated by Fr(U) around a segment.

Proof. (i) Consider first the case $P = [\![x, y]\!]$. For $\varepsilon > 0$, observe that

$$P_\varepsilon = \{z \in \mathbb{R}^2 : d(z, P) < \varepsilon\}.$$

Then P_ε is clearly a connected open set. When ε is sufficiently small, P_ε only meets Fr(U) in radii $[\![x, x_i]\!]$ of the disk $D(x, \varepsilon)$, so that $T_\varepsilon = P_\varepsilon \cap U$ has the form of a truncated tube (Figure 5.10).

Let $u, v \in U \setminus P$, and let γ be an arc joining u to v in U.

- If γ *avoids* P, then γ connects u to v in $U \setminus P$;
- otherwise, choose $\varepsilon < \min(d(u, P); d(v, P))$ so that $u, v \notin \overline{T}_{\varepsilon/2}$; thus γ enters $\overline{T}_{\varepsilon/2}$ at a point u' and exits at a point v'. We have $u', v' \in T_\varepsilon \setminus P$, and $T_\varepsilon \setminus P$ is connected (it is even a star-shaped set). Therefore u' can be connected to v' inside $T_\varepsilon \setminus P$. Then, by concatenation, we obtain an arc from u to v in $U \setminus P$.

In the general case of a polygonal line P, we use induction on the number of segments of P.

(ii) The polygonal line between x and y on Fr(U) allows us to close the arc P, which becomes a polygonal arc P' in $\mathbb{R}^2$. By Theorem 5.2.6, we obtain two connected components U_1', U_2' for $\mathbb{R}^2 \setminus P'$, and we have $P' = \mathrm{Fr}(U_i)$, $i = 1, 2$. Then, setting $U_i = U_i' \cap U$, $i = 1, 2$, we obtain exactly two connected components for $U \setminus \mathring{P}$.

Let $z \in \mathring{P}$. If $z \notin \mathrm{Fr}(U_2)$, then there exists $r > 0$ such that $D(z, r) \subset U$ and $D(z, r) \cap U_2 = \emptyset$. Thus $D(z, r) \cap U_2' = \emptyset$. But this contradicts the fact that $z \in P' = \mathrm{Fr}(U_2')$, so we have $\mathring{P} \subset \mathrm{Fr}(U_2)$, then similarly, $\mathring{P} \subset \mathrm{Fr}(U_1)$. As the boundaries $\mathrm{Fr}(U_i)$, $i = 1, 2$, are closed and $x, y \in P = \overline{\mathring{P}}$, we conclude that $P \subset \mathrm{Fr}(U_1) \cap \mathrm{Fr}(U_2)$.

(iii) Let X and Y be the connected components of Fr(U) containing x and y respectively. We have $X \cup Y \subset \mathrm{Fr}(U)$, $X \cap Y = \emptyset$, $x \in X$, $y \in Y$.

We use induction on the number m of segments of Y:

(α) For $m = 0$, we have $Y = \{y\}$; $U' = U \cup \{u\}$ is *visibly* open and connected and we can apply (i) to it: we find that $U' \setminus P = U \setminus P$ is connected.

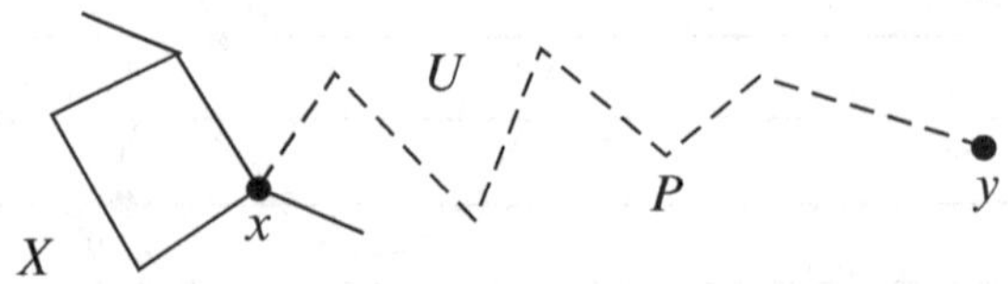

(β) Let $m = 1$, so $Y = [\![a, b]\!]$ (we may have $y = a$ or b); a and b are terminal points of $\mathrm{Fr}(U)$, so $U' = U \cup [\![a, b]\!]$ is a neighbourhood of a and of b (i.e. U' contains open disks centred at a and b); the *immediate neighbours* of a on $[\![a, b]\!]$ are points immersed in U, and we see easily that the set I of points of $]\!]a, b[\![$ that are immersed is an open set in $]\!]a, b[\![$; similarly, the set $]\!]a, b[\![\setminus I$ of boundary points of U in $]\!]a, b[\![$ is an open set of $]\!]a, b[\![$, so $I =]\!]a, b[\![$ and consequently U' is a neighbourhood of each of its points, so it is open.

We decompose $P = \bigcup_{i=0}^{n} [\![x_i, x_{i+1}]\!]$ with $x_0 = x$ and $x_{n+1} = y$. Successively, we have $U' \setminus [\![x_0, x_1]\!] = U'_1$ is open and connected by (i), then $U'_1 \setminus [\![x_1, x_2]\!] = U'_2$ is again open and connected by (i), and so forth until $U'_n \setminus [\![a, y]\!]$ which is again open and connected.

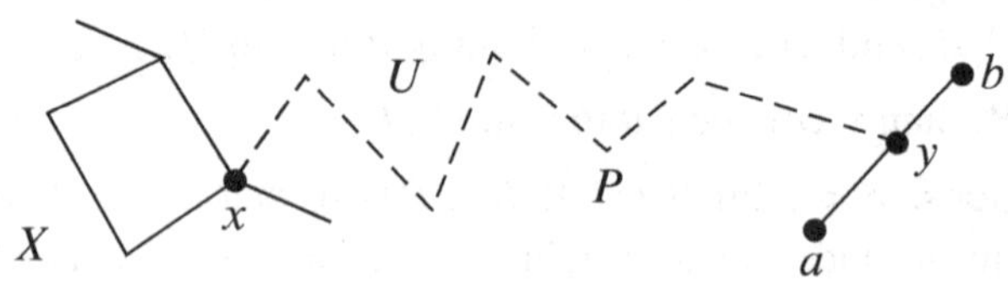

(γ) $m \curvearrowright m + 1$: y belongs to a segment $[\![a, b]\!]$ of Y.

Case 1: there exists a point $z \in Y$, $z \neq a, b$ which is a terminal point of $\mathrm{Fr}(U)$ corresponding to the segment $S = [\![t, z]\!]$. The points of $]\!]t, z[\![$ are then immersed, and as we saw for $m = 1$, we find that $U' = U \cup]\!]t, z]\!]$ is open, and the connected component of its boundary contained in Y has one less segment. By the induction hypothesis, then $U'' = U' \setminus P$ is open and connected, $z \in U''$, $t \in \mathrm{Fr}(U'')$: then by (i), $U'' \setminus [\![t, z]\!] = U \setminus P$ is open and connected.

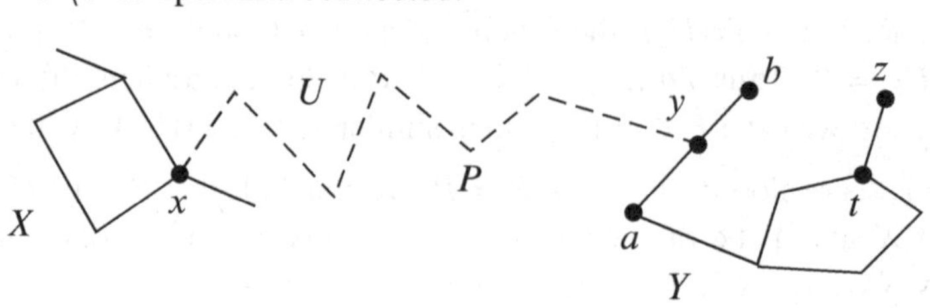

Case 2: Otherwise, Y contains a closed polygonal line L which is a Jordan arc: $\mathbb{R}^2 \setminus L$ consists of two connected components C_1 and C_2 of which one, say C_1, contains U. Let $S = [\![a', b']\!]$ be a segment of L adjacent to U (it is easy to see that S exists!):

- $U' = U \cup \mathring{S} \cup C_2$ is connected; indeed, as $\mathring{S} \subset \overline{U}$ and $\mathring{S} \subset \overline{C_2}$, $U' = (U \cup \mathring{S}) \cup (C_2 \cup \mathring{S})$ is the union of two intersecting connected sets;

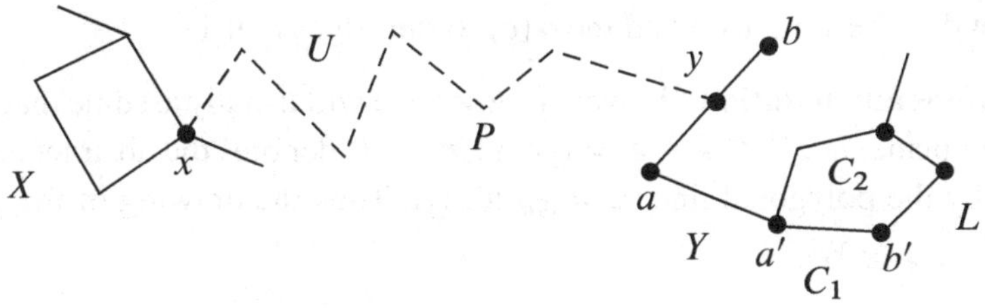

- U' is open since the points of $\mathring{S}$ are immersed points on the boundary of $U \cup C_2$. The connected component of the boundary of U' contained in Y has one less segment. By the induction hypothesis, then, $U'' = U' \setminus P$ is open and connected; we have $U \setminus P = U'' \setminus ([\![a', b']\!] \cup C_2)$ with $a', b' \in \mathrm{Fr}(U'')$ in the same connected component $Y \setminus]\!]a', b'[\![$ of $\mathrm{Fr}(U'')$. By (ii), $U'' \setminus [\![a', b']\!]$ has two connected components: C_2 is necessarily one of them, so the other is exactly $U'' \setminus (C_2 \cup]\!]a', b']\!])$ and this set is precisely $U \setminus P$. ∎

The properties of Proposition 5.2.7 are illustrated in Figure 5.11.

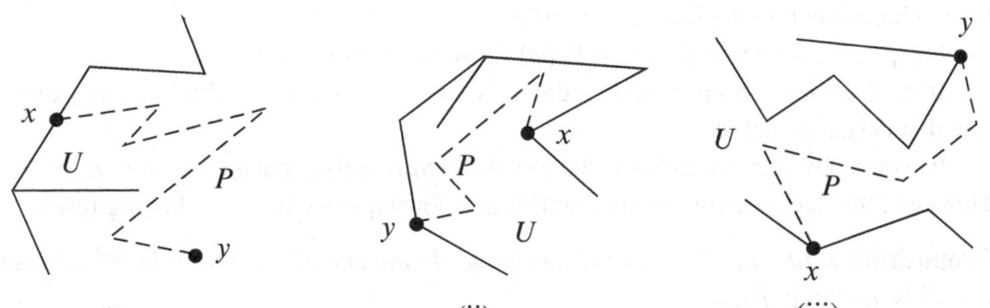

Figure 5.11. Illustration of the three alternatives described in Proposition 5.2.7.

5.2.3 Embedded graphs

Let $\Gamma = (V; E, N)$ be a planar graph and $\sigma: \Gamma \to \mathbb{R}^2$ an arbitrary polygonal embedding:

$$\sigma: V \sqcup E \to \mathbb{R}^2 \sqcup \mathcal{C}_{\mathbb{R}^2}.$$

The curves $\sigma(e)$ are thus polygonal lines that are disjoint (except at the extremities); the graph is visible from its drawing $F = \sigma(V) \cup \bigcup_{e\in E} \sigma(e)$ since the $\sigma(v)$ correspond to the vertices v and the $\sigma(e)$ to the edges e of Γ.

Conventions and notation. In what follows, we will usually assume that Γ is *embedded* in the plane, i.e. $V \subset \mathbb{R}^2$; $e = ([x, y], n) \in E$ denotes the abstract edge which is realised by the polygonal line $\gamma_e = \xi_e([0, 1])$. Thus the drawing of the graph is just $F = V \cup \bigcup_{e\in E} \gamma_e$.

5.2.4 Faces

Let $\Gamma = (V; E, N)$ be an embedded graph and F the corresponding drawing in $\mathbb{R}^2$. As F is a compact and thus closed subset of the plane, $\mathbb{R}^2 \setminus F$ is an open set, which decomposes into connected components (which are open, see Section 5.2.1), which we call the *faces* of F, or of the embedded graph Γ: thus we can write

$$\mathbb{R}^2 \setminus F = \bigsqcup_{\varphi\in\Phi} \varphi,$$

where $\Phi = \Phi(\Gamma)$ is the set of faces of Γ. Furthermore, F is bounded (i.e. contained in a disk $D(0, r)$ for sufficiently large r), so there is one and only one unbounded face, which is called the *infinite face* of F.

For $\varphi \in \Phi$, we write $\overline{\varphi} \cap F = \mathrm{Fr}(\varphi)$ for the *boundary* of φ.

If $\gamma_e \subset \mathrm{Fr}(\varphi)$, we say that the edge e is *adjacent* to φ, or on the boundary of φ, and that φ is adjacent to e.

It seems obvious that $\mathrm{Fr}(\varphi)$ is the union of *entire* polygonal arcs γ_e, $e \in E_\varphi \subset E$. However, this fact needs to be justified! It is a consequence of the following result.

Proposition 5.2.8. *Let $\Gamma = (V; E, N)$ be a planar embedded graph in $\mathbb{R}^2$ and let $e \in E$ be an edge. Then*

(i) *if e lies on a cycle of Γ, then γ_e is adjacent to exactly two faces φ_1, φ_2 and $\gamma_e \subset \mathrm{Fr}(\varphi_1) \cap \mathrm{Fr}(\varphi_2)$;*

(ii) *if e does not lie on a cycle, then γ_e is adjacent to a single face φ: $\gamma_e \subset \mathrm{Fr}(\varphi)$.*

Proof. (i) If the edge $e = ([x, y], n)$ lies on a cycle, it lies on an elementary cycle C, which we may assume of minimal length, so having no chords. Then γ_e is drawn in a face U of $\Gamma \setminus e$. Since C is a cycle, x and y lie in the same connected component of

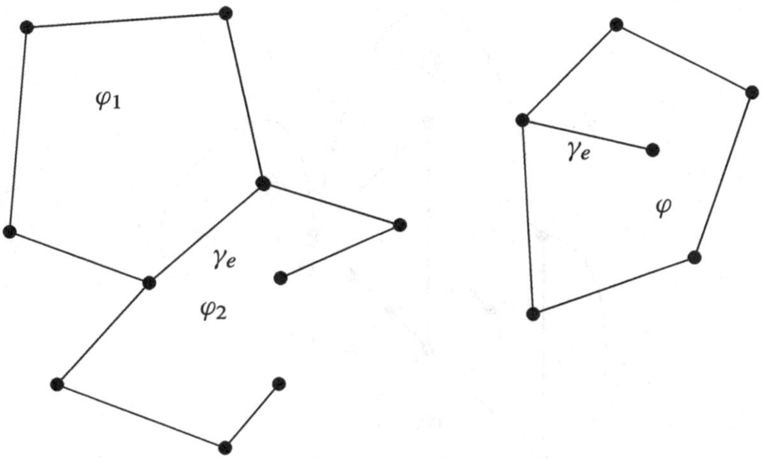

Figure 5.12. Left-hand figure: γ_e is adjacent to two faces φ_1, φ_2. Right-hand figure: γ_e is adjacent to one face φ.

$\mathrm{Fr}(U)$. But $C \backslash e$ is contained in $\mathrm{Fr}(U)$, thus by Proposition 5.2.7 (ii), $U \backslash \gamma_e$ has two connected components φ_1 and φ_2 which are the two faces adjacent to γ_e.

(ii) $e = ([x, y], n)$ is an isthmus; γ_e is drawn in a face $U = \varphi$ of $\Gamma \backslash e = \Gamma_1 + \Gamma_2$; x and y are in two distinct connected components of $\mathrm{Fr}(U)$: $x \in V(\Gamma_1)$, $y \in V(\Gamma_2)$, so by Proposition 5.2.7 (iii), γ_e does not disconnect φ, so it is adjacent to a single face.

These two cases are shown in Figure 5.12. ■

This proposition shows that for every face φ of F, $\mathrm{Fr}(\varphi)$ consists of polygonal lines γ_e, $e \in E_\varphi$: thus $\mathrm{Fr}(\varphi)$ is the drawing in the plane of a subgraph of Γ, called the *boundary* of φ and written $\partial\varphi = (V_\varphi; E_\varphi)$ where $E_\varphi = \{e \in E : \gamma_e \subset \mathrm{Fr}(\varphi)\}$ and V_φ is the set of extremities of all the edges $e \in E_\varphi$.

It seems clear (but again requires proof!) that for a graph of finite order, the number of faces in any of its possible drawings is finite.

Proposition 5.2.9. *Let $\Gamma = (V; E, N)$ be an embedded planar graph in $\mathbb{R}^2$. The set Φ of faces of Γ is finite.*

Proof. We use induction on the number of edges of Γ. If $|E| = 0$, there is only one face, and the drawing of Γ is the plane with a finite number of points removed. If $E = \{e\}$, then Γ admits at most two faces: if e is a loop, there are two faces by Theorem 5.2.6; otherwise γ_e is a non-closed polygonal arc and there is only one face by Proposition 5.2.7 (i).

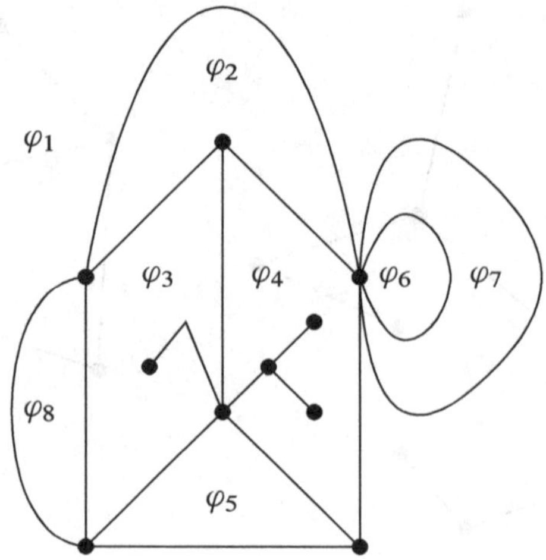

Figure 5.13. Planar graph with 10 vertices and 8 faces; the face φ_1 is the infinite face.

Suppose now that Γ has $m + 1$ edges. Let e be one of them. Then $\mathring{\gamma}_e$ is contained in a single face φ of $\Gamma\backslash e$, so its extremities x, y lie in $\overline{\varphi}$; the other faces are not modified when we add γ_e to the drawing of $\Gamma\backslash e$.

- if y (or x) is in φ, then adding the curve γ_e to the drawing of $\Gamma\backslash e$ does not disconnect φ (cf. Proposition 5.2.7 (i)), so that Γ has the same number of faces as $\Gamma\backslash e$;
- if $x, y \in \mathrm{Fr}(\varphi)$, then Proposition 5.2.7 (ii) and (iii) show that adding γ_e separates φ into at most two connected components: the number of faces thus increases by at most one. ■

We observe that in general, the boundary of a face is not necessarily connected. It suffices to consider the graph composed of two disjoint cycles, for example two concentric circles of different radii. This graph has 3 faces, and the boundary of the intermediate face (the annulus) is not connected.

We now show that a tree, or a forest, is always planar.

Proposition 5.2.10. *Every finite forest* Γ *can be embedded in the plane, and its drawing possesses only one face.*

Proof. We use induction on $|E|$. If the forest has no edge, i.e. if $|E| = 0$, the result is clear.

Assume now that the forest has $m + 1$ edges. Remove one of its edges e and embed $\Gamma\backslash e$, which is still a forest, into the plane via an embedding $\sigma: V \sqcup E\backslash e \to \mathbb{R}^2 \cup \mathcal{C}_{\mathbb{R}^2}$. Let F be the corresponding drawing of $\Gamma\backslash e$. Then by the induction

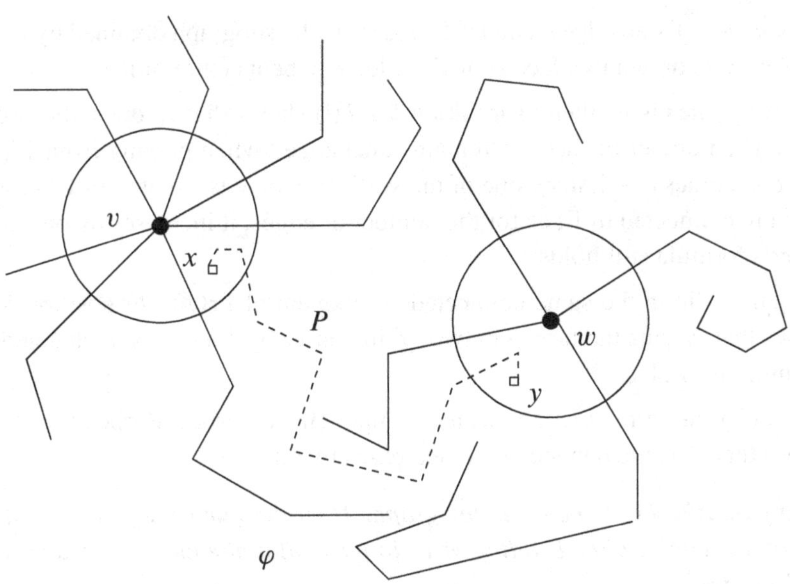

Figure 5.14. Forest embedded in the plane.

hypothesis, $\varphi := \mathbb{R}^2 \setminus F$ is the unique face of F, so it is a open set connected by polygonal arcs.

Let x and y denote the extremities of e. Then x (resp. y) belongs to a connected component C (resp. C') of F, and $C \neq C'$ since otherwise there would be a cycle. By Proposition 5.2.7 (iii), we obtain the result, since $x, y \in F = \mathrm{Fr}(\varphi)$. ■

5.2.5 Euler's formula

In 1750, Leonhard Euler discovered and proved a remarkable formula relating the number of faces, the number of vertices and the number of edges of a graph embedded in $\mathbb{R}^2$.

Theorem 5.2.11 (Euler's formula). *Let $\Gamma = (V; E, N)$ be a graph of finite order embedded in $\mathbb{R}^2$. Set $n = |V|$, $m = |E|$ and let $f = |\Phi|$ be the number of faces of Γ. Then if Γ is connected, we have $n - m + f = 2$. More generally, if Γ possesses exactly c connected components, we have $n - m + f = c + 1$.*

Proof. Let us prove the general case by induction on the number of edges of Γ.

If $m = 0$, then $n = c$ and $f = 1$ (the plane with s points removed is connected), so $n - m + f = c - 0 + 1 = c + 1$.

$m \curvearrowright m+1$: fix an edge e and let $\Gamma \backslash e$ denote the subgraph obtained by removing the edge e; γ_e is drawn in a face φ of $\Gamma \backslash e$; let x, y be its extremities.

- If x (or y) lies in φ, then Proposition 5.2.7 (i) shows that γ_e does not disconnect $\Gamma \backslash e$. The number of faces f remains unchanged when passing from $\Gamma \backslash e$ to Γ, and c becomes $c - 1$ since one of the vertices x or y is isolated in $\Gamma \backslash e$, whereas both are connected in Γ; as for the number of edges, it increases by one, and thus Euler's formula still holds.
- If x and y lie in the same connected component of $\mathrm{Fr}(\varphi)$, Proposition 5.2.7 (ii) shows that γ_e cuts the face φ in two: f increases by 1 and c is unchanged, so the formula still holds.
- If x and y lie in distinct connected components of $\mathrm{Fr}(\varphi)$, Proposition 5.2.7 (iii) shows that f is unchanged and c decreases by one. ■

Corollary 5.2.12. *Let Γ be a planar graph. Then the number of faces f does not depend on the choice of embedding of Γ. In particular, the number of faces is equal to 1 if Γ is a tree.*

Proof. Euler's formula gives $f = c + 1 - n + m$, and f depends only on Γ. ■

In general, the boundaries $\partial\varphi$ of faces may contain cycles. The following result shows that forests are the only exception.

Proposition 5.2.13. *Let Γ be a graph embedded in the plane.*

(i) *If Γ is a forest, there is a single face φ and $\partial\varphi$ has no cycle.*

(ii) *If there exists a face φ whose boundary $\partial\varphi$ has no cycle, then Γ is a forest and φ is its only face.*

Proof. Point (i) was already proved in Proposition 5.2.10.

(ii) Since $\partial\varphi$ has no cycle, it is a subgraph of Γ, which is a forest, so its drawing has a single face: $U = \mathbb{R}^2 \backslash \mathrm{Fr}(\varphi)$ is connected. It follows that Γ has only one face, since if ψ were another face we would have $\psi \subset U$; taking $x \in \psi$, $y \in \varphi$, we could draw an arc α from x to y *in U* (since U is arc-connected); the arc α would then meet both φ and $\mathbb{R}^2 \backslash \varphi$ since $\psi \subset \mathbb{R}^2 \backslash \varphi$, so it would intersect $\mathrm{Fr}(\varphi)$, but then we would have $U \cap \mathrm{Fr}(\varphi) \neq \emptyset$, which is impossible (Figure 5.15). But then Γ has no cycle, i.e. it is a forest, since if an edge e lies on a cycle, then it is adjacent to two faces (see Proposition 5.2.8). ■

Recall that the *girth* of a simple graph $\Gamma = (V; E)$, denoted $\mu(\Gamma)$, is the length of the smallest cycle contained in Γ (see Chapter 1). We write $\mu(\Gamma) = +\infty$ when Γ has no cycle (so is a forest); otherwise $|E| \geqslant \mu(\Gamma) \geqslant 3$. The number $\mu(\Gamma)$ is related to the numbers $m = |E|$, $n = |V|$, $f = |\Phi|$ as we will see in Proposition 5.2.16 below.

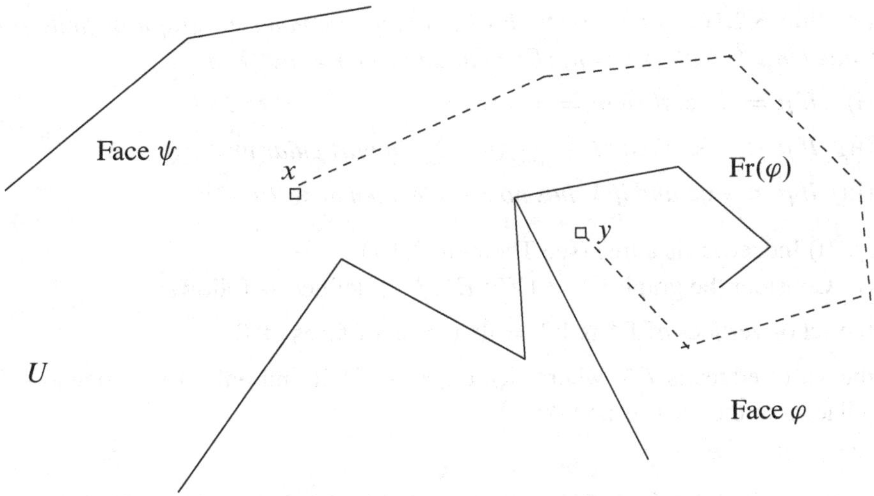

Figure 5.15. The dotted polygonal line connects $x \in \psi$ to $y \in \varphi$, while remaining in U.

Proposition 5.2.14. *Let $\Gamma = (V; E)$ be a simple planar graph embedded in $\mathbb{R}^2$. If $\mu(\Gamma) < +\infty$, then every face φ admits a boundary $\partial\varphi$ consisting of at least $\mu(\Gamma)$ edges.*

Proof. Let $\partial\varphi = (V'; E')$. If we had $|E'| < \mu(\Gamma)$, then $\partial\varphi$ would have no cycle, so by Proposition 5.2.13, Γ would be a forest whose unique face would be φ, and thus we would have $\mu(\Gamma) = +\infty$. ∎

Proposition 5.2.15. *Let $\Gamma = (V; E, N)$ be a planar graph embedded in $\mathbb{R}^2$ with $|E| \geqslant 3$, and let φ be a face. Then*

(i) *if φ is adjacent to exactly one edge e, then e is a loop;*

(ii) *if φ is adjacent to exactly two edges e and e′ which are not loops, then e and e′ form a double edge;*

(iii) *if Γ is simple, then every face is adjacent to at least* 3 *edges.*

Proof. (i) If e is not a loop, then $\partial\varphi$ has no cycle, so by Proposition 5.2.13, the graph Γ is a forest with unique face φ and a unique edge.

(ii) If e and e' do not form a cycle, then $\partial\varphi$ has no cycle. As above, we deduce that Γ is a forest with unique face φ, which thus has two edges.

(iii) We distinguish two cases: if the graph is a forest, it has one face, and since $|E| \geqslant 3$ this face is adjacent to at least 3 edges; otherwise $\mu(\Gamma) < +\infty$ and we apply Proposition 5.2.14. ∎

Proposition 5.2.16. *Let $\Gamma = (V; E)$ be a simple connected graph of finite order, embedded in $\mathbb{R}^2$, with $|V| = n$, $|E| = m$ and $\mu(\Gamma) = \mu$. Then*

(i) *If $\mu = +\infty$, then $m = n - 1$.*

(ii) *If $\mu < +\infty$, then $m \leqslant \frac{\mu}{\mu-2}(n-2)$; in particular $m \leqslant 3(n-2)$.*

(iii) *If $\mu < +\infty$ and if Γ has no 3-cycle, then $m \leqslant 2n - 4$.*

Proof. (i) Indeed, Γ is a tree (see Theorem 2.2.1).

(ii) Consider the graph $\Gamma^* = (V^*; E^*, N^*)$ defined as follows:

- the set of vertices of Γ^* is $V^* = \Phi$, the set of faces of Γ;
- the set of edges is E^*, where $([\varphi, \psi], e) \in E^*$ if and only if the edge $e \in E$ is adjacent to the faces φ and ψ;
- $N^* = E$.

This graph, called the *dual* of Γ, will be studied further below; it may have loops and double edges. We have

- $\sum_{\varphi \in \Phi} d(\varphi) = 2|E^*|$, where $d(\varphi)$ is the degree of the vertex φ of Γ^* (see Lemma 1.1.9);
- $|E^*| \leqslant m$ since the map $f : E \to E^*$ which associates $([\varphi, \psi], e) \in E^*$ to e (where φ and ψ are the faces adjacent to e) is surjective;
- for $\varphi \in \Phi$, the degree $d(\varphi)$ is at least equal to the number of edges adjacent to φ (this quantity is twice the number of loops in Γ^*, plus the number of non-loop edges); this number is at least equal to μ by Proposition 5.2.14.

It follows that

$$\mu \cdot f \leqslant \sum_{\varphi \in \Phi} d(\varphi) = 2|E^*| \leqslant 2m.$$

We apply Euler's formula (with $c = 1$):

$$m = n + f - 2 \leqslant n + \frac{2m}{\mu} - 2,$$

so $m \leqslant \frac{\mu}{\mu-2}(n-2)$. Since $\mu \geqslant 3$, we have $\frac{\mu}{\mu-2} \leqslant 3$, which gives the second upper bound.

(iii) In this case $\mu \geqslant 4$ and $\frac{\mu}{\mu-2} \leqslant 2$; the result then follows from (ii) above. ∎

Exercise 5.2. Show that the upper bound of (ii) in Proposition 5.2.16 is optimal.

This proposition confirms the intuition that a planar graph may not have "too many" edges with respect to its number of vertices.

Corollary 5.2.17. *The graphs K_5 and $K_{3,3}$ are not planar.*

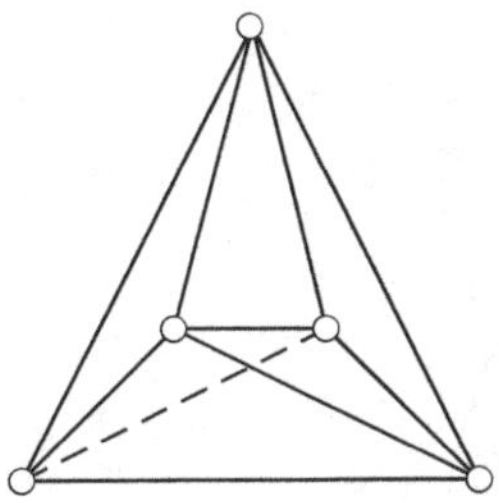

K_5

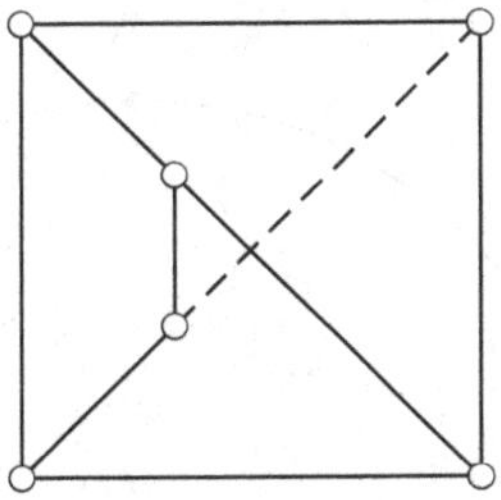

$K_{3,3}$

Figure 5.16. The complete graph K_5 is not planar; the bipartite complete graph $K_{3,3}$ is not planar.

Proof. These graphs are connected, and they are not trees. For each of them, let n, m and μ respectively denote its number of vertices, its number of edges and its girth.

For K_5, we have $n = 5, m = 10$ and $\mu = 3$, so $\frac{\mu}{\mu-2}(n-2) = 3(n-2) = 9$, which contradicts the upper bound (ii) of Proposition 5.2.16: thus K_5 cannot be planar.

For $K_{3,3}$, we have $n = 6, m = 9$ and $\mu = 4$: hence $\frac{\mu}{\mu-2}(n-2) = 2(n-2) = 8$ and $K_{3,3}$ also cannot be planar. ■

Corollary 5.2.18. *Every simple planar graph $\Gamma = (V; E)$ has at least one vertex whose degree is $\leqslant 5$.*

Proof. With no loss of generality, we may assume that the graph is connected and has at least three vertices.

(i) If $\mu(\Gamma) = +\infty$, then $|E| = |V| - 1$ by Proposition 5.2.16 (i), so $\delta(\Gamma) = 1$.

(ii) If $\mu(\Gamma) < +\infty$ and $d(x) \geqslant 6$ for all $x \in V$, then by Lemma 1.1.9 we have $2m = \sum_{x \in V} d(x)$ so $2m \geqslant 6n$, so $3n \leqslant m$. But by Proposition 5.2.16 (ii) we have $m \leqslant 3n - 6$, which is a contradiction. ■

Exercise 5.3. Show that the statement of Corollary 5.2.18 becomes false if Γ is not simple.

5.2.6 Planar graphs and blocks

Lemma 5.2.19. *For any given face of a planar graph, there exists a graph embedding in the plane such that the given face becomes the infinite face.*

Proof. Let x, y, u denote the coordinates of a point M in $\mathbb{R}^3$, and consider the unit sphere $S_2 = \{(x, y, u) : x^2 + y^2 + u^2 = 1\}$. Equipped with the topology induced by that of $\mathbb{R}^3$, the sphere S_2 is a compact space. Let N be the point of S_2 of coordinates $(0, 0, 1)$ (the *north pole* of the sphere). Consider the stereographic projection ν of

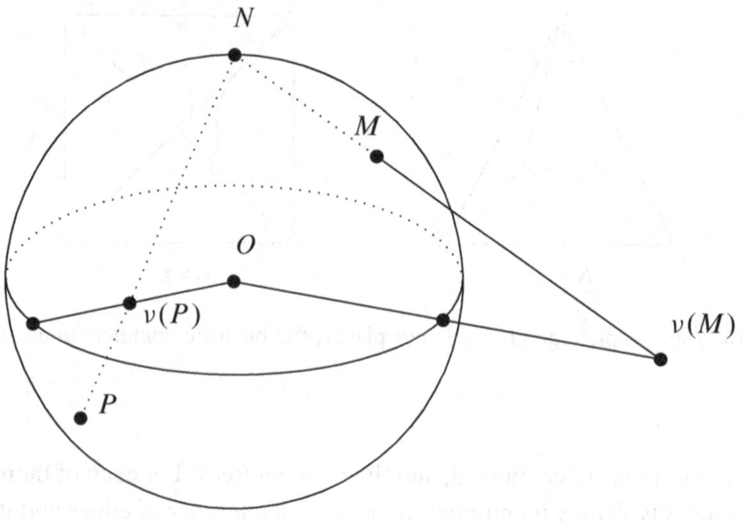

Figure 5.17. The stereographic projection. The northern hemisphere is mapped to the exterior of the unit disk, the equator is invariant, the southern hemisphere is sent to the interior of the unit disk.

pole N. It associates to every point M of S_2 distinct from N the point in the plane $u = 0$ aligned with N and M. It is given as a complex number z by

$$z = \frac{x + iy}{1 - u},$$

where x, y, u are the coordinates of M. The map $\nu\colon (x, y, u) \mapsto z$ is a homeomorphisme of $S_2 \backslash N$ to $\mathbb{C}$, where $\mathbb{C}$ is identified with $\mathbb{R}^2$ (Figure 5.17).

A planar graph Γ with drawing F corresponds to a drawing $F' = \nu^{-1}(F)$ on $S_2 \backslash N$; we see that via a well-chosen rotation ρ of the sphere, we can place the point N in the interior of a face $\nu^{-1}(\varphi)$ of F'. Then, applying ν, we obtain $F'' = \nu \circ \rho \circ \nu^{-1}(F)$ which is homeomorphic to F; it is another drawing of Γ, but for which the face $\nu \circ \rho \circ \nu^{-1}(\varphi)$ corresponding to φ is now the infinite face. ■

Corollary 5.2.20. *A graph is planar if and only if it can be embedded on the sphere.*

Theorem 5.2.21. *A graph $\Gamma = (V; E, N)$ is planar if and only if all of its blocks are planar.*

Proof. One direction is obvious. For the other, we use induction on the number of blocks of Γ. It suffices to show that each connected component of Γ is planar. By Proposition 5.2.5, we may assume that Γ is simple.

If the graph has only one block, there is nothing to prove. Let Γ be a connected simple graph with $k+1$ blocks, $k \geqslant 1$. The block graph $\flat(\Gamma)$ is then a tree (see Proposition 4.2.5). Therefore Γ possesses a block B with a unique articulation vertex x (it suffices to take a leaf of $\flat(\Gamma)$). By the induction hypothesis and Lemma 5.2.19, the subgraphs B and $\Gamma \backslash B$ can be embedded into $\mathbb{R}^2$ in such as way that x lies on the boundary of the infinite face of each of the graphs B and $\Gamma \backslash B$. More precisely, B and $\Gamma \backslash B$ can be drawn separately in two complementary half-planes of $\mathbb{R}^2$, with the articulation vertex x lying on the boundary line between them. ∎

This theorem indicates the importance of 2-connected graphs in studying planarity. We will investigate it in more detail in the following paragraph.

5.2.7 Two-connected planar graphs

Proposition 5.2.22. *In a 2-connected planar graph Γ, the boundaries of faces are elementary cycles.*

Proof. We use induction on the number of edges of the 2-connected graph Γ. If Γ is an elementary cycle, the result clearly holds. If Γ has 3 edges, it is a triangle which is an elementary cycle and which bounds the two faces of Γ.

Since every 2-connected graph is obtained by the chain method (see Proposition 4.2.8), if Γ is not a cycle, then we can write $\Gamma = \Gamma' \cup P$ where Γ' is a 2-connected graph and P is an elementary chain connecting two vertices of Γ' and not passing through any other vertex of Γ'.

By the induction hypothesis, the boundaries of the faces of Γ' are elementary cycles. By the construction of P, there exists a unique face φ' of Γ' containing $\mathring{P}$. Since Γ is 2-connected, the extremities x, y of the chain P lie on the boundary of φ'. By Proposition 5.2.7 (ii) it follows that P separates the face φ' into two faces φ_1, φ_2 of Γ, and the boundaries of these two faces are both elementary cycles containing the chain P. The other faces of Γ are also faces of Γ', so by the induction hypothesis their boundaries are also elementary cycles. ∎

5.3 Maximal planar graphs and polyhedra

5.3.1 Maximal planar graphs

A planar graph is said to be *maximal* if adding an edge between any two non-neighbouring vertices produces a non-planar graph. For example K_4 and $K_5 \backslash e$ where e is an arbitrary edge of K_5 are maximal planar graphs. We also note that every maximal graph is connected (otherwise we could add an edge between two vertices belonging to different connected components without contradicting planarity), and that a planar graph is maximal if and only if the associated simplified graph is maximal. Consequently, we will only consider simple graphs.

Proposition 5.3.1. *Every maximal simple planar graph with $n \geqslant 3$ vertices is 2-connected.*

The proof of this result is technical, and relies on an argument of connectedness which makes it possible to construct a polygonal line between two vertices without *crossing* any other edges. Before explaining this reasoning, we need to introduce the notion of a regular boundary point. Let U be an open set of $\mathbb{R}^2$ and $x \in \mathrm{Fr}(U)$. We say that x is *regular* if there exists a point $y \in U$ such that $]\!]x, y]\!] \subset U$. Thus, by Lemma 5.2.3, if U is connected in $\mathbb{R}^2$, any two points x, y in the set $\tilde{U} := U \cup \{\text{regular boundary points of } U\}$ can be connected by a polygonal line in $U \cup \{x, y\}$.

Lemma 5.3.2. *Let U be a connected open set of $\mathbb{R}^2$ such that* $\mathrm{Fr}(U)$ *is polygonal. Then $\tilde{U} = \overline{U}$.*

Proof. It suffices to show that every point of $\mathrm{Fr}(U)$ is regular. Let $x \in \mathrm{Fr}(U)$. There exists $r > 0$ such that $D(x, r) \cap U$ is an open non-empty angular section of $D(x, r)$. Choose y in this section to conclude. ∎

Proof of Proposition 5.3.1. Let Γ be a connected embedded planar graph. Assume that it is not 2-connected. Then there exists a vertex x such that the subgraph $\Gamma' = \Gamma - x$ consists of $k \geqslant 2$ connected components $\Gamma_1, \Gamma_2, \ldots, \Gamma_k$. Up to modifying the embedding, we may assume that x lies on the infinite face of Γ'.

Let φ denote the infinite face of Γ. We have $x \in \mathrm{Fr}(\varphi)$. Since Γ' is not connected, unlike Γ, there exist two vertices $y \in V(\Gamma_1) \cap \mathrm{Fr}(\varphi)$ and $z \in V(\Gamma_2) \cap \mathrm{Fr}(\varphi)$ that are not neighbours but are both connected to x in Γ. However, φ has polygonal boundary, so by Lemma 5.3.2 we can connect y to z by a polygonal line e in $\varphi \cup \{y, z\}$. We conclude that $\Gamma \cup \{e\}$ is planar; thus Γ is not maximal planar. ∎

A triangle in a graph is a 3-cycle; by extension, a face of a planar graph is called a *triangle* or said to be *triangular* if, for a given embedding $\sigma\colon \Gamma \to \mathbb{R}^2$, the inverse image $\sigma^{-1}(\mathrm{Fr}(\varphi))$ is a 3-cycle of Γ.

A graph is *triangulated* if every cycle of length strictly greater than 3 admits a chord, i.e. an edge connecting two non-consecutive vertices in the cycle. We also say that the graph is a *triangulation*.

Proposition 5.3.3. *A planar graph is maximal if and only if its faces are triangles.*

Proof. By Proposition 5.2.22, the boundaries of the faces in the drawing F of Γ via an arbitrary embedding σ are elementary cycles. Suppose there exists a face φ

of F which is bounded by a k-cycle $\sigma(x_1), \sigma(x_2), \ldots, \sigma(x_k)$ with $k \geqslant 4$. Since this face is a connected open set of $\mathbb{R}^2 \setminus F$ with polygonal boundary, by Lemma 5.3.2 we may connect two non-consecutive vertices of this cycle, say $\sigma(x_1)$ and $\sigma(x_3)$, by a polygonal line whose interior does not intersect any other polygonal line of F. This shows that Γ is not maximal, which is a contradiction.

We leave the converse as an exercise. ∎

Recall that the degree $d(\varphi)$ of a face φ is the number of edges in its boundary.

Corollary 5.3.4. *In a simple maximal planar graph $\Gamma = (V; E)$ with $n \geqslant 3$ vertices, m edges, and f faces, we have*

(i) $m = 3(n-2)$;

(ii) $f = 2n - 4$.

Proof. In an embedded maximal planar graph, each edge is adjacent to 2 faces: otherwise there would be a leaf, so we could add an edge to it without destroying the planarity. Thus, letting Φ denote the set of faces of (the embedding of) Γ, we have $\sum_{\varphi \in \Phi} d(\varphi) = 2|E|$. By the preceding proposition, every face has exactly 3 edges in its boundary; we thus conclude from the preceding formula that $3f = 2m$, which gives $f = 2 + m - n = 2 + 3(n-2) - n = 2n - 4$ by Euler's formula, taking account of the fact that Γ is necessarily connected. ∎

Exercise 5.4. Show the following properties (Figure 5.16):

- $K_5 \setminus e$ is maximal planar and its faces are triangles.
- $K_{3,3} \setminus e$ is non-maximal planar, and the boundary of each of its faces is an elementary 4-cycle.
- The number of edges that must be added to $K_{3,3} \setminus e$ to make it maximal is equal to 4.

Proposition 5.3.5. *A triangulation with $n \geqslant 4$ vertices is 3-connected.*

Proof. Let x, y be two arbitrary vertices of the triangulation Γ. We must show that $\Gamma' = \Gamma - \{x, y\}$ is connected. Let z and t be two distinct vertices of x and y. There exists an elementary chain P between z and t in Γ. If P avoids x and y, then P connects z to t in Γ'. If x lies on P, let x_1 denote the vertex preceding x on P. The edge $\{x_1, x\}$ is adjacent to exactly two faces, that are both triangular. Let x_2 denote the third vertex of one of the faces adjacent to $\{x_1, x\}$. The edge $\{x, x_2\}$ is adjacent to the triangle xx_1x_2 and to another triangle xx_2x_3. We thus define a sequence $x_1, x_2, \ldots, x_k$ of neighbours of x such that x_k is the successor of x on P. If the chain $x_1x_2 \ldots x_k$ avoids y, we then obtain another chain from z to t avoiding x. If one of x_i, $2 \leqslant i \leqslant k-1$, is the vertex y, then we choose the other face adjacent to the edge

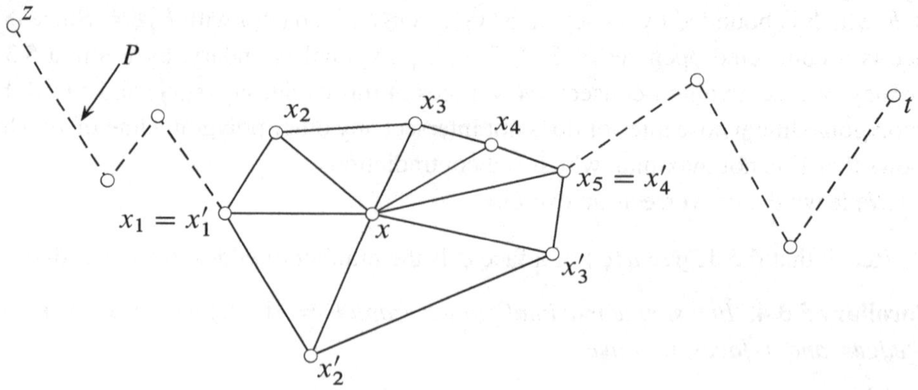

Figure 5.18. Avoiding the vertex x in a triangulation.

$\{x_1, x\}$ from the start, so as to avoid x on the other side. We can thus avoid y in the same way if y is neither a predecessor nor a successor of x on P.

It remains to deal with the case where $\{x, y\}$ is an edge of P. We avoid the two vertices and the edge in question similarly, by defining step by step the vertices of triangles neighbouring x, then those of y. ■

5.3.2 Polyhedra

A *convex polygon* P is the *convex envelope* of a family of $n \geqslant 3$ distinct cocyclic points, called vertices of P. Its boundary consists of exactly n segments, namely the sides of P, whose extremities are the vertices. The centre of the circle containing the vertices is called the centre of the polygon.

A convex polygon is said to be *regular* if the segments of its boundary all have the same length.

Remark 5.3.6. The n points and the n segments bounding a regular convex polygon can thus be viewed as the vertices and the edges of a planar graph, which is nothing other than the elementary cycle C_n.

A *convex polyhedron* T is the convex envelope of a family of $n \geqslant 4$ points, called the vertices of T, that lie on a sphere in the 3-dimensional space $\mathbb{R}^3$ but are not coplanar. The boundary of T consists of (planar) *faces* whose sides are the sides of T.

A convex polyhedron T is said to be *regular* if all its faces are regular (convex) polygons that can be superimposed on one another, and if all its vertices are neighbours of the same number of faces. The centre of the sphere containing the n vertices is called the centre of T.

A polyhedron can be viewed as the drawing of a graph in $\mathbb{R}^3$. The vertices are the extremities of the segments bounding the faces, and the edges are exactly these segments. A regular convex polyhedron with n vertices can thus be seen as a d-regular graph, with $d \geqslant 3$.

Let us give some of the combinatorial properties of this representation.

Proposition 5.3.7. *The graph associated to a convex regular polyhedron P is planar.*

Proof. This follows from Corollary 5.2.20 by considering the sphere in which the polyhedron is inscribed, taking care to choose for the pole a point of the sphere that is not a vertex. ■

As $\sum_{\varphi\in\Phi} d(\varphi) = 2|E| = \sum_{x\in V} d(x)$, we see that a convex regular polyhedron is determined by the number p of edges per face and by the degree d of each vertex: this is the *Schläfli* symbol written (p, d).

Theorem 5.3.8. *There are only five Schläfli symbols (p, d). Indeed, there exist exactly five convex regular polyhedra: they are the Platonic solids T_4, T_6, T_8, T_{12}, T_{20}, with respective Schläfli symbols* (3, 3), (4, 3), (3, 4), (5, 3), (3, 5). *See Table* 5.1.

		f	m	n	Geometry of faces
T_4	Tetrahedron	4	6	4	Triangle
T_6	Hexahedron (cube)	6	12	8	Square
T_8	Octahedron	8	12	6	Triangle
T_{12}	Dodecahedron	12	30	20	Pentagon
T_{20}	Icosahedron	20	30	12	Triangle

Table 5.1. The five convex regular polyhedra.

Proof. Let P be a convex regular polyhedron with n vertices, m edges, f faces, p edges per face and vertices of degree d. We have $\sum_{\varphi\in\Phi} d(\varphi) = 2m$, so $pf = 2m$. Furthermore, $\sum_{x\in V} d(x) = 2m$, so $dn = 2m$, i.e. $f = \frac{2m}{p}$ and $n = \frac{2m}{d}$. Since the graph is connected and planar, we can apply Euler's formula, obtaining $\frac{2m}{d} - m + \frac{2m}{p} = \frac{2m}{p}$; thus $\frac{1}{d} + \frac{1}{p} = \frac{1}{m} + \frac{1}{2}$, so $\frac{1}{d} + \frac{1}{p} > \frac{1}{2}$. But $p \geqslant 3$, so $d \leqslant 5$, which leads to the combinations indicated in the statement. The existence of these polyhedra is shown in Figures 5.19, 5.20 and 5.21. ■

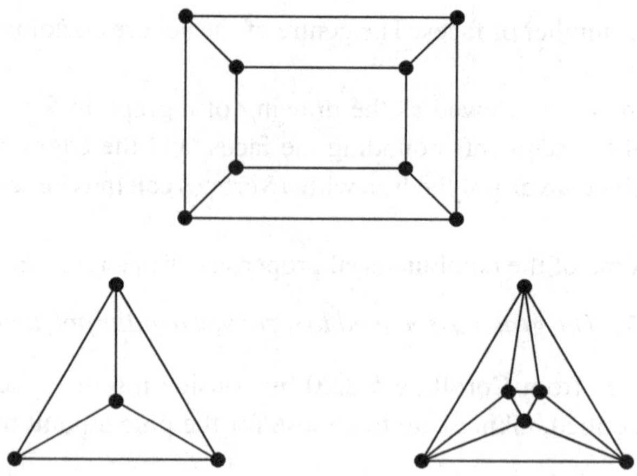

Figure 5.19. The planar hexahedron T_6 (cube), the planar tetrahedron T_4 (pyramid) and the planar octahedron T_8.

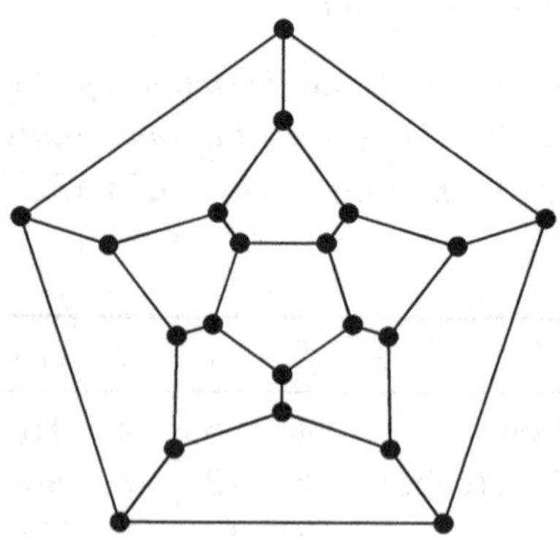

Figure 5.20. The planar dodecahedron T_{12}.

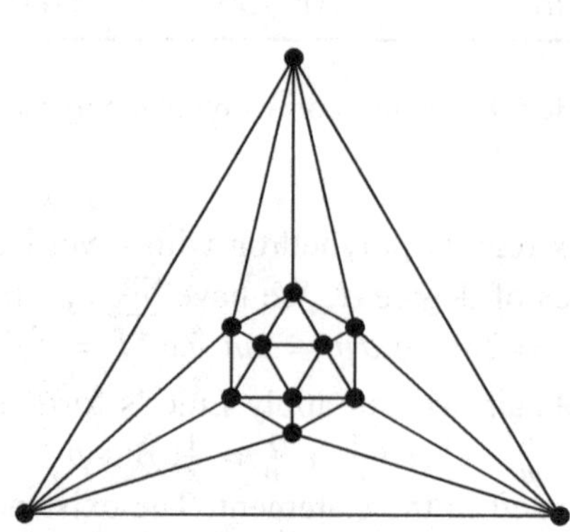

Figure 5.21. The planar icosahedron T_{20}.

5.4 Comparison of embeddings

Let $\Gamma = (V; E, N)$ be a planar graph, and consider two embeddings

$$\sigma, \sigma': V \sqcup E \to \mathbb{R}^2 \sqcup \mathscr{C}_{\mathbb{R}^2}$$

associated to drawings $F = \sigma(V) \cup \bigcup_{e \in E} \sigma(e)$, $F' = \sigma'(V) \cup \bigcup_{e \in E} \sigma'(e)$ whose sets of faces are respectively Φ, Φ'. The two graphs $\Gamma^\sigma = (\sigma(V); \sigma(E), N)$ and $\Gamma^{\sigma'} = (\sigma'(V); \sigma'(E), N)$ are thus embedded in the plane (and isomorphic to Γ). Thus there is an isomorphism of graphs, denoted for simplicity $\tau = \sigma' \circ \sigma^{-1}$:

$$\begin{aligned} \tau: \Gamma^\sigma &\to \Gamma^{\sigma'}, \\ v \in V &\mapsto \sigma'(\sigma^{-1}(v)), \\ e \in E &\mapsto \sigma'(\sigma^{-1}(e)). \end{aligned}$$

Furthermore, we have $|\Phi| = |\Phi'|$, by Euler's theorem.

We say that τ is a *planar isomorphism* between Γ^σ and $\Gamma^{\sigma'}$ if we can extend τ to a bijection $\Phi \to \Phi'$ that preserves adjacency:

$$e \text{ adjacent to } \varphi \iff \tau(e) \text{ adjacent to } \tau(\varphi).$$

In this case we say that the planar embeddings σ and σ' are *equivalent*.

Example 5.4.1. If Γ is a forest, all its planar embeddings are equivalent since there is only a single face. In Figure 5.22, the two planar embeddings are not equivalent.

Exercise 5.5. (i) Show that there are 2 isomorphisms τ and τ' between the two embeddings Γ^σ and $\Gamma^{\sigma'}$ drawn in Figure 5.22.

(ii) Show that neither of these isomorphisms can be extended to a planar isomorphism.

We will prove, however, that when the graph is *sufficiently connected*, then all its planar embeddings are equivalent.

Theorem 5.4.2 (Whitney, 1932). *Let Γ be a simple connected planar graph of finite order that is 3-connected. Then all its planar embeddings are equivalent.*

Proof. Consider two embeddings: $\sigma: \Gamma \to \Gamma^\sigma = (V; E)$ and $\sigma': \Gamma \to \Gamma^{\sigma'} = (V'; E')$ of the 3-connected planar graph Γ, associated to drawings F and F' with sets of faces Φ and Φ' respectively. Then there is an isomorphism $\tau = \sigma' \circ \sigma^{-1}$ between Γ^σ and $\Gamma^{\sigma'}$. We need to show that if φ is a face of Γ^σ, then $\beta := \tau(\partial\varphi)$ is the boundary of a face of $\Gamma^{\sigma'}$. Firstly, β is an elementary cycle since $\partial\varphi$ is one, as the graph is 2-connected (cf. Proposition 5.2.22). Thus β corresponds to a Jordan curve of interior U

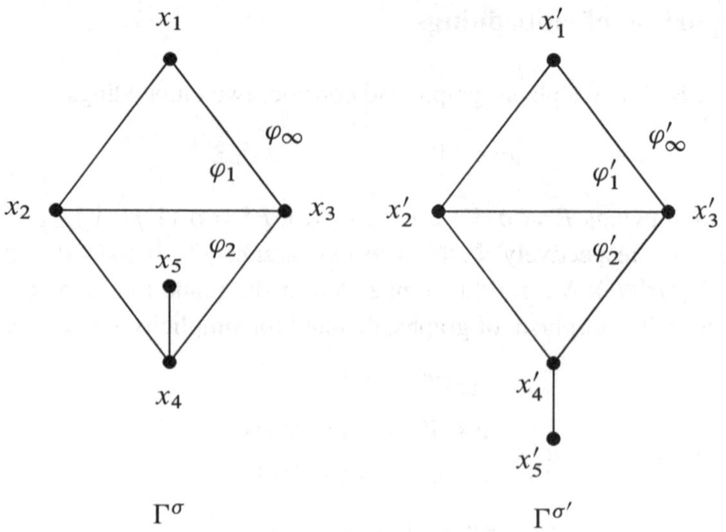

Figure 5.22. Two planar embeddings that are not equivalent.

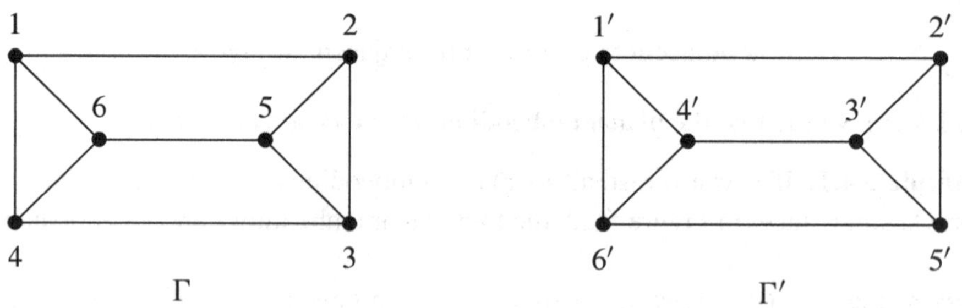

Figure 5.23. Planar isomorphism.

and exterior W. Since Γ is 3-connected, $F \setminus \partial\varphi$ is connected (cf. Menger's theorem), so $F' \setminus \beta$ is connected. Thus $F' \setminus \beta$ is contained in either U or W. Suppose for example that $F' \setminus \beta \subset U$; then $F' \subset U \cup \beta \subset \overline{U}$ and W contains no vertex of $\Gamma^{\sigma'}$, so W is a face ψ of $\Gamma^{\sigma'}$ and $\partial\psi = \beta$. ■

Exercise 5.6. Consider the two embedded graphs Γ and Γ' in Figure 5.23.

(i) Show that Γ and Γ' are 3-connected.

(ii) Set $\tau(i) = i'$, for $1 \leqslant i \leqslant 3$; show that τ extends to a unique isomorphism $\tau\colon \Gamma \to \Gamma'$.

(iii) Show that τ is a planar isomorphism that illustrates the proof of Whitney's theorem.

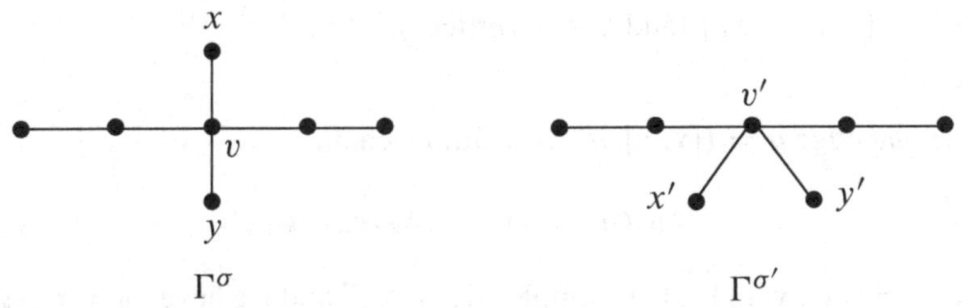

Figure 5.24. Two non-topologically equivalent embeddings.

Remark 5.4.3. Let Γ be a planar graph embedded in the plane, and let e be an edge of Γ. Let Γ_e denote the subgraph induced by the vertices *in contact* with the faces to which e is adjacent. Then Euler's formula applied to Γ_e shows that the number of these faces does not depend on which embedding is chosen within a class of equivalent planar embeddings.

There exists another, more topological notion of isomorphism, which corresponds quite well to intuition. Keeping the same notation, consider two embeddings $\sigma: \Gamma \to \Gamma^\sigma = (V; E)$ and $\sigma': \Gamma \to \Gamma^{\sigma'} = (V'; E')$ of the planar graph Γ, and let F and F' be the associated drawings, with sets of faces Φ and Φ' respectively. There is an isomorphism of graphs $\tau = \sigma' \circ \sigma^{-1}$ between Γ^σ and $\Gamma^{\sigma'}$; we say that the two embeddings are *topologically equivalent* if τ can be extended to a homeomorphism $\tau^*: \mathbb{R}^2 \to \mathbb{R}^2$ such that $\tau^*|_F: F \to F'$ is a homeomorphism. Clearly this notion is stronger than mere planar equivalence.

Exercise 5.7. Consider the two embeddings σ, σ' of a graph Γ inducing the embedded graphs $\Gamma^\sigma, \Gamma^{\sigma'}$ in Figure 5.24, with associated drawings F, F'.

(a) Show that the planar embeddings σ and σ' are equivalent.

(b) Let $\pi: \mathbb{R}^2 \to \mathbb{R}^2$ be a homeomorphism such that $\pi(F) = F'$.

  (i) Show that $\pi(v) = v'$.

  (ii) Show that $\pi(\{x, y\}) = \{x', y'\}$.

  (iii) Considering the segments $[\![x', y']\!]$ and $\pi^{-1}([\![x', y']\!])$, show that this leads to an impossibility: the embeddings are not topologically equivalent.

5.5 Kuratowski's theorem and Wagner's theorem

An *elementary subdivision* of a graph $\Gamma = (V; E, N)$ is a graph $\Gamma' = (V'; E', N')$ defined as follows:

- $V' = V \sqcup \{x_1, \ldots, x_k\}$ (add $k \geqslant 1$ vertices);
- $N' = N$;
- replace *one* edge $a = ([x, y], n)$ by a simple chain:

$$(x_0, e_0, x_1, e_1, \ldots, x_k, e_k, x_{k+1}),$$

where $e_i = (\{x_i, x_{i+1}\}, n)$ are simple edges of Γ' and we have $x_0 = x$, $x_{k+1} = y$;
- $E' = (E \backslash \{a\}) \cup \{e_i,\ 0 \leqslant i \leqslant k\}$.

A *subdivision* of a graph $\Gamma = (V; E, N)$ is a finite sequence of elementary subdivisions. Observe that the operation of subdivision preserves the degree of the vertices of the initial graph; the degree of the newly created vertices is equal to 2.

Let Γ_0 and Γ be two graphs. We will say that a graph Γ *contains a subdivision of* Γ_0 if Γ contains a subgraph Γ' isomorphic to a subdivision of Γ_0. Figure 5.25 shows an example.

The inverse operation of subdivision is *contraction*, which is defined in the framework of multigraphs. Let $\Gamma = (V; E, N)$ be a multigraph and $e = ([x, y], n_e) \in E$.

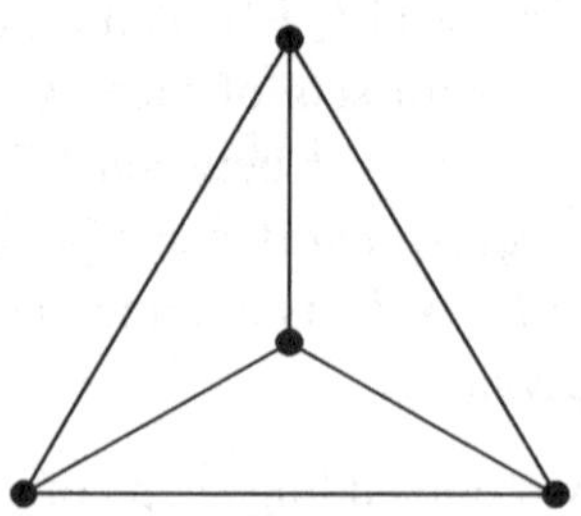

K_4

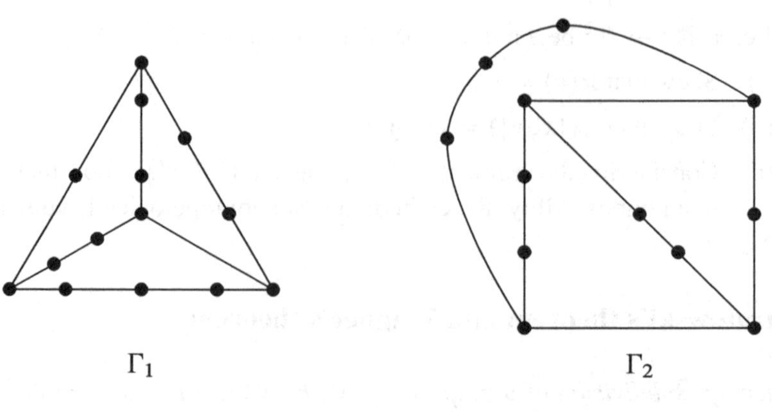

Γ_1 Γ_2

Figure 5.25. Γ_1 and Γ_2 are subdivisions of K_4.

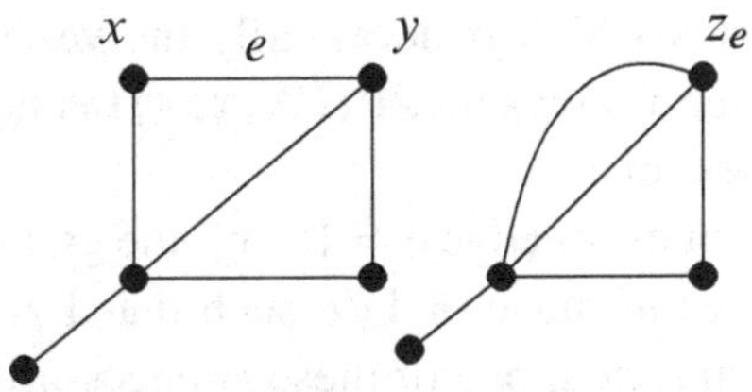

Figure 5.26. Example of a contraction.

We delete the edge e and *fuse* x and y into a single (new) vertex z_e. The new graph is written $\Gamma/e = (V'; E', N')$. More precisely,

- $V' = (V \backslash \{x, y\}) \sqcup \{z_e\}$;
- for each $t \in V$, we define $t' \in V'$ as follows: $x' = y' = z_e, t' = t$ for $t \neq x, y$;
- $E' = \{([u', v'], n) : ([u, v], n) \in E \backslash \{e\}\}$;
- $N' = N$.

In this contraction, a p-edge between x and y ($p \geqslant 2$) becomes a $(p-1)$-loop at z_e, and multi-edges appear for each vertex t adjacent to both x and y.

We say that a multigraph Γ is *contractible to* Γ_0 if Γ_0 can be obtained from Γ by a finite number of contractions. We also say that a graph Γ' is a *minor* graph of a graph Γ if it can be obtained (up to isomorphism) from Γ by deletion of vertices and edges and contraction of edges. In other words, a minor of Γ arises from applying contractions to a subgraph of Γ.

Exercise 5.8. Show that if a graph Γ' is a subdivision of a graph Γ, then Γ is a contraction of Γ' (use elementary subdivision).

This notion allowed K. Wagner to give another characterisation of planarity, that we will prove below. As we saw in Corollary 5.2.17, the graphs $K_{3,3}$ and K_5 are not planar; it turns out that they are the only obstructions to planarity, as stated in a famous theorem by Kuratowski that we will also prove below. In what follows, we consider only simple graphs.

Lemma 5.5.1. *Let $\Gamma = (V; E)$ be a 3-connected graph of order $n \geqslant 5$. Then*

(i) *for every edge e of Γ, the graph Γ/e is 2-connected.*

(ii) *Furthermore, there exists an edge e such that Γ/e is 3-connected.*

Proof. (i) Observe first that contraction is an operation which preserves connectedness; the graph Γ/e is connected for every edge e. Assume that there exists an edge $e = \{x, y\}$ such that Γ/e is not 2-connected. As the graph Γ/e is connected, it possesses an articulation vertex which is necessarily the vertex z_e that arose from the contraction of e. But it then follows that $\Gamma(V\backslash\{x,y\})$ is not connected, which contradicts the 3-connectedness of Γ.

(ii) Assume now that for every edge $e=\{x,y\}$, the graph Γ/e is not 3-connected. Then there are two vertices u and v of Γ/e such that $\Gamma/e-\{u,v\}$ is not connected. Since Γ is itself 3-connected, one of these is necessarily the vertex z_e obtained by contraction of e. In other words, there exists a vertex $z\in V\backslash\{x,y\}$ such that $\Gamma(V\backslash\{x,y,z\})$ is not connected.

Choose a connected component C with minimal cardinality among all those of the induced graphs $\Gamma(V\backslash\{x,y,z\})$ when $\{x,y\}$ runs over E and z runs over V. This choice fixes an edge $e=\{x,y\}$ and a vertex z. As Γ is 3-connected, $\Gamma(V\backslash\{x,y\})$ is connected: if C' is another connected component of $\Gamma(V\backslash\{x,y,z\})$, there exists a chain between C and C' that necessarily passes through z in Γ, so also through a neighbour u of z belonging to C. Set $e'=\{u,z\}$. Now, Γ/e' is not 3-connected, so as before there exists $w\in V$ such that the graph $\Gamma(V\backslash\{u,w,z\})$ is not connected. Let D be one of its connected components not containing the vertices x,y. Such a D exists since x and y remain adjacent in $\Gamma(V\backslash\{u,w,z\})$ which admits at least two connected components. As before, since $\Gamma(V\backslash\{z,w\})$ is connected, there is a neighbour $v\in D$ of u which is neither x nor y nor z. Consequently, $v\in C$ and D is connected in $\Gamma(V\backslash\{x,y,z\})$: hence $D\subset C$. But $u\notin D$, so $|D|<|C|$, which contradicts the minimality of C. ∎

Proposition 5.5.2. *Let $\Gamma=(V;E)$ be a connected graph containing no subdivision of $K_{3,3}$ or of K_5. Then for every edge e, Γ/e contains no subdivision of $K_{3,3}$ or of K_5.*

Proof. Let us show the contrapositive of this statement: if Γ/e contains a subdivision of $K_{3,3}$ or of K_5, then Γ also contains a subdivision of $K_{3,3}$ or of K_5 (which is not necessarily the same).

Let $\Gamma=(V;E)$, $e=\{x,y\}\in E$, and let z_e the vertex obtained by contracting e. Let $\Gamma'\subseteq\Gamma/e$ be a subdivision of $K_{3,3}$ or of K_5. If $z_e\notin V(\Gamma')$, then $\Gamma'\subseteq\Gamma$ is a subdivision of $K_{3,3}$ or of K_5 in Γ. So we can assume that $z_e\in V(\Gamma')$. The degree $d(z_e)$ of z_e in Γ' can be equal to 2, 3 or 4. There are several cases to deal with, according to the value of this degree. If $d(z_e)=3$ (resp. 4), we will say that z_e is a *principal vertex* of $K_{3,3}$ (resp. of K_5); if $d(z_e)=2$, we will say that z_e is a vertex of the subdivision.

(i) If $d(z_e)=2$, then z_e is a subdivision vertex of Γ'. Since the graph Γ is connected, replacing z_e by the edge $e=\{x,y\}$ yields a subdivision of $K_{3,3}$ or of K_5 as a subgraph of Γ.

(ii) If $d(z_e)=3$, then z_e is a principal vertex of $K_{3,3}$. Let

$$\Gamma'(z_e)=(\Gamma(x)\cup\Gamma(y))\backslash\{x,y\}=(\Gamma(x)\backslash\{y\})\cup(\Gamma(y)\backslash\{x\}).$$

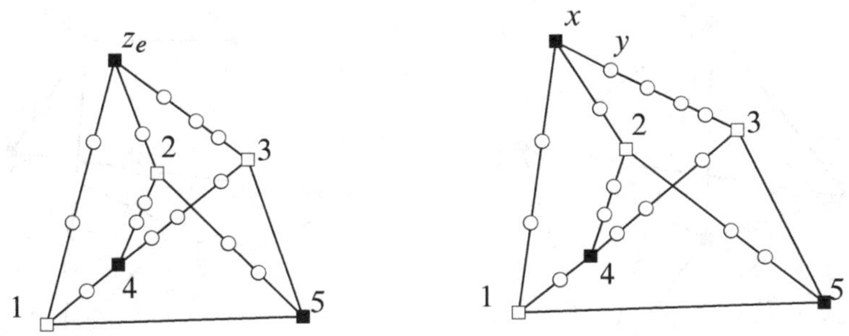

Figure 5.27. On the left, the subdivision of $K_{3,3}$ in Γ/e, considering the bipartition $\{z_e, 4, 5\} \sqcup \{1, 2, 3\}$; on the right the associated subdivision of $K_{3,3}$ in Γ, considering the bipartition $\{x, 4, 5\} \sqcup \{1, 2, 3\}$.

As the graph is connected, we can assume that $|\Gamma(x)\backslash\{y\}| = 2$ and $|\Gamma(y)\backslash\{x\}| = 1$ up to exchanging x and y. Let $u \in \Gamma'(z_e)$ with $u \in \Gamma(y)\backslash\{x\}$. If we replace the vertex z_e by the vertex x and add the edge e, then since the vertex y is again adjacent to u, we obtain a subdivision of $K_{3,3}$ which is a subgraph of Γ (Figure 5.27).

(iii) If $d(z_e) = 4$, then z_e is a principal vertex of K_5. There are essentially two cases. One is the asymmetric situation where $|\Gamma(x)\backslash\{y\}| = 3$ and $|\Gamma(y)\backslash\{x\}| = 1$; then replacing z_e by x and adding the edge e incident to the vertex y, we obtain a subdivision of K_5 as a subgraph of Γ (Figure 5.28). The other is when $|\Gamma(x)\backslash\{y\}| = |\Gamma(y)\backslash\{x\}| = 2$; we obtain a subdivision of $K_{3,3}$ as a subgraph of Γ (Figure 5.29). ■

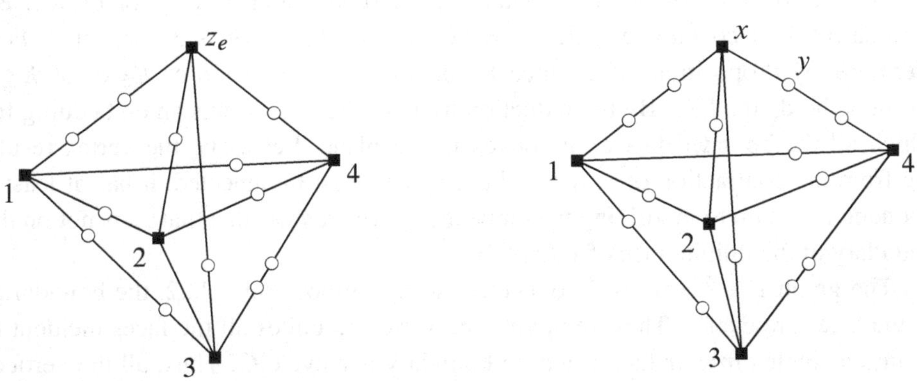

Figure 5.28. First situation: on the left the subdivision of K_5 in Γ/e, considering the vertices $\{z_e, 1, 2, 3, 4\}$; on the right the associated subdivision of K_5 in Γ, considering the vertices $\{x, 1, 2, 3, 4\}$.

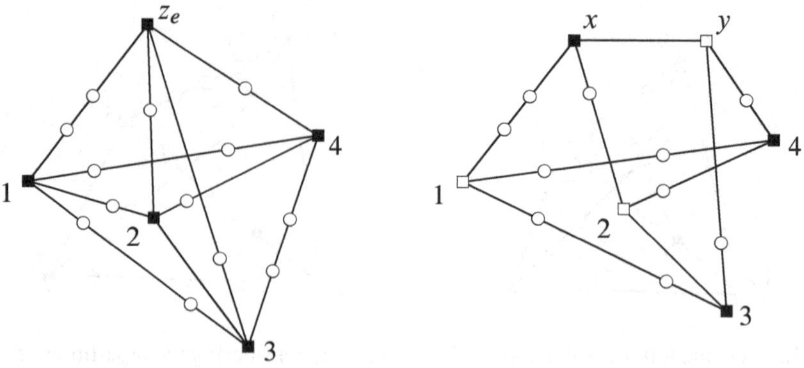

Figure 5.29. Second situation: on the left, the subdivision of K_5 in Γ/e considering the vertices $\{z_e, 1, 2, 3, 4\}$; on the right, the associated subdivision of $K_{3,3}$ in Γ considering the bipartition of vertices $\{y, 1, 2\} \sqcup \{x, 3, 4\}$.

Theorem 5.5.3. *Let* $\Gamma = (V; E)$ *be a* 3*-connected graph containing no subdivision of* $K_{3,3}$ *or* K_5*. Then* Γ *is planar and admits an embedding all of whose bounded faces are convex.*

Proof. We use induction on the number of edges.

The *smallest* 3-connected graph is K_4, which contains no subdivision of $K_{3,3}$ or of K_5: it is easy to draw it in the plane in such a way that the bounded faces are convex in the plane (Figure 5.1).

Now let $n \geqslant 5$, and assume that the assertion holds for all 3-connected graphs of order $n - 1$ containing no subdivision of $K_{3,3}$ or of K_5. Let $\Gamma = (V; E)$ be a 3-connected graph of order n containing no subdivision of $K_{3,3}$ or of K_5. By Lemma 5.5.1, it contains an edge $e = \{x, y\}$ such that Γ/e is 3-connected. Furthermore, by Proposition 5.5.2, since Γ contains no subdivision of K_5 or of $K_{3,3}$, the same holds for Γ/e. By the induction hypothesis, Γ/e admits an embedding for which all the bounded faces are convex in the plane. Let z_e be the vertex resulting from the contraction of e. Since the graph Γ/e is 3-connected, it has at least 2 bounded faces; up to modifying the embedding, we may assume that z_e is not on the boundary of the infinite face of Γ/e.

The graph $\Gamma/e - \{z_e\}$ is 2-connected, so by Proposition 5.2.22, the boundaries of the faces are cycles. Thus, removing the vertex z_e unites all the faces incident to z_e into a single *circular* face φ whose boundary is a cycle C. Thus, all the vertices neighbouring z_e in Γ/e lie on C.

Observe first that C cannot contain a subdivision that is an *alternating chain* (such as x_1, y_1, x_2, y_2 where $x_1, x_2 \in \Gamma(x)$ and $y_1, y_2 \in \Gamma(y)$): indeed, otherwise, connecting y_2 to x_1 along C would give a subdivision of $K_{2,2}$ along the chain

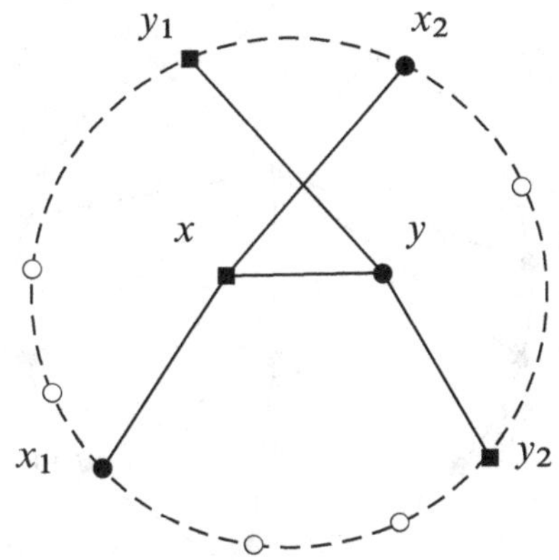

Figure 5.30. Configuration in which there exists a chain of the type x_1, y_1, x_2, y_2 with $x_1, x_2 \in \Gamma(x)$ and $y_1, y_2 \in \Gamma(y)$ in Γ: the latter contains a subdivision of $K_{3,3}$.

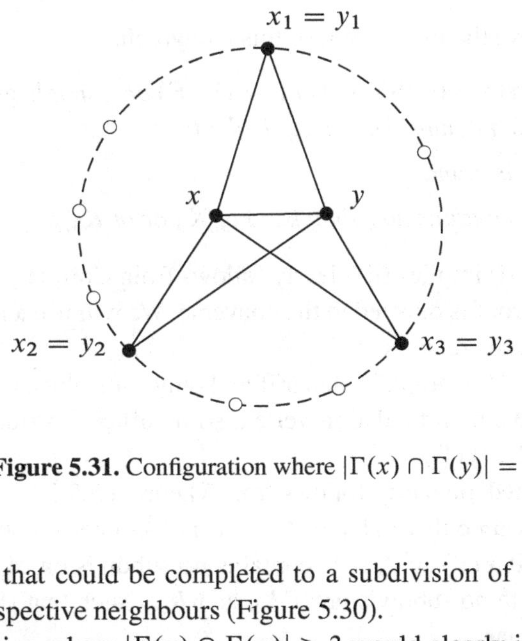

Figure 5.31. Configuration where $|\Gamma(x) \cap \Gamma(y)| = 3$.

x_1, y_1, x_2, y_2, x_1, that could be completed to a subdivision of $K_{3,3}$ by connecting x and y to their respective neighbours (Figure 5.30).

Then the situation where $|\Gamma(x) \cap \Gamma(y)| \geqslant 3$ would clearly induce a subdivision of K_5 (Figure 5.31).

It thus follows that C is a cycle of the form $x_1, \dots, x_r, y_1, \dots, y_s, x_1$, allowing $x_r = y_1$ and $y_s = x_1$.

We then take x and y on the perpendicular bisector of $[\![x_1, x_r]\!]$ as in Figure 5.32 in such a way that we can connect x and y to their respective neighbours by the segments $[\![x, x_i]\!]$, $[\![y, y_j]\!]$, $i = 1, \dots, r$, $j = 1, \dots, s$. The induced faces are clearly convex since the initial face φ was. ∎

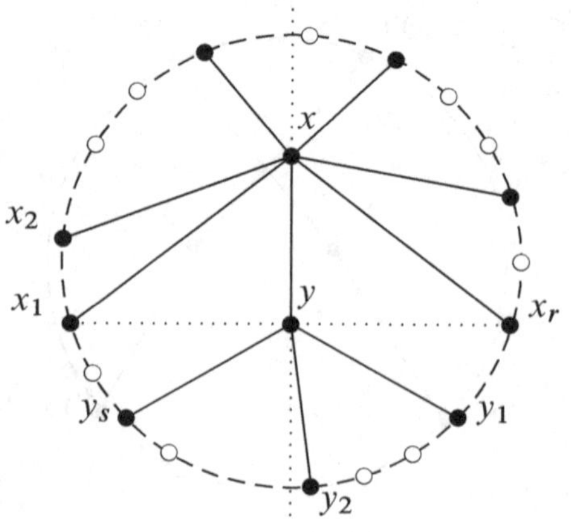

Figure 5.32. Tiling of φ with convex faces of Γ.

We can now prove the main result of this paragraph.

Theorem 5.5.4 (Kuratowski, 1930). *Let $\Gamma = (V; E)$ be a simple graph of finite order. Then the two following conditions are equivalent:*

(i) *the graph Γ is planar;*

(ii) *the graph Γ contains no subdivision of K_5 or of $K_{3,3}$.*

Proof. The fact that (i) implies (ii) clearly follows from Corollary 5.2.17.

The rest of the proof is devoted to the converse. We will use an argument depending on the vertex-connectedness of Γ.

By Theorem 5.2.21, a graph is planar if and only if its blocks are planar. Furthermore, the blocks have no articulation vertex, so it suffices to study the planarity of graphs Γ such that $2 \leqslant \kappa(\Gamma)$.

If Γ is 3-connected, planarity follows from Theorem 5.5.3.

Thus we may assume that $\kappa(\Gamma) = 2$, i.e. Γ is 2-connected but not 3-connected. We will prove that if $\kappa(\Gamma) = 2$, if Γ contains no subdivision of K_5 and $K_{3,3}$, and if every graph Γ' with no subdivision of K_5 and $K_{3,3}$ such that $|V(\Gamma')| < |V(\Gamma)|$ is planar, then Γ is planar.

As $\kappa(\Gamma) = 2$, there exist $x, y \in V$ such that $\Gamma - \{x, y\}$ is not connected. Thus we can write $V = V_1 \sqcup V_2 \sqcup \{x, y\}$ where V_1 and V_2 are not connected in $\Gamma - \{x, y\}$. Let $\Gamma_1 = \Gamma(V_1 \cup \{x, y\})$ and $\Gamma_2 = \Gamma(V_2 \cup \{x, y\})$ be the induced subgraphs on $V_1 \cup \{x, y\}$ and $V_2 \cup \{x, y\}$. Every chain from V_1 to V_2 passes through x or y. Fix $v_1 \in V_1, v_2 \in V_2$: then there are two vertex-disjoint elementary chains C_1 and C_2 between v_1 and v_2, of which one necessarily passes through x and the other through y (Figure 5.33).

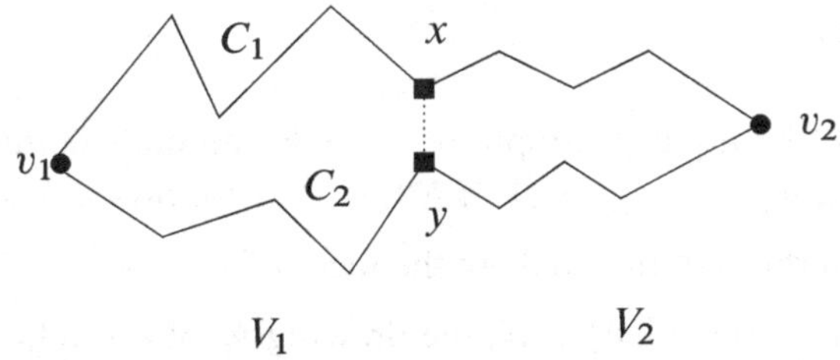

Figure 5.33. The two chains C_1 and C_2 between v_1 and v_2 in the proof of Theorem 5.5.4.

These two chains then give two elementary chains C_1', C_2' between x and y, one in Γ_1 and the other in Γ_2. Let $a = \{x, y\}$; we denote by $\Gamma_i \cup \{a\}$, $i = 1, 2$, the graph obtained by adding the edge a to the set of edges of Γ_i. Then $\Gamma_1 \cup \{a\}$ and $\Gamma_2 \cup \{a\}$ have no subdivision of K_5 or of $K_{3,3}$. Indeed, if for example $\Gamma_1 \cup \{a\}$ had such a subdivision S, then a would necessarily be among the edges of S (since Γ_1 has no subdivision of K_5 or $K_{3,3}$), but then $S' = S \backslash a \cup C_2'$ (the notation is clear) would be a subdivision of K_5 or $K_{3,3}$ satisfying $S' \subset \Gamma$: contradiction.

As $|V(\Gamma_1 \cup \{a\})|, |V(\Gamma_2 \cup \{a\})| < |V(\Gamma)|$, the hypotheses imply that $\Gamma_1 \cup \{a\}$ and $\Gamma_2 \cup \{a\}$ are planar. We take embeddings σ_1 and σ_2 of $\Gamma_1 \cup \{a\}$ and $\Gamma_2 \cup \{a\}$ in such a way that a lies on the boundary of the infinite face of two (disjoint) drawings, and then, given some intuitively possible deformations, we join $\sigma_1(\Gamma_1 \cup \{a\})$ and $\sigma_2(\Gamma_2 \cup \{a\})$ by gluing the drawings along a, which yields an embedding of $\Gamma \cup \{a\}$ and thus of Γ (note that a may lie in E). ■

The following theorem gives another characterisation of planar graphs.

Theorem 5.5.5 (Wagner, 1937). *A graph is planar if and only if it contains no subgraph contractible to K_5 or $K_{3,3}$.*

Proof. Suppose that Γ is not planar. Then by Kuratowski's theorem it contains a subdivision of K_5 or of $K_{3,3}$. By contracting in this subdivision all the edges between vertices with degree 2, we obtain a minor isomorphic to K_5 or $K_{3,3}$. Clearly, deletion of an edge or a vertex in a graph preserves planarity. Proposition 5.5.2 shows that the contraction of an edge in a graph also preserves planarity. Consequently, a planar graph cannot contain any minor isomorphic to K_5 or $K_{3,3}$, which concludes the proof. ■

There exists yet another characterisation of planarity, which is formulated using linear algebra and the cycle space (see MacLane's theorem, Theorem 6.4.5).

5.6 Dual graphs

Let $\Gamma = (V; E, N)$ be a planar graph, $\sigma: \Gamma \to \mathbb{R}^2$ an embedding and Φ the set of faces. Define the *dual graph* $\Gamma^*_\sigma = (V^*; E^*, N^*)$ as follows:

- $V^* = \Phi$ (the vertices of the dual are the faces of Γ);
- for each edge $e = ([x, y], n) \in E$, the drawing γ_e of e is adjacent to one or two faces $\varphi, \psi \in \Phi$; set $e^* = ([\varphi, \psi], n) \in E^*$ (note that e^* is a loop when $\varphi = \psi$);
- $N^* = N$.

Note that $|E^*| = |E|$ and $|V^*| = |\Phi|$.

The fascinating idea of duality is very old: it can be found already in the 15th book of Euclid's Elements!

If we consider another planar embedding σ' of Γ equivalent to σ, the property of adjacency of faces and edges is preserved, so the dual graphs Γ^*_σ and $\Gamma^*_{\sigma'}$ are actually *identical*. This property does not hold in general if the planar embeddings are not equivalent.

Exercise 5.9. Consider the simple graph $\Gamma = (V; E)$, where

$$V = \{1, 2, 3, 4, 5\} \quad \text{and} \quad E = \{12, 23, 31, 14, 15\}$$

(where ij means $\{i, j\}$)). Two embeddings of this graph in the plane are shown in Figure 5.34. Prove that the associated duals Γ^*_σ and $\Gamma^*_{\sigma'}$ are not isomorphic.

Example 5.6.1. (i) If Γ is a forest with m edges and $\sigma: \Gamma \to \mathbb{R}^2$ is an embedding, then Γ^*_σ is an m-loop.

(ii) For every embedding of an m-loop, the corresponding dual is a chain with m edges.

(iii) Whatever the embedding, the dual of a cycle C_m is always an m-edge (i.e. an edge of multiplicity m). Conversely, the dual of an m-edge (which is not an m-loop) is always a cycle of length m.

(iv) Every embedding of K_4 induces a dual of K_4 isomorphic to K_4!

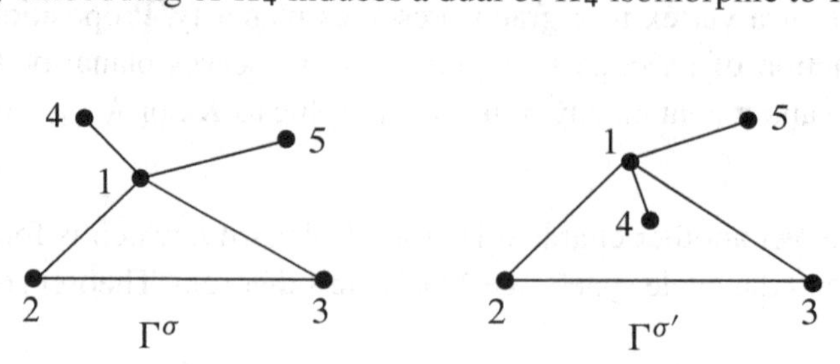

Figure 5.34. Two embeddings of Γ.

(v) If e is an isthmus in a graph, then e^* is a loop, and if e is a loop, then e^* is an isthmus.

Proposition 5.6.2. *Let* $\Gamma = (V; E, N)$ *be a non-empty planar graph together with an embedding* $\sigma\colon \Gamma \to \mathbb{R}^2$. *Then* Γ^*_σ *is connected.*

Proof. We use induction on $|E|$. If $E = \emptyset$, then Γ^*_σ is an isolated vertex, so it is connected.

Now let Γ be a planar graph having $m + 1$ edges. If Γ is a forest, then Γ^*_σ is an $(m + 1)$-loop, so it is connected. Otherwise, choose an edge e that is not an isthmus: thus γ_e is adjacent to two faces φ, ψ. If we delete e, then φ and ψ now form a single face, so in the dual Γ^*_σ, everything takes place as if we had simply contracted the edge e^*:

$$(\Gamma\backslash e)^*_{\tilde{\sigma}} \simeq \Gamma^*_\sigma/e^*,$$

where $\tilde{\sigma}$ is the restriction of the embedding σ to $\Gamma\backslash e$.

By the induction hypothesis, we deduce that Γ^*_σ/e^* is connected. Since the number of connected components in a graph remains unchanged under contractions, contraction of an edge in a graph does not modify the connectedness, so we see that Γ^*_σ is connected. ■

As application of the dual, we have the following result.

Corollary 5.6.3. *Let* Γ *be an embedded planar graph. If every face* φ *of* Γ *has connected boundary* $\mathrm{Fr}(\varphi)$*, then* Γ *is connected.*

Proof. Let Φ denote the set of faces of Γ. We have

$$\Gamma = \bigcup_{\varphi\in\Phi} \mathrm{Fr}(\varphi).$$

Let x and y lie on Γ. As the dual Γ^* is connected (cf. Proposition 5.6.2), there exists a sequence of faces φ_i, $i = 1, \dots, j$, such that $\mathrm{Fr}(\varphi_i) \cap \mathrm{Fr}(\varphi_{i+1}) \neq \emptyset$, $i = 1, \dots, j-1$, with $x \in \mathrm{Fr}(\varphi_1)$ and $y \in \mathrm{Fr}(\varphi_j)$. As the boundaries $\mathrm{Fr}(\varphi_i)$ are connected, the union $\bigcup_{i=1}^{j} \mathrm{Fr}(\varphi_i)$ is connected, so x and y are connected in $\Gamma = \bigcup_{\varphi\in\Phi} \mathrm{Fr}(\varphi)$. ■

Proposition 5.6.4. *Let* $\Gamma = (V; E, N)$ *be a planar graph together with an embedding* $\sigma\colon \Gamma \to \mathbb{R}^2$. *Then* Γ^*_σ *is planar.*

Proof. For each face φ, choose a point $O_\varphi \in \varphi$. Then for each edge e of the drawing γ_e adjacent to φ and ψ, draw a polygonal arc in φ from O_φ to a point of γ_e that we extend to O_ψ by a polygonal arc in ψ. This gives an arc ξ_e that cuts γ_e at a single point. Intuitively, we see that this can be done in such a way that the two arcs ξ_e and $\xi_{e'}$ have (at least) one common extremity, so O_φ and O_ψ intersect only at this (or these) extremities. This construction is illustrated in Figure 5.35. ■

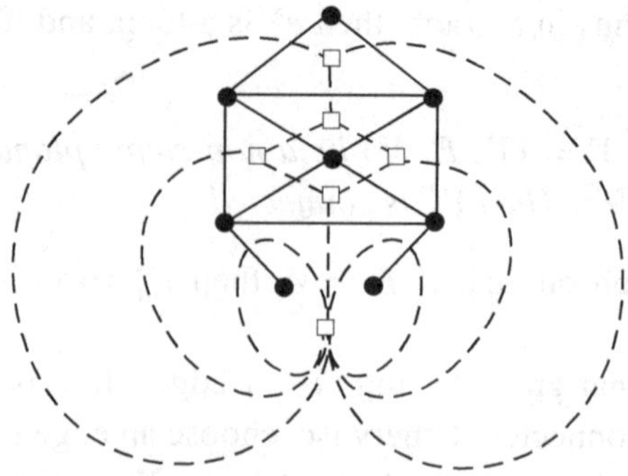

Figure 5.35. A planar graph (with edges shown in solid lines) and its dual (with edges shown in dotted lines).

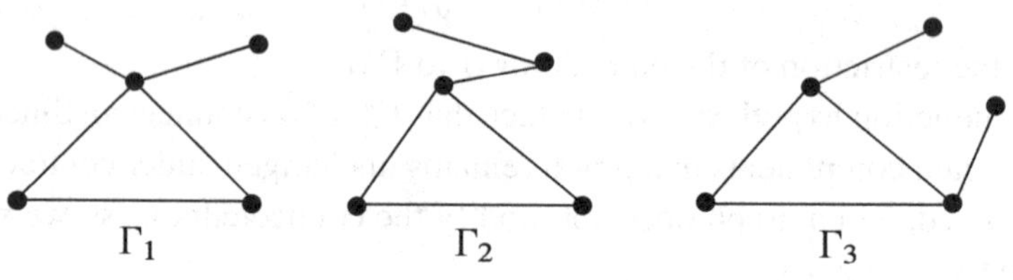

Figure 5.36. Three non-isomorphic graphs with the same dual.

By construction, the number of vertices and the number of edges of the dual graph of Γ do not depend on the choice of embedding. This is also the case for the number of faces.

Corollary 5.6.5. *Let $\Gamma = (V; E, N)$ be a planar graph and $\sigma\colon \Gamma \to \mathbb{R}^2$ an embedding. Let c denote the number of connected components of Γ. Then the dual graph Γ^*_σ possesses exactly $|V| - c + 1$ faces. In particular, if Γ is connected, then Γ^*_σ possesses exactly $|V|$ faces.*

Proof. Let n, m, f (resp. n^*, m^*, f^*) denote the numbers of vertices, edges and faces of Γ (resp. Γ^*_σ). Euler's formula applied to Γ gives $n - m + f = c + 1$. As Γ^*_σ is planar and connected, Euler's formula gives $n^* - m^* + f^* = 2$. But $n^* = f$ and $m^* = m$, which gives the result. ∎

Exercise 5.10. Consider the three planar graphs $\Gamma_1, \Gamma_2, \Gamma_3$ drawn in Figure 5.36.

(i) Show that these graphs are pairwise non-isomorphic.

(ii) Show that they have the same dual Γ^*.

(iii) Show that Γ^* can be embedded in the plane in three different ways, and that $\Gamma_1, \Gamma_2, \Gamma_3$ are the three associated duals.

Example 5.6.6. The five regular polyhedra in $\mathbb{R}^3$ (see Paragraph 5.3.2) are spectacular examples of duality. Let P be any one of the five: the cube T_6 (6 faces), the tetrahedron T_4 (4 faces), the octahedron T_8 (8 faces), the dodecahedron T_{12} (12 faces) and the icosahedron T_{20} (20 faces). We realise the dual P^* of P by placing a point in the centre of each face of P and connecting two of these points with an edge whenever the corresponding faces are adjacent. An effort at visualisation shows that $T_6^* \simeq T_8$, $T_8^* \simeq T_6$ and $T_4^* \simeq T_4$; we can also see that $T_4 \simeq K_4$. It is a little harder to check that T_{12} and T_{20} are actually duals of each other.

We will see below that the notion of dual allows us to give another formulation of planarity (Whitney's theorem, cf. Theorem 9.2.5).

5.7 Crossings, thickness and genus of a graph

An *invariant* of a graph is a quantity or concept that is identical for every isomorphism class of graphs, such as for example the average degree, the girth, the number of edges, the number of connected components, or the number of faces for a planar graph. In this paragraph we will study three other important invariants of graphs.

5.7.1 Crossings and thickness

The *number of crossings* of a simple graph $\Gamma = (V; E)$ is the minimal number of crossings of edges that can occur in a drawing of Γ in the plane (counting 1 for every pair of edges that intersect). We denote it by $\mathrm{Cr}(\Gamma)$. We have for example $\mathrm{Cr}(K_5) = 2$. This parameter essentially gives a way to measure the non-planarity of a graph.

The *thickness* of a simple connected graph $\Gamma = (V; E)$ is the smallest number k of connected partial planar graphs with no common edge, which are called thickness factors $\Gamma_i = (V; E_i)$, $i = 1, \ldots, k$, necessary to construct the graph: $E = \bigsqcup_{1 \leqslant i \leqslant k} E_i$. This invariant is denoted by $\mathrm{Th}(\Gamma)$. An example is shown in Figure 5.37.

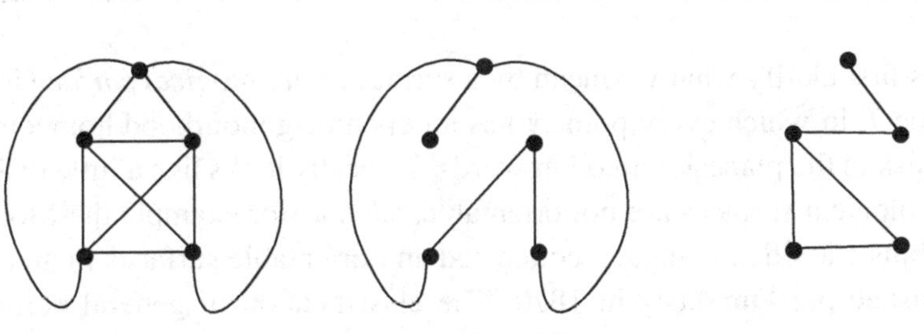

Figure 5.37. Thickness factors of K_5. The complete graph K_5 (left-hand figure) has two thickness factors (the two right-hand figures), so $\mathrm{Th}(K_5) = 2$.

Like the number of crossings, the thickness of a graph measures the (default of) planarity of a graph. Proposition 5.7.1 yields a lower bound for the thickness of a graph as a function of the numbers of its vertices and of edges.

Proposition 5.7.1. *Let $\Gamma = (V; E)$ be a simple graph having $n \geqslant 3$ vertices and m edges. Then the thickness of Γ satisfies:*

$$\mathrm{Th}(\Gamma) \geqslant \left\lceil \frac{m}{3n-6} \right\rceil.$$

Proof. By Proposition 5.2.16, the number of edges of each subplanar graph of Γ is at most $3n - 6$. We thus have

$$m \leqslant (3n-6)\mathrm{Th}(\Gamma),$$

which gives the result. ■

Exercise 5.11. Prove the stronger inequality $\mathrm{Th}(\Gamma) \geqslant \lfloor \frac{m+3n-7}{3n-6} \rfloor$.

Exercise 5.12. Give a lower bound for the thickness of K_n, $K_{m,m}$, for $n = 5, 6, 7, 8, 9, 10$ and $m = 3, 4, 5$. Draw their thickness factors.

Exercise 5.13. Let $K_{m,n}$ be a complete bipartite graph. Show that

$$\mathrm{Th}(K_{m,n}) \geqslant \left\lceil \frac{mn}{2m+2n-4} \right\rceil.$$

5.7.2 Genus of a graph

Even though a graph is not necessarily planar, we have an intuitive impression that a graph is locally 2-dimensional, since a vertex with the beginnings of its incident edges resembles a bicycle wheel minus its outer rim, with the vertex at the centre and the *germs* of edges appearing as spokes. But of course, being (or not being) planar depends on a more subtle reality, namely the *global* geometric nature of the plane. This makes it natural to ask whether graphs can be embedded in other types of surfaces.

Let us first clarify what we mean by a surface. A *topological surface* is a topological space T in which every point x has an open neighbourhood homeomorphic to an open disk in the plane $\mathbb{R}^2$; in other words, T locally looks like a little disk $D(a, r)$. Some topological surfaces are not orientable, such as for example the Möbius band. A. F. Möbius classified compact, connected and orientable surfaces in an article that was published posthumously in 1870. The classification of general non-orientable surfaces was not completed until 1925.

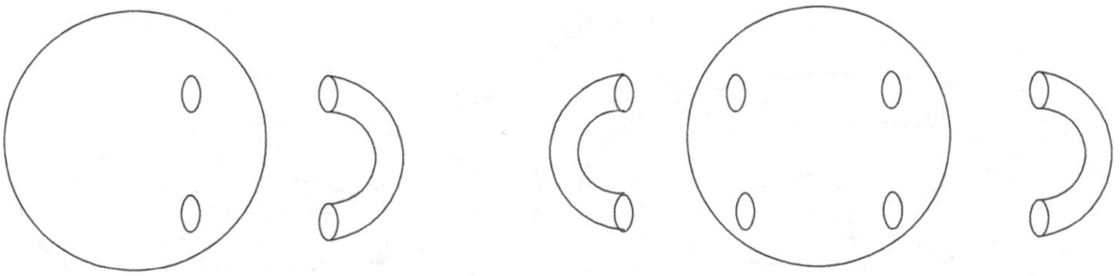

Figure 5.38. Left-hand figure: creating $\mathbb{S}_1$. Right-hand figure: creating $\mathbb{S}_2$.

The oriented surfaces are classified by a remarkably simple invariant, namely the genus, which is the number of *holes*; a surface of genus g, $\mathbb{S}_g$, can be obtained by starting from the sphere $\mathbb{S}_0$ in $\mathbb{R}^3$ and adding g "handles" as follows: for each handle, one removes 2 small disks from the surface of the sphere and connects the pair by adding a cylinder whose ends are glued to the holes (Figure 5.38).

Theorem 5.7.2 (Möbius, 1870). *Every compact and connected orientable topological surface is homeomorphic to a surface $\mathbb{S}_g$, $g \geqslant 0$.*

Another way of visualising $\mathbb{S}_g$, $g \geqslant 1$ is as follows: start with the torus $\mathbb{S}_1$ (a tire), then rotate a vertical circle around the z-axis, and create $\mathbb{S}_2$ by gluing two tori, etc. (Figure 5.39).

The *genus* $\gamma(\Gamma)$ of a graph $\Gamma = (V; E, N)$ is the smallest integer g such that Γ can be embedded in $\mathbb{S}_g$. Proposition 5.2.19 shows that the planar graphs are exactly those that can be embedded on the sphere; in other words, the planar graphs are the graphs of genus 0. Note that since we know that $K_{3,3}$ and K_5 are not planar (by Corollary 5.2.17), we must have $\gamma(K_{3,3})$, $\gamma(K_5) \geqslant 1$. In fact, they are of genus 1; the presence of a hole on the torus gives just enough room to draw both K_5 and $K_{3,3}$, as shown in Figure 5.40.

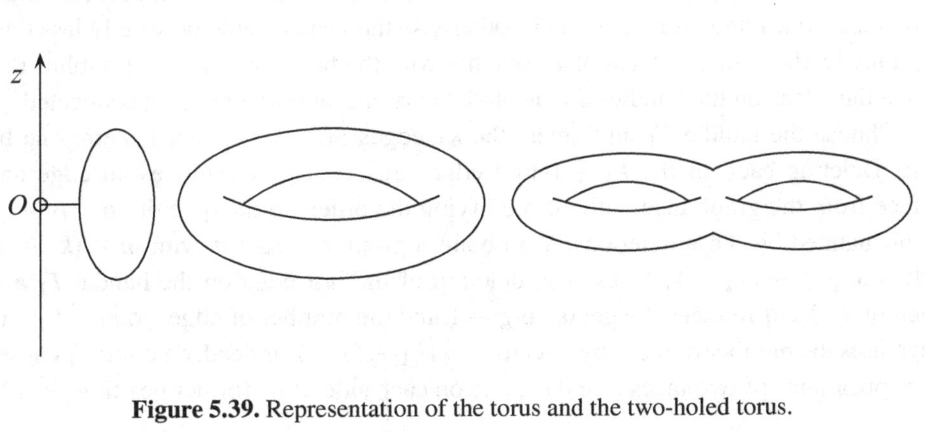

Figure 5.39. Representation of the torus and the two-holed torus.

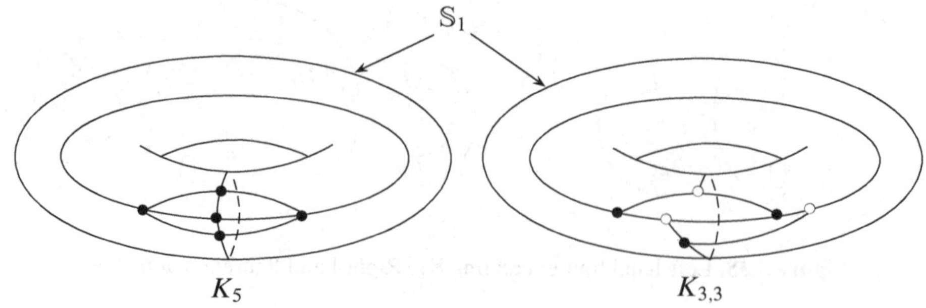

Figure 5.40. K_5 and $K_{3,3}$ are of genus 1.

Proposition 5.7.3. *The genus of a graph is less than or equal to the number of crossings: $\gamma(\Gamma) \leqslant \mathrm{Cr}(\Gamma)$.*

Proof. Suppose that $\mathrm{Cr}(\Gamma) = g$, and draw the graph on the sphere. At each crossing, add a handle, and replace one of the edges by a bridge along the handle. By creating g handles, we thus eliminate all of the crossings. ∎

For an embedded graph on $\mathbb{S}_g$, we define the faces to be the connected components of $\mathbb{S}_g \setminus \Gamma$. A face is thus homeomorphic to an open disk of $\mathbb{R}^2$. Euler's formula generalises amazingly to this situation.

Theorem 5.7.4. *Let $\Gamma = (V; E, N)$ be a connected graph of genus g with n vertices and m edges, embedded in $\mathbb{S}_g$. Let f denote the number of faces of the corresponding drawing. Then f does not depend on the embedding, and $n - m + f = 2 - 2g$.*

Sketch of the proof. The fact that Γ is connected and of genus g indicates that it can be embedded in $\mathbb{S}_g$, which is the sphere with g added handles $T_1, T_2, \ldots, T_g$. We may assume that all the vertices of Γ are drawn outside of the handles, so that each handle has at least one edge running along it (otherwise the genus would be strictly less than g). Finally, the connectedness of Γ together with the fact that $\gamma(\Gamma) = g$ implies that when the edges on the handles are deleted, the resulting graph remains connected.

Choose the handle T_1 and delete the k_1 edges $e_1, \ldots, e_{k_1}$ along them, one by one. Deleting each of the $k_1 - 1$ first edges $e_1, \ldots, e_{k_1-1}$ removes an edge and a face from the graph face without modifying the order, so the quantity $n - m + f$ is unchanged by these deletions. We obtain a graph of order n with $m - (k_1 - 1)$ edges and $f - (k_1 - 1)$ faces. The deletion of the last edge on the handle T_1 and then of T_1 itself reduces the genus to $g - 1$ and the number of edges to $m - k_1$, but increases the number of faces by one, to $f - (k_1 - 1) + 1$: indeed, deleting T_1 causes the appearance of two faces φ and φ', one on each side of its former position.

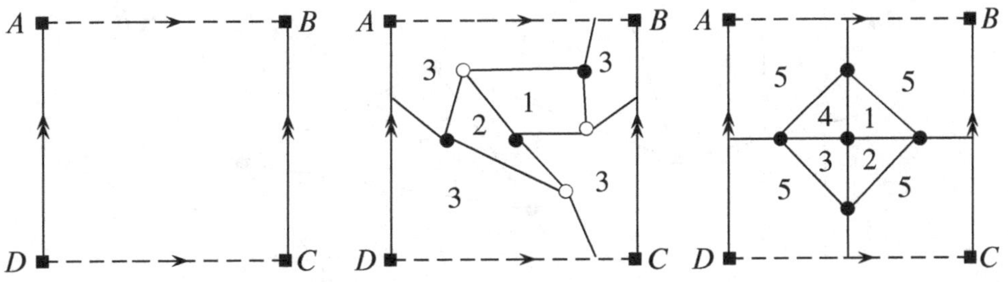

Figure 5.41. The graphs $K_{3,3}$ and K_5 on the torus. Left-hand figure: a representation of the torus. Middle figure: $K_{3,3}$ with its 3 faces; we see the 8 edges adjacent to the face 3. Right-hand figure: K_5 with its 5 faces.

We repeat this process successively on each handle T_i. The final graph Γ' is thus connected, planar and has $n' = n$ vertices, $m' = m - \sum_{i=1}^{g} k_i$ edges and $f' = f - \sum_{i=1}^{g}(k_i - 2)$. Euler's simple formula (Theorem 5.2.11) for planar graphs gives

$$2 = n' - m' + f' = n - m + \sum_{i=1}^{g} k_i + f - \sum_{i=1}^{g}(k_i - 2),$$

which after simplification gives $2 = n - m + f + 2g$, the desired result. ■

We can check the formula of Theorem 5.7.4 for K_5 and $K_{3,3}$ with Figure 5.40, but it is not very easy, in general, to count the number of faces, which is why we use a third realisation of $\mathbb{S}_g$. For $g = 1$, starting from a square $ABCD$, we identify AB with CD respecting the orientation of the arrows ($\overrightarrow{AB} \equiv \overrightarrow{DC}$), which gives a cylinder, and then identify $\overrightarrow{BC}$ with $\overrightarrow{AD}$, which gives a torus. Figure 5.41 gives another view of the embeddings of K_5 and $K_{3,3}$ on the torus.

Example 5.7.5. The two parts of Figure 5.42 use the representation of the torus $\mathbb{S}_1$ explained above, and the representation of the sphere $\mathbb{S}_0$ as a cylinder after gluing along the segment $[A, B]$. In this figure, we have embedded a genus 1 graph on the torus $\mathbb{S}_1$ (that has been cut as explained above). Gluing the faces on $\mathbb{S}_0$ produces a planar graph, represented in the lower figure. We can check the validity of Euler's formula on both the initial graph and its planar version. On the first one we find 8 vertices, 13 edges and 5 faces; after cutting and gluing, we find 16 vertices, 25 edges and 11 faces.

In the case of a simple graph, there exists a lower bound for the genus that uses only the number of vertices and the number of edges. Let $\lceil x \rceil$ denote the smallest integer greater than or equal to the real number x (its *ceiling*).

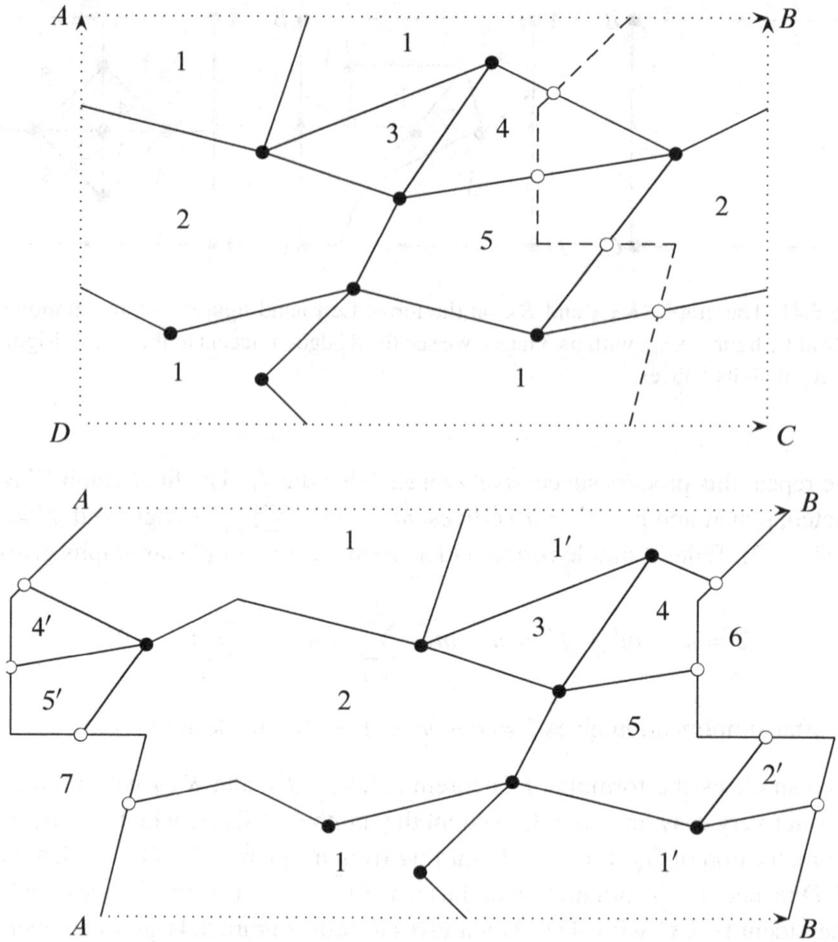

Figure 5.42. An embedded graph on $\mathbb{S}_1$ (upper figure), cut along the boldface dotted line and glued on $\mathbb{S}_0$ (lower figure). The faces $1, 2, 4, 5$ are doubled in $1', 2', 4', 5'$, to which we add the lateral faces 6 and 7.

Proposition 5.7.6. *Let $\Gamma = (V; E)$ be a simple connected graph of finite order, of genus $\gamma(\Gamma) = g$, with $n \geqslant 4$ vertices and m edges. Then*

$$g \geqslant \left\lceil \frac{m-3n}{6} + 1 \right\rceil.$$

Proof. By Proposition 5.2.15, every face is adjacent to at least 3 edges, and by Proposition 5.2.8, every edge is adjacent to either one or two faces; consequently $3f \leqslant 2m$. By Theorem 5.7.4, we obtain $n - m + f = 2 - 2g$, so $2g = m - n - f + 2 \geqslant m - n - \frac{2}{3}m + 2 = \frac{1}{3}m - n + 2$. Thus $g \geqslant \frac{m-3n}{6} + 1$. ∎

Example 5.7.7. We know the genus of some classical graphs:

(i) $\gamma(K_{m,n}) = \lceil \frac{(m-2)(n-2)}{4} \rceil$ (Ringel 1965): in particular, we have $\gamma(K_{4,5}) = 2$ and $\gamma(K_{5,5}) = 3$.

(ii) $\gamma(K_n) = \lceil \frac{(n-3)(n-4)}{12} \rceil$ (Ringel–Youngs 1968): in particular, $\gamma(K_7) = 1$ and $\gamma(K_8) = 2$.

Observe that Euler's formula in Theorem 5.7.4 concerns graphs of genus exactly equal to g, not graphs whose true genus is less than g, even if they can also be embedded in $\mathbb{S}_g$. For example, we can embed a cycle C_4 of length 4 on a torus, and we have $n = 4$, $m = 4$, but the number of faces f of the drawing on the torus is equal to 1 or 2 according to the embedding. Thus $n - m + f$ may be equal to either 1 or 2 according to the embedding, while $2 - 2g = 0$. Thus there exists no embedding of C_4 on the torus ($g = 1$) such that the formula of Theorem 5.7.4 holds.

Example 5.7.8. The surface $\mathbb{S}_2$ is represented as an octagon whose (oriented) sides are glued in pairs, respecting the orientations (Figure 5.43). Figure 5.44 contains the graph K_8 on $\mathbb{S}_2$. The edges *crossing* the sides of the octagon are labelled to help the visualisation, with the edge connecting the vertex x to the vertex y denoted by xy or yx.

Remark 5.7.9. Let Γ be a connected graph of genus g_0. Then there exists $g \geqslant g_0$ and an embedding σ of Γ in $\mathbb{S}_g$ whose drawing has only one face. Indeed, it suffices to add a handle between one face and each of the others.

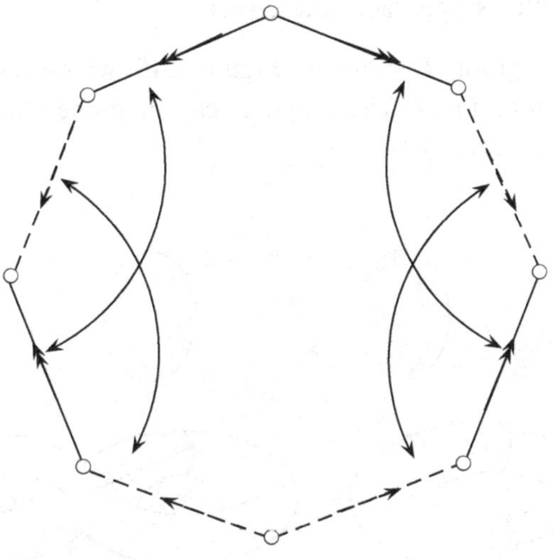

Figure 5.43. A representation of $\mathbb{S}_2$. The sides of the octogon are identified with each other pairwise, respecting the directions of the arrows.

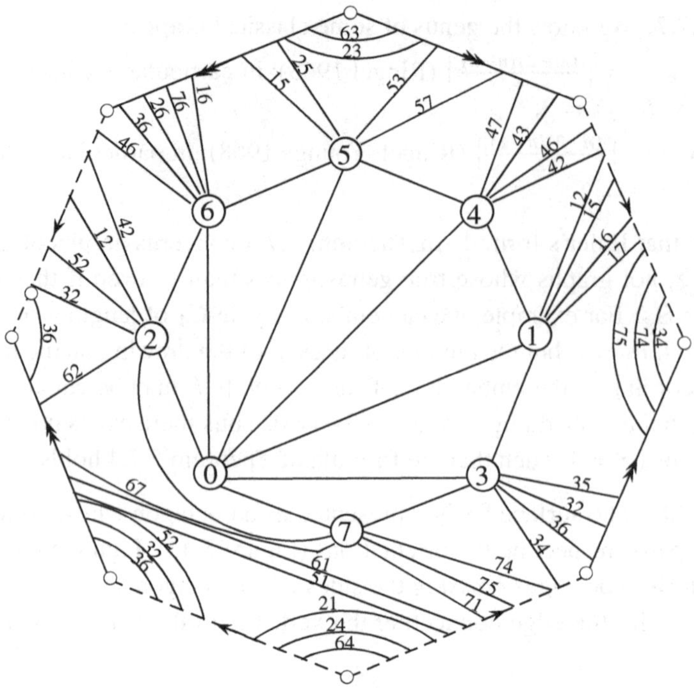

Figure 5.44. Embedding of K_8 in $\mathbb{S}_2$.

Example 5.7.10. The cycle C_n has genus zero and two faces in $\mathbb{S}_0$. However, it can be drawn in on $\mathbb{S}_1$ with a single face (exercise!).

Example 5.7.11. The graph $\mathcal{L}$ shown in Figure 5.45, which resembles a pince-nez, has genus zero and three faces. The graph $\mathcal{L}$ can be embedded in $\mathbb{S}_1$ (resp. in $\mathbb{S}_2$) with two faces (resp. one face).

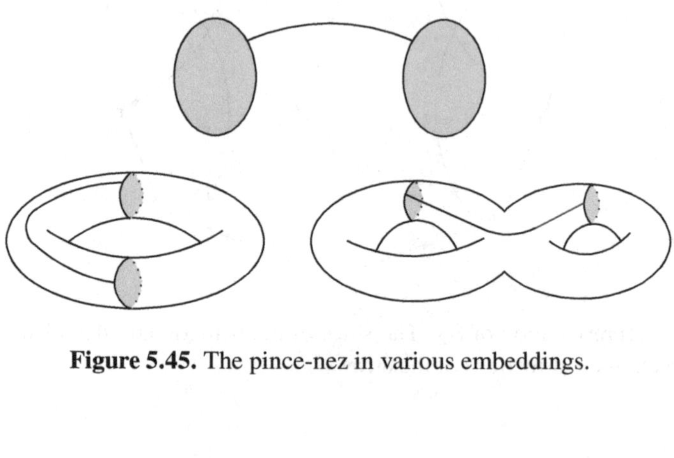

Figure 5.45. The pince-nez in various embeddings.

5.8 Elements of plane topology and geometry

5.8.1 Elements of topology

Let us recall the essential elements of topology that we have used in this chapter, adapting the notions we introduce to the particular framework of metric spaces.

Let (X, d) be a metric space. For $x \in X$ and $r > 0$, we consider

- $\mathcal{B}(x,r) = \{t \in X : d(x,t) < r\}$, the *open ball* of centre x and radius r;
- $\mathcal{B}'(x,r) = \{t \in X : d(x,t) \leqslant r\}$, the *closed ball* of centre x and radius r;
- $\mathcal{C}(x,r) = \{t \in X : d(x,t) = r\}$, the circle of centre x and radius r.

A sequence $(x_n)_{n\geqslant 0}$ of points of X is *convergent* (in X) if there exists $x \in X$ (called a limit point or just a limit) such that

$$\forall r > 0,\ \exists N > 0 : \big(n \geqslant N \Longrightarrow x_n \in \mathcal{B}(x,r)\big).$$

A convergent sequence has a unique limit.

An open set is a subset $U \subset X$ satisfying

$$\forall x \in X,\ \exists r > 0 : \mathcal{B}(x,r) \subset U.$$

Open balls are open sets.

For $A, B \subset X$ and $x \in X$, define the distance from x to A by

$$d(x, A) = \inf\{d(x,y),\ y \in A\},$$

and the distance from B to A by

$$d(B, A) = \inf\{d(z, A),\ z \in B\} = \inf\{d(a,b),\ a \in A,\ b \in B\}.$$

The closure $\overline{A}$ in X is the set of $x \in X$ such that

$$\forall r > 0 : \mathcal{B}(x,r) \cap A \neq \emptyset.$$

We always have $\overline{\mathcal{B}}(x,r) \subset \mathcal{B}'(x,r)$. In fact, when $X = \mathbb{R}^k$, we have the equality $\overline{\mathcal{B}}(x,r) = \mathcal{B}'(x,r)$.

A subset A is said to be *closed* in X if its complement $X \setminus A$ is an open set of X. Thus A is closed if and only if $\overline{A} = A$. Note that X and $\emptyset$ are closed sets.

Closed balls and circles are closed sets.

Lemma 5.8.1. (i) *A is closed if and only if the limit of every convergent sequence $(a_n)_{n\geqslant 0}$ of points of A also lies in A.*

(ii) *The singletons $\{x\}$, and more generally the finite sets of X, are closed sets.*

(iii) *Every finite union of closed sets is closed.*

(iv) *If A is closed and $x \notin A$, then $d(x, A) > 0$.*

Define the *boundary* of A by

$$\mathrm{Fr}(A) = \overline{A} \cap \overline{X \backslash A}.$$

The metric space (X, d) is *connected* if there exists no non-trivial partition into open subsets of X, i.e.

$$X = U \sqcup V,\ U, V \text{ open sets} \implies U = \emptyset \text{ or } V = \emptyset.$$

Thus (X, d) is connected if and only if the only non-empty subset of X that is both open and closed is X itself. A subset A is *connected* if (A, d_A) is connected, where d_A is the distance induced by d on A.

Lemma 5.8.2. (i) *If $A \subset X$ is connected, then $\overline{A}$ is connected. In fact, if A is connected and $A \subset B \subset \overline{A}$, then B is connected.*

(ii) The passing-customs theorem. *If a connected set C intersects both A and $X \backslash A$, then C intersects* $\mathrm{Fr}(A)$.

For every $x \in X$, the connected component $C(x)$ of x is the largest connected subset of X containing x. The connected components $C(x)$ are closed, and form a partition of X into disjoint subsets: $X = \bigsqcup_{x \in \tilde{X}} C(x)$ for a suitable set of elements $\tilde{X} \subset X$. If U is an open set of $\mathbb{R}^2$, the connected components of U are open subsets of U and $\mathbb{R}^2$, but they are also closed subsets of U.

Let $f: (X, d) \to (X', d')$. We say that f is *continuous at $x \in X$* if

$$\forall \varepsilon > 0,\ \exists \eta > 0 \big(d(t, x) < \eta \implies d'(f(t), f(x)) < \varepsilon\big).$$

The function f is said to be *continuous* (on X) if f is continuous at every point $x \in X$, and f is a *homeomorphism* if it is continuous and bijective and f^{-1} is also continuous.

Lemma 5.8.3. *The following properties are equivalent:*

(i) *f is continuous;*

(ii) *for every open set U of X', $f^{-1}(U)$ is an open set of X.*

Equip $\mathbb{R}$ with the distance associated to the absolute value, $d(x, y) = |x - y|$, and equip $\mathbb{R}^2$ with the distance associated to the Euclidean norm $d(x, y) = \|x - y\|$, where $\|(x_1, x_2)\| = \sqrt{x_1^2 + x_2^2}$. In the case of $\mathbb{R}^2$, the ball $\mathcal{B}(x, r)$ is called a *disk* denoted by $D(x, r)$.

Lemma 5.8.4. (i) *The connected subsets of* $\mathbb{R}$ *are the open, closed and semi-open intervals, bounded or not.*

(ii) Bolzano's theorem. *Let* $f: (X, d) \to (X', d')$ *be continuous. If* X *is connected, then* $f(X)$ *is connected.*

Lemma 5.8.5. *Let* (X, d) *be a connected metric space and* $\mathcal{U}$ *an open cover of* X *(i.e. each* $U \in \mathcal{U}$ *is open and* $X = \bigcup_{U \in \mathcal{U}} U$*). Then for all* $x, y \in X$ *there exists a finite simple chain of elements of* $\mathcal{U}$ *connecting* y *to* x*, i.e. there exists* $U_1, \dots, U_m \in \mathcal{U}$ *such that* $y \in U_1$ *and* y *is not in any other* U_i*,* $x \in U_m$ *and* x *is not in any other* U_i*, and* $U_i \cap U_j \neq \emptyset$ *if and only if* $|i - j| \leqslant 1$ *(each* U_i *intersects only the preceding* U_{i-1} *or the following* U_{i+1}*, but no others).*

Proof. Let $x \in X$ be fixed and let A be the set of points of X (including x) that can be connected to x by a finite simple chain of elements of $\mathcal{U}$. We have $A \neq \emptyset$ since $x \in A$.

(i) A is open: let $y \in A$, with $y \in U_1, \dots, U_m \ni x$ and let $t \in U_1$.

- If $t \in U_1 \backslash U_2$, then $U_1, \dots, U_m$ forms a simple chain from t to x.
- If $t \in U_1 \cap U_2$, then $U_2, \dots, U_m$ forms a simple chain from t to x.

Thus $U_1 \subset A$, so A is open. Indeed, as U_1 is open, for every $t \in U_1$ there exists an open disk centred at t and contained in U_1, so also contained in A.

(ii) $X \backslash A$ is open: let $b \in X \backslash A$. Since $\mathcal{U}$ is a cover, there exists $U \in \mathcal{U}$ containing b; we necessarily have $U \cap A = \emptyset$. Indeed, assume that $y \in U \cap A$, with a simple chain $y \in U_1, \dots, U_m \ni x$. We then create a simple chain from b to x as follows. Let k denote the largest index for which $U \cap U_k \neq \emptyset$:

- if $k = m$: either U, U_m or U or U_m forms a simple chain from b to x, according to whether $b \notin U_m$ and $x \notin U$ or $x \in U$ or $b \in U_m$;
- if $k < m$, then $x \notin U$ (since $x \in U_m$ and $U \cap U_m = \emptyset$) and
  - either $b \notin U_k$ and $U, U_k, \dots, U_m$ is a simple chain from b to x,
  - or $b \in U_k$ and $U_k, \dots, U_m$ is a simple chain from b to x (in this case $b \notin U_i, k + 1 \leqslant i \leqslant m$ since $U \cap U_i = \emptyset$).

This shows that $U \cap A = \emptyset$, i.e. $U \subset X \backslash A$. Just as in (i), we conclude that $X \backslash A$ is open.

Thus A is an open, closed, non-empty connected subset of X, so $A = X$, which concludes the proof (cf. [27]). ∎

Lemma 5.2.3 follows immediately from this result.

Proof of Lemma 5.2.3. For all $x \in U$, choose an open disk $D_x = D(x, r_x)$ contained in U, and take $\mathcal{U} = \{D_x, x \in U\}$. This is an open cover of U, so by Lemma 5.8.5, for all $x, y \in U$, there exists a simple chain $x \in D_{x_1}, \dots, D_{x_m} \ni y$; thus we have a polygonal arc $P = [\![x, x_1]\!] \cup [\![x_1, x_2]\!] \cup [\![x_{m-1}, x_m]\!] \cup [\![x_m, y]\!] \subset U$. ■

The metric space (X, d) is *compact* if it is possible to extract a finite open cover from every open cover:

$$X = \bigcup_{i \in I} U_i,\ U_i \text{ open sets} \implies \exists J \subset I, J \text{ finite such that } X = \bigcup_{j \in J} U_j.$$

A subset K of X is said to be *compact* if the subspace (K, d_K) is compact (d_K denotes the distance induced on K by d). One can show that the compact subsets of $\mathbb{R}$ and $\mathbb{R}^2$ are the closed and bounded subsets (here, *bounded* means that there exists $r > 0$ such that $K \subset D(O, r)$, where O is any given point). In particular, the segment $[a, b]$ is a compact set of $\mathbb{R}$, and the closed disk $D'(x, r)$ is a compact set of $\mathbb{R}^2$.

Lemma 5.8.6. (i) Weierstrass' theorem. *Let $f : (X, d) \to (X', d')$ be continuous. If X is compact, then $f(X)$ is compact.*

(ii) *A compact subset is closed.*

(iii) *A finite union of compact subsets is compact.*

5.8.2 Proof of Jordan's *polygonal* theorem

We recall the statement of Theorem 5.2.6.

Theorem 5.2.6. *Let P be a polygonal arc in the plane $\mathbb{R}^2$. If P is closed, then the open set $\mathbb{R}^2 \setminus P$ has exactly 2 connected components U_1, U_2 of which only one is unbounded, and* $\mathrm{Fr}(U_1) = \mathrm{Fr}(U_2) = P$.

Proof of Theorem 5.2.6. *A little topological stroll along the customs officers' path, or the art of avoiding polygons!*

(1) Let P also denote the map $[0, 1] \to \mathbb{R}^2$ used to parameterise the arc P. For each $p = P(t) \in P$, there exists an open disk $D_p = D(p, r)$ of radius $r = r(p)$ sufficiently small for its closure $\overline{D_p}$ to cut only one small arc $\{P(t + u) : |u| < \varepsilon(p)\}$ of P, and no other part of P. Furthermore, we can suppose that the disk $\overline{D_p}$ does not contain any angular point other than p itself (if p is angular). Then $D_p \cap P$ is either a *diameter* or an *angle* (Figure 5.46). Let $\mathbb{A}$ be the (finite) set of angular points of P and Σ the (finite) set of segments that form P.

(2) $\mathbb{R}^2 \setminus P$ is an open set in the plane since P is a finite union of closed segments; it is not bounded, so it has at least one unbounded connected component. Furthermore, since P is compact, $\mathbb{R}^2 \setminus P$ has exactly one unbounded connected component. Indeed, P is contained in a disk $D = D((0, 0), R)$ for $R > 0$ sufficiently large, and $\mathbb{R}^2 \setminus D$ is connected, so it is contained in a certain connected component of $\mathbb{R}^2 \setminus P$; the other components are contained in D, so they are bounded.

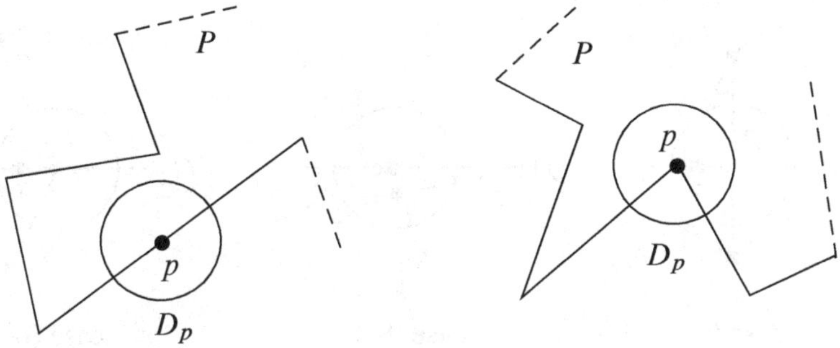

Figure 5.46. Left-hand side: D_p, p not angular. Right-hand side: D_p, p angular.

Fix a point O in this connected component, in such a way that O does not lie on any of the lines generated by the segments of P: this is possible since P contains only a finite number of segments. Thanks to this, for any x in the plane, $[\![x, O]\!] \cap P$ is finite.

(3) Let $x \notin P$. For $p \in [\![x, O]\!] \cap P$, the disk D_p with the line (Ox) removed consists of two open half-disks U_1 and U_2. Set

- $\tau(p) = 1$ if P cuts U_1 and U_2, i.e. $[\![x, O]\!]$ *crosses* P;
- $\tau(p) = 0$ if P only cuts open half-disks U_1, U_2, i.e. $[\![x, O]\!]$ *does not cross* P.

See Figure 5.47 for getting more intuition. We conclude by setting $\tau(p) = 0$ if $p \in [\![x, O]\!] \setminus P$.

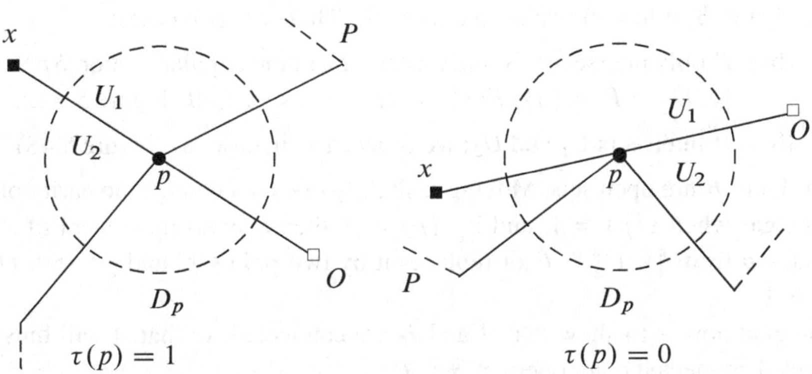

Figure 5.47. On the left: $[\![x, O]\!]$ crosses P. On the right: $[\![x, O]\!]$ does not cross P.

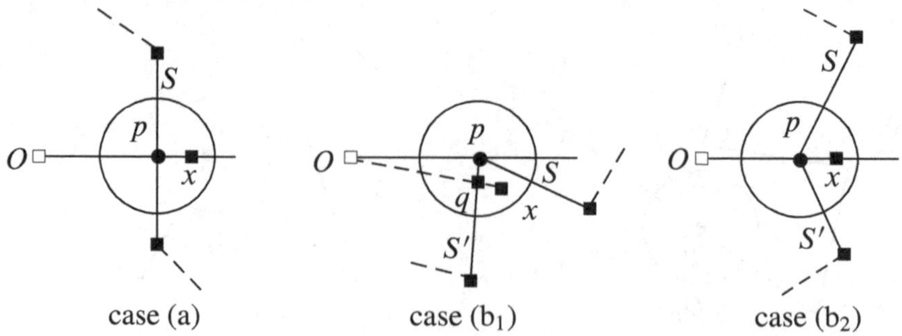

Figure 5.48. B is not empty.

The quantity $T(x) = \sum_{p\in[\![x,O]\!]} \tau(p)$ counts the *crossings* of P when going from x to O.

(4) Let

$$A = \{x \notin P : T(x) \text{ is even}\}, \quad B = \{x \notin P : T(x) \text{ is odd}\}.$$

We have $\mathbb{R}^2 \setminus P = A \sqcup B$ and $O \in A$, so $A \neq \emptyset$. It is less obvious that B is non-empty. Let $p \in P$ be a point that realises the minimum $\min_{x\in P} d(x, O)$.

(a) If $p \notin \mathbb{A}$ (i.e. is not an angular point), it lies on a unique segment S (*perpendicular* to the line (Op)); we define x by $\overrightarrow{Ox} = \lambda \overrightarrow{Op}$ where $\lambda = 1 + \frac{r}{2\|p-O\|}$ and $r = r(p)$ is the number defined in step (1), so that $\|x - p\| = \frac{r}{2}$ and $T(x) = \tau(p) = 1$. Thus $x \in B$ (Figure 5.48).

(b) If $p \in \mathbb{A}$, it lies on two segments S, S'. There are two cases:

(b_1) P only intersects U_2; indeed let x be in the angular section $\widehat{SpS'}$ of D_p: $[\![x, O]\!] \cap P = \{q\}$; $T(x) = \tau(q) = 1$, so $x \in B$ (Figure 5.48);

(b_2) P intersects U_1 and U_2; we proceed as in case (a) (Figure 5.48).

(5) A and B are open sets. Moving x slightly does not change the parity of $T(x)$. This is clear when $\tau(p) = 1$, and if $\tau(p) = 0$, then a small movement of x either eliminates p from $[\![x, O]\!] \cap P$ or replaces it by two points p' and p'' with $\tau(p') = \tau(p'') = 1$.

The goal now is to show that A and B are connected, so that A will thus be the unbounded connected component of $\mathbb{R}^2 \setminus P$.

(6) A is connected. For this, we show that all $x \in A$ can be connected to O by an arc γ in A. It suffices to construct such a γ which does not cut P. Indeed, since A and B are open sets, we have $\mathrm{Fr}(A), \mathrm{Fr}(B) \subset P$, and if γ intersected B, then since γ is a connected subset of $\mathbb{R}^2$ (being the image under a continuous function of the

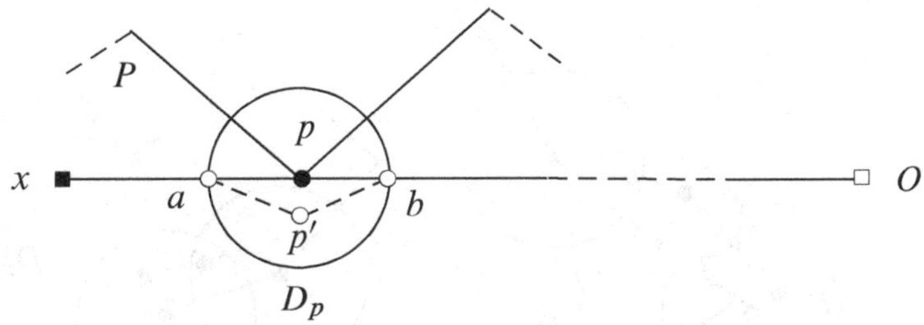

Figure 5.49. The art of avoiding a point p such that $\tau(p) = 0$. We start from x, arrive at a, then pass to p', then to b, and we continue until we reach the next point to be avoided.

connected interval $[0, 1]$ in $\mathbb{R}$, see Lemma 5.8.4), it would also intersect $\mathrm{Fr}(A)$, and thus also P (see Lemma 5.8.2):

- when $[\![x, O]\!] \cap P = \emptyset$, the segment $[\![x, O]\!]$ works;
- otherwise, let p be the point of P on $[\![x, O]\!]$ nearest to x ($[\![x, p[\![\cap P = \emptyset$); if $\tau(p) = 0$ a small movement, for example by $\frac{r(p)}{2}$ in D_p, allows us to avoid p (Figure 5.49). We continue this way along the segment $[\![x, O]\!]$, always avoiding the points p such that $\tau(p) = 0$.
- Let p' be the first point encountered such that $\tau(p') = 1$ (if p' did not exist, we would have an *almost rectilinear* arc from x to O that avoids P). Then there exists at least one other point $q \in [\![x, O]\!]$ such that $\tau(q) = 1$ (since $T(x)$ is even); let q' denote the one which is nearest to 0.
- We will now move along P from p' to q' without crossing it. We write $p' = P(t_1)$, $q' = P(t_2)$ and assume that $0 \leqslant t_1 < t_2 \leqslant 1$. Then $K = P([t_1, t_2])$ is a connected subset of the plane since P is continuous and $[t_1, t_2]$ is connected (see Lemma 5.8.4). Note that $\mathcal{U} = \{D_z \cap K, z \in K\}$ is an open cover of K. By Lemma 5.8.5, there exists a finite simple chain $p' \in D_{z_1}, \dots, D_{z_m} \ni q'$, in which each disk only intersects the next one. We move forward in $D_{z_1} \cap [\![x, O]\!]$ until $p'' \notin P$ before p', then pass to p_1 in D_{z_2} via the biconvex lens-shaped intersection $D_{z_1} \cap D_{z_2}$, then take a point p_2' of D_{z_2} near the centre of D_{z_2} and connect it by a segment to a point p_2 of the biconvex lens $D_{z_2} \cap D_{z_3}$ and so forth. We end with $p_{m-1} \in D_{z_{m-1}} \cap D_{z_m}$, which we connect to q'', an arbitrary neighbour of q' on $[\![q', O]\!]$ (Figure 5.50). It can happen that in this trajectory, we meet $[\![x, O]\!]$, but this makes no difference; the main thing is that we avoid P.
- There now remain between q' and O only points $q \in [\![x, O]\!] \cap P$ such that $\tau(q) = 0$, which we avoid in the same way as the points p at the beginning.
- Finally, we obtain an arc γ from x to O that does not intersect P.

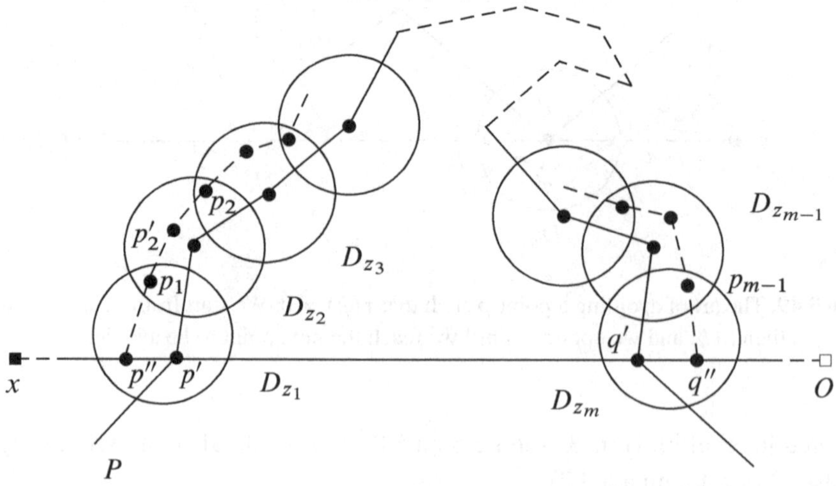

Figure 5.50. The art of avoiding P between p' and q'.

Observe that this procedure is similar to the method used in the proof of Proposition 5.2.4 to double an edge.

(7) B is connected. The method is the same, but this time we take two points x, y of B that must be connected without meeting P. We start from x and move along the segment $[\![x, O]\!]$, avoiding the points p of $[\![x, O]\!] \cap P$ such that $\tau(p) = 0$ as above (Figure 5.49). Let p' be the first point encountered such that $\tau(p') = 1$ (such a point p' exists since $T(x)$ is odd). Similarly, starting from y along $[\![y, O]\!]$ and avoiding the points q of $[\![y, O]\!] \cap P$ such that $\tau(q) = 0$, we necessarily encounter a point q' such that $\tau(q') = 1$. As above, thanks to the connectedness of the arc of P connecting p' to q', we create an arc from p'', a neighbour of p, to q'', a neighbour of q, which avoids P. The union of these three arcs gives an arc γ from x to y that does not meet P (Figure 5.51).

(8) We saw that $\mathrm{Fr}(A), \mathrm{Fr}(B) \subset P$. If $p \in P$, every disk $D(p, \varepsilon)$ for ε sufficiently small is contained in $\overline{D}_p$ (defined at the beginning of this proof) and is cut by P into two connected sections $U_1, U_2 \subset A \cup B$ (Figure 5.47). If for example $U_1 \subset A$, then $U_2 \subset B$ (by applying Lemma 5.8.2), so $p \in \mathrm{Fr}(A) \cap \mathrm{Fr}(B)$. Thus $\mathrm{Fr}(A) = \mathrm{Fr}(B) = P$. ∎

Returning to Figure 5.7, this shows that in order to decide whether a point p lies in the interior of the curve P (i.e. in the bounded connected component of $\mathbb{R}^2 \setminus P$) or in the exterior (i.e. in the unbounded connected component), it suffices to fix O sufficiently far from P and to consider the parity of the number of times that the segment $[\![p, O]\!]$ intersects P (Figure 5.52).

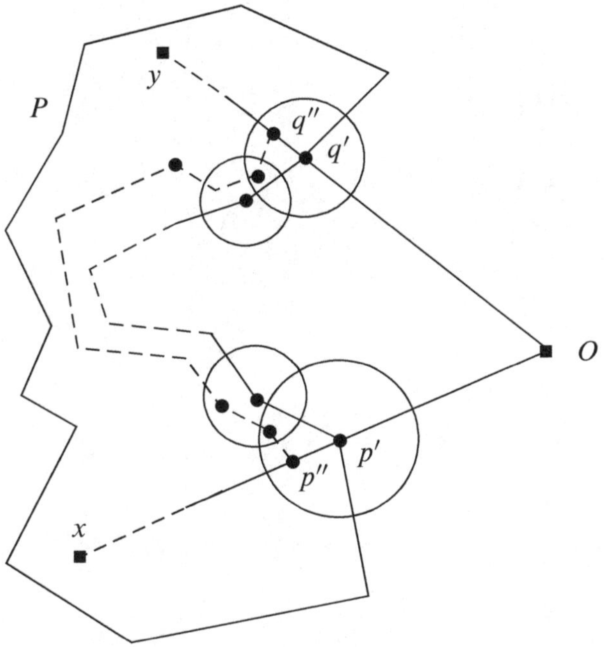

Figure 5.51. B is connected.

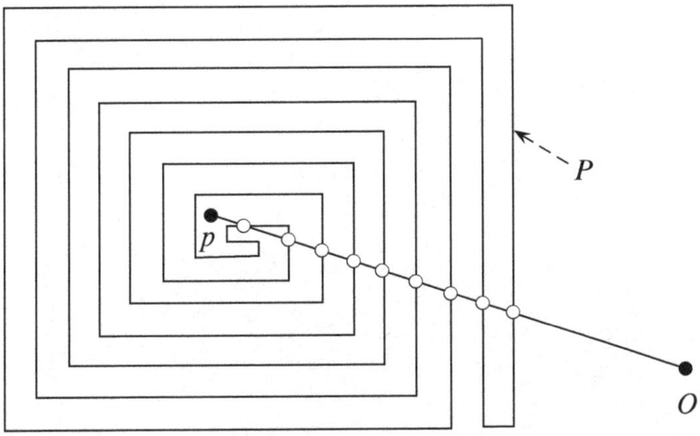

Figure 5.52. Position of the point p with respect to the closed spiral line P. The segment $[\![O, p]\!]$ cuts P an odd number of times (equal to 9), so the point p lies in the interior of P, i.e. in the bounded connected component of $\mathbb{R}^2 \setminus P$.

Chapter 6

Graphs and linear algebra

We saw in Chapter 1 that several types of matrix can be naturally associated to a graph. These representations, which are not always much used in practice since they can be algorithmically quite time-consuming, can nevertheless be extremely useful in the theoretical study of graphs. The idea of a matrix is a theoretically important concept, since it allows us to harness all the power of linear algebra to study graph structures. Furthermore, as we will see, many interesting and useful practical properties of graphs can also be addressed using algebraic methods.

Throughout this chapter, all graphs are assumed to be *finite*.

6.1 Graphs and matrices

Let $\Gamma = (V; E, N)$ be a graph with $V = \{x_1, x_2, \ldots, x_n\}$ and $E = \{e_1, e_2, \ldots, e_m\}$. Recall that the *adjacency matrix* of Γ is the symmetric square matrix $A = (a_{i,j}) \in \mathcal{M}_{n\times n}(\mathbb{Z})$, the set of $n \times n$ matrices with coefficients in $\mathbb{Z}$, defined by

$$a_{i,j} = \text{the number of edges between the vertices } x_i \text{ and } x_j.$$

Similarly, when $\vec{\Gamma} = (V; \vec{E}, N)$ is a digraph,

$$a_{i,j} = \text{the number of arcs from } x_i \text{ to } x_j.$$

Figures 6.1 and 6.2 illustrate these notions.

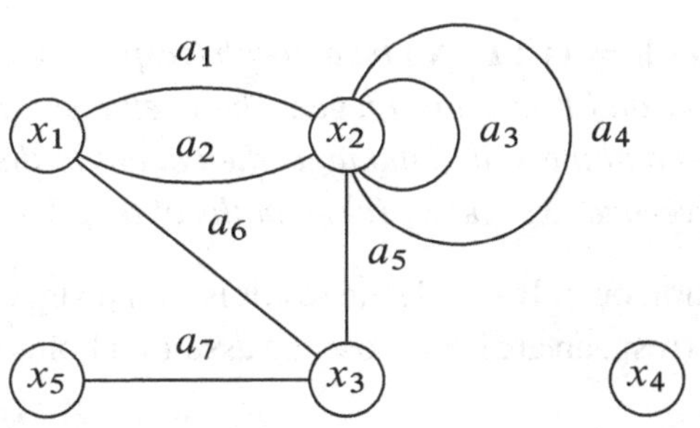

Figure 6.1. Graph with adjacency matrix A_1.

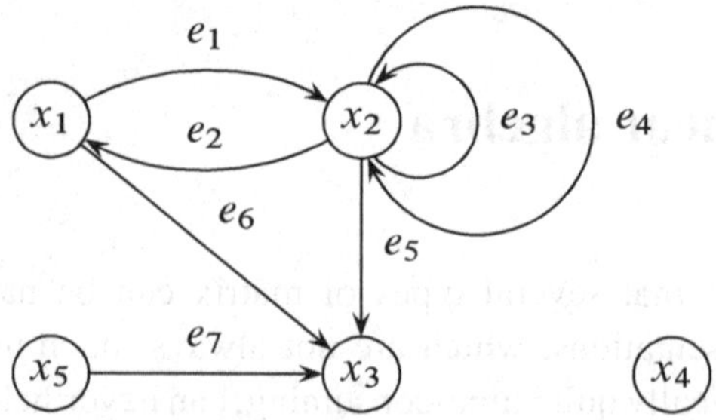

Figure 6.2. Digraph whose adjacency matrix is A_2.

Example 6.1.1. The adjacency matrix of the graph shown in Figure 6.1 is given by

$$A_1 = \begin{array}{c} \\ x_1 \\ x_2 \\ x_3 \\ x_4 \\ x_5 \end{array} \begin{array}{c} \begin{array}{ccccc} x_1 & x_2 & x_3 & x_4 & x_5 \end{array} \\ \begin{pmatrix} 0 & 2 & 1 & 0 & 0 \\ 2 & 2 & 1 & 0 & 0 \\ 1 & 1 & 0 & 0 & 1 \\ 0 & 0 & 0 & 0 & 0 \\ 0 & 0 & 1 & 0 & 0 \end{pmatrix} \end{array}$$

The adjacency matrix of the oriented graph shown in Figure 6.2 is given by

$$A_2 = \begin{array}{c} \\ x_1 \\ x_2 \\ x_3 \\ x_4 \\ x_5 \end{array} \begin{array}{c} \begin{array}{ccccc} x_1 & x_2 & x_3 & x_4 & x_5 \end{array} \\ \begin{pmatrix} 0 & 1 & 1 & 0 & 0 \\ 1 & 2 & 1 & 0 & 0 \\ 0 & 0 & 0 & 0 & 0 \\ 0 & 0 & 0 & 0 & 0 \\ 0 & 0 & 1 & 0 & 0 \end{pmatrix} \end{array}$$

Proposition 6.1.2. *Let $\Gamma = (V; E, N)$ be a graph (resp. $\vec{\Gamma} = (V; \vec{E}, N)$ a digraph), A its adjacency matrix and $r \geqslant 1$ an integer. The coefficient $a_{i,j}^{(r)}$ of the matrix A^r, located in the i-th row and the j-th column, is then equal to the number of chains of length r between the vertices x_i and x_j (resp. paths of length r from x_i to x_j).*

Proof. We use induction on r. If $r = 1$, the result is trivial since a chain (resp. a path) of length 1 is an edge (resp. an arc). Suppose the assertion holds up to $r - 1$, for $r \geqslant 2$. We have

$$a_{i,j}^{(r)} = \sum_{k=1}^{n} a_{i,k}^{(r-1)} a_{k,j}^{(1)}.$$

We know there are $a_{k,j}^{(1)}$ edges (resp. arcs) between x_k and x_j. But by the induction hypothesis, $a_{i,k}^{(r-1)}$ is the number of chains (resp. paths) of length $r-1$ between the vertex x_i and the vertex x_k. Thus for $k = 1, \dots, n$, each term $a_{i,k}^{(r-1)} a_{k,j}^{(1)}$ counts the number of chains (resp. paths) of length r between x_i and x_j that pass through x_k at the $(r-1)$-th step. By adding them together for $k = 1, \dots, n$, this shows that $a_{i,j}^{(r)}$ counts the total number of chains (resp. paths) of length r between x_i and x_j. ■

Proposition 6.1.3. *Let $\vec{\Gamma} = (V; \vec{E}, N)$ be a digraph and $1 \leqslant i \leqslant |V|$, $1 \leqslant j \leqslant |V|$ two integers. Then the sum of the entries in the i-th row of the adjacency matrix of Γ is equal to the out degree of the vertex x_i and the sum of coefficients of the j-th column of the adjacency matrix of Γ is equal to the in degree at x_j.*

Proof. The sum of the coefficients of the i-th row (which corresponds to the vertex x_i) indicates the number of arcs starting from x_i, so by definition it is equal to $d^+(x_i)$. The sum of coefficients of the j-th column (which corresponds to the vertex x_j) indicates the number of arcs having terminal vertex x_j, so it is $d^-(x_j)$. ■

The *incidence matrix of a graph* is the rectangular matrix $J = (\alpha_{i,j}) \in \mathcal{M}_{n\times m}(\mathbb{Z})$ defined by

$$\alpha_{i,j} = \begin{cases} 1 & \text{if } x_i \text{ is one of the two vertices of the edge } a_j, \\ 2 & \text{if } x_i \text{ is the vertex of the loop } a_j, \\ 0 & \text{otherwise.} \end{cases}$$

We define the *incidence matrix of a digraph* as the rectangular matrix $J = (\alpha_{i,j}) \in \mathcal{M}_{n\times m}(\mathbb{Z})$ defined by

$$\alpha_{i,j} = \begin{cases} -1 & \text{if } x_i \text{ is the initial vertex of the arc } e_j, \\ 1 & \text{if } x_i \text{ is the terminal vertex of the arc } e_j, \\ 2 & \text{if } x_i \text{ is the vertex of the loop } e_j, \\ 0 & \text{otherwise.} \end{cases}$$

Example 6.1.4. The two matrices J_1, J_2 are the incidence matrices of the (di)graphs shown in Figures 6.1 and 6.2 respectively.

$$J_1 = \begin{array}{c} \\ x_1 \\ x_2 \\ x_3 \\ x_4 \\ x_5 \end{array} \begin{array}{c} \begin{array}{ccccccc} a_1 & a_2 & a_3 & a_4 & a_5 & a_6 & a_7 \end{array} \\ \begin{pmatrix} 1 & 1 & 0 & 0 & 0 & 1 & 0 \\ 1 & 1 & 1 & 2 & 1 & 0 & 0 \\ 0 & 0 & 0 & 0 & 1 & 1 & 1 \\ 0 & 0 & 0 & 0 & 0 & 0 & 0 \\ 0 & 0 & 0 & 0 & 0 & 0 & 1 \end{pmatrix} \end{array}$$

$$J_2 = \begin{array}{c} \\ x_1 \\ x_2 \\ x_3 \\ x_4 \\ x_5 \end{array} \begin{array}{c} \begin{array}{ccccccc} e_1 & e_2 & e_3 & e_4 & e_5 & e_6 & e_7 \end{array} \\ \begin{pmatrix} -1 & 1 & 0 & 0 & 0 & -1 & 0 \\ 1 & -1 & 2 & 2 & -1 & 0 & 0 \\ 0 & 0 & 0 & 0 & 1 & 1 & 1 \\ 0 & 0 & 0 & 0 & 0 & 0 & 0 \\ 0 & 0 & 0 & 0 & 0 & 0 & -1 \end{pmatrix} \end{array}$$

Proposition 6.1.5. *Let $\vec{\Gamma} = (V; \vec{E}, N)$ be a loopless digraph. Then each column of its incidence matrix sums to zero.*

Proof. The proof is left as an exercise. ■

Let $\Gamma = (V; E, N)$ be a graph or a digraph, with $V = \{x_1, x_2, \dots, x_n\}$ and $E = \{e_1, e_2, \dots, e_m\}$. The *matrix of degrees* of Γ is the diagonal matrix $D = (d_{i,j}) \in \mathcal{M}_{n\times n}(\mathbb{Z})$ defined by

$$d_{i,j} = \begin{cases} d(x_i) & \text{if } i = j, \\ 0 & \text{otherwise.} \end{cases}$$

For a digraph, we have $d(x) = d^+(x) + d^-(x)$.

The following proposition relates the incidence matrix of a (di)graph to its adjacency matrix.

Proposition 6.1.6. *Let Γ be a loopless (di)graph, A its adjacency matrix, J its incidence matrix and D its matrix of degrees. Let ${}^tJ$ be the transpose of J; then $J\,{}^tJ \in \mathcal{M}_{n\times n}(\mathbb{Z})$ and*

$$J\,{}^tJ = \begin{cases} A + D & \text{if } \Gamma \text{ is a graph,} \\ -A + D & \text{if } \Gamma \text{ is a digraph.} \end{cases}$$

Proof. (i) *Non-oriented case*: if $J = (\alpha_{i,j})_{\substack{1\leqslant i\leqslant n \\ 1\leqslant j\leqslant m}}$, then $J\,{}^tJ = (\beta_{i,j})_{1\leqslant i,j\leqslant n}$ where

$$\beta_{i,j} = \sum_{k=1}^{m} \alpha_{i,k}\alpha_{j,k}, \quad 1 \leqslant i, j \leqslant n.$$

- $\beta_{i,i} = \sum_{1\leqslant k\leqslant m} \alpha_{i,k}^2$ is the number of edges incident to x_i, so it is the degree of x_i. But $d(x_i) = d_{i,i} = a_{i,i} + d_{i,i}$ since $a_{i,i} = 0$, as the graph is loopless;
- for $j \neq i$, we have $d_{i,j} = 0$. We distinguish two cases:
  - if x_j is not adjacent to x_i, then $a_{i,j} = 0$ and for all $1 \leqslant k \leqslant m$, we have either $\alpha_{i,k} = 0$ or $\alpha_{j,k} = 0$ (otherwise x_i and x_j would be incident to the edge e_k and thus adjacent). Therefore $\beta_{i,j} = 0 = a_{i,j} + d_{i,j}$;

- if x_j is adjacent to x_i, then

$$\alpha_{i,k}\alpha_{j,k} = \begin{cases} 1 & \text{if } e_k \text{ is incident to } x_i \text{ and } x_j, \\ 0 & \text{otherwise,} \end{cases}$$

so $\beta_{i,j}$ is the number of edges incident to x_i and x_j, which is given by $a_{i,j} = a_{i,j} + d_{i,j}$.

(ii) *Oriented case*: the only difference with the preceding case is when x_j is adjacent to x_i: if e_k is oriented from x_i to x_j, then $\alpha_{i,k}\alpha_{j,k} = (-1)\cdot(+1) = -1$, and if e_k is oriented from x_j to x_i, then $\alpha_{i,k}.\alpha_{j,k} = (+1)\cdot(-1) = -1$, so $\beta_{i,j}$ is the opposite of the number of edges incident to x_i and x_j, i.e. $\beta_{i,j} = -a_{i,j} = -a_{i,j} + d_{i,j}$. This concludes the proof. ∎

Exercise 6.1. Let $\Gamma = (V; E)$ be a (simple) graph, and let J be its incidence matrix and $L(\Gamma)$ its line graph. Let $A(L(\Gamma))$ denote the adjacency matrix of $L(\Gamma)$ and I the identity matrix of order $m = |E|$. Show that

$${}^tJJ = 2I + A\big(L(\Gamma)\big).$$

Hint: Observe that the scalar product of two column vectors of J corresponding to two edges is non-zero if and only if the two corresponding edges have a common vertex.

We say that a simple digraph $\vec{\Gamma} = (V; \vec{E})$ is a *forest* (resp. a *tree*) if the underlying graph of $\vec{\Gamma}$ is a forest (resp. a tree).

Theorem 6.1.7. *Let $\vec{\Gamma} = (V; \vec{E})$ be a simple digraph, and let J denote its incidence matrix. Then the digraph $\vec{\Gamma}$ is a forest if and only if the column vectors of J are linearly independent over $\mathbb{Z}$.*

Proof. Set $m = |\vec{E}|$. For each arc $e_j = (x, y) \in \vec{E}$, let $a_j = \{x, y\}$ denote the underlying edge in the underlying graph of $\vec{\Gamma}$.

If the digraph is not a forest, the underlying graph possesses an elementary cycle

$$C = (x_1, a_1, x_2, a_2, \ldots, x_t, a_t, x_1)$$

of length $t \leqslant m$. Let $c_1, c_2, \ldots, c_t$ denote the column vectors of J corresponding to the arcs whose underlying edges are $a_1, a_2, \ldots, a_t$. Let us show that these vectors are linearly dependent in $\mathbb{Z}^m$. For each i, define the orientation coefficient u_i of the edge a_i in the cycle C by

$$u_i = \begin{cases} 1 & \text{if } e_i = (x_i, x_{i+1}), \\ -1 & \text{if } e_i = (x_{i+1}, x_i). \end{cases}$$

Let $\vec{\Gamma}' = (V'; \vec{E}')$ be the subdigraph induced by the arcs and vertices corresponding to the columns $c_1, c_2, c_3, \ldots, c_t$. Since each x_i, $i = 1, \ldots, t$, is an extremity of two edges of the cycle C, each row of the incidence matrix J' of $\vec{\Gamma}'$ contains exactly two non-zero coefficients. These two non-zero coefficients of the i-th row of the matrix J' (corresponding to the vertex x_i) depend on the two orientation coefficients of the only two edges incident to x_i in the elementary cycle C. Up to reordering the vertices and arcs of $\vec{\Gamma}$, we can write

$$J' = \begin{array}{c} \\ x_1 \\ x_2 \\ x_3 \\ x_4 \\ \vdots \\ x_{t-1} \\ x_t \end{array} \begin{array}{c} \begin{array}{ccccccc} e_1 & e_2 & e_3 & e_4 & \ldots & e_{t-1} & e_t \end{array} \\ \begin{pmatrix} u_1 & 0 & 0 & 0 & \ldots & 0 & -u_t \\ -u_1 & u_2 & 0 & 0 & \ldots & 0 & 0 \\ 0 & -u_2 & u_3 & 0 & \ldots & 0 & 0 \\ 0 & 0 & -u_3 & u_4 & \ldots & 0 & 0 \\ & & & & & & \\ 0 & 0 & 0 & 0 & & u_{t-1} & 0 \\ 0 & 0 & 0 & 0 & & -u_{t-1} & u_t \end{pmatrix} \end{array}.$$

Since $u_i^2 = 1$ for all i, we have $u_1c_1 + u_2c_2 + \cdots + u_tc_t = 0$, so the column vectors of J' are linearly dependent. The same obviously holds for the column vectors of J.

Conversely, suppose that the column vectors of the matrix J are linearly dependent. Then there exist complex numbers $u_1, u_2, u_3, \ldots, u_t$, not all zero, such that $u_1c_1 + u_2c_2 + \cdots + u_tc_t = 0$. Let $\vec{\Gamma}' = (V'; \vec{E}')$ be the subdigraph generated by the vertices of arcs corresponding to the columns $c_1, c_2, c_3, \ldots, c_t$, and let J' denote its incidence matrix. As $u_1c_1 + u_2c_2 + \cdots + u_tc_t = 0$ with $u_i \neq 0$, $i = 1, \ldots, t$, each line of J' admits at least 2 non-zero coefficients. Consequently, no vertex in the underlying graph of $\vec{\Gamma}'$ can be of degree 1. Therefore the underlying graph of $\vec{\Gamma}'$ must contain a cycle (see Proposition 2.2.5). This cycle is a fortiori a cycle of the underlying graph of $\vec{\Gamma}$, and thus $\vec{\Gamma}$ is not a forest. ∎

Remark 6.1.8. The linear dependence of the column vectors of the incidence matrix of a digraph which is a forest would thus consist of linear combinations with coefficients in $\{-1, 0, 1\}$; it follows that the linear independence over $\mathbb{Z}$ of columns of the incidence matrix is equivalent to their linear independence over $\mathbb{Q}$, $\mathbb{R}$ or $\mathbb{C}$.

We will now show that the incidence matrix of a digraph possesses a very particular structure. A matrix M with coefficients in $\mathbb{Z}$ is said to be *totally unimodular* if every square matrix extracted from M has determinant equal to 1, -1 or 0.

6.1.1 The oriented case

Theorem 6.1.9. *The incidence matrix of a loopless digraph $\vec{\Gamma} = (V; \vec{E}, N)$ is totally unimodular.*

Proof. We use induction on the size of a square matrix extracted from the incidence matrix J of $\vec{\Gamma}$. If $J' = (u)$ is of order 1, then $\det J' = u$. As u is a coefficient of the matrix J, we have $u \in \{-1, 0, 1\}$ since $\vec{\Gamma}$ is loopless. Let J' be a square matrix of order k with $k \geqslant 2$. There are three cases.

(i) If one of the column vectors of J' is identically zero, we obviously have $\det J' = 0$.

(ii) If all the column vectors of J' have at least two non-zero coefficients, the matrix J' defines a subdigraph $\vec{\Gamma}'$ of $\vec{\Gamma}$ with k vertices and k arcs, and thus the underlying graph of $\vec{\Gamma}'$ is not a forest (see Corollary 2.2.3), so it has a cycle. Then by Theorem 6.1.7, the column vectors of the matrix J' are linearly dependent, so $\det J' = 0$.

(iii) If there exists a column of J' with exactly one non-zero coefficient, then this coefficient is equal to ± 1. By expanding the determinant of J' with respect to this column, we obtain $\det J' = \pm \det J''$, where J'' is a square matrix of order $k - 1$ extracted from J. By the induction hypothesis, we have $\det J'' \in \{-1, 0, 1\}$, which gives the result. ∎

Proposition 6.1.10. *Let $\vec{\Gamma} = (V; \vec{E}, N)$ be a loopless digraph with n vertices, m edges and k connected components. Then the rank of its incidence matrix J satisfies*

$$\operatorname{rank}(J) = n - k \leqslant m.$$

Proof. Set $r = \operatorname{rank}(J)$, so r is the maximum number of linearly independent column vectors of J. The edges of the underlying graph corresponding to these columns define a partial graph F, which is a forest by Theorem 6.1.7; F possesses n vertices, c connected components and r edges. Moreover, we have $c = n - r$ (see Corollary 2.2.3).

Let us show that $c = k$. We clearly have $c \leqslant k$ since the vertices of F are those of $\vec{\Gamma}$. If we had $c > k$, there would exist two connected components of F that we could connect by an arc in $\vec{\Gamma}$. The digraph obtained this way would still be a forest, so its incidence matrix would have $r + 1$ linearly independent column vectors, contradicting the maximality of r. We conclude that $m \leqslant r = n - k$, which gives the desired result. ∎

6.1.2 The non-oriented case

Theorem 6.1.11. *Let $\Gamma = (V; E)$ be a graph and J its incidence matrix. The matrix J is a totally unimodular matrix if and only if Γ is bipartite.*

Proof. Assume that $\Gamma = (V; E)$ is a bipartite graph: $V = V_1 \sqcup V_2$. Let J' be a square matrix of order t extracted from J. If $t = 1$, we simply have $J' = (u)$ with $u \in \{-1, 0, 1\}$ since a bipartite graph is loopless. Suppose $t > 1$. We then have two cases.

(i) If each column of J' contains exactly two non-zero coefficients (so both equal to 1), the t vertices corresponding to the rows of J' can be partitioned into two sets V_1' and V_2' such that $V_1' \subseteq V_1$ and $V_2' \subseteq V_2$. Each column of J' corresponds to an edge of which one extremity lies in V_1' and the other in V_2'. Thus the sum of row vectors corresponding to V_1' is equal to the sum of row vectors corresponding to V_2'. Therefore the row vectors of J' are linearly dependent, so $\det J' = 0$.

(ii) Suppose now that there exists at least one column of J' with at most one non-zero coefficient, equal to 1. Expand the determinant of J' with respect to this column. Up to sign, we obtain the determinant of a square matrix of order $(t-1)$, which gives the result by the induction hypothesis.

Conversely, assume that Γ is not bipartite. In this case Γ contains a cycle of odd length; write it as $2k+1$ with $k \geqslant 1$. It is easy to check (exercise) that the determinant of the square matrix J' corresponding to the $2k+1$ vertices and $2k+1$ edges of this cycle is equal to 2, which shows that J is not totally unimodular. ∎

Proposition 6.1.12. *Let $\Gamma = (V; E, N)$ be a loopless graph with n vertices, m edges, and k connected components. Let J be the incidence matrix of Γ. Then the rank of J is equal to $n - k \leqslant m$.*

Exercise 6.2. Prove this proposition.

The *rank* of Γ is the integer $\rho(\Gamma) = n - k$ that gives the number of edges of a *spanning forest* (i.e. of a partial (di)graph which is a forest). For connected graphs, we recover the notion of a spanning tree introduced in Proposition 2.2.6. The *nullity* of the graph Γ is the number $\nu(\Gamma) = m - n + k$. Observe that $\rho(\Gamma) \geqslant 0$ and $\rho(\Gamma) + \nu(\Gamma) = m$. Furthermore, $\rho(\Gamma) \leqslant m$ (so $\nu(\Gamma) \geqslant 0$) since for each connected component C_i the number of edges is at least equal to the number of vertices minus 1 by Proposition 1.4.2.

6.2 Vector spaces and graphs

In this paragraph, we will associate vector spaces to graphs. This will allow us to subsequently introduce or reintroduce some properties of graphs in an algebraic form. In this section, we assume that all (di)graphs are *finite and loopless*.

6.2.1 The case of oriented graphs

Let δ_{ij} denote the *Kronecker symbol*:

$$\delta_{ij} = 1 \text{ if } i = j, \quad \delta_{ij} = 0 \text{ if } i \neq j.$$

Let $\vec{\Gamma} = (V; \vec{E}, N)$ be a digraph and write $V = \{x_1, x_2, \ldots, x_n\}$. Represent each vertex x_i by the vector $\xi_i = (\delta_{ij})_{1 \leqslant j \leqslant n} \in \mathbb{C}^n$ in such a way that $\mathcal{B}_n = \{\xi_1, \xi_2, \ldots, \xi_n\}$ is the canonical basis of $\mathbb{C}^n$. Observe that changing the numbering of vertices comes down to permuting the elements of $\mathcal{B}_n$. Let $C_0(\vec{\Gamma})$ denote the vector space $\mathbb{C}^n$ equipped with this basis; we call it the *vertex space of the digraph* (recall that every vector space of finite dimension n over a field K is isomorphic to K^n).

Similarly, let $\vec{E} = \{e_1, e_2, \ldots, e_m\}$ be a numbering of the set of arcs; we represent each arc e_i by $\varepsilon_i = (\delta_{ij})_{1 \leqslant j \leqslant m} \in \mathbb{C}^m$. Then $\mathcal{B}'_m = \{\varepsilon_1, \varepsilon_2, \ldots, \varepsilon_m\}$ is the canonical basis of $\mathbb{C}^m$; it is convenient to identify the arcs e_i with the vectors ε_i. Let $C_1(\vec{\Gamma})$ denote the vector space $\mathbb{C}^m$ equipped with this basis: we call it the *arc space of the digraph*.

In the language of graphs, the basis $\mathcal{B}_n$ of $C_0(\vec{\Gamma})$ is called the *standard basis of the vertex space of the digraph*, and the basis $\mathcal{B}'_m$ of $C_1(\vec{\Gamma})$ is called the *standard basis of the arc space of the digraph*.

Let $\Gamma = (V; E, N)$ be the underlying graph of $\vec{\Gamma}$ and let $a_i \in E$ denote the edge underlying the arc $e_i \in \vec{E}$, so that $E = \{a_1, \ldots, a_m\}$. Let C be a cycle of Γ given by

$$C = (x_0, a_0, x_1, a_1, \ldots, x_n, a_n, x_0).$$

This cycle possesses a *proper orientation*, namely the cyclic orientation according to the ordering $x_0, x_1, \ldots, x_n$. Such a cycle will be called an *oriented cycle*; every cycle obviously has two possible orientations. We will see in what follows that the distinction between these two orientations is not essential, even though we do need to fix a choice.

We can now compare this orientation to that of the corresponding arcs of the digraph by associating to C the vector $\mu_C = (c_1, \ldots, c_m) \in C_1(\vec{\Gamma})$ defined by

$$c_i = \begin{cases} 1 & \text{if } e_i = ((x_i, x_{i+1}), n_i) : e_i \text{ is oriented as in } C, \\ -1 & \text{if } e_i = ((x_{i+1}, x_i), n_i) : e_i \text{ is oriented in the opposite direction to } C, \\ 0 & \text{if } e_i \notin C. \end{cases}$$

We can consider μ_C as a map from $\vec{E}$ to $\mathbb{C}$ by setting $\mu_C(e_i) = c_i$.

Example 6.2.1. In the digraph represented in Figure 6.3, the vector associated to the cycle

$$C = (x_6, a_6, x_3, a_3, x_4, a_4, x_5, a_7, x_6)$$

is $\mu_C = (0, 0, -1, -1, 0, 1, -1, 0) \in C_1(\vec{\Gamma})$.

Clearly distinct cycles are associated to distinct vectors. More precisely, the two vectors associated to two distinct oriented cycles are also distinct, and conversely, two distinct vectors defined as above must be associated to two distinct oriented

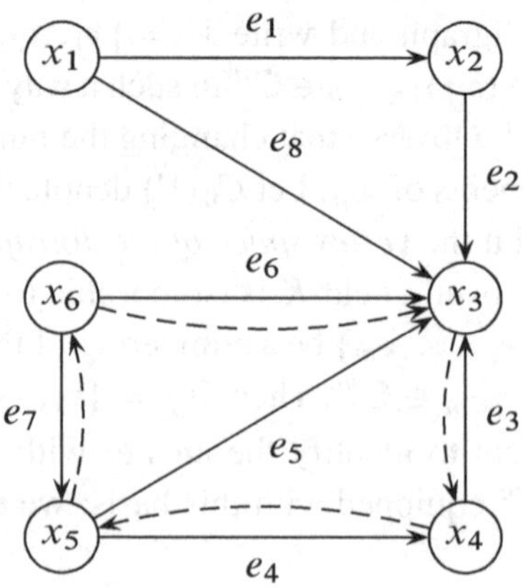

Figure 6.3. Cycle C in a digraph.

cycles. The *cycle space* is the vector subspace of $C_1(\vec{\Gamma})$ generated by the set $\{\mu_C \mid C$ elementary cycle of $\Gamma\}$. It is denoted by $\mathcal{C}$.

Let (V_1, V_2) be a partition of the vertices of a connected graph $\vec{\Gamma}$. The set $C^o = C^o(V_1, V_2)$ of arcs between V_1 and V_2, or between V_2 and V_1, is called a *cocycle*. Just as we saw for a cycle, we can associate to a cocycle C^o an element $\mu_{C^o} = (\mu_{C^o}(e_1), \mu_{C^o}(e_2), \ldots, \mu_{C^o}(e_m))$ of $C_1(\vec{\Gamma})$ defined by

$$\mu_{C^o}(e_i) = \begin{cases} 1 & \text{if } e_i \text{ is oriented from } V_1 \text{ to } V_2, \\ -1 & \text{if } e_i \text{ is oriented from } V_2 \text{ to } V_1, \\ 0 & \text{if } e_i \notin C^o. \end{cases}$$

The *cocycle space* is the vector subspace of $C_1(\Gamma)$ generated by the set $\{\mu_{C^o} \mid C^o$ cocycle of $\vec{\Gamma}\}$; it is denoted by $\mathcal{C}^o$.

Certain cycles play a particular role. Let $\vec{\Gamma} = (V; \vec{E}, N)$ be a connected digraph with $|V| = n$, $|\vec{E}| = m$, and let A be a spanning tree of the underlying graph $\Gamma = (V; E, N)$. The tree A then has $n-1$ edges (see Theorem 2.2.1). Let

$$R = \{e_i \in \vec{E} \mid 1 \leqslant i \leqslant m \text{ and } a_i \notin A\}.$$

If $R = \emptyset$, then Γ is a tree, otherwise $|R| = m - (n-1)$. For each $e \in R$, we observe that adding the edge a (corresponding to the arc e) to the tree A creates exactly one elementary cycle (see Proposition 2.2.6), which is called the *fundamental cycle* and denoted C_e; it corresponds to a vector μ_{C_e}. Thus there are $m - (n-1)$ fundamental cycles attached to the spanning tree A.

Similarly, certain cocycles also play a particular role. Here, instead of adding an edge to the tree A, we remove one. By removing an edge a of A corresponding to an

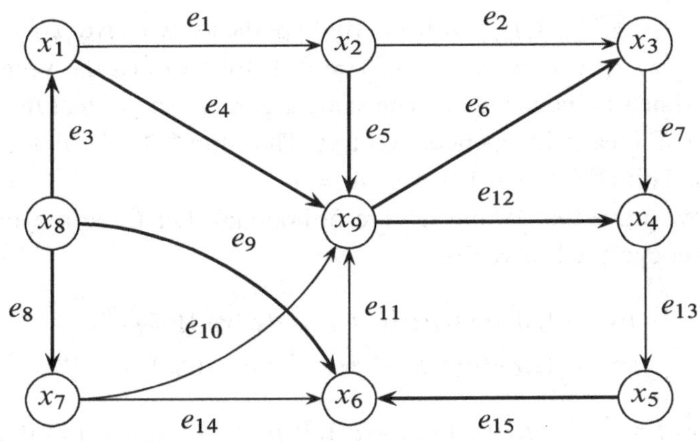

Figure 6.4. Fundamental cycles and cocycles in a digraph.

arc e of $\vec{\Gamma}$, we obtain two connected components, which yields a partition (V_1, V_2) of V, and therefore a cocycle C_e^o called the *fundamental cocycle*. Thus there are $n-1$ fundamental cocycles attached to the spanning tree A.

Example 6.2.2. In the digraph shown in Figure 6.4, the spanning tree A is shown in boldface arcs. The vector of the cycle $C_{e_{11}}$ is $\mu_{C_{e_{11}}} = e_{11} - e_4 - e_3 + e_{10}$. The cocycle associated to the arc e_4 has vector $\mu_{C^o_{e_4}} = e_1 + e_4 + e_9 + e_{11} - e_{13}$.

We have the following fundamental result.

Theorem 6.2.3. *Let $\vec{\Gamma} = (V; \vec{E}, N)$ be a connected loopless digraph, and let A be a spanning tree of the underlying graph Γ. Then*

(i) *the set of $m-n+1$ fundamental cycles $\{\mu_{C_e}\}$ attached to the tree A forms a basis of the cycle space and*

$$\dim_{\mathbb{C}} \mathcal{C} = m - (n-1);$$

(ii) *the set of $n-1$ fundamental cocycles $\{\mu_{C_e^o}\}$ attached to A forms a basis of the cocycle space and*

$$\dim_{\mathbb{C}} \mathcal{C}^o = n - 1.$$

(iii) *The $\mathbb{C}$-subspaces $\mathcal{C}$ and $\mathcal{C}^o$ are supplementary, and orthogonal (for the canonical scalar product on $\mathbb{C}^n$) in $C_1(\vec{\Gamma})$.*

Proof. Let $(C_i)_{n \leqslant i \leqslant m}$ be the fundamental cycles associated to A. Up to reordering the arcs, fix

$$R = \{e_n, \dots, e_m\} = \{e \in \vec{E} \mid e \text{ does not induce an edge of } A\}.$$

Suppose that $\mu = \sum_{i=n}^{m} \lambda_i \mu_{C_i} = 0_{\mathbb{C}^m}$. By hypothesis, we have $\mu(e_j) = 0$, so $0 = \mu(e_j) = \sum_{i=n}^{m} \lambda_i \mu_{C_i}(e_j) = \pm\lambda_j$, so $\lambda_j = 0$. It follows that the vectors μ_{C_i}, $n \leqslant i \leqslant m$, are $\mathbb{C}$-linearly independent. The same argument shows that the $\mu_{C_e^o}$, $e \in A$, form a system of linearly independent vectors. Thus $\dim \mathcal{C} \geqslant m - n + 1$ (since $|R| = m - n + 1$) and $\dim \mathcal{C}^o \geqslant n - 1$ (since $|A| = n - 1$).

Let us now prove that $\mathcal{C}$ and $\mathcal{C}^o$ are orthogonal. Let C be a cycle and $C^o = C^o(V_1, V_2)$ a cocycle. Observe that

$$\mu_C = (\mu_C(e_1), \mu_C(e_2), \ldots, \mu_C(e_m)) \in \mathbb{C}^m,$$
$$\mu_{C^o} = (\mu_{C^o}(e_1), \mu_{C^o}(e_2), \ldots, \mu_{C^o}(e_m)) \in \mathbb{C}^m.$$

Then $\langle \mu_C, \mu_{C^o} \rangle = \sum_{i=1}^{m} \mu_C(e_i) \cdot \mu_{C^o}(e_i)$. If $\mu_C(e_i) \cdot \mu_{C^o}(e_i) \neq 0$, then we have $e_i \in C \cap C^o$, and this number is equal to 1 or -1 according to whether the arc e_i goes from V_1 to V_2 or from V_2 to V_1; thus the number of arcs belonging to $C \cap C^o$ is even, because since C is a cycle, every time we pass from V_1 to V_2 we then have to pass back from V_2 to V_1. Thus $\langle \mu_C, \mu_{C^o} \rangle$ is simply the number of arcs in C from V_1 to V_2 minus the number of arcs from V_2 to V_1, so $\langle \mu_C, \mu_{C^o} \rangle = 0$. Thus

$$\dim \mathcal{C} + \dim \mathcal{C}^o \leqslant m = m - n + 1 + n - 1.$$

Hence, by the above, $\dim \mathcal{C} = m - n + 1$ and $\dim \mathcal{C}^o = n - 1$. It follows that $\{\mu_{C_e} \mid e \notin A\}$ forms a basis of the cycle space and $\{\mu_{C_e^o} \mid e \in A\}$ forms a basis of the cocycle space. ∎

Corollary 6.2.4. *Let $\vec{\Gamma} = (V; \vec{E}, N)$ be a loopless digraph with k connected components. Then*

$$\dim_{\mathbb{C}} \mathcal{C} = \mu(\Gamma) = m - n + k, \quad \dim_{\mathbb{C}} \mathcal{C}^o = \rho(\Gamma) = n - k.$$

Exercise 6.3. Prove Corollary 6.2.4.

6.2.2 The case of non-oriented graphs

We can develop an analogous theory for non-oriented graphs simply by deleting the signs; Then, since the coefficients 1 and -1 no longer compensate for each other, we replace the field $\mathbb{C}$ by the field with two elements $\mathbb{F}_2 = \mathbb{Z}/2\mathbb{Z}$.

Let $\Gamma = (V; E, N)$ be a loopless graph; write $V = \{x_1, x_2, \ldots, x_n\}$. We represent each vertex x_i by the vector $\xi_i = (\delta_{ij})_{1 \leqslant j \leqslant n} \in \mathbb{F}_2^n$ by using the Kronecker symbols δ_{ij} in such a way that $\mathcal{B}_n = \{\xi_1, \xi_2, \ldots, \xi_n\}$ is the canonical basis of $\mathbb{F}_2^n$. Identifying the vertices with their corresponding vectors, we see that changing their numbering comes down to permuting the basis vectors. Let $C_0(\Gamma)$ denote this $\mathbb{F}_2$-vector space they generate; we call it the *vertex space of the graph*.

Similarly, let $E = \{a_1, a_2, \ldots, a_m\}$ be a numbering of the edges. We represent each edge a_i by $\varepsilon_i = (\delta_{ij})_{1 \leqslant j \leqslant m} \in \mathbb{F}_2^m$. Again, $\mathcal{B}'_m = \{\varepsilon_1, \varepsilon_2, \ldots, \varepsilon_m\}$ is the canonical basis of $\mathbb{F}_2^m$. We identify a_i with ε_i and write $C_1(\Gamma)$ for the $\mathbb{F}_2$-vector space they generate, which is called the *edge space of the graph.*

The vector $\mu_C = (c_1, c_2, \ldots, c_m) \in C_1(\Gamma)$ associated to the cycle:

$$C = (x_0, a_0, x_1, a_1, \ldots, x_n, a_n, x_0)$$

is defined by

$$c_i = \begin{cases} 1 & \text{if } a_i \in C, \\ 0 & \text{if } a_i \notin C. \end{cases}$$

Clearly two distinct cycles are associated to two distinct vectors. The *cycle space* $\mathcal{C}$ is the $\mathbb{F}_2$-vector subspace of $C_1(\Gamma)$ generated by the set $\{\mu_C \mid C \text{ cycle of } \Gamma\}$.

Exercise 6.4. For every cycle C, let E_C denote the set of edges of C. Let C and C' be two cycles having a single common edge a. Show that the vector $\mu_C + \mu_{C'}$ is the vector associated to a cycle C''. What happens when C and C' have exactly two common edges?

Exercise 6.5. (i) Show that in a graph of finite order, every cycle is an edge-disjoint union of elementary cycles (use induction).

(ii) Deduce that the cycle space is generated by the elementary cycles.

An *algebraic cycle* is an element μ of $C_1(\Gamma)$. If (V_1, V_2) is a partition of the vertices of a connected graph Γ, then the set $C^o = C^o(V_1, V_2)$ of edges between V_1 and V_2 is called a *cocycle*, to which we associate a vector μ_{C^o} of $C_1(\Gamma)$ defined by

$$\mu_{C^o}(a_i) = \begin{cases} 1 & \text{if } a_i \in C^o, \\ 0 & \text{if } a_i \notin C^o. \end{cases}$$

The *cocycle space*, denoted $\mathcal{C}^o$, is the $\mathbb{F}_2$-vector subspace of $C_1(\Gamma)$ generated by the vectors μ_{C^o}, with C^o a cocycle of Γ.

When A is a spanning tree of Γ, we can similarly define fundamental cycles and cocycles for digraphs attached to this tree, denoted respectively C_a for the edges a that are not on the tree A and C_a^o for the a which are on A: C_a is the unique cycle induced by adding the edge a to the tree A, and C_a^o is the cocycle relative to the partition of $A \backslash a$ into two connected components obtained by removing the edge a from A.

Theorem 6.2.5. *Let $\Gamma = (V; E, N)$ be a connected loopless graph, and A a spanning tree of Γ. Then*

(i) *the set* $\{\mu_{C_a} \mid a \in E \setminus A\}$ *forms a basis of the* $\mathbb{F}_2$*-vector cycle space, and*

$$\dim_{\mathbb{F}_2} \mathcal{C} = m - (n-1);$$

(ii) *the set* $\{\mu_{C_a^o} \mid a \in A\}$ *forms a basis of the* $\mathbb{F}_2$*-vector cocycle space, and*

$$\dim_{\mathbb{F}_2} \mathcal{C}^o = n - 1.$$

(iii) *The spaces* $\mathcal{C}$ *and* $\mathcal{C}^o$ *are supplementary in the* $\mathbb{F}_2$*-vector space* $C_1(\Gamma)$ *and orthogonal for the canonical scalar product.*

The number $m - n + 1$ is called the *cyclomatic number* of Γ.

Exercise 6.6. Write out the details of the proof of Theorem 6.2.5.

In the following exercise, we refer back to Remark 6.3.2.

Exercise 6.7. (i) Show that $\mathcal{C}^o = \mathcal{C}^\perp$, the space of vectors orthogonal to $\mathcal{C}$ for the scalar product.

(ii) Let $\varphi_{{}^tJ}$ be the linear map associated to ${}^tJ$. Show that $\operatorname{Im} \varphi_{{}^tJ} = \mathcal{C}^o$.

(iii) Does this hold in the oriented case?

6.3 Circulation

In Chapter 4, we introduced the notion of a flow in a network associated to a digraph $\vec{\Gamma} = (V; \vec{E}, N)$. In a network, we need to consider two particular vertices s and p, the source and the sink. However it is sometimes convenient not to fix a source and a sink: this gives rise to the similar notion of a *circulation* in a digraph $\vec{\Gamma} = (V; \vec{E}, N)$, which is a map $f : \vec{E} \to \mathbb{C}$ satisfying

$$\forall x \in V, \quad \sum_{\substack{e \in \vec{E} \\ i(e) = x}} f(e) = \sum_{\substack{e \in \vec{E} \\ t(e) = x}} f(e).$$

With the conventions above ($\vec{E} = \{e_1, e_2, \dots, e_m\}$, the arc e_j being identified with the j-th vector of the canonical basis of $\mathbb{C}^m$), we can represent f in the form of a column vector of $\mathbb{C}^m$:

$$\tilde{f} = {}^t\big(f(e_1)\ f(e_2)\ \dots\ f(e_m)\big).$$

Proposition 6.3.1. *Let* $\vec{\Gamma} = (V; \vec{E}, N)$ *be a loopless digraph and* J *its incidence matrix. A map* $f : \vec{E} \to \mathbb{C}$ *is a circulation if and only if*

$$J\tilde{f} = 0_{\mathbb{C}^n}.$$

Proof. Let $f\colon \vec{E} \to \mathbb{C}$. The k-th component of $J\tilde{f}$ can be written

$$(J\tilde{f})_k = \sum_{j=1}^{m} \alpha_{k,j} f(e_j).$$

For fixed k, we have $\alpha_{k,j} = 1, -1$ or 0 according to whether $t(e_j) = x_k$ (i.e. x_k is the terminal vertex of the arc e_j), $i(e_j) = x_k$ (i.e. x_k is the initial vertex of the arc e_j) or e_j is not incident to x_k. Thus

$$(J\tilde{f})_k = \sum_{\substack{j=1 \\ t(e_j)=x_k}}^{m} f(e_j) - \sum_{\substack{j=1 \\ i(e_j)=x_k}}^{m} f(e_j),$$

so if f is a circulation, we have

$$(J\tilde{f})_k = \sum_{t(e_j)=x_k} f(e_j) - \sum_{i(e_j)=x_k} f(e_j) = 0,$$

and hence $J\tilde{f} = 0_{\mathbb{C}^n}$.

Conversely, assume that $J\tilde{f} = 0_{\mathbb{C}^n}$. Then for all $k = 1, \dots, n$,

$$0 = (J\tilde{f})_k = \sum_{j=1}^{m} \alpha_{k,j} f(e_j) = \sum_{t(e_j)=x_k} f(e_j) - \sum_{i(e_j)=x_k} f(e_j).$$

Thus $\sum_{i(e)=x} f(e) = \sum_{t(e)=x} f(e)$ for all $x \in V$, so f is a circulation. ∎

Remark 6.3.2. The matrix J corresponds to a linear map $\varphi_J\colon \mathbb{C}^m \to \mathbb{C}^n$: the j-th column of J consists of components of $\varphi_J(e_j)$ in the basis of ξ_k, $1 \leqslant k \leqslant n$. In this column, we find the coefficient 1 if $t(e_j) = x_k$, -1 if $i(e_j) = x_k$ and 0 otherwise. Thus $J\tilde{f} = 0_{\mathbb{C}^n}$ indicates that $\tilde{f} \in \operatorname{Ker}\varphi_J$. Then Proposition 6.3.1 shows that $\operatorname{Ker}\varphi_J$ is equal to $\widetilde{\mathcal{C}}$, where $\widetilde{\mathcal{C}}$ denotes the set of circulations of $\vec{\Gamma}$, so this set $\widetilde{\mathcal{C}}$ is a vector subspace of $\mathbb{C}^m$. Furthermore, we have the isomorphism $\mathbb{C}^m/\operatorname{Ker}\varphi_J \cong \operatorname{Im}\varphi_J$, so we have the equality of dimensions $m - \dim\widetilde{\mathcal{C}} = \dim\operatorname{Im}\varphi_J = \operatorname{rank}(J) = n - k$ (see Proposition 6.1.10), and this is then equal to $m - \dim\mathcal{C}$ by Corollary 6.2.4. Thus $\dim\widetilde{\mathcal{C}} = \dim\mathcal{C}$. Actually, we even have the equality $\widetilde{\mathcal{C}} = \mathcal{C}$, as asserted in the following theorem.

Theorem 6.3.3. *Let $\vec{\Gamma} = (V; \vec{E}, N)$ be a loopless digraph. Then $\widetilde{\mathcal{C}} = \mathcal{C}$, i.e. the space of circulations is equal to the cycle space.*

Proof. By Remark 6.3.2, it suffices to show that $\mathcal{C} \subset \widetilde{\mathcal{C}}$ (or the inverse). Let

$$C = (x_0, a_0, x_1, a_1, \dots, x_n, a_n, x_0)$$

be a cycle in the underlying graph associated to the vector $\mu_C = (c_1, \dots, c_m)$ of $\mathcal{C}$. Let us compute $J\mu_C$. The k-th line of the matrix J is associated to the vertex x_k. If this vertex is not on the cycle C, then $(J\mu_C)_k = 0$. Otherwise x_k is incident to exactly two edges of C, and these edges correspond to two arcs. Considering the orientation of these arcs and their incidence at x_k, particularly whether x_k is the initial or terminal extremity of the arcs, we find that in every case, we have $(J\mu_C)_k = 0$. Thus μ_C corresponds to a circulation f, and $f(e_j) = c_j$ for all $j = 1, \dots, m$. ∎

Corollary 6.3.4. (i) *The kernel of the map* $\varphi_J : C_1(\vec{\Gamma}) \to C_0(\vec{\Gamma})$ *is* $\operatorname{Ker} \varphi_J = \mathcal{C}$.

(ii) *If the underlying graph of* $\vec{\Gamma}$ *is a tree, then* $\widetilde{C} = \{0_{\mathbb{C}^m}\}$.

Proof. (i) is clear. For (ii), observe that if Γ is a tree, we have $\mathcal{C} = \{0_{\mathbb{C}^m}\}$, so $\widetilde{\mathcal{C}} = \{0_{\mathbb{C}^m}\}$. ∎

The final result of this paragraph shows that the arcs of a digraph can be partitioned in such a way that they belong either to a cycle or to a cocycle. We first prove a lemma.

Lemma 6.3.5 (*Coloured arcs lemma*, Minty, 1960). *Let* $\vec{\Gamma} = (V; \vec{E}, N)$ *be a digraph with no loops whose arcs are coloured arbitrarily in black, red or green. Let* e_1 *be a black arc. Then one and only one of the following two assertions holds.*

(1) *There exists an (oriented) elementary cycle* C *containing* e_1 *all of whose arcs are either black or red and such that all the black arcs are oriented in the same direction.*

(2) *There exists a cocycle* C^o *containing* e_1 *all of whose arcs are either black or green and such that all the black arcs are oriented in the same direction.*

Proof. Let $e_1 = ((x_1, y_1), n_1)$. We will mark all the vertices successively, starting from the vertex y_1:

(a) mark y_1;

(b) if the vertex u is already marked and if v is a vertex that is not yet marked, we mark v if and only if

- either there exists a black arc from u to v,
- or there exists a red arc from u to v or from v to u.

The process stops when no more vertices can be marked by rule (b). We distinguish two cases.

(i) Suppose we have marked the vertex x_1. Then rule (b) implies that there exists an oriented elementary chain from y_1 to x_1 containing only red or black arcs, all of whose black arcs are oriented in the same direction. Adding the arc e_1, we obtain a cycle all of whose arcs are either black or red, such that all the black arcs are oriented in the same direction, so (1) is satisfied.

(ii) Suppose that the vertex x_1 was not marked in the procedure. Let X denote the set of vertices that were already marked by the procedure. Then $y_1 \in X$ and $x_1 \in V\backslash X$. The cocycle C^o associated to this partition contains no black arc whose initial vertex lies in X, and C^o contains no red arc. Consequently, C^o is a cocycle all of whose black arcs are oriented in the same direction (towards X) and whose other arcs are coloured green. Thus C^o satisfies (2).

Let us show now that (1) and (2) are mutually exclusive. Suppose there is a cycle C satisfying (1) and a cocycle $C^o = C^o(X, V\backslash X)$ satisfying (2). We then have $e_1 \in C \cap C^o$, so C^o must contain another arc of C since $y_1 \in X$ and $x_1 \in V\backslash X$. Let e_2 denote this arc. By (2), it cannot be red. Suppose e_2 is black; then e_2 has the same direction as e_1 in C^o. But then e_2 is oriented in the opposite direction to e_1, which is impossible by (1). Thus e_2 is green, which is impossible since $e_2 \in C$, and we have obtained a contradiction. ■

A cocycle is a *cocircuit* if all its arcs are oriented in the same direction.

Corollary 6.3.6. *Every arc of a digraph belongs either to a circuit or to a cocircuit.*

Proof. Colour all the arcs black and apply Lemma 6.3.5. ■

6.4 Planar graphs

We introduced the property of planarity of graphs in Chapter 5, essentially using topological and combinatorial tools. We will now study the same property using linear algebra.

A non-empty cocycle $C^o = C^o(V_1, V_2)$ is said to be *minimal* if C^o contains no non-empty proper subset defining a cocycle. This concept is illustrated in Figure 6.5.

Exercise 6.8. Show that in a connected graph, $C^o(V_1, V_2)$ is minimal if and only if V_1 and V_2 are connected.

For example, the C_e^o associated to a spanning tree A are minimal cocycles; by Theorems 6.2.3 and 6.2.5, these minimal cocycles generate the cocycle space $\mathcal{C}^o$.

Example 6.4.1. Figure 6.5 shows the same graph twice. In figure (A), the cocycle $C^o(X, \overline{X})$ consisting of the edges a_1, a_2, a_3 is not minimal, since in figure (B) we have a cocycle $C^o(Y, \overline{Y})$ consisting of a_1, a_2, which is included in a_1, a_2, a_3.

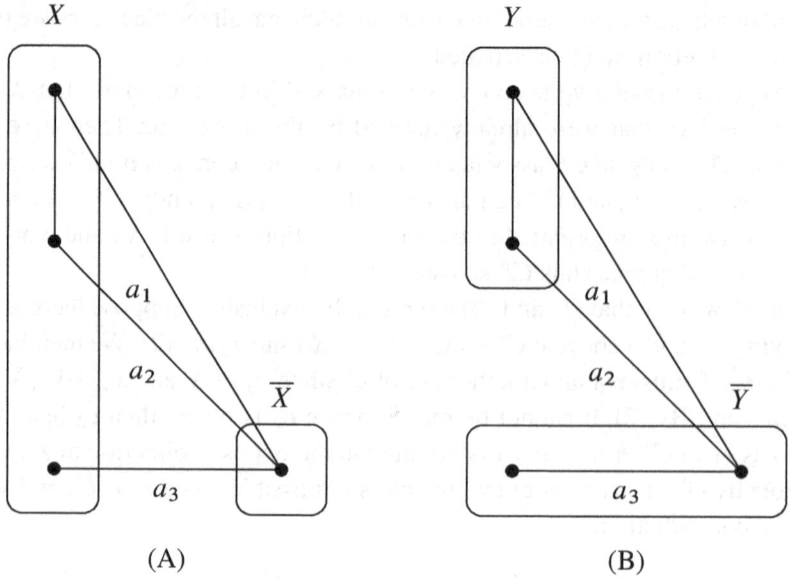

Figure 6.5. Minimal cocycle in a graph.

Exercise 6.9. Show that if $\kappa(\Gamma) \geqslant 2$, then for all $x \in V$ the set $E(x)$ of edges adjacent to x is a minimal cocycle (here graphs are assumed loopless). Compare this situation to the one shown in Figure 6.5.

Let $\Gamma = (V; E, N)$ be a connected planar embedded graph in $\mathbb{R}^2$ and $\Gamma^* = (V^*; E^*, N^*)$ its topological dual (see Section 5.6). Let C be an elementary cycle of Γ; it is a Jordan curve (i.e. a simple closed curve) that divides the plane into two connected components. The faces in the interior of C form a subset V_1 of Γ^*, and the exterior faces correspond to V_2. Consequently, to the cycle C there corresponds the minimal cocycle $C^o(V_1, V_2)$ of Γ^*. Now let C be a set of edges of Γ of which no subset forms a cycle. Then the graph induced by C is a forest and possesses only a single face (see Proposition 5.2.10). Thus each pair of vertices of Γ^* (i.e. each pair of faces of Γ) can be connected by a chain of which no edge (in Γ^*) corresponds to an edge of C. It follows that no non-empty subset of edges of C corresponds to a minimal cocycle of Γ^*.

This proves the following proposition.

Proposition 6.4.2. *Let $\Gamma = (V; E, N)$ be a connected planar graph embedded in $\mathbb{R}^2$ and $\Gamma^* = (V^*; E^*, N^*)$ its topological dual. Then a set of edges C forms a cycle of Γ if and only if the set of edges C^* of Γ^* corresponding to the edges of C forms a cocycle.*

Thus by identifying the edges of Γ and Γ^*, since $\mathfrak{C}$ is generated by these cycles and $\mathfrak{C}^o(\Gamma^*)$ is generated by the minimal cocycles, we have $\mathfrak{C}^o(\Gamma^*) = \mathfrak{C}(\Gamma)$.

The correspondence above is even more precise, since the elementary cycles of Γ are exactly the minimal cocycles of Γ^*. This suggests the following definition. Let $\Gamma = (V; E, N)$ be a connected loopless graph. We will say that $\Gamma^c = (V^c; E^c, N^c)$ is a *combinatorial dual* of Γ if there exists a bijection $f: E \to E^c$ such that

$$f(\mathfrak{C}(\Gamma)) = \mathfrak{C}^o(\Gamma^c),$$

where this equality is justified by the fact that an elementary cycle C is entirely determined by the set of its edges $E(C)$. Thus when Γ is planar, Γ^* is the combinatorial dual of Γ.

The combinatorial dual possesses the following properties.

Lemma 6.4.3. (i) $|E^c| = |E|$, $|V^c| = |E| - |V| + 2$;

(ii) $\delta(\Gamma^c) \geqslant \mu(\Gamma)$;

(iii) *every partial graph of* Γ *has a combinatorial dual;*

(iv) *every subdivision of* Γ *has a combinatorial dual;*

(v) *every contraction of* Γ *has a combinatorial dual.*

Proof. (i) By Theorem 6.2.5, $\dim \mathfrak{C}^o(\Gamma^c) = |V^c| - 1 = \dim \mathfrak{C}(\Gamma) = |E| - |V| + 1$.

(ii) Let $x \in V(\Gamma^c)$ be of degree $\delta(\Gamma^c)$. The set of edges adjacent to x is a cocycle, to which there corresponds a cycle of Γ with $\delta(\Gamma^c)$ edges, so $\mu(\Gamma) \leqslant \delta(\Gamma^c)$.

(iii) If we remove the edge e from Γ and contract the edge $f(e)$ of Γ^c, then f yields a bijection between $E(\Gamma \backslash e)$ and $E(\Gamma^c / f(e))$ under which the cycles of $\Gamma \backslash e$ are sent onto the cycles of $\Gamma^c / f(e)$ (we have $\dim \mathfrak{C}(\Gamma \backslash e) = \dim \mathfrak{C}(\Gamma) - 1$ and $\dim \mathfrak{C}^o(\Gamma^c / f(e)) = |V^c| - 2 = \dim \mathfrak{C}^o(\Gamma^c) - 1$).

(iv) If we add a vertex to the middle of the edge e, this replaces e by two edges e' and e''. In Γ^c, we double the edge $f(e)$ in such a way that if Γ_+ and Γ_+^c denote the graphs thus obtained, then $\dim \mathfrak{C}(\Gamma_+) = \dim \mathfrak{C}(\Gamma)$ and $\dim \mathfrak{C}^o(\Gamma_+^c) = \dim \mathfrak{C}^o(\Gamma)$. We deduce from f an obvious bijection $E(\Gamma_+) \to E(\Gamma_+^c)$ sending cycles to cocycles. It is even possible to show that doubling an edge e in Γ corresponds to a subdivision in Γ^c.

(v) If we contract e in Γ and remove $f(e)$ from Γ^c, f again gives a bijection between $E(\Gamma / e)$ and $E(\Gamma^c \backslash f(e))$, under which cycles of Γ / e are sent to cocycles of $\Gamma^c \backslash f(e)$ (we have $\dim \mathfrak{C}(\Gamma / e) = \dim \mathfrak{C}(\Gamma)$ and $\dim \mathfrak{C}^o(\Gamma^c \backslash f(e)) = \dim \mathfrak{C}^o(\Gamma^c)$). ∎

Let us now give the main result of this paragraph, a new characterisation of planarity that is algebro-combinatorial in nature.

Theorem 6.4.4 (Whitney, 1933). *Let $\Gamma = (V; E, N)$ be a connected loopless graph. Then Γ is planar if and only if it has a combinatorial dual.*

Proof. Let $\Gamma = (V; E, N)$ be a connected loopless graph. If Γ is planar, we saw that Γ^* is a combinatorial dual of Γ.

Conversely, let Γ^c be a combinatorial dual of Γ, and suppose that Γ is not planar. Then by Wagner's theorem (Theorem 5.5.5), Γ can be contracted to a graph that contains $K_{3,3}$ or K_5. By Lemma 6.4.3 (v), it suffices to prove that neither $K_{3,3}$ nor K_5 admits a combinatorial dual.

If $\Gamma^c = (V^c; E^c, N^c)$ is a combinatorial dual of K_5, we have

$$|V^c| = |E| - |V| + 2 = 10 - 5 + 2 = 7 \quad \text{and} \quad \mu(K_5) = 3.$$

But since

$$\sum_{x \in V^c} d(x) = 2|E^c|,$$

it follows that $\delta(\Gamma^c) \times |V^c| \leqslant 2|E^c|$. But then with Lemma 6.4.3 (ii), we have

$$|V^c| \leqslant \frac{2|E^c|}{\delta(\Gamma^c)} \leqslant \frac{2|E^c|}{\mu(\Gamma)} = \frac{2|E|}{3} = \frac{20}{3} < 7 = |V^c|,$$

which is absurd. If Γ^c is a combinatorial dual of $K_{3,3}$, then

$$|V^c| = |E| - |V| + 2 = 9 - 6 + 2 = 5 \quad \text{and} \quad \mu(K_{3,3}) = 4;$$

just as above, we have

$$|V^c| \leqslant \frac{2|E^c|}{\delta(\Gamma^c)} \leqslant \frac{2|E^c|}{\mu(\Gamma)} = \frac{2|E|}{4} = \frac{18}{4} < 5 = |V^c|,$$

which is again absurd. This concludes the proof. ∎

There exists another criterion of planarity which is even more algebraic in nature, related to the following observation. The cocycle space $\mathscr{C}^o(\Gamma)$ of a graph Γ is generated by the set of cocycles $E(x) = C^o(\{x\}, V \backslash \{x\})$, $x \in V$ (exercise). When Γ is planar, we can apply this observation to Γ^*: its cocycles $E(\varphi)_{\varphi \in \Phi}$ are such that every edge belongs to at most two $E(\varphi)$, since every edge is adjacent to at most two faces (see Proposition 5.2.8). Since on the one hand $\mathscr{C}(\Gamma) = \mathscr{C}^o(\Gamma^*)$ and on the other, we can extract a basis from the generating set $E(\varphi)_{\varphi \in \Phi}$, this proves the following theorem.

Theorem 6.4.5 (MacLane, 1937). *Let $\Gamma = (V; E, N)$ be a connected loopless graph. Then Γ is planar if and only if the cycle space possesses a simple basis, i.e. a basis such that every edge of Γ is adjacent to at most two basis elements.*

6.5 Background in linear algebra

6.5.1 Vector spaces

A *ring* is a non-empty set A equipped with two internal composition laws, written as addition and multiplication, satisfying the following properties:

- the set A forms an abelian group (cf. the annex of Chapter 1) for the additive law, whose neutral element is written 0;
- the multiplication law is associative;
- the multiplication law is right and left distributive with respect to the addition law.

If the multiplication law of A admits a neutral (or unit) element 1, we say that A is a *unitary ring*. A unitary ring is *commutative* if the multiplication law is commutative. If all the non-zero elements of a unitary ring A are invertible, then A is a *field*. The fields that we will encounter in this book are mainly $K = \mathbb{R}$, $\mathbb{C}$ or $\mathbb{F}_2$ (the field with only two elements).

Let K be a field. A set E is a K*-vector space* if it is equipped with an internal composition law, written additively, and an external law $K \times E \to E$, $(a, v) \mapsto av$, written multiplicatively, that satisfy the following properties:

- the set A is an abelian group for the internal additive law, whose neutral element is written 0_E;
- if 1 is the unit element of K, then $1v = v$ for all $v \in E$;
- $a(bv) = (ab)v$ for all $a, b \in K$ and all $v \in E$;
- $(a + b)v = av + bv$ for all $a, b \in K$ and all $v \in E$;
- $a(u + v) = au + av$ for all $a \in K$ and all $u, v \in E$.

The elements of K are called *scalars*, and K is called the *base field* of the vector space E. The elements of E are called *vectors*.

The simplest examples are $\mathbb{R}^n$, $\mathbb{C}^n$ and $\mathbb{F}_2^n$, which are vector spaces over $\mathbb{R}$, $\mathbb{C}$ and $\mathbb{F}_2$ respectively.

Let E be a K-vector space and let $F \subseteq E$ be a non-empty subset. Then F is a *vector subspace* of E if the restriction to F of the laws of E makes F into a vector space over K.

A family $F = \{v_1, v_2, v_3, \ldots, v_k\}$ of elements of a K-vector space E is a *generating family* if for every $u \in E$, there exist $a_1, a_2, \ldots, a_k \in K$ such that

$$u = \sum_{i=1}^{k} a_i v_i .$$

This type of expression is called a K*-linear combination* of vectors of E.

A family $F = \{v_1, v_2, v_3, \ldots, v_k\}$ of vectors in a K-vector space E is *linearly independent* if

$$\left.\begin{array}{r} \exists a_1, a_2, \ldots, a_k \in K \\ \text{such that } \displaystyle\sum_{i=1}^{k} a_i v_i = 0_E \end{array}\right\} \Longrightarrow a_1 = a_2 = \cdots = a_k = 0.$$

A *basis* of a K-vector space E is a family that is both linearly independent and generating. Thus, a family $B = \{v_1, v_2, v_3, \ldots, v_k\}$ or $(v_1, v_2, v_3, \ldots, v_k)$ if the order of which the n vectors are listed needs to be fixed, of a K-vector space E is a basis if for all $u \in E$, there exists a *unique* k-tuple of scalars $a_1, a_2, \ldots, a_k \in K$ such that

$$u = \sum_{i=1}^{k} a_i v_i.$$

Let E be a K-vector space equipped with a basis containing a finite number of elements n. Then every basis of E contains exactly n elements. This number is called the *dimension* of E over K, and is denoted $\dim_K E$.

Let E and F be two K-vector spaces and let f be a map from E to F. Then f is a *K-linear map* if

- $f(u + v) = f(u) + f(v)$ for all $u, v \in E$;
- $f(au) = af(u)$ for all $a \in K$ and all $u \in E$.

The set $\mathcal{A}(E, F)$ of maps from E to F forms a K-vector space, and the subset of K-linear maps, denoted $\mathcal{L}(E, F)$, is a vector subspace of $\mathcal{A}(E, F)$. A K-linear map from E to E is called an *endomorphism* of E. If a K-linear map f from E to F is also a bijection, we say that f is an *isomorphism* of K-vector spaces, and that E and F are *isomorphic*. Any two K-vector spaces of the same dimension (over K) are isomorphic.

Let $f \in \mathcal{L}(E, F)$; the *image* of f is the set

$$\mathrm{Im}(f) = \{f(u) \mid u \in E\}.$$

It is a vector subspace of F. The *rank* of f is the dimension of $\mathrm{Im}(f)$.

The *kernel* of f is the set

$$\mathrm{Ker}(f) = \{u \in E \mid f(u) = 0\}.$$

It is also a vector subspace of E.

If F is a vector subspace of E, then the structure of E induces a K-vector space structure on the quotient E/F: $(u + F) + \lambda(v + F) = (u + \lambda v) + F$ for all $u, v \in E$ and $\lambda \in K$. There is an isomorphism $E/\mathrm{Ker}\, f \simeq \mathrm{Im}\, f$. The image and the kernel of a group morphism are defined similarly.

Let E_1 and E_2 be two vector subspaces of a K-vector space E. Define the *sum* of E_1 and E_2 by

$$E_1 + E_2 = \{u_1 + u_2 \mid u_1 \in E_1, u_2 \in E_2\}.$$

If for all $u \in E_1 + E_2$ the decomposition $u = u_1 + u_2$ with $u_1 \in E_1$ and $u_2 \in E_2$ is unique, we say that the sum is *direct*, and write $E_1 \oplus E_2$. The vector subspaces E_1 and E_2 are said to be *supplementary* in E if $E = E_1 \oplus E_2$.

Let E_1 and E_2 be two vector subspaces of a K-vector space E. Then $E_1 + E_2$ is a direct sum if and only if $E_1 \cap E_2 = \{0_E\}$. In this case, we have the identity $\dim_K E_1 + \dim_K E_2 = \dim_K E$.

If $f \in \mathcal{L}(E, F)$, we have the *rank formula*

$$\dim_K E = \dim_K \operatorname{Im}(f) + \dim_K \operatorname{Ker}(f).$$

6.5.2 Matrices

A *matrix* of type $m \times n$ over a field K is an ordered set of n column vectors of length m or m row vectors of length n, with coefficients in the base field K. Thus, a matrix is a table of $m \times n$ elements of K. A matrix is written as follows:

$$A = (a_{i,j})_{\substack{1\leqslant i\leqslant m\\ 1\leqslant j\leqslant n}} = \begin{pmatrix} a_{1,1} & a_{1,2} & a_{1,3} & \dots & a_{1,n} \\ a_{2,1} & a_{2,2} & a_{2,3} & \dots & a_{2,n} \\ a_{3,1} & a_{3,2} & a_{3,3} & \dots & a_{3,n} \\ \vdots & \vdots & \vdots & & \vdots \\ a_{m,1} & a_{m,2} & a_{m,3} & \dots & a_{m,n} \end{pmatrix}.$$

The elements $a_{i,j}$ are the *coefficients* of the matrix A. The set of $m \times n$ matrices is denoted $K^{m,n}$ or else $\mathcal{M}_{m,n}(K)$. When $n = 1$, we shall often make use of the identification of n-tuples (that is elements of K^m) with the corresponding column vector of $K^{m,1}$. So if $x = (x_1, \dots, x_n) \in K^n$ and $A \in K^{m,n}$, then the product Ax means the m-tuple $\big(\sum_{j=1}^n a_{k,j} x_j\big)_{1\leqslant k\leqslant m}$.

If $m = n$, we say that the matrix A is *square*. The *transpose matrix* of the matrix A is the matrix

$${}^tA = (a_{j,i})_{\substack{1\leqslant j\leqslant n\\ 1\leqslant i\leqslant m}} = \begin{pmatrix} a_{1,1} & a_{2,1} & a_{3,1} & \dots & a_{m,1} \\ a_{1,2} & a_{2,2} & a_{3,2} & \dots & a_{m,2} \\ a_{1,3} & a_{2,3} & a_{3,3} & \dots & a_{m,3} \\ \vdots & \vdots & \vdots & & \vdots \\ a_{1,n} & a_{2,n} & a_{3,n} & \dots & a_{m,n} \end{pmatrix}.$$

The transpose is thus an element of $K^{n,m}$.

Let A be a matrix of $K^{m,n}$. A matrix is said to be *extracted* from A if it is obtained by deleting whole rows and/or whole columns of A.

Let $A = (a_{i,j}) \in K^{m,n}$ and $B = (b_{i,j}) \in K^{n,p}$. The *product matrix* $C = AB = (c_{i,j}) \in K^{m,p}$ is defined by

$$c_{i,j} = \sum_{k=1}^{n} a_{i,k} b_{k,j}, \text{ for } 1 \leqslant i \leqslant m \text{ and } 1 \leqslant j \leqslant p.$$

Observe that the matrix A^tA is an $m \times m$ square matrix.

Let E be a K-vector space of dimension n and F a K-vector space of dimension m. Let $U = (u_1, u_2, \dots, u_n)$ be a basis of E, and let $V = (v_1, \dots, v_m)$ be a basis of F. Let f be a linear map from E to F. Define matrix coefficients $a_{i,j}$, $i = 1, \dots, m$, $j = 1, \dots, n$, by denoting $a_{i,j}$ the component of $f(u_j)$ along v_i, $1 \leqslant j \leqslant n$, $1 \leqslant i \leqslant m$:

$$f(u_j) = \sum_{i=1}^{m} a_{i,j} v_i, \quad 1 \leqslant j \leqslant n.$$

The matrix $A = (a_{i,j}) \in K^{m,n}$ formed by the coefficients $a_{i,j}$ is called the *matrix associated to the linear map* f relative to the bases U and V. This matrix completely determines the linear map f.

Conversely, let $A \in K^{m,n}$ be a matrix. Then the map $K^{n,1} \to K^{m,1}$, $x \mapsto Ax$ (the matrix product of A by the column vector x) defines a linear map. Observe that the j-th column of A is Ae_j, where

$$e_j = {}^t(\underbrace{0\,0\,0 \dots 0}_{j-1}\,1\,\underbrace{0 \dots 0}_{n-j}) \in K^{n,1}.$$

The vectors $e_1, e_2, e_3, \dots, e_n$ form a basis called the *canonical basis* of $K^{n,1}$ which can be identified with K^n.

Let $(f_1, f_2, f_3, \dots, f_m)$ be the canonical basis of $K^{m,1}$. Then

$$Ae_j = \sum_{k=1}^{n} a_{k,j} f_k, \quad j = 1, \dots, m.$$

Thus, if $x = {}^t(x_1\, x_2\, \dots\, x_n) \in K^{n,1}$, then the x_i, $i = 1, \dots, n$, are the coordinates of the vector x in the basis $(e_1, e_2, e_3, \dots, e_n)$. If $y = Ax = {}^t(y_1\, y_2\, \dots\, y_m) \in K^{m,1}$, then the y_i are the coordinates of the vector Ax in the basis $(f_1, f_2, f_3, \dots, f_m)$.

Let E be a K-vector space of dimension n and $U = (u_1, u_2, \dots, u_n)$ a basis of E. Let $V = (v_1, v_2, \dots, v_n)$ be another basis of E. We can write

$$u_j = \sum_{i=1}^{m} p_{i,j} v_i, \quad j = 1, \dots, n, \quad \text{where } p_{i,j} \in K.$$

The matrix $P = (p_{i,j})_{1\leqslant i,j\leqslant n}$ is called the *transition matrix* from the basis V to the basis U. If $x \in E$ is represented in the basis U by X_U and in the basis V by X_V, then we have $X_V = PX_U$.

The *determinant of a square matrix* is defined recursively as follows:

- the determinant of the 1×1 matrix $A = (a)$ is $\det(a) = a$;
- the determinant of the 2×2 square matrix

$$A = \begin{pmatrix} a_{1,1} & a_{1,2} \\ a_{2,1} & a_{2,2} \end{pmatrix}$$

  is

$$\det(A) = \begin{vmatrix} a_{1,1} & a_{1,2} \\ a_{2,1} & a_{2,2} \end{vmatrix} = a_{1,1}a_{2,2} - a_{2,1}a_{1,2};$$

- the determinant of the $n \times n$ square matrix, $n \geqslant 1$,

$$A = (a_{i,j}) = \begin{pmatrix} a_{1,1} & a_{1,2} & a_{1,3} & \dots & a_{1,n} \\ a_{2,1} & a_{2,2} & a_{2,3} & \dots & a_{2,n} \\ a_{3,1} & a_{3,2} & a_{3,3} & \dots & a_{3,n} \\ \vdots & \vdots & \vdots & & \vdots \\ a_{n,1} & a_{n,2} & a_{n,3} & \dots & a_{n,n} \end{pmatrix}$$

  is given by

$$\det(A) = \sum_{j=1}^{n} (-1)^{j+1} a_{1,j} \Delta_{1,j},$$

  where $\Delta_{i,j}$ is the determinant of the extracted square matrix $A_{i,j}$ obtained by deleting the i-th row and the j-th column from A. The determinant of a matrix can be expanded starting from any row or column of A:

$$\det(A) = \sum_{j=1}^{n} (-1)^{i+j} a_{i,j} \Delta_{i,j} = \sum_{i=1}^{n} (-1)^{i+j} a_{i,j} \Delta_{i,j}.$$

For $n \geqslant 1$, let I denote the diagonal square matrix of order n whose diagonal contains only 1's. This matrix is called the *identity matrix* of order n. A square matrix A of order n, $n \geqslant 1$, is said to be *invertible* if there exists a square matrix B of order n such that

$$AB = BA = I,$$

i.e. $\sum_{k=1}^{n} a_{i,k} b_{k,j} = \sum_{k=1}^{n} b_{i,k} a_{k,j} = \delta_{ij}$ (the Kronecker symbol). The matrix B is called the *inverse* of A, and denoted A^{-1}.

We also have the property $\det(AB) = \det(A) \times \det(B)$, so that the map det defines a group homomorphism from the group of invertible square matrices of order n to the multiplicative group $K\backslash\{0\}$. In particular, if P is an invertible square matrix, we have $\det(I) = 1$ and $\det(P^{-1}) = (\det(P))^{-1}$.

Thus, by fixing a basis $U = (u_1, u_2, \dots, u_n)$ of E, we can associate a square matrix A_U to every endomorphism f of E, called the matrix of f relative to the basis U: A_U is the matrix $(a_{i,j})_{1\leqslant i,j\leqslant n}$ defined by

$$f(u_j) = \sum_{i=1}^{n} a_{i,j} u_i, \quad j = 1, \dots, n.$$

To f we can thus also associate the determinant $\det(A_U)$. One can show that this determinant does not depend on the chosen basis of E, but only on f, so we write it as $\det(f)$. If U and V are two bases of E and P is the transition matrix from V to U, and if A_U denotes the matrix of f relative to U and A_V the matrix of f relative to V, then we have $A_U = P^{-1} A_V P$, and passing to the determinant, we have $\det(A_U) = \det(A_V)$.

6.5.3 Scalar products

We can equip $\mathbb{C}^n$ with the *canonical scalar product* defined as follows: for $z = (z_1, \dots, z_n)$ and $z' = (z'_1, \dots, z'_n) \in \mathbb{C}^n$, set

$$\langle z, z' \rangle = \sum_{i=1}^{n} z_i \overline{z'_i} \quad \text{(where } \overline{z'_i} \text{ denotes the complex conjugate of } z'_i\text{)}.$$

It satisfies

(i) $\langle z, z' + z'' \rangle = \langle z, z' \rangle + \langle z, z'' \rangle, \forall z, z', z'' \in \mathbb{C}^n$;

(ii) $\langle \lambda z, z' \rangle = \lambda \langle z, z' \rangle, \forall z, z' \in \mathbb{C}^n, \forall \lambda \in K$;

(iii) $\langle z', z \rangle = \overline{\langle z, z' \rangle}, \forall z, z' \in \mathbb{C}^n$ (*Hermitian symmetry*);

(iv) if $\langle z, w \rangle = 0$ for all $w \in \mathbb{C}^n$, then $z = 0_{\mathbb{C}^n}$ (*non-degeneracy*).

This scalar product defines a norm (the *Hermitian norm* on $\mathbb{C}^n$) via the formula

$$\|z\| = \sqrt{\langle z, z \rangle} = \Big(\sum_{i=1}^{n} |z_i|^2\Big)^{1/2};$$

the associated distance $d(x, y) = \|x - y\|$ gives $\mathbb{C}^n$ the structure of a metric space. The linear space $\mathbb{C}^n$ equipped with its canonical scalar product is said to be a *Hermitian space*.

In $\mathbb{R}^n$, we proceed similarly but without complex conjugation: if $x = (x_1, \dots, x_n)$ and $x' = (x'_1, \dots, x'_n) \in \mathbb{R}^n$, we set

$$\langle x, x' \rangle = \sum_{i=1}^{n} x_i x'_i.$$

Properties (i), (ii), (iii) and (iv) still hold, and the Hermitian symmetry (iii) becomes

(iii′) $\langle x', x\rangle = \langle x, x'\rangle$ (*symmetry*).

The norm induced by this scalar product is the *Euclidean norm* on $\mathbb{R}^n$.

In the case of $\mathbb{F}_2^n$, the scalar product is defined as for $\mathbb{R}^n$, and properties (i), (ii), (iii′) and (iv) still hold.

For an arbitrary field K and a subset A of K^n equipped with the canonical scalar product, the *orthogonal* set of A is the set

$$A^{\perp} = \{x \in K^n \mid \forall a \in A,\ \langle x, a\rangle = 0\}.$$

It forms a vector subspace of K^n. Let $B = (e_1, e_2, \dots, e_n)$ be a basis of K^n. Then B is said to be *orthonormal* if

$$\langle e_i, e_j\rangle = \begin{cases} 1 & \text{if } i = j, \\ 0 & \text{if } i \neq j. \end{cases}$$

The canonical basis of K^n is an orthonormal basis. Note that, when

$$x = \sum_{i=1}^{n} x_i e_i \in K^n,$$

then $\langle x, e_i\rangle = x_i$, $1 \leqslant i \leqslant n$.

6.5.4 Endomorphisms of the Hermitian space $\mathbb{C}^n$

We suppose that $\mathbb{C}^n$ is equipped with its canonical scalar product. Let $u \in \mathcal{L}(\mathbb{C}^n)$ be an endomorphism of $\mathbb{C}^n$. The *characteristic polynomial* of u is defined by

$$\Phi(u, X) = \det(XI_n - A),$$

where A is the matrix of u in an arbitrary basis of $\mathbb{C}^n$ and I_n denotes the identity matrix of $\mathbb{C}^n$ (one can show that this determinant does not depend on the basis in which u is represented). It is a monic polynomial of degree n, and thus by d'Alembert's theorem, it has n roots in $\mathbb{C}$ (counting multiplicities).

A value $\lambda \in \mathbb{C}$ is an *eigenvalue* of u if there exists $x \neq 0_{\mathbb{C}^n}$ such that $u(x) = \lambda x$; such a vector $x \in \mathbb{C}^n$ is called an *eigenvector* associated to the eigenvalue λ. The *eigenspace* associated to λ is $E_\lambda = \{z \in \mathbb{C}^n \mid u(z) = \lambda z\}$; if there are p distinct eigenvalues $\lambda_1, \dots, \lambda_p$ with respective multiplicities $m_1, \dots, m_p$, then $n = \sum_{1 \leqslant i \leqslant p} m_i$. The set of eigenvalues is often called the *spectrum* of u and denoted $\mathrm{Sp}(u)$. The n roots of the characteristic polynomial are exactly the eigenvalues of u.

The *adjoint* of the endomorphism $u \in \mathcal{L}(\mathbb{C}^n)$ is the endomorphism u^* defined by

$$\forall x, y \in \mathbb{C}^n, \quad \langle u(x), y\rangle = \langle x, u^*(y)\rangle.$$

Clearly, we have $(u^*)^* = u$.

An endomorphism u is a *Hermitian operator* if $u^* = u$. Let $\mathcal{H} = \mathcal{H}(\mathbb{C}^n)$ denote the set of Hermitian endomorphisms of $\mathbb{C}^n$. The Hermitian endomorphism u is said to be *positive* if $\langle u(x), x\rangle \geqslant 0$ for all $x \in \mathbb{C}^n$: we write $u \geqslant 0$.

When we fix an orthonormal basis $\mathfrak{B} = (b_1, \dots, b_n)$ of $\mathbb{C}^n$ and represent u in this basis by the matrix $A = (a_{i,j})$, then u^* is represented by $A^* = (\overline{a}_{j,i})$, called the *adjoint* matrix of A; in other words $A^* = {}^t\overline{A}$. Indeed, since $u(b_j) = \sum_{1\leqslant i\leqslant n} a_{i,j} b_i$, we have

$$\langle u(b_j), b_k\rangle = \Big\langle \sum_{1\leqslant i\leqslant n} a_{i,j} b_i, b_k\Big\rangle = \sum_{1\leqslant i\leqslant n} a_{ij}\langle b_i, b_k\rangle = a_{k,j},$$

and similarly, writing $u^*(b_j) = \sum_{1\leqslant i\leqslant n} a^*_{i,j} b_i$, we obtain

$$\langle u^*(b_j), b_k\rangle = \Big\langle \sum_{1\leqslant i\leqslant n} a^*_{i,j} b_i, b_k\Big\rangle = \sum_{1\leqslant i\leqslant n} a^*_{i,j}\langle b_i, b_k\rangle = a^*_{k,j},$$

so

$$\overline{a^*_{j,k}} = \overline{\langle u^*(b_j), b_k\rangle} = \langle b_k, u^*(b_j)\rangle = \langle u(b_k), b_j\rangle = a_{j,k}.$$

Thus $\overline{A^*} = {}^tA$, so $A^* = {}^t\overline{A}$.

It follows that if u is Hermitian, it is represented in an orthonormal basis by a *Hermitian matrix*: $A = {}^t\overline{A} = A^*$. Conversely, if A is a Hermitian matrix (i.e. if $A = A^*$) and if we take an orthonormal basis $\mathfrak{B}$, then A can be seen as the matrix in this basis of a Hermitian endomorphism u defined as follows: for $x = \sum_{1\leqslant i\leqslant n} x_i b_i$, we set $u(x) = A\,{}^t(x_1 \dots x_n)$, so

$$\langle x, u(y)\rangle = \langle x, A\,{}^ty\rangle = \langle x, A^*\,{}^ty\rangle = \langle x, u^*(y)\rangle,$$

but then $u = u^*$ and u is a Hermitian endomorphism.

Remark 6.5.1. Every symmetric matrix A with real coefficients is Hermitian: $A = {}^tA = \overline{{}^tA} = A^*$.

The Hermitian endomorphisms have several remarkable properties, which we collect together in the following *spectral theorem.*

Theorem 6.5.2. *Let $u \in \mathcal{H}(\mathbb{C}^n)$ be a Hermitian endomorphism of $\mathbb{C}^n$. Then*

(i) *the eigenvalues of u are real;*

(ii) *the eigenspaces E_λ, $\lambda \in \mathrm{Sp}(u)$, are pairwise orthogonal;*

(iii) *u is diagonalisable, i.e. there exists a basis of $\mathbb{C}^n$ consisting of eigenvectors of u.*

Proof. (i) If $u(x) = \lambda x$, $x \neq 0_{\mathbb{C}^n}$, then $\langle u(x), x\rangle = \langle \lambda x, x\rangle = \lambda\langle x, x\rangle$. Also, we have $\langle u(x), x\rangle = \langle x, u^*(x)\rangle = \langle x, u(x)\rangle = \langle x, \lambda x\rangle = \overline{\lambda}\langle x, x\rangle$. Since $x \neq 0_{\mathbb{C}^n}$, we have $\langle x, x\rangle \neq 0$, so $\lambda = \overline{\lambda}$, i.e. $\lambda \in \mathbb{R}$.

(ii) Let $\lambda \neq \mu$ be two eigenvalues of u; suppose $\mu \neq 0$ and let $x \in E_\lambda$ and $y \in E_\mu$. Then $\langle u(x), u(y)\rangle = \lambda\overline{\mu}\langle x, y\rangle = \lambda\mu\langle x, y\rangle$ by (i), but also

$$\langle u(x), u(y)\rangle = \langle u(x), \mu y\rangle = \langle x, u^*(\mu y)\rangle = \langle x, u(\mu y)\rangle = \langle x, \mu^2 y\rangle = \mu^2\langle x, y\rangle,$$

so $\mu(\lambda - \mu)\langle x, y\rangle = 0$, and finally, $\langle x, y\rangle = 0$ since $\lambda \neq \mu \neq 0$.

(iii) See Exercise 6.10 below. ∎

Exercise 6.10. Prove that if $u \in \mathcal{H}(\mathbb{C}^n)$ is a Hermitian endomorphism of $\mathbb{C}^n$, there exists a basis of $\mathbb{C}^n$ consisting of orthonormal eigenvectors of u, i.e.:

$$\mathbb{C}^n = \bigoplus_{\lambda \in \mathrm{Sp}(u)} E_\lambda.$$

Hint: use induction on n; to pass from n to $n+1$, observe that the characteristic polynomial $\Phi(u, x)$ has at least one root λ, then show that the space of vectors that are orthogonal to E_λ is stable under u, so that $u(E_\lambda) \subset E_\lambda$.

To a Hermitian endomorphism u, it is convenient to associate its Rayleigh quotient:

$$0_{\mathbb{C}^n} \neq x \mapsto R(u, x) = \frac{\langle u(x), x\rangle}{\langle x, x\rangle} \in \mathbb{R}.$$

This value $R(u, x)$ is a real number since $\langle x, x\rangle \in \mathbb{R}$ and

$$\langle u(x), x\rangle = \langle x, u(x)\rangle = \overline{\langle u(x), x\rangle}.$$

The Rayleigh quotient has the following property.

Proposition 6.5.3. *Let $\lambda_{\min}, \lambda_{\max}$ denote the smallest and the greatest eigenvalue of u respectively. Then*

$$\forall x \neq 0_{\mathbb{C}^n}, \quad \lambda_{\min} \leqslant R(u, x) \leqslant \lambda_{\max}.$$

Proof. Let $\{e_1, \dots, e_n\}$ be a basis consisting of eigenvectors of u associated to the eigenvalues $\lambda_1 \leqslant \dots \leqslant \lambda_n$, which we may assume orthonormal (it suffices to choose an orthonormal basis for each eigensubspace). For $x \in \mathbb{C}^n$, we have $x = \sum_{1 \leqslant i \leqslant n} x_i e_i$, hence

$$\begin{aligned}\langle u(x), x\rangle &= \Big\langle \sum_{1\leqslant i\leqslant n} x_i u(e_i), \sum_{1\leqslant i\leqslant n} x_i e_i\Big\rangle = \Big\langle \sum_{1\leqslant i\leqslant n} \lambda_i x_i e_i, \sum_{1\leqslant i\leqslant n} x_i e_i\Big\rangle \\ &= \sum_{1\leqslant i\leqslant n} \lambda_i x_i \overline{x_i}.\end{aligned}$$

Observing that $\lambda_{\min} \leqslant \lambda_i \leqslant \lambda_{\max}$ for all i, we have

$$\frac{\sum_{1\leqslant i\leqslant n} \lambda_i x_i \overline{x_i}}{\sum_{1\leqslant i\leqslant n} x_i \overline{x_i}} \leqslant \frac{\sum_{1\leqslant i\leqslant n} \lambda_{\max} x_i \overline{x_i}}{\sum_{1\leqslant i\leqslant n} x_i \overline{x_i}} = \frac{\lambda_{\max} \sum_{1\leqslant i\leqslant n} x_i \overline{x_i}}{\sum_{1\leqslant i\leqslant n} x_i \overline{x_i}} = \lambda_{\max}.$$

We reason analogously for $\lambda_{\min}$. ∎

Observe that the values $\lambda_{\min}$ and $\lambda_{\max}$ are attained by taking the associated eigenvectors.

Exercise 6.11. Prove that a Hermitian endomorphism is

(i) positive if and only if its eigenvalues are all positive,

(ii) positive definite if and only if its eigenvalues are all strictly positive.

If u, v are Hermitian endomorphisms, we write $u \leqslant v$ or $v \geqslant u$ when $v - u$ is positive.

Exercise 6.12. (i) Show that $\leqslant$ is an order relation on the set of Hermitian endomorphisms.
Hint: we can use the fact that u is diagonalisable.

(ii) If u is Hermitian, show that $u^2 = u \circ u$ is positive.

Chapter 7

Colouring

Problem A. How many colours are needed to colour a geographic map in such a way that no two countries sharing a frontier have the same colour?

Problem B. The department of mathematics and computer science at the university of X proposes a certain number of different classes. Students must choose to take several of these. For the scheduling, the head of the department and the teaching staff are confronted with the following problem: two classes which have both been chosen by the same student cannot take place at the same time. We can model this problem by a graph in the following way: the vertices of the graph represent the classes, and two vertices are adjacent if there is at least one student who has chosen both of them. To know the number of different time slots (colours) needed in order for all students to be able to take all their chosen classes, we need to colour the vertices of the graph in such a way that if two vertices are adjacent, then they have different colours.

This chapter is devoted to the topic of graph colourings, which arises frequently in graph theory. We will see some applications and prove some classical theorems, among which Brooks' theorem and Vizing's theorem. In the final paragraph, we will introduce some more recent results. Throughout this chapter, all graphs will be considered to be *finite and loopless.*

7.1 Vertex colourings

7.1.1 General properties

A *vertex colouring* of a graph $\Gamma = (V; E, N)$ is a map

$$f : V \to \mathbb{C}$$

such that $f(x) \neq f(y)$ when x is adjacent to y. A *k-vertex colouring* or *k-colouring* is a vertex colouring that uses k colours, i.e. $|f(V)| = k$; we often take $f(V) = \{1, 2, \dots, k\}$. If there is no ambiguity about the type of colouring, we simply refer to a *k-colouring*. The smallest k such that the graph Γ admits a k-colouring is called the *chromatic number* of Γ, and denoted by $\chi(\Gamma)$. A graph Γ such that $\chi(\Gamma) \leqslant k$ (resp. $\chi(\Gamma) = k$) is said to be *k-colourable* (resp. *k-chromatic*). Figure 7.1 illustrates this concept.

Clearly, for a complete graph K_n we have $\chi(K_n) = n$; thus for every graph Γ we have $\chi(\Gamma) \geqslant \omega(\Gamma)$, where $\omega(\Gamma)$ is the maximum cardinality of a clique.

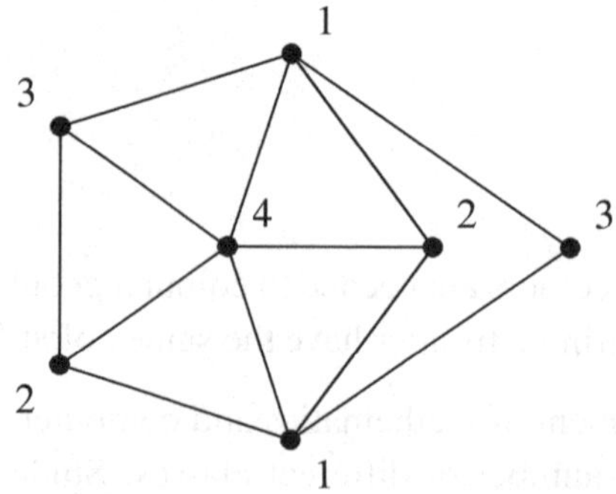

Figure 7.1. Graph admitting a 4-colouring, with chromatic number equal to $\chi(\Gamma) = 4$.

	9			**2**				**7**
				9		**2**	**6**	
5					**7**		**9**	
	1	**9**	**6**			**3**		
					9		**1**	
6	**8**	**5**		**1**				
	7		**4**	**8**	**1**			
1			**9**					
9							**4**	**1**

8	**9**	1	5	**2**	6	4	3	**7**
4	3	7	1	**9**	8	**2**	**6**	5
5	6	2	3	4	**7**	1	**9**	8
7	**1**	**9**	**6**	5	4	**3**	8	2
3	2	4	8	7	**9**	5	**1**	6
6	**8**	**5**	2	**1**	3	9	7	4
2	**7**	3	**4**	**8**	**1**	6	5	9
1	4	8	**9**	6	5	7	2	3
9	5	6	7	3	2	8	**4**	**1**

Figure 7.2. A problem of *sudoku* and its solution.

Example 7.1.1. A square *sudoku* grid consists of 9 square subgrids each containing 9 squares. Starting from a partially filled grid, the game consists in trying to fill it in completely by placing a digit from 1 to 9 in each square in such a way that every row, every column and every subgrid contains all the digits from 1 to 9, once each. Every sudoku (i.e. partially filled grid) possesses a unique solution; there is only one way to complete it. A sudoku is usually modelled by a graph Γ with 81 vertices representing the squares, in which two vertices are connected by an edge if they lie on the same row, in the same column or in the same subgrid. This graph is 20-regular. Solving the sudoku comes down to a solving a colouring problem in which the role of the colours is now played by the digits. The subgrids are all K_9 graphs, so a minimum of 9 colours is indeed necessary, so $\chi(\Gamma) \geqslant 9$. But, actually, in a properly constructed sudoku, we have $\chi(\Gamma) = 9$. Solving a sudoku is essentially the same thing as completing a partial 9-colouring. Figure 7.2 gives an example of a partially filled sudoku that, like all sudokus, possesses a unique solution (9-colouring).

Let Γ' be a subgraph of Γ. All the colours used for the vertices of Γ' in a minimal colouring of Γ can be used for a k'-colouring of Γ'. Since $k' \geqslant \chi(\Gamma')$, we have $\chi(\Gamma) \geqslant \chi(\Gamma')$.

Exercise 7.1. Let C_n be the elementary cycle with n vertices. Prove that

$$\chi(C_n) = \begin{cases} 2 & \text{if } n \geqslant 2 \text{ is even,} \\ 3 & \text{if } n \geqslant 3 \text{ is odd.} \end{cases}$$

Recall that the simplification Γ_s of the graph $\Gamma = (V; E, N)$ is the simple graph obtained by deleting the loops and replacing each p-edge by a simple edge in Γ. Clearly $\chi(\Gamma) = \chi(\Gamma_s)$. In what follows, we will consider only simple graphs.

Let $\Gamma = (V; E)$ be a graph. A subset of vertices V' of V is called an *independent set* (or a *stable set*) if no pair of vertices $x, y \in V'$ ever corresponds to an edge. An independent set V' is *maximum* if for every independent set V'', we have $|V''| \leqslant |V'|$. This means that for every $x \notin V'$, there exists an edge from x to a vertex in V'. The cardinality of a maximum independent set will be written $\alpha(\Gamma)$. We can also define the notion of an independent set in a multigraph $\Gamma = (V; E, N)$. A subset of vertices V' of V is an independent set (or stable set) if $\Gamma(V')$ contains no edge.

A k-colouring of a graph is a partition of its vertices into stable sets $V_1, V_2, \ldots, V_k$.

Proposition 7.1.2. *Let $\Gamma = (V; E)$ be a simple graph. Then*

$$\chi(\Gamma) \geqslant \left\lceil \frac{|V|}{\alpha(\Gamma)} \right\rceil \quad \textit{and} \quad \chi(\Gamma) + \alpha(\Gamma) \leqslant |V| + 1.$$

Proof. Let $\chi(\Gamma) = k$, with the partition of V into stable sets $\{V_1, V_2, \ldots, V_k\}$. We have $|V_i| \leqslant \alpha(\Gamma)$, so $|V| = |V_1 \cup V_2 \cup \cdots \cup V_k| = \sum_{1 \leqslant i \leqslant k} |V_i| \leqslant k\alpha(\Gamma)$. This gives the first inequality.

If we colour a maximum independent set V' with one colour, we will need at most $|V| - \alpha(\Gamma)$ colours to colour $V \setminus V'$. Thus we have $\chi(\Gamma) \leqslant (|V| - \alpha(\Gamma)) + 1$. ■

Proposition 7.1.3. *Let $\Gamma = (V; E)$ be a simple graph. Then*

$$\chi(\Gamma) \leqslant \Delta(\Gamma) + 1.$$

Proof. We use induction on the number n of vertices of the graph. If $n = 1$, the result is trivial. Suppose the assertion holds for every graph with at most $n - 1$ vertices, $n \geqslant 2$, and let Γ be a graph with n vertices. If we delete a vertex x, the resulting graph Γ' has maximal degree at most equal to $\Delta(\Gamma)$. By the induction hypothesis, this graph is $(\Delta(\Gamma') + 1)$-colourable. A $(\Delta(\Gamma) + 1)$-colouring of Γ is realised by colouring the vertex x with a colour that is different from the ones already used for the vertices adjacent to x of which there are at most $\Delta(\Gamma)$. ■

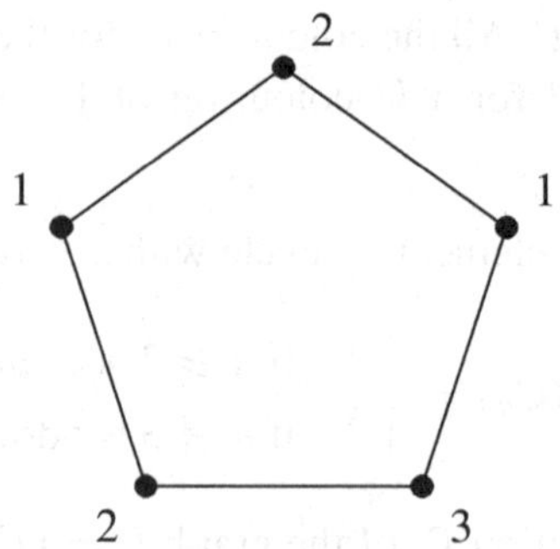

Figure 7.3. Graph such that $\chi(\Gamma) = 3 = \Delta(\Gamma) + 1$.

Combining Proposition 7.1.2 and Proposition 7.1.3, we obtain bounds for the chromatic number of a graph.

Corollary 7.1.4. *Let* $\Gamma = (V; E)$ *be a simple graph. Then*

$$\left\lceil \frac{|V|}{\alpha(\Gamma)} \right\rceil \leqslant \chi(\Gamma) \leqslant \Delta(\Gamma) + 1.$$

Remark 7.1.5. The first inequality can be an equality, as for $\Gamma = C_4$. The second inequality can also be an equality; Figure 7.3 shows an example.

It is possible to develop a simple algorithm to colour a graph. Suppose that the vertices are ordered as $x_1, x_2, x_3, \dots, x_n$. We will colour the vertices one by one by always using the *smallest* possible colour, where colours are represented by integers $1, 2, 3, \dots$

Algorithm 7.1 is a colouring algorithm with polynomial complexity. However, this algorithm does not give the chromatic number. In fact, the number of colours used to colour a graph using this algorithm depends on the chosen ordering of the vertices, and there are $n!$ possible orderings! So it is not very likely that this algorithm actually produces a colouring with a minimum number of colours. In fact, we have the following problem:

Data: a graph Γ and an integer k.

Problem: Is it possible to colour the vertices of Γ with k colours?

This problem is $\mathcal{NP}$-complete for $k \geqslant 3$ (why $k \geqslant 3$?). Therefore there exists no algorithm with polynomial complexity that can solve the above problem, unless $\mathcal{P} = \mathcal{NP}$.

```
Function Colour(Γ: graph) : colouring
  Data: A_{x_1}, A_{x_2}, ..., A_{x_n} (adjacency lists of vertices);
  Data: x_1, x_2, x_3, ..., x_n (ordering on the vertices of Γ);
  Output: f: V(Γ) → ℕ (a colouring of Γ);
  begin
    f(x_1) := 1;
    for every i from 1 to n do
      f(x_i) := min{j ≠ f(x_t), x_t ∈ A_{x_i}, t = 1,2,...,i − 1};
    end
  end
EndFunction
```

Algorithm 7.1. Colouring.

In our algorithm the vertex x_i to be coloured admits at most $\Delta(\Gamma)$ adjacent vertices in the induced subgraph on the vertices $x_1, x_2, x_3, \ldots, x_i$; more precisely, it has at most $\Delta(\Gamma)$ predecessors when it is coloured in the course of the **for** loop, so there are at most $\Delta(\Gamma)$ colours with which we cannot colour it. Thus one needs at most $\Delta(\Gamma) + 1$ colours to colour the graph using the algorithm.

Exercise 7.2. Use Algorithm 7.1 to give another proof of Proposition 7.1.3.

A slightly finer bound is given in the following proposition.

Proposition 7.1.6. *Let $\Gamma = (V; E)$ be a simple graph. Then*

$$\chi(\Gamma) \leqslant 1 + \max\{\delta(\Gamma') \mid \Gamma' \textit{ induced subgraph of } \Gamma\} \leqslant 1 + \Delta(\Gamma).$$

Proof. Recall that $\delta(\Gamma)$ is the minimal degree of vertices. Let

$$k = \max\{\delta(\Gamma'), \Gamma' \text{ induced subgraph of } \Gamma\} \quad \text{and} \quad n = |V|.$$

The graph Γ contains a vertex x_n such that $d(x_n) = \delta(\Gamma) \leqslant k$. If we delete this vertex, we obtain the graph $\Gamma_{n-1} = \Gamma(V \backslash \{x_n\})$. Let x_{n-1} be a vertex of Γ_{n-1} such that $d(x_{n-1}) = \delta(\Gamma_{n-1}) \leqslant k$. We continue the process until we obtain the graph $\Gamma_1 = \Gamma(V \backslash \{x_n, x_{n-1}, \ldots, x_2\})$ consisting of only the vertex x_1. In this way, we have *ordered* the vertices $x_1, x_2, \ldots, x_n$ in such a way that each x_i, $1 \leqslant i \leqslant n$, has at most k neighbours in Γ_i, since $\delta(\Gamma_i) \leqslant k$ is the degree of x_i in Γ_i. Thus we need at most $1 + k$ colours to colour the graph. ■

Corollary 7.1.7. *Let $\Gamma = (V; E)$ be a simple connected graph that is not regular. Then $\chi(\Gamma) \leqslant \Delta(\Gamma)$.*

Proof. Suppose that $\chi(\Gamma) = \Delta(\Gamma) + 1$. By applying Proposition 7.1.6, there exists an induced subgraph Γ' such that $\delta(\Gamma') = \Delta(\Gamma)$. But then Γ' is an induced regular subgraph, which cannot be connected to a vertex exterior to Γ' since it is regular of degree $\Delta(\Gamma)$. Since Γ is connected, we have $\Gamma = \Gamma'$, but Γ' is regular, contradiction. ∎

One interesting question is determining which graphs satisfy $\chi(\Gamma) = \Delta(\Gamma) + 1$. These graphs are necessarily regular, and K_n and C_{2n+1} are among them. The next result, Brooks' theorem, shows that there are no others.

7.1.2 Brooks' theorem

Lemma 7.1.8. *Let $\Gamma = (V; E)$ be a simple connected graph with no articulation vertex (i.e. 2-connected), not complete, satisfying $\delta(\Gamma) \geqslant 3$ (so $|V| \geqslant 4$). Then there exist two non-adjacent vertices x_1 and x_2 having a common neighbour, such that $\Gamma(V \backslash \{x_1, x_2\})$ is connected.*

Proof. There are two cases to study.

First case: Γ is 3-connected. There exists a vertex x having two non-adjacent neighbours. Indeed, if we suppose that any two neighbours x_1, x_2 of each vertex x are adjacent, we can show that the graph is then complete. To do this, let $x, y \in V$ be two arbitrary distinct vertices. Since the graph is connected, there exists a chain $(x, t_1, t_2, \dots, t_{n-1}, y)$. Since x and t_2 are both neighbours of t_1, they are adjacent; then t_1 and t_3 are both neighbours of t_2, so they are adjacent; then x and t_3 are both neighbours of t_1, so they are also adjacent, and thus we show step by step that y is a neighbour of x. Thus there exists an x having two non-adjacent neighbours x_1, x_2. But since the graph is 3-connected, $\Gamma(V \backslash \{x_1, x_2\})$ is connected.

Second case: Γ is 2-connected but not 3-connected. (1) *There exists a vertex x of Γ such that $\Gamma(V \backslash \{x\})$ (denoted $\Gamma - x$) possesses at least one articulation vertex.* By Corollary 4.2.6 of Proposition 4.2.5, $\Gamma - x$ admits at least two blocks, each containing just one articulation vertex. Let B_1 and B_2 be these blocks, and let b_1 and b_2 be their articulation vertices (note that we may have $b_1 = b_2$).

(2) *B_1 and B_2 each have at least two vertices:* if $\{a\}$ is a block, then it is an isolated vertex, but this is not possible since $\Gamma - x$ is connected with at least two vertices.

(3) *Let $e = \{a, b\}$ be an edge coming out of B_1, i.e. $a \in B_1$, $b \notin B_1$.* Then either $b = x$ or $a = b_1$, and similarly for B_2. Indeed, if $b \neq x$, then e is an edge of $\Gamma - x$, so it belongs to a block B_1', and $a \in B_1 \cap B_1'$ is an articulation vertex of $\Gamma - x$ (Lemma 4.2.3). But since $a \in B_1$, we must have $a = b_1$.

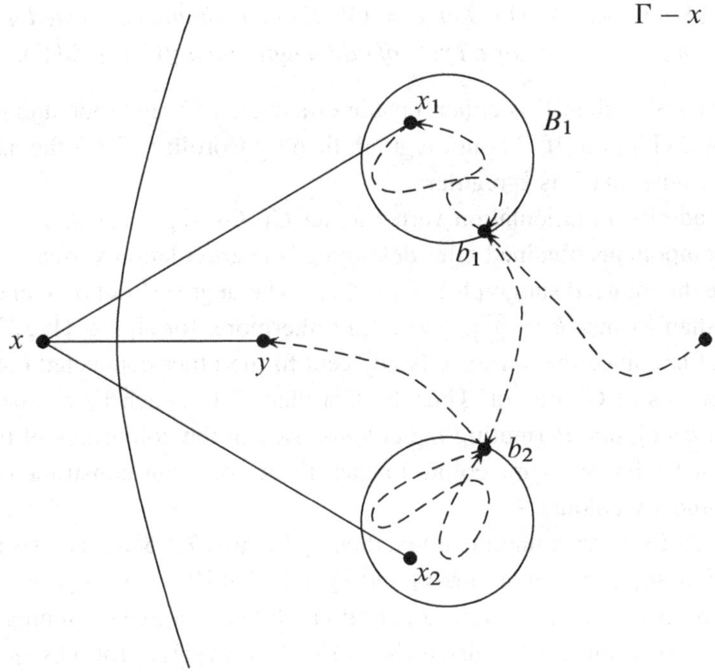

Figure 7.4. Case $y \notin B_1 \cup B_2$ of Lemma 7.1.8.

(4) *The vertex x has two neighbours x_1 and x_2 which are not adjacent.* Let $t_1 \in B_1, t_1 \neq b_1$. Then since Γ is 2-connected, there exists a chain from t_1 to x that avoids b_1. This chain leaves B_1 along an edge $e = \{a, b\}$ with $a \neq b_1$. By point 3 above, we have $b = x$. Set $x_1 = a$; then x is adjacent to $x_1 \neq b_1, x_1 \in B_1$, and similarly, x is adjacent to $x_2 \neq b_2, x_2 \in B_2$. But x_1, x_2 are not adjacent.

(5) *Finally,* $\Gamma(V \backslash \{x_1, x_2\})$ *is connected.* Firstly, since the degree of x is at least 3, there exists $y \in \Gamma - x$ adjacent to x, $y \neq x_1, x_2$. It suffices to show that every vertex t of $\Gamma - x$ can be connected to y inside $\Gamma - x$ with a path that avoids x_1 and x_2.

- If $y \notin B_1 \cup B_2$, let C be a chain from t to y in $\Gamma - x$. If C passes through x_1, it enters into B_1, so it passes through b_1 (by 3), and it must come out again, also by b_1; thus we can shorten C by deleting the part in B_1, so that it avoids x_1 completely. We ensure that C also avoids x_2 in the same way (Figure 7.4).
- If $y \in B_1$, let C be a chain from t to y in $\Gamma - x$. Then it enters into B_1 through b_1 (again by 3 above), then goes from b_1 to y in B_1. If it passes through x_1, then as B_1 is 2-connected, there is another vertex-disjoint chain from b_1 to y which avoids x_1. We also ensure that it avoids x_2 by deleting the part in B_2.
- If $y \in B_2$, the reasoning is analogous. ■

Theorem 7.1.9 (Brooks, 1941). *Let* $\Gamma = (V; E)$ *be a simple connected graph. If* Γ *is neither a complete graph nor a cycle of odd length, then* $\chi(\Gamma) \leqslant \Delta(\Gamma)$.

Proof. If $\Delta(\Gamma) \leqslant 2$, then Γ is either a chain or a cycle of even order, and it is finite. Suppose that $\Delta(\Gamma) \geqslant 3$. If Γ is not regular, then by Corollary 7.1.7 the assertion is true. Suppose now that Γ is k-regular.

(1) If Γ admits an articulation vertex x, let $C_1, C_2, C_3, \ldots, C_p$, $p \geqslant 2$, be the connected components obtained after deletion of the articulation vertex x. For all i, let Γ_i denote the induced subgraph $\Gamma(C_i \cup \{x\})$. The degree $d_i(x)$ of x in each Γ_i is strictly less than k since $k = \sum_{i=1}^{p} d_i(x)$. Furthermore, for all $i \in \{1, 2, 3, \ldots, p\}$, Γ_i is not regular, since the vertex x is adjacent to the other connected components while the vertices of C_i are not. Thus, by Corollary 7.1.7, each Γ_i can be coloured with at most k colours. Permuting the colours used in the colourings of each Γ_i in such a way that x has the same colour in each Γ_i, we can thus construct a colouring of Γ with at most k colours.

(2) If Γ admits no articulation vertex, then by Lemma 7.1.8 there exists a vertex x having two non-adjacent neighbours x_1 and x_2 such that $\Gamma(V \backslash \{x_1, x_2\})$ is connected. Thus by Proposition 2.2.6, it admits a partial graph that is a tree (spanning tree). We order the vertices of this tree (so also those of $\Gamma(V \backslash \{x_1, x_2\})$) as follows: $t_0 = x$ is the largest element, then we (arbitrarily) order the neighbours of x: $t_1 > t_2 > \cdots > t_\alpha$, then the vertices neighbouring t_i in Γ: $t_{\alpha+1} > t_{\alpha+2} > \cdots > t_\beta$, etc. from t_1 to t_n; finally, we complete the ordering by forcing $t_n > t_{n+1} = x_1 > t_{n+2} = x_2$. Thus $V(\Gamma) = V$ is ordered as follows:

$$x = t_0 > t_1 > t_2 > \cdots > t_n > t_{n+1} = x_1 > t_{n+2} = x_2$$

in such a way that

$$\forall i, 1 \leqslant i \leqslant n+2,\ \exists t_j > t_i \text{ such that } t_i \text{ is adjacent to } t_j.$$

We can now construct a colouring $f\colon V \to \{1, 2, \ldots, k\}$ as follows: $f(t_{n+2}) = f(t_{n+1}) = 1$, $f(t_n) = 2$; this is possible since x_1 and x_2 are not adjacent; then t_n has a neighbour greater than itself, so at most $k-1$ smaller neighbours (here these are at most t_{n+1} and t_{n+2}), so we can colour it with one of k colours; generally, each t_i, having a larger neighbour, can have at most $k-1$ smaller neighbours, and thus we can colour it with one of k colours with no contradiction; we eventually arrive at t_1, then finally at $t_0 = x$ which has k neighbours, of which t_1, t_{n+1}, t_{n+2}. But the last two have the same colour, so the neighbours of x require at most $k-1$ colours and we can colour it with no contradiction. ∎

7.1.3 Zykov's theorem

We saw that the colouring problem is $\mathcal{NP}$-complete for $k \geqslant 3$, meaning that there exists no polynomial algorithm to solve the colouring problem. Nevertheless, the colouring problem is still *decidable* in the algorithmic sense. The next theorem allows us to elaborate a procedure (with exponential complexity) to compute the chromatic number of a graph.

Let $\Gamma = (V; E)$ be a graph. The *fusion* of vertices x and y is the *identification* of vertices x and y to form a single vertex: the neighbours of this vertex are the neighbours of both x and y in the graph Γ. Let $\Gamma/(x = y)$ denote the fused graph obtained this way.

For two non-adjacent vertices x and y, we denote by $\Gamma + xy$ the (simple) graph obtained by adding the edge $\{x, y\}$ to Γ. Note that $(\Gamma + xy)/xy$ coincides with $\Gamma/(x = y)$.

Theorem 7.1.10 (Zykov, 1949). *Let $\Gamma = (V; E)$ and x, y be two non-adjacent vertices. Then*

$$\chi(\Gamma) = \min\{\chi(\Gamma/(x = y)), \chi(\Gamma + xy)\}.$$

Proof. Let $\mathcal{F}(\Gamma)$ denote the set of colourings of vertices of Γ. Then

$$\begin{aligned}\chi(\Gamma) &= \min\{|f(V)| \mid f \in \mathcal{F}(\Gamma)\}\\ &= \min\big\{\min\{|f(V)| \mid f(x) = f(y)\}, \min\{|f(V)| \mid f(x) \neq f(y)\}\big\}\\ &= \min\{\chi(\Gamma/(x = y)), \chi(\Gamma + xy)\}.\end{aligned}$$

∎

Iteratively applying Theorem 7.1.10, we obtain a *Zykov tree* of Γ ; it is binary, and its root is the graph Γ whose two sons are $\Gamma/(x = y)$ and $\Gamma + xy$ for a certain pair x, y of non-adjacent vertices. The process stops when we obtain cliques; thus the leaves are complete graphs. This tree is not unique; it depends on successive choices of non-adjacent vertices. However, this procedure allows us to set up an algorithm (which is obviously not polynomial).

7.1.4 Critical graphs

In this paragraph, we introduce a fundamental notion concerning colourings: critical graphs.

A graph $\Gamma = (V; E)$ is *critical* if $\chi(\Gamma') < \chi(\Gamma)$ for every subgraph $\Gamma' = \Gamma - x$ of Γ. A graph is k*-critical* if it is k-chromatic and critical. Clearly we have

(i) every k-chromatic graph contains a k-critical subgraph;

(ii) every critical graph is connected (exercise).

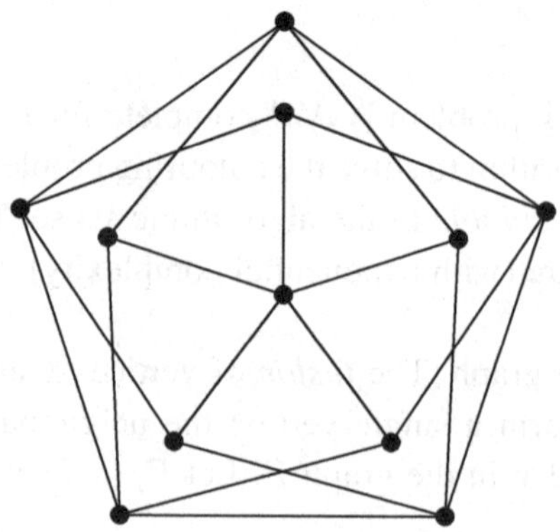

Figure 7.5. Grötzsch's graph is a 4-critical graph with minimal degree 3.

Proposition 7.1.11. *Let $\Gamma = (V; E)$ be a critical graph. Then $\chi(\Gamma) \leqslant \delta(\Gamma) + 1$.*

Proof. Suppose $k = \chi(\Gamma) > \delta(\Gamma) + 1$, and let $x \in V$ be such that $d(x) = \delta(\Gamma)$. Then $\Gamma - x$ is $(k-1)$-colourable. Let $f : V \backslash \{x\} \to \{1, 2, \ldots, k-1\}$ be a $(k-1)$-colouring of $\Gamma - x$. Since $d(x) < k - 1$, there exists a colour i not incident to x with which we can colour x. But this is a contradiction. ■

Corollary 7.1.12. *Let Γ be a k-chromatic graph. Then Γ has at least k vertices of degree at least $k - 1$.*

This result allows us to recover the bound $\chi(\Gamma) \leqslant \Delta(\Gamma) + 1$ that was given in Proposition 7.1.6. Indeed, by Corollary 7.1.12, Γ contains a vertex x satisfying $\Delta(\Gamma) \geqslant d(x) \geqslant k - 1 \geqslant \chi(\Gamma) - 1$.

Proposition 7.1.13. *Every k-critical graph $\Gamma = (V; E)$ is $(k-1)$-edge-connected.*

Proof. Let X be a set of edges that disconnect the graph, and assume that $|X| < k - 1$. Let $\Gamma(V_1)$ and $\Gamma(V_2)$ be two connected components resulting from deleting X. By hypothesis, $\Gamma(V_1)$ and $\Gamma(V_2)$ are $(k-1)$-colourable.

Let $C_i = \{V_{i,1}, V_{i,2}, V_{i,3}, \ldots, V_{1,k-1}\}$, $i = 1, 2$, be $(k-1)$-colourings of $\Gamma(V_1)$ and $\Gamma(V_2)$ respectively. Let us construct a bipartite graph $\Gamma' = (V'; E')$ as follows:

- $V' = \{V_{1,1}, V_{1,2}, V_{1,3}, \ldots, V_{i,k-1}\} \sqcup \{V_{2,1}, V_{2,2}, V_{2,3}, \ldots, V_{2,k-1}\}$;
- the pair $\{V_{1,i}, V_{2,j}\}$ is an edge in Γ' by definition if there is no edge between $V_{1,i}$ and $V_{2,j}$.

Let $\overline{\Gamma'}$ be the complement of Γ': it has the same vertices, and there is an edge between $V_{1,i}$ and $V_{2,j}$ if there exists a vertex of $V_{1,i}$ adjacent to a vertex of $V_{2,j}$.

Every edge of Γ between a vertex of $V_{1,i}$ and a vertex of $V_{2,j}$ belongs to X, so the graph $\overline{\Gamma'}$ has at most $(k-2)$ edges. Thus the graph Γ' has at least $(k-1)^2 - (k-2) = k^2 - 3k + 3$ edges.

Every vertex of Γ' is adjacent to at most $k-1$ vertices, so each vertex is incident to at most $k-1$ edges. Consequently, a set of $k-2$ vertices is incident to at most $(k-2)(k-1) = k^2 - 3k + 2$ edges. Therefore every transversal set T of Γ' (a transversal set is a set containing at least one extremity of each edge (see chapter 8)), necessarily possesses at least $k-1$ vertices. By König's theorem (Theorem 8.1.5), the transversality number $\beta(\Gamma')$ is equal to the cardinality of the maximum matching $\alpha'(\Gamma')$, i.e. to $k-1 \leqslant \beta(\Gamma') = \alpha'(\Gamma') \leqslant k-1$; the last inequality follows from $|V'| = 2(k-1)$. Hence $\beta(\Gamma') = \alpha'(\Gamma') = k-1$, which implies that Γ' possesses a matching of cardinality $k-1$, which is necessarily perfect since $|V'| = 2(k-1)$.

Let C be a perfect matching, and let $a = \{V_{1,i}, V_{2,j}\} \in C$. Since there is no edge in Γ between $V_{1,i}$ and $V_{2,j}$, we can colour $V_{1,i} \cup V_{2,j}$ with the same colour. Repeating the process for every edge of C, we obtain a $(k-1)$-colouring of Γ, which is a contradiction. ■

7.1.5 Constructible graphs

In this section we give a rather surprising theorem, which allows us to characterise graphs with a chromatic number $\chi(\Gamma) \geqslant k$ thanks to an elementary construction.

Definition 7.1.14. A graph is *k-constructible* if it can be constructed using the three following procedures:

(i) the complete graph K_k is k-constructible;

(ii) if $\Gamma = (V; E)$ is k-constructible and if $x, y \in V$ are such that $\{x, y\} \notin E$, then $\Gamma/(x = y)$ is k-constructible;

(iii) if $\Gamma = (V; E)$ and $\Gamma' = (V'; E')$ are two simple k-constructible graphs such that $V \cap V' = \{x\}$, $\{x, y\} \in E$ and $\{x, y'\} \in E'$, then the graph $(\Gamma - xy) \cup (\Gamma' - xy') + yy'$ is k-constructible (*Hajós construction*).

The following theorem allows us to show that every graph with chromatic number $\chi(\Gamma) \geqslant k$ can be obtained recursively starting from the complete graph K_k by

- adding vertices,
- adding edges,
- fusing non-adjacent vertices,
- applying the Hajós construction (iii).

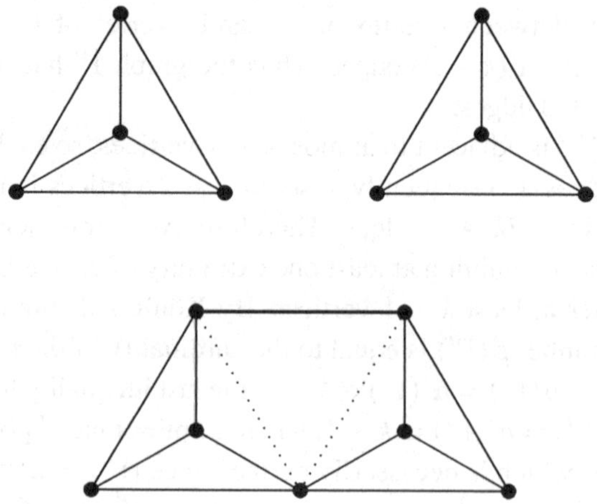

Figure 7.6. Application of the Hajós construction to two copies of K_4 : the dotted edges have been deleted; the boldface edge has been added.

Theorem 7.1.15 (Hajós' theorem). *Let* $\Gamma = (V; E)$ *be a graph. Then* $\chi(\Gamma) \geqslant k$ *if and only if* Γ *contains a* k*-constructible subgraph.*

Proof. When x and y are not adjacent, the graph $\Gamma/(x = y)$ coincides with $(\Gamma + xy)/xy$, the graph obtained by adding an edge between the vertices x and y and then contracting it. We see that $\chi(\Gamma) = \chi\big((\Gamma + xy)/xy\big)$: the colour of the fused vertex $x = y$ in $(\Gamma + xy)/xy$ can be attributed to x and y in Γ.

(1) Suppose that Γ contains a subgraph Γ' which is k-constructible. Then $\chi(\Gamma) \geqslant \chi(\Gamma')$. We proceed by induction on the number of transformations necessary to pass from K_k to Γ'. We consider the following elementary transformations.

- If $\Gamma' = K_k$, then $\chi(\Gamma') = k$.
- If Γ' is obtained starting from a graph Γ'' such that $\chi(\Gamma'') \geqslant k$ by adding a vertex, then $\chi(\Gamma') = \chi(\Gamma'') \geqslant k$.
- If Γ' is obtained starting from a graph Γ'' such that $\chi(\Gamma'') \geqslant k$ by adding an edge, then $\chi(\Gamma') \geqslant \chi(\Gamma'') \geqslant k$.
- If Γ' is of the form $(\Gamma'' + xy)/xy$ with $\chi(\Gamma'') \geqslant k$, then $\chi(\Gamma') = \chi(\Gamma'') \geqslant k$ (cf. preliminary remark).
- Finally, suppose that Γ' is of the form $(\Gamma_1 - xy_1) \cup (\Gamma_2 - xy_2) + y_1y_2$ with $V(\Gamma_1) \cap V(\Gamma_2) = \{x\}$ and $\chi(\Gamma_i) \geqslant k$, $i = 1, 2$. In every colouring of Γ', the two vertices y_1 and y_2 have different colours, so one of them, say y_1, must have a different colour than x. Thus we obtain a colouring of Γ_1, so $\chi(\Gamma') \geqslant \chi(\Gamma_1) \geqslant k$.

(2) Let us now show that if $\chi(\Gamma) \geqslant k$, then Γ contains a k-constructible subgraph.

If $k = 1$, then it suffices to take the subgraph N_1. If $k = 2$, the graph has at least one edge xy; we consider the corresponding subgraph K_2.

Assume now that $k \geqslant 3$, and suppose there exists a graph $\Gamma = (V; E)$ with no k-constructible subgraph and such that $\chi(\Gamma) \geqslant k$. We may also assume that this graph is maximal for this property in terms of $|E|$; thus $\Gamma + xy$ contains a k-constructible subgraph for every pair of non-adjacent vertices x, y. We see that Γ is not complete r-partite, otherwise we would have $r \geqslant \chi(\Gamma) \geqslant k$ and Γ would contain at least a K_r, so also a K_k, in which case it would be k-constructible (by adding vertices and edges to this K_k). It follows that Γ contains three vertices $x, y_1, y_2 \in V$ such that $\{y_1, y_2\} \in E$ and $\{x, y_1\}, \{x, y_2\} \notin E$; indeed it suffices to observe that the relation $v \mathfrak{R} w \Leftrightarrow \{v, w\} \notin E$ is not transitive in a non-r-partite graph with $r \geqslant 3$.

By the maximality of Γ, the graphs $\Gamma + xy_i$, $i = 1, 2$, each contain a k-constructible subgraph Γ_i. Naturally, we have $xy_i \in E(\Gamma_i)$. We consider the map $y \mapsto y' = \varphi(y)$ such that $\varphi(y) = y$ for all $y \in V(\Gamma_1) \setminus V(\Gamma_2)$ and y' is a copy of y if $y \in V(\Gamma_1) \cap V(\Gamma_2)$. The vertices y' form a graph Γ_1' isomorphic to Γ_1; it is also k-constructible, and $V(\Gamma_1') \cap V(\Gamma_2) = \{x\}$. Since $xy_1' \in E(\Gamma_1')$ and $xy_2 \in E(\Gamma_2)$, the Hajós construction applied to Γ_1' and Γ_2 implies that $(\Gamma_1' - xy_1') \cup (\Gamma_2 - xy_2) + y_1'y_2$ is k-constructible.

It remains to fuse each vertex $y \in V(\Gamma_1) \cap V(\Gamma_2)$ with the corresponding vertex $\varphi(y) \in V(\Gamma_1') \setminus V(\Gamma_2)$. We obtain a k-constructible graph Γ' (the fusion of non-adjacent vertices preserves k-constructibility). Then Γ' is effectively a subgraph of Γ: it suffices to consider the edges between x and y_1, y_2 in Γ' and to observe that we have $xy_i \notin E(\Gamma')$, $i = 1, 2$. ■

7.1.6 Planar graphs and maps

Historically, the four-colour theorem is related to the colouring of geographic maps: the mapmaker wishes to impose the condition that two countries having a common frontier must have different colours, and he wishes to accomplish this using a minimal number of different colours. But what, exactly, is a geographic map? From the point of view of graph theory, it is a planar graph (i.e. a graph of genus 0 using the terminology of Section 5.7). These planar graphs are not necessarily connected (the Vatican is an enclave (loop) inside Italy), but this situation does not create any colouring problem (since once Italy has been coloured, it suffices to choose any other colour for the Vatican). Furthermore, planar graphs corresponding to maps may have double edges (as illustrated in the situation of the country Andorra situated on the frontier between France and Spain).

Here, also, there is no colouring problem: once France and Spain have been coloured, we can choose any other colour for Andorra. Note that in a realistic map,

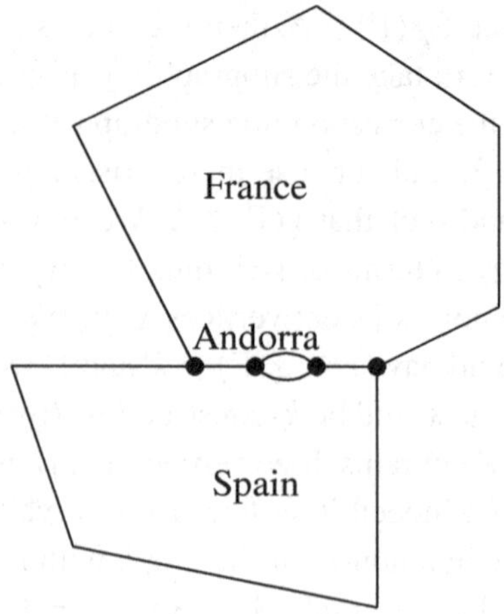

Figure 7.7. Graph of the frontier between France and Spain.

Figure 7.8. A planar graph showing a rather unrealistic geographic situation.

there are no hanging edges: the planar graph in Figure 7.8 is not geographically very plausible, but it is still legitimate.

A *map* is defined to be an embedded planar graph that is simple and 2-connected. A map is said to be k*-face colourable* if it can be coloured with k different colours in such a way that two adjacent faces are never of the same colour. This association of maps with planar graphs allows us to reduce the colouring problem to a purely combinatorial question on graphs.

Proposition 7.1.16. *A map Γ is k-face colourable if and only if its dual Γ^* is k-vertex colourable.*

Proof. The faces of Γ are the vertices of Γ^*, so that every colouring of the faces of Γ yields a colouring of the vertices of Γ^* and vice versa. ■

For planar graphs, there exist rather fine bounds for colourings. The first result is due to P. J. Heawood.

Theorem 7.1.17 (Heawood, 1890). *Every planar graph* Γ *is 5-vertex colourable.*

Proof. We may assume that Γ is simple. We will use induction on the number n of vertices. If $n \leqslant 5$, the result is trivial. Assume the result holds until $n - 1 \geqslant 5$, and let $\Gamma = (V; E)$ have n vertices.

The graph Γ has a vertex x of degree at most equal to 5 (Corollary 5.2.18). Deleting this vertex gives a planar graph with $n - 1$ vertices. By the induction hypothesis, this graph is 5-colourable. The goal now is to show that we can colour x with one of these 5 colours.

(1) If $d(x) < 5$, we are done.

(2) If $d(x) = 5$, let x_1, x_2, x_3, x_4, x_5 be the vertices adjacent to x. Two of these vertices, say x_1 and x_2, are not adjacent, since otherwise Γ would contain K_5 and it would be thus not be planar by Kuratowski's theorem (Theorem 5.5.4). Let us contract the two edges $\{x, x_1\}$ and $\{x, x_2\}$ in Γ: this means that we contract first $a_1 = \{x, x_1\}$, giving Γ/a_1, with new vertex z_{a_1}, then we contract the edge $a_2 = \{z_{a_1}, x_2\}$ in Γ/a_1, which creates the new vertex z_{a_2}, which we now denote u_x. We obtain a planar graph Γ' having less than n vertices, and thus 5-colourable by the induction hypothesis. The vertices adjacent to u_x are: firstly, the vertices adjacent to x, then the vertices adjacent to x_1, and finally the vertices adjacent to x_2. Let $f' \colon V' \to \{1, \dots, 5\}$ be a 5-colouring of Γ'. Set

- $f(y) = f'(y)$ for $y \neq x, x_1, x_2$;
- $f(x_1) = f(x_2) = f'(u_x)$;
- $f(x) = c \notin \{f'(x_3), f'(x_4), f'(x_5), f'(u_x)\}$.

This gives a 5-colouring of Γ. ■

Corollary 7.1.18. *Every map is 5-face colourable.*

Proof. It is an immediate consequence of Proposition 7.1.16 and Heawood. ■

Exercise 7.3. Draw the dual graph associated to each of the graphs in Figures 7.8 and 7.9, after placing vertices at each intersection.

Experiments show that all maps appear to be 4-colourable, and that 4 colours are in general necessary, as shown in the examples in Figure 7.9.

Theorem 7.1.19 (Appel and Haken, 1976). *Let* $\Gamma = (V; E, N)$ *be a planar graph. Then* Γ *is 4-vertex colourable.*

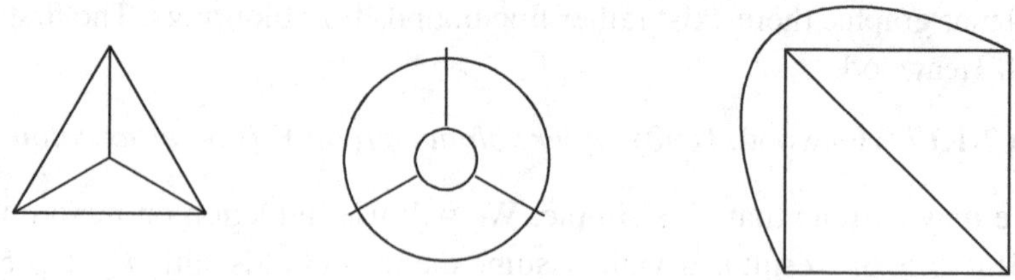

Figure 7.9. Three maps requiring 4 colours.

An immediate consequence (see Proposition 7.1.16) is the following extremely famous result.

Corollary 7.1.20 (*Four-colour theorem*). *Every map can be coloured with* 4 *colours.*

We will not give the proof of this theorem: it is very long, occupying several dozen pages, and very technical. The idea of the proof is to bring all graphs back to a finite number of configurations (precisely 1478), then to colour these with the help of a computer programme. An improvement on this proof was subsequently given by N. Robertson, D. Sanders, P. D. Seymour and R. Thomas in 1997 [30].

We saw in Chapter 5 that K_4 is planar but K_5 is not: this suggests that 4 may be the minimal number of colours, since n colours are needed to colour the vertices of K_n.

Here is a theorem that shows how to restrict the four-colour problem to a particular class of planar graphs. A simple graph is said to be *cubic* if it is regular of degree 3. Every simple planar graph Γ such that $\delta(\Gamma) \geqslant 3$ can be transformed into a simple cubic planar graph as follows. Let σ be an embedding of a planar graph with polygonal edges.

- Take a compass, place the point on a vertex $\sigma(x)$, and draw a small circle that cuts all the edges incident to $\sigma(x)$, but only those;
- Add a new vertex at each intersection point of the circle with an edge.
- Take an eraser and erase everything in the interior of the contained in the circle.
- Each arc of the circle between two new vertices becomes a new edge.
- Start the process again for each vertex.

We obtain a connected cubic planar graph. Figure 7.10 shows this construction.

Theorem 7.1.21. *To prove the four-colour theorem, it is sufficient to prove that every cubic map is* 4*-face colourable.*

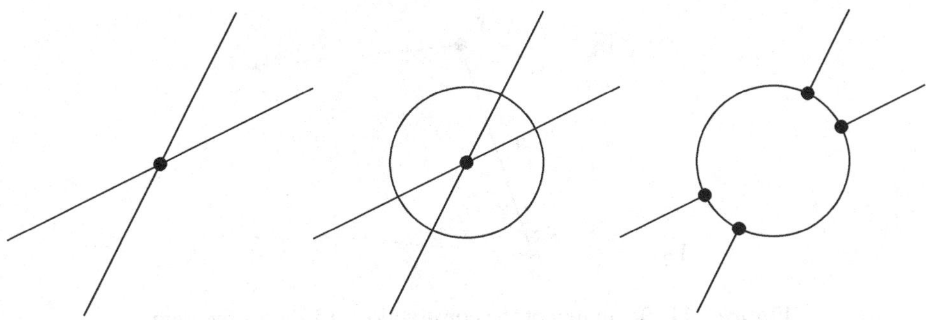

Figure 7.10. Cubic transformation.

Proof. Let Γ be a map. Thanks to the construction above, we obtain a cubic map that is 4-face colourable. A 4-face colouring of Γ is then obtained by applying the inverse transformation to the one above, fusing the vertices on the small circles. ■

It is natural to ask what can be said about map colourings for graphs of genus $g \geqslant 1$ (see Section 5.7). For example, K_5 is of genus 1, so it can be embedded in the torus $\mathbb{S}_1$: how many colours are needed to colour the resulting *toric map*? One can prove that 7 colours are sufficient to colour all maps drawn on a torus. Note that by the Ringel–Youngs identities (see Example 5.7.7), K_7 can be embedded in the torus but not K_8, which already suggests that 7 may be the desired number; indeed, if K_n is embedded in $\mathbb{S}_m$, then n colours are needed for its vertex colouring. A further remarkable result, due to Heawood in 1890, is that for the vertex colouring of all genus g graphs, $\lfloor \frac{7+\sqrt{1+48g}}{2} \rfloor$ colours suffice.

7.2 Edge colourings

The question of colouring edges is in some sense *dual* to the question of colouring vertices, but as we will see, it is dealt with by entirely different methods. An *edge colouring* of a graph $\Gamma = (V; E, N)$ is a map $f: E \to \mathbb{C}$ such that $f(a_1) \neq f(a_2)$ whenever a_1 is incident to a_2. A k*-edge colouring* is an edge colouring with k colours, i.e. $|f(E)| = k$ (which is why we often take $f(E) = \{1, 2, \ldots, k\}$). If there is no ambiguity, we will simply speak of a k*-colouring*. A k-edge colouring gives a partition of the set of edges into k subsets. The smallest k such that the graph Γ admits a k-edge colouring is called the *chromatic index*, and we denote it by $\chi'(\Gamma)$.

A graph Γ such that $\chi'(\Gamma) \leqslant k$ is said to be k*-edge colourable*. We will say that *a colour is incident to a vertex* if there exists an edge of that colour incident to the given vertex.

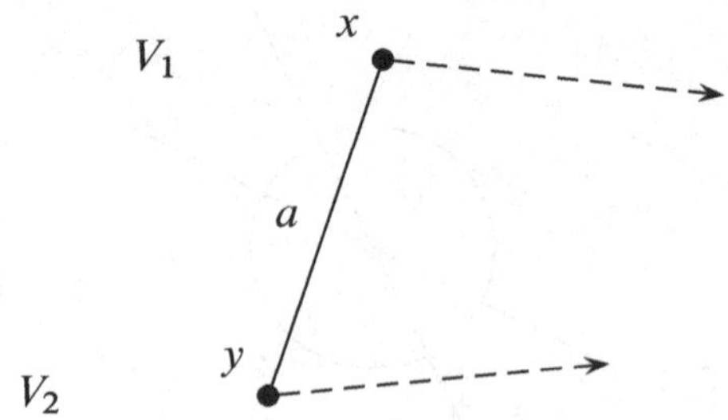

Figure 7.11. Beginning of the construction in König's theorem.

The colouring is related to the maximal degree $\Delta(\Gamma)$ of the graph, as we will see below.

Theorem 7.2.1 (König, 1916). *Let $\Gamma = (V; E)$ be a simple bipartite graph. Then $\chi'(\Gamma) = \Delta(\Gamma)$.*

Proof. It suffices to prove the result when Γ is connected. We use induction on the number of edges of the graph. If $m = |E| = 0$ or 1, it is trivial. For $m = 2$ the graph consisting of 2 edges and 3 vertices satisfies $\Delta = 2$ and $\chi' = 2$. Suppose the assertion holds for $|E| = m$ ($m \geqslant 2$) and let $\Gamma = (V_1 \sqcup V_2; E)$ be a bipartite graph with $m + 1$ edges. Let $a = \{x, y\}$ be an edge of Γ, with $x \in V_1, y \in V_2$. Then by the induction hypothesis, the bipartite graph $\Gamma' = \Gamma \backslash a$ obtained by deleting this edge admits a $\Delta(\Gamma')$-edge colouring. Since $\Delta(\Gamma') \leqslant \Delta(\Gamma)$, there exists a colouring $\chi\colon E\backslash a \to \mathbb{C}$ with $\Delta(\Gamma)$ colours.

(a) If there exists a colour c not incident to x nor to y, we colour a with it, which gives a colouring of Γ.

(b) Otherwise every colour is incident to either x or y. Since we have deleted a, we see that $d_{\Gamma'}(x) < d_{\Gamma}(x) \leqslant \Delta(\Gamma)$, so there exists a colour c_1 not incident to x; similarly, there exists a colour $c_2 \neq c_1$ not incident to y. It follows that c_1 is incident to y and c_2 incident to x (Figure 7.11).

Now let $B = \Gamma'_{c_1,c_2}$ be the subgraph of Γ' whose edges are the edges of Γ' coloured with the colours c_1 and c_2; the vertices of B are the extremities of these edges, so $x, y \in V(B)$. Each vertex of B is of degree 1 or 2, so the connected components of B are cycles (of even length) or chains whose edges are coloured alternatively by c_1 and c_2. By construction, $d_B(x) = d_B(y) = 1$; the connected component C in B containing x is an elementary chain (otherwise the colour c_1 or c_2 would be incident twice to a vertex of C). If there existed a chain from x to y, then as c_2 is incident to x and c_1 is incident to y, it would be of the form $(x, c_2, x_1, c_1, \dots, x_n, c_1, y)$, so it would require an even number of edges, so an odd number of vertices, but since B is bipartite, we would have to have $y \in V_1$, which is not the case. Thus the connected component C of x does not contain y, and ends at a vertex z to which the colour c_2 is not incident. Consequently, we can exchange the colours c_1 and c_2 in C: this operation makes c_2 no longer incident to x, and as c_2 is not incident to y, we can colour a with c_2 to obtain a colouring of Γ. ■

Exercise 7.4. Extend the notion of bipartite graph to multi-graphs. How would one adapt the above proof for multi-graphs?

Theorem 7.2.2 (Vizing, 1964). *Let $\Gamma = (V; E)$ be a simple graph. Then*

$$\Delta(\Gamma) \leqslant \chi'(\Gamma) \leqslant \Delta(\Gamma) + 1.$$

Proof. The first inequality is obvious. Indeed, let x be a vertex such that $d(x) = \Delta(\Gamma)$; it takes at least $\Delta(\Gamma)$ colours to colour the edges incident to x.

We will prove the second inequality by induction on the number of edges. If $|E| = 0, 1$ or 2, it is trivial. For $|E| = 3$ and $\Gamma = K_3$, we have $\Delta = 2$, $\chi' = 3$; the other cases are also simple.

Suppose now that the assertion holds for $|E| = m$ ($m \geqslant 3$). Let $\Gamma = (V; E)$ be a graph with $|E| = m + 1$ edges. To simplify the notation, let us write $a = xy$ for an edge $a = \{x, y\}$. Fix an edge $a = xy_1$. By the induction hypothesis, we can colour the edges of $\Gamma \backslash a$ (as in the proof of König's theorem):

$$\chi\colon E\backslash a \to \mathbb{C}$$

with the colours $c_1, \ldots, c_{\Delta+1}$, where $\Delta = \Delta(\Gamma)$.

The degrees of vertices of $\Gamma\backslash a$ are at most $\Delta(\Gamma)$, and we use $\Delta + 1$ colours to colour the graph. Consequently, there exists at least one colour not incident to each vertex: *if the same colour is not incident to x and to y_1*, then we colour the edge with this colour and we are done. *If the missing colours at x and y_1 are different*, we construct a sequence $(xy_1; c_1), (xy_2; c_2), (xy_3; c_3), \ldots, (xy_k; c_k)$, $k \geqslant 2$, satisfying the following conditions:

(a) c_i is a colour missing at y_i; the edge xy_{i+1}, $1 \leqslant i < k$ is coloured with c_i;

(b) the sequence above is maximal, i.e. :

(b$_1$) either there is no edge incident to x of colour c_k (there exists no xy_{k+1} coloured with c_k),

(b$_2$) or the colour c_k is already present in the set $c_1, \ldots, c_{k-1}$ of preceding colours in the sequence.

Let c denote the colour missing at x. By hypothesis, by (a) we have $c \neq c_i$ for $1 \leqslant i \leqslant k - 1$. There are two cases.

Case 1: *The colour c_k is not incident to x.* We shift all the colours: the edge xy_1 is coloured with the colour c_1, the edge xy_2 is coloured with the colour c_2, the edge xy_3

is coloured with the colour c_3, ..., the edge xy_k is coloured with the colour c_k; this is possible since c_i is not incident either to x nor to y_i by hypothesis and construction. We thus obtain a $(\Delta(\Gamma)+1)$-edge colouring of Γ.

Case 2: *The colour c_k is incident to x.* By (b_2) above, there exists a j such that xy_j is coloured with c_k. Then by construction of the sequence above we have $c_{j-1} = c_k$.

Here also, we shift the colours: xy_i is coloured with the colour c_i for all $i < j$. For $i > j$, xy_i remains with the colour c_{i-1}. The edge xy_j temporarily has no colour.

Consider the subgraph B of Γ whose edges are coloured with c or $c_k = c_{j-1}$ (recall that c is a colour that is not incident to x and $c_k = c_{j-1}$ is not incident to y_k). The vertices of B are the extremities of the edges in question: each vertex of B is thus of degree 1 or 2, so that the connected components of B are either chains or cycles (of even length).

We have $d_B(x) = 1$ since c is not incident to x but c_k is present, and we have $d_B(y_k) \leqslant 1$ since c_k is not incident to y_k but c can be incident to y_k, and finally $d_B(y_j) \leqslant 1$ since we have removed the colour of xy_j which was $c_{j-1} = c_k$. We distinguish two cases.

(a) *If x and y_j are in two different connected components of B*:

- If $d_B(y_j) = 0$, i.e. if c is missing at y_j, then it suffices to colour xy_j with c and we are done.
- If $d_B(y_j) = 1$, then c is incident to y_j. The connected component of x in B is a string of alternating colours c and c_k ending at x (since $d_B(x) = 1$) with the colour c_k since c is not incident to x. The connected component of y_j is a chain ending at y_j (since $d_B(y_j) = 1$) with an edge of colour c. Since the other extremity u of the chain is incident either to c or to c_k, one of the colours c or c_k is thus not incident to u, so we can exchange c and c_k in this final connected component. Neither of the vertices x and y_j are now of the colour c, and therefore we can colour xy_j with c.

(b) *If x and y_j are in the same connected component of B*, we stop shifting the colours: xy_i is coloured with c_i for all $1 \leqslant i \leqslant k-1$, and xy_k is not coloured. Note that this operation does not modify B, and that the edge xy_{j-1} of course remains coloured with $c_k = c_{j-1}$.

- If $d_B(y_k) = 0$, i.e. if c is absent at y_k, then it suffices to colour xy_k with c and we are done.
- If $d_B(y_k) = 1$, the connected component of y_k ends at y_k with an edge coloured with c. But y_j is in the same connected component in B as x. This connected component, which is a chain from x to y_j, is not the one containing y_k since $d_B(y_k) = 1$. So it suffices to permute the colours c and c_k in the connected component of y_k in B, and to colour xy_k with the colour c. ■

Determining whether $\chi'(\Gamma) = \Delta(\Gamma)$ or $\Delta(\Gamma) + 1$ is an $\mathcal{NP}$-complete problem. However, the proof above is effective: it is possible to create an algorithm that colours the edges of Γ with at most $\Delta + 1$ colours. The above theorem separates simple graphs into two classes: *class 1* contains the graphs such that $\chi' = \Delta$ and *class 2* contains the graphs satisfying $\chi' = \Delta + 1$. We see that $\chi'(C_{2n+1}) = 3 = \Delta + 1$ (class 2) and $\chi'(C_{2n}) = 2 = \Delta$ (class 1). The following proposition deals with the case of complete graphs.

Proposition 7.2.3. *Let $n \geqslant 2$.*

(i) *If n is odd, then K_n is of class 2: $\chi'(K_n) = n = \Delta(K_n) + 1$.*

(ii) *If n is even, then K_n is of class 1: $\chi'(K_n) = n - 1 = \Delta(K_n)$.*

Proof. (i) Suppose n odd. K_n is not $(n-1)$-edge colourable: the maximum number of pairs of vertices having empty intersection is equal to $\frac{n-1}{2}$, so the maximum number of edges of the same colour is $\frac{n-1}{2}$; we have

$$|E(K_n)| = \binom{n(n-1)}{2} \leqslant \frac{n-1}{2}\chi'(K_n),$$

hence $\chi'(K_n) \geqslant n$, then $\chi'(K_n) = n$ by Vizing's theorem. As $2|E| = \sum_{x \in V} d(x) = n(n-1) = \chi'(K_n)(n-1)$, we deduce that each colour is incident to exactly $n-1$ vertices of K_n and thus that there is a bijection between the n colours of this edge colouring corresponding to $\chi'(K_n)$ and the n vertices of K_n.

(ii) Suppose n even. By (i), we can colour the subgraph K_{n-1} with $n-1$ colours. The remaining vertex x is adjacent to the $n-1$ vertices of K_{n-1}. Each vertex u of K_{n-1} is of degree $n-2$, so there is a colour c_u that is not incident to u in K_{n-1}. By the above observation, the map $u \mapsto c_u$ is an injection. Thus, for each vertex u of K_{n-1}, we can colour the edge xu by c_u. ∎

We can also give a direct proof of (i) without using Vizing's theorem.

Exercise 7.5. (i) Let n be odd, and represent K_n by the vertices of a regular polygon, labelled $0, 1, 2 \ldots, n-1$ (modulo n); each edge $\{k, l\}$ is represented by the segment $[\![k, l]\!]$. Colour the edges in the boundary $\{k, k+1\}$, $0 \leqslant k \leqslant n-1$ with the colour c_k, $0 \leqslant k \leqslant n-1$. Show that every edge $\{i, i+\lambda\}$ of K_n is parallel to an edge $\{k, k+1\}$.

Hint: $\{i, j\}$ is parallel to $\{i-1, j+1\}$ (modulo n). We will need to solve an equation of degree one in the ring $\mathbb{Z}/n\mathbb{Z}$, which has only a single solution since n odd implies that 2 is invertible (more precisely, if we write $n = 2m - 1$, then the inverse of 2 is m). Conclude that $\chi'(K_n) = n$.

Figure 7.12 illustrates this method for $n = 5$.

(ii) Why does the technique above fail when n is even?

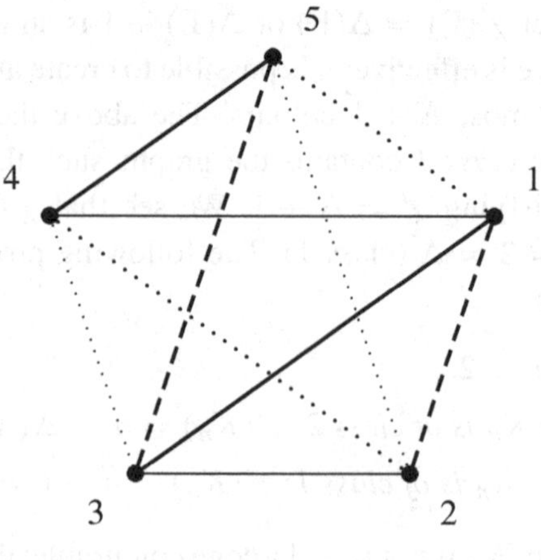

Figure 7.12. Edge colouring of K_5.

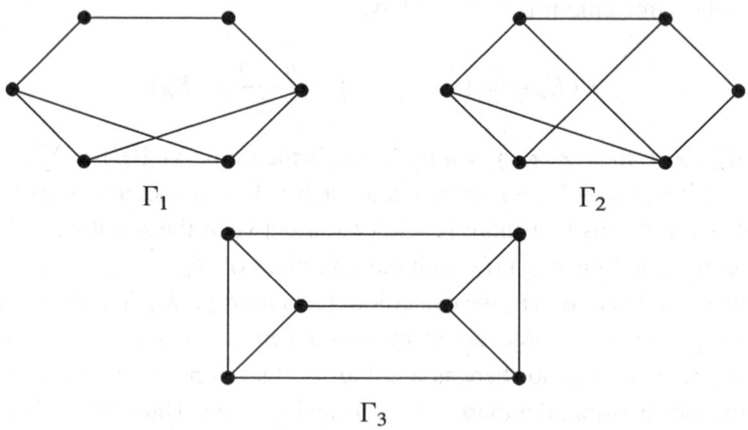

Figure 7.13. What are their chromatic indices?

Exercise 7.6. Find the chromatic indices of the graphs in Figure 7.13.

To conclude this section, we will study the chromatic index of Petersen's graph, shown in Figure 7.14. This is the 3-regular graph with 10 vertices $V = \{0, 1, 2, 3, 4, 5, 6, 7, 8, 9\}$ and 15 edges $E = \{01, 12, 23, 34, 40, 56, 67, 78, 89, 95, 05, 18, 26, 39, 47\}$.

Proposition 7.2.4. *We have* $\chi'(P) = 4$.

Proof. We first show that the graph P is 4-edge colourable. Number the successive vertices of the external cycle C of P from 0 to 4, and those of the internal cycle C' from 5 to 9. Then partition $E(P)$ as $\{01, 23, 56, 78\} \sqcup \{12, 34, 67, 89\} \sqcup \{05, 18, 26, 39, 47\} \sqcup \{04, 59\}$. This yields a 4-edge colouring of P.

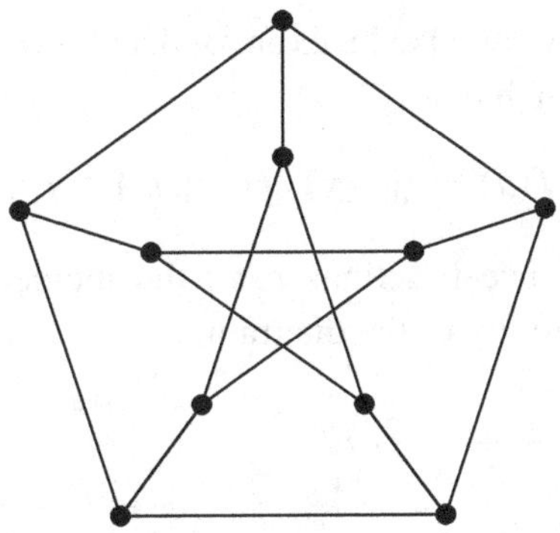

Figure 7.14. Petersen's graph.

We now show that P is not 3-edge colourable. Suppose that $\chi'(P) \leqslant 3$. The cycle $C = (0, 1, 2, 3, 4)$ is not 2-edge colourable, so it requires 3 colours. Let r be any colour, and let xy be an edge of C of colour r. The two vertices x' and y' on C' are connected to x and y respectively, defining two edges of colours different from r. Since $d(x') = 3$, there must exist another vertex $x'' \neq x$ on C' such that $x'x''$ has the colour r; the same holds for an edge $y'y''$ of C'. Consequently, C' has at least two different edges of colour r. Since this holds for the 3 colours of the colouring, we obtain a contradiction, since C' is of length 5. ■

As an immediate consequence, we see that P is not Hamiltonian. Indeed, if C is a Hamiltonian cycle of P, then C is 2-edge colourable. But since P is 3-regular, we can thus attribute to each of the 5 other edges connecting the vertices pairwise the same third colour, which is a contradiction.

7.3 Graph morphisms

The notion of a morphism concerns all graphs, whether or not they are simple and whether or not they have loops. Let $\Gamma = (V; E, N)$ and $\Gamma' = (V'; E', N')$ be two graphs. A *homomorphism* or *morphism* $f : \Gamma \to \Gamma'$ is a map

$$f : V \sqcup E \to V' \sqcup E'$$

satisfying

- $f(V) \subseteq V'$;
- $f(E) \subseteq E'$;
- if $a \in E$ is incident to $x \in V$, then $f(a) \in E'$ is incident to $f(x)$.

The *incidence property* can also be expressed as follows: if $a = ([x, y], n) \in E$, then there exists $n' \in N'$ such that

$$f(a) = ([f(x), f(y)], n') \in E'.$$

Or, we can use the incidence functions ε, ε'; the incidence property can then be expressed by the commutativity of the diagram

$$\begin{array}{ccc} E & \xrightarrow{\;f\;} & E' \\ \downarrow{\scriptstyle \varepsilon} & & \downarrow{\scriptstyle \varepsilon'} \\ P_2(V) & \xrightarrow{\;f\;} & P_2(V') \end{array} \qquad \varepsilon' \circ f = f \circ \varepsilon$$

We see that a multi-edge can become simple under a morphism, but a loop remains a loop. When Γ' is simple, giving $f|_V$ together with the incidence property suffices to determine f.

Exercise 7.7. (i) Show that the composition of two morphisms is a morphism.

(ii) Show that Id_Γ, the identity map of Γ, is a morphism.

(iii) Let $\Gamma = (V; E, N)$ be a graph and $\Gamma' = (V'; E')$ a simple graph, and suppose that $\varphi: V \to V'$ satisfies the incidence property. Show that there exists a unique morphism $f: \Gamma \to \Gamma'$ such that $f|_V = \varphi$.

Let $f: \Gamma \to \Gamma'$ be a morphism. If $\Delta = (V; E, N)$ is a subgraph of Γ, define the *image graph* of Γ by f, denoted $f(\Delta)$, as the subgraph of Γ' whose set of vertices is $f(V)$ and whose set of edges is $f(E)$. We have the following result on the conservation of connectedness in topological spaces under continuous maps.

Exercise 7.8. Let $f: \Gamma \to \Gamma'$ be a morphism and Δ a subgraph of Γ; prove that if Δ is connected, then $f(\Delta)$ is connected.

We say that a morphism $f: \Gamma \to \Gamma'$ is an *isomorphism* if there exists a morphism $g: \Gamma' \to \Gamma$ such that $f \circ g = \mathrm{Id}_{\Gamma'}$ and $g \circ f = \mathrm{Id}_\Gamma$.

Exercise 7.9. (i) Show that if

$$f: \Gamma = (V; E, N) \to \Gamma' = (V'; E', N')$$

is an isomorphism, we have

- $f|_V: V \to V'$ is bijective and
- $f|_E: E \to E'$ is bijective.

(ii) Conversely, if $f|_V$ and $f|_E$ are bijective and $\varepsilon' \circ f = f \circ \varepsilon$, show that f is an isomorphism.

The last property shows the coherence of these definitions with the notion of isomorphism introduced in Section 1.9.

We see thus that if f is an *injective morphism*, i.e. if $f|_V$ and $f|_E$ are injective, then f gives an isomorphism between Γ and $f(\Gamma)$; we also say that f is an *embedding* or an *immersion*.

These notions can be illustrated by interpreting the number $\omega(\Gamma)$ (Section 1.1.3), which is the maximum cardinality of a clique, in the following way. Let $\Gamma = (V; E)$ be a graph with $|V| = n$ and $|E| = m$, and define the complete graph with t vertices (for $t \geqslant 1$) as follows: $K_t = (W; A)$ where $W = \{1, 2, \dots, t\}$ and A is the set of all subsets of W with two elements if $t \geqslant 2$; $A = \emptyset$ if $t = 1$. Then $\omega(\Gamma)$ is the largest integer t for which there exists an embedding of K_n in Γ.

7.3.1 Graph quotients

Just as in topology, the notion of a quotient corresponds to the idea of a *gluing* (or identifying) described via equivalence relations. If $\mathcal{R}$ is an equivalence relation on X and $\mathcal{R}(x)$ denotes the class of x, then in the quotient set $X/\mathcal{R}$, all the elements of $\mathcal{R}(x)$ are glued together (or identified) to form a single element.

Let $\Gamma = (V; A, \varepsilon)$ be a graph (see Section 1.8). We can always glue vertices together, or edges together, but these gluings must be coherent to define a quotient. This coherence is expressed in the following definition.

Let $\mathcal{R}$ be an equivalence relation on $V \sqcup A$ such that

- $\mathcal{R}_{|V\times V}$ (denoted $\mathcal{R}_V$) is an equivalence relation on V;
- $\mathcal{R}_{|A\times A}$ (denoted $\mathcal{R}_A$) is an equivalence relation on A.

We require this equivalence relation to be *compatible*, which means that $a\mathcal{R}_A a'$ implies that the extremities of $\varepsilon(a)$ are *equivalent* to the extremities of $\varepsilon(a')$ in the following precise sense: if $\varepsilon(a) = [x, y]$ and $\varepsilon(a') = [x', y']$, then

- either $x\mathcal{R}_V x'$ and $y\mathcal{R}_V y'$,
- or $x\mathcal{R}_V y'$ and $y\mathcal{R}_V x'$

(and this holds even when a or a' are loops). Define $\Gamma/\mathcal{R} = (\overline{V}; \overline{A}, \overline{\varepsilon})$ by

- $\overline{V} = V/\mathcal{R}_V$;
- $\overline{A} = A/\mathcal{R}_A$;
- $\overline{\varepsilon}$: let $\overline{a} = \mathcal{R}_A(a)$ (i.e. the class of a) with $\varepsilon(a) = [x, y]$; set

$$\overline{\varepsilon}(\overline{a}) = [\mathcal{R}_V(x), \mathcal{R}_V(y)].$$

It is easy to see (Exercise 7.10) that $\overline{\varepsilon}$ is well defined; $\Gamma/\mathcal{R}$ is called the *graph quotient of* Γ *by* $\mathcal{R}$. Furthermore, the map $p\colon \Gamma \to \Gamma/\mathcal{R}$ defined by

- $p(x) = \mathcal{R}_V(x), x \in V$, and
- $p(a) = \mathcal{R}_A(a), a \in A$,

is a surjective graph morphism: $\overline{\varepsilon} \circ p = p \circ \varepsilon$.

Exercise 7.10. (i) Show that $\overline{\varepsilon}$ is well defined.

(ii) Show that $\overline{\varepsilon} \circ p = p \circ \varepsilon$.

(iii) Let $\Gamma = (V; A, \varepsilon)$ with $V = \{1,2,3,4\}$, $A = \{a, a', b, b', b'', c, d, e, f\}$ with $\varepsilon(a) = \varepsilon(a') = \varepsilon(d) = \{1,2\}$, $\varepsilon(b) = \varepsilon(b') = \varepsilon(b'') = \{2,3\}$, $\varepsilon(c) = \{1,4\}$, $\varepsilon(e) = \{1,3\}$, $\varepsilon(f) = \{3,4\}$.
Show that the equivalence $\mathcal{R}$ which *glues* 1 to 3, a to a', b to b' and b'' to d is compatible. Draw Γ and $\Gamma/\mathcal{R}$.

7.3.2 Morphisms and quotients of simple graphs

(1) We first need to adapt the notion of a morphism, since if Γ and Γ' are simple, it can happen that $\mathrm{Hom}(\Gamma, \Gamma')$ is very small, due to the absence of loops; also, in simple graphs, the information is reduced to adjacency or non-adjacency. We will say that $f: \Gamma \to \Gamma'$ is a morphism if

(i) $f: V \to V'$ is a map;

(ii) if $x, y \in V$ are adjacent, then either $f(x) = f(y)$, or $f(x)$ and $f(y)$ are adjacent.

(2) For the quotient, we construct $\Gamma/\mathcal{R}$ as we did for multigraphs; the resulting graph can have multiple loops and edges, so we consider its simplification. We check that the canonical map p is a morphism of simple graphs.

This comes down to accepting that a vertex can be adjacent to itself.

Exercise 7.11. Show that if $m \geqslant n$, then K_n and C_n are quotients of K_m and C_m respectively.

Exercise 7.12. (i) Let $\Gamma = (V; E)$ with $V = \{1,2,3,4,5,6\}$ and $E = \{12, 23, 34, 45, 56, 61\}$ (where kl denotes the edge $\{k, l\}$). Show that the graph quotients obtained by gluing 23 and 45 in two different ways ($2 \equiv 4$, $3 \equiv 5$ or else $2 \equiv 5$, $3 \equiv 4$) are not isomorphic.

(ii) Same question, by gluing 12 and 56, with

$$V = \{0,1,2,3,4,5,6,7\} \quad \text{and} \quad E = \{01, 12, 23, 34, 45, 56, 67\}.$$

Exercise 7.13. Let $L_\infty = (V; E)$ with $V = \mathbb{Z}$, $E = \{(k, k+1) \mid k \in \mathbb{Z}\}$ (an infinite chain) and $n \geqslant 3$ a fixed integer. Let $\mathcal{R}$ be the equivalence defined by $u\mathcal{R}v$ if and only if $n \mid (u - v)$ in $\mathbb{Z}$. Show that $L_\infty/\mathcal{R} \simeq C_n$.

7.3.3 Morphisms and colouring

The notion of a k-vertex colouring is related to the notion of a morphism by the following proposition.

Proposition 7.3.1. *Let $\Gamma = (V; E, N)$ be a loopless graph.*

(i) *If $f: V \to \{1, \dots, k\}$ is a k-vertex colouring of Γ, then f extends to a morphism $\tilde{f}: \Gamma \to K_k$ (where K_k is a complete graph with k vertices).*

(ii) *Conversely, if $f: \Gamma \to K_k$ is a morphism, $f|_V: V \to \{1, \dots, k\}$ is a k-vertex colouring of Γ.*

Proof. (i) Let $a = ([x, y], n) \in E$. Then we have $f(x) \neq f(y)$, which implies that $\{f(x), f(y)\}$ is an edge of K_k. Set $\tilde{f}(a) = \{f(x), f(y)\}$; this defines a morphism.

(ii) If x and y are adjacent, there exists an edge $a = ([x, y], n) \in E$, so $f(a)$ is an edge of K_k incident to $f(x)$ and $f(y)$, which implies that $f(x) \neq f(y)$; thus $f|_V$ is a k-vertex colouring. ∎

Corollary 7.3.2. *If $f: \Gamma \to \Gamma'$ is a morphism of loopless graphs, then $\chi(\Gamma) \leqslant \chi(\Gamma')$.*

Proof. Suppose that $\chi(\Gamma') = k$, so it is a k-vertex colouring of Γ'. By Proposition 7.3.1, we can see this k-vertex colouring as a morphism $h: \Gamma' \to K_k$; but then $h \circ f$ is a k-vertex colouring of Γ. Thus $\chi(\Gamma) \leqslant \chi(\Gamma')$. ∎

Let

$$f: V \to \{1, 2 \dots, k\}$$

be a k-vertex colouring of a loopless graph $\Gamma = (V; E, N)$. We can associate to f a partition of V:

$$V = \bigsqcup_{i \in f(V)} f^{-1}(i) \quad \text{(we take only the } f^{-1}(i) \text{ that are non-empty).}$$

It is easy to check that the sets $f^{-1}(i)$ are independent sets (since Γ is loopless). Conversely, to each partition of V into independent sets $V = V_1 \sqcup V_2 \sqcup \dots \sqcup V_k$, we can associate a k-vertex colouring by attributing the colour i to all the vertices of V_i.

More generally, consider a morphism $f: \Gamma_1 = (V_1; E_1, N_1) \to \Gamma_2 = (V_2; E_2, N_2)$, where Γ_2 is loopless. The partition $V_1 = \bigsqcup_{y \in V_2} f^{-1}(y)$ consists of independent sets (since Γ_2 is loopless), so that Γ_1 is k-partite with $k = |f(V_1)|$. This partition can be interpreted as a sort of kernel. To understand this, consider the following general situation. Let $\Gamma = (V; E, N)$ be a graph and $\mathcal{R}$ an equivalence relation on V. Let $\Gamma/\mathcal{R} = (\overline{V}; \overline{E}, \overline{N})$ denote the quotient graph, and $p: \Gamma \to \Gamma/\mathcal{R}$ the canonical surjection. Then we have the following *universal quotient proposition*.

Theorem 7.3.3. *If $f:\Gamma \to \Gamma'$ is a morphism that is constant on the classes modulo $\mathcal{R}$ (i.e. $x\mathcal{R}y \Rightarrow f(x) = f(y)$), then there exists a unique morphism $\tilde{f}:\Gamma/\mathcal{R} \to \Gamma'$ such that $\tilde{f} \circ p = f$:*

$$\begin{array}{ccc} \Gamma & \xrightarrow{\;p\;} & \Gamma/\mathcal{R} \\ & {\searrow}^{f} & \downarrow \tilde{f} \\ & & \Gamma' \end{array}$$

Proof. The existence and the uniqueness of $\varphi: V/\mathcal{R} \to V(\Gamma')$ satisfying $\varphi \circ p|_V = f|_V$ are immediate. It remains to define $\psi: \overline{E} \to E(\Gamma')$. If $a = ([\mathcal{R}(x), \mathcal{R}(y)], e) \in E(\Gamma/\mathcal{R})$, set $\psi(a) = f(e)$; then $\psi \circ p|_E = f|_E$. Taking $\tilde{f}|_{\overline{V}} = \varphi$ and $\tilde{f}|_{\overline{E}} = \psi$, we obtain $\tilde{f} \circ p = f$. ■

Let us apply this theorem to the following situation: let Γ and Γ' be two simple graphs, and let $f:\Gamma \to \Gamma'$ be a morphism. Consider the following equivalence relation associated to f: $x\mathcal{R}_f y$ if and only if $f(x) = f(y)$. By Theorem 7.3.3, there exists a unique $g = \tilde{f}$ such that $g \circ p = f$. Moreover, here $g|_{\overline{V}}$ and $g|_{\overline{E}}$ are injective (exercise). But then $\Gamma/\mathcal{R}_f$ is isomorphic to $f(\Gamma)$, which shows that $\mathcal{R}_f$ plays the role of $\operatorname{Ker}(f)$ (the *kernel* of the morphism f), and the classes modulo $\mathcal{R}_f$ are the $f^{-1}(y)$.

7.4 Perfect graphs

In this paragraph, we will assume that all graphs are simple. Let $\Gamma = (V; E)$ be a graph. The *index of partition into cliques* is the minimal number of cliques that are necessary to provide a partition of the set of vertices; it is written $\theta(\Gamma)$.

We have $\theta(C_4) = 2$ and $\theta(C_5) = 3$ (exercise).

The following lemma gives relations between the parameters ω (the maximum cardinality of a clique), χ (the chromatic number), θ, and α (the maximum cardinality of a stable set).

Lemma 7.4.1. *Let $\Gamma = (V; E)$ be a simple graph. Then*

$$\chi(\Gamma) \geqslant \omega(\Gamma), \quad \alpha(\Gamma) \leqslant \theta(\Gamma),$$
$$\alpha(\Gamma) = \omega(\overline{\Gamma}), \quad \theta(\Gamma) = \chi(\overline{\Gamma}).$$

Proof. Left to the reader. ■

It can happen that χ is much bigger than ω. Using non-effective probabilistic methods, P. Erdős proved the following result.

Theorem 7.4.2 (Erdős). *Let $g \geqslant 3$ and $k \geqslant 2$ be two positive integers. There exists a graph Γ of order k^{3g} such that $\chi(\Gamma) \geqslant k$ and $\mu(\Gamma) \geqslant g$, where $\mu(\Gamma)$ is the girth of Γ.*

Perfect graphs are the opposite of this situation: a graph is said to be *perfect* if for every induced subgraph Γ' of Γ we have $\chi(\Gamma') = \omega(\Gamma')$; in other words, the number of colours $\omega(\Gamma')$ trivially necessary to colour Γ' is also sufficient.

Let Γ be an arbitrary graph such that $\chi(\Gamma) \leqslant k$. Is it possible to give an effective way to colour Γ with k colours? This problem is $\mathcal{NP}$-complete for $k \geqslant 3$ (the case $k = 2$ is left to the reader). However, if the graph is perfect, then it is in the class $\mathcal{P}$, i.e. there exists an algorithm that can solve the problem in polynomial time.

Perfect graphs were introduced by C. Berge, who proposed the following conjecture.

Conjecture 7.4.3 (*Conjecture of perfect graphs*). *A graph is perfect if and only if its complement is perfect.*

This conjecture was proven by L. Lovász.

Theorem 7.4.4 (*Theorem of perfect graphs*, Lovász, 1972). *A simple graph $\Gamma = (V; E)$ is perfect if and only if its complement $\overline{\Gamma}$ is perfect.*

Proof. We will not give this delicate proof here. ∎

Recall that C_n denotes the cycle of length n. Observe that $\chi(C_3) = \omega(C_3) = 3$, but if $n \geqslant 5$ is odd, then $\chi(C_n) = 3 > \omega(C_n) = 2$. Thus if a graph, or its complement, contains such a cycle C_n, it cannot be perfect (by Lovász' theorem).

Are odd cycles the only obstruction to perfection? That was the audacious question behind the following conjecture.

Conjecture 7.4.5 (*Strong conjecture on perfect graphs*). *A graph is perfect if and only if neither it nor its complement contains a cycle of odd length* $\geqslant 5$.

First proposed by C. Berge in 1966, this conjecture was proven in 2002 by Chudnovsky, Robertson, Seymour and Thomas.

Exercise 7.14. Show that bipartite graphs are perfect graphs.

Here is another class of perfect graphs: a *comparability graph* is a graph associated to a preorder relation (namely a binary relation that is reflexive and transitive) in the following way. Let $\leqslant$ be a preorder on the set X, and define the simple graph $\Gamma = (V; E)$ by taking

- $V = X$;
- $\{x, y\} \in E$ if and only if $x < y$ or $y < x$ (i.e. x and y are distinct and comparable).

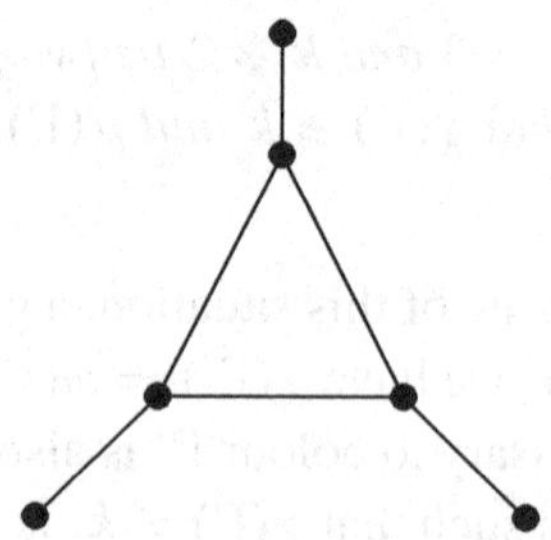

Figure 7.15. A graph that is not a comparability graph.

Example 7.4.6. The graph shown in Figure 7.15 is not a comparability graph: no matter which order is chosen on the vertices of the triangle, one of the edges adjacent to this triangle cannot be oriented correctly: the transitivity edge will be missing.

Theorem 7.4.7. *Comparability graphs are perfect.*

Proof. As every induced subgraph of a comparability graph is also a comparability graph, it suffices to see that $\chi(\Gamma) = \omega(\Gamma)$ for a comparability graph $\Gamma = (V; E)$ associated to the preorder $\leqslant$. Let $x \in V$, and let $r(x)$ denote the largest integer r for which there exist r vertices x_i, $i = 1, \dots, r$, such that $x_1 \leqslant x_2 \leqslant \cdots \leqslant x_r = x$. Note that the transitivity of the preorder relation $\leqslant$ implies that the subgraph induced by the vertices x_i is complete, so it is a clique of Γ. Let $k = \max_x r(x)$, and take $y \in V$ such that $r(y) = k$. The size of a maximum clique is then $\omega(\Gamma) = k$, and the map

$$r: V \to \{1, 2, 3, \dots, k\}$$

is a colouring of Γ. If $x < y$, then clearly $r(x) < r(y)$, so if $\{x, y\} \in E$ we have $r(x) \neq r(y)$. Thus $\chi(\Gamma) \leqslant \omega(\Gamma)$, but then $\chi(\Gamma) = \omega(\Gamma)$ by Lemma 7.4.1. ∎

Remark 7.4.8. In a comparability graph, only the *ordered chains* of length k induce a clique of size k.

Let $\{I_1, I_2, \dots, I_n\}$ be a set of intervals of $\mathbb{R}$, and associate to it a simple graph $\Gamma = (V; E)$:

- $V = \{I_1, I_2, \dots, I_n\}$;
- for $i \neq j$, $\{I_i, I_j\} \in E$ if and only if $I_i \cap I_j \neq \emptyset$.

This graph is called an *interval graph.*

Proposition 7.4.9. *The complement of an interval graph is a comparability graph.*

Proof. Define a preorder on $V = \{I_1, I_2, \dots, I_n\}$ as follows:

$$I_i \leqslant I_j \text{ if and only if } i = j \text{ or } \forall x \in I_i, \forall y \in I_j : x < y.$$

This preorder is associated to a comparability graph $\Delta = (V; E')$. Let $i \neq j$. Then

- if $\{I_i, I_j\} \in E'$ we have $I_i \leqslant I_j$ or $I_j \leqslant I_i$, so $I_i \cap I_j = \emptyset$ and $\{I_i, I_j\} \notin E$;
- if $\{I_i, I_j\} \notin E'$, then $I_i \cap I_j \neq \emptyset$. But as $\mathbb{R}$ is totally ordered, we have $\forall x \in I_i$, $\forall y \in I_j$, $x < y$ or the contrary, i.e. $I_i \leqslant I_j$ or $I_j \leqslant I_i$. Hence $\{I_i, I_j\} \in E'$, so $\overline{\Gamma} = \Delta$ is a comparability graph. ■

Corollary 7.4.10. *Interval graphs are perfect graphs.*

Proof. It follows from Theorem 7.4.4, Theorem 7.4.7 and Proposition 7.4.9. ■

Algorithmic remarks. Determining the invariants $\omega(\Gamma)$, $\chi(\Gamma)$ and $\alpha(\Gamma)$ of a graph Γ is an $\mathcal{NP}$-hard problem, which means that every $\mathcal{NP}$-problem can be reduced in polynomial time to the problem of computing one of the parameters above (see Section 1.7.2). Thus it is not very probable (unless $\mathcal{P} = \mathcal{NP}$) that polynomial algorithms exist to compute ω, χ and α. However, all these parameters can be computed in polynomial time for perfect graphs, which makes this class particularly interesting from the algorithmic point of view. Further details on these graphs and the associated algorithmics can be found in [2].

7.5 List colourings

List colourings are more general versions of vertex colourings: they are in fact vertex colourings in which colours must be chosen from a specific list of colours associated to each vertex.

Associate to each vertex x of a graph $\Gamma = (V; E, N)$ a list $L(x)$ of k available colours for that vertex. A *list colouring* is a colouring f such that $f(x)$ must be chosen in $L(x)$.

A graph is *k-list colourable* if for every family $L(x)_{x \in V}$ of sets with k elements, where $k = |L(x)|$, there exists a vertex colouring f such that for all x we have $f(x) \in L(x)$. The smallest integer k such that Γ is k-list colourable is called the *list chromatic number* of Γ, and denoted $\chi_\ell(\Gamma)$.

Similarly, we can define the notion of a list edge colouring. The *list chromatic index* is denoted $\chi'_\ell(\Gamma)$.

Clearly $\chi_\ell(\Gamma) \geqslant \chi(\Gamma)$. A k-colouring comes down to choosing a list $L(x) = \{1, 2, \dots, k\}$ for each x. Similarly, we have $\chi'_\ell(\Gamma) \geqslant \chi'(\Gamma)$.

Example 7.5.1. The chromatic number of $K_{2,2}$ is equal to 2 (it is a bipartite graph), and its list chromatic number is equal to 2 (Figure 7.16). If $\{a_1, b_1\} \cap \{a_2, b_2\} \neq \emptyset$ or $\{a_3, b_3\} \cap \{a_4, b_4\} \neq \emptyset$, we suppose for example that $a_1 = a_2$ and we colour the

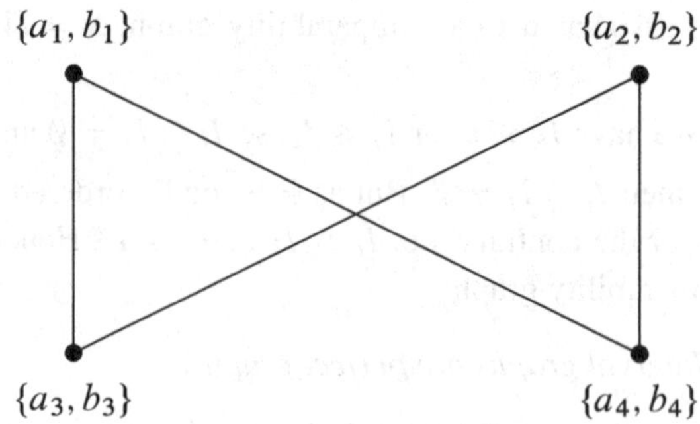

Figure 7.16. List chromatic number of $K_{2,2}$.

upper vertices (at the top of the figure) with a_1. For the lower vertices (at the bottom), it suffices to take a different colour from a_1 (there exists at least one such in each list).

If $\{a_1, b_1\} \cap \{a_2, b_2\} = \emptyset$ and $\{a_3, b_3\} \cap \{a_4, b_4\} = \emptyset$, we may assume that $\{a_1, b_1, a_2, b_2\} = \{a_3, b_3, a_4, b_4\}$ up to replacing the colour of the lower vertices by a colour already present in the lists of the upper vertices. Thus, up to exchanging the roles of two lower vertices and the colours in each list, we find two possible configurations:

- either $(a_1, b_1) = (a_3, b_3)$ and $(a_2, b_2) = (a_4, b_4)$: we colour the vertices with a_1, a_2, b_1, b_2 respectively;
- or $(a_1, a_2) = (a_3, a_4)$ and $(b_1, b_2) = (b_4, b_3)$: we colour the four vertices with a_1, a_2, b_2, b_1 respectively.

Example 7.5.2. We have $\chi(K_{3,3}) = 2$ (it is a bipartite graph), but the choice of lists indicated in Figure 7.17 shows that $\chi_\ell(K_{3,3}) \geqslant 3 > \chi(K_{3,3})$.

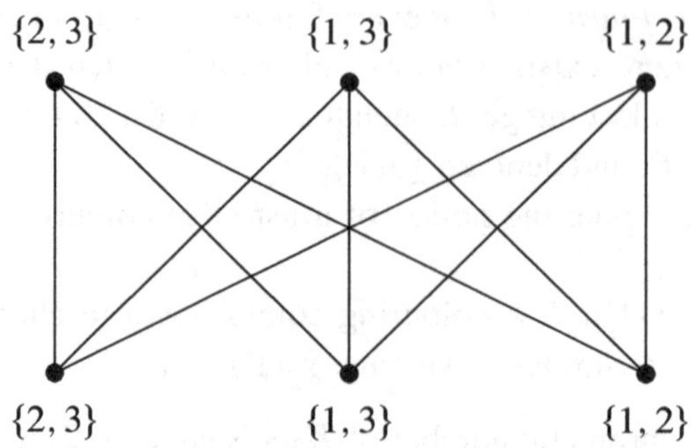

Figure 7.17. List chromatic number of $K_{3,3}$.

The following theorem yields a particularly important result.

Theorem 7.5.3 (Thomassen, 1994). *Every planar graph is 5-list colourable.*

For the list chromatic index, the situation seems to be less complicated, as indicated by the following conjecture.

Conjecture 7.5.4 (*On list edge colourings*). *Every simple graph* $\Gamma = (V; E)$ *satisfies* $\chi'_\ell(\Gamma) = \chi'(\Gamma)$.

At present, this conjecture has only been proven for bipartite graphs.

Theorem 7.5.5 (Galvin, 1995). *For every simple bipartite graph* $\Gamma = (V; E)$, *we have* $\chi'_\ell(\Gamma) = \chi'(\Gamma)$.

We conclude this paragraph with the description of an application of list colourings to telecommunications. In order to avoid interference during mobile phone calls, the frequencies used by the cell phone towers covering overlapping territories must be chosen in such a way that a given frequency cannot be used simultaneously by two towers whose territories overlap. We model this problem as follows. The towers are the vertices of the graph, and we place an edge between two towers if their territories overlap. Each tower has a fixed number k of distinct available frequencies; in other words, each vertex possesses a list of "colours" $L(x) = \{f_1, f_2, \ldots, f_k\}$. At each moment, the telephone company must ensure that if the frequency (or colour) i is being used by the vertex x, then the neighbouring vertices cannot receive the frequency i (i.e. each neighbouring vertex must be coloured with a colour from its list different from i). Thus this problem is actually a list colouring problem.

7.6 Chromatic polynomials

We conclude this chapter with some further elementary notions. In this section, we admit graphs with loops.

We saw how to colour the vertices of a graph with k colours. This leads to the question of how many distinct ways there are to obtain such a colouring.

Let $\Gamma = (V; E, N)$ be a multigraph with at least one vertex, and let $k \geqslant 1$. Let $P_\Gamma(k)$ denote the number of possible colourings

$$f : V(\Gamma) \to \{1, 2, 3, \ldots, k\}$$

of Γ with at most k colours. It is natural to set $P_\Gamma(k) = 0$ when Γ possesses a loop, since a vertex that is adjacent to itself cannot be coloured.

We see that P_Γ is an increasing function with integer values: $P_\Gamma\colon \mathbb{N}^* \to \mathbb{N}$. Clearly,

- if $k < \chi(\Gamma)$, then $P_\Gamma(k) = 0$,
- if $k \geqslant \chi(\Gamma)$, then $P_\Gamma(k) > 0$

(where $\chi(\Gamma)$ is the chromatic number of Γ, see Chapter 7).

Consequently $\chi(\Gamma)$ is the smallest integer k such that $P_\Gamma(k) \neq 0$.

Theorem 7.6.1. *Let $\Gamma = (V; E, N)$ be a multigraph. Then for every edge a, we have*

$$P_\Gamma(k) = P_{\Gamma\backslash a}(k) - P_{\Gamma/a}(k), \quad k \geqslant 1.$$

Proof. Let $\varepsilon(a) = [x, y]$.

If a is a 1-loop, then $P_\Gamma(k) = 0$ and clearly $P_{\Gamma\backslash a}(k) = P_{\Gamma/a}(k)$. If there is a p-loop ($p \geqslant 2$) between x and y, then $P_\Gamma(k) = P_{\Gamma\backslash a}(k) = P_{\Gamma/a}(k) = 0$.

If a is a 1-edge, the number of k-colourings of $\Gamma\backslash a$ such that x and y have different colours is the same as the number of k-colourings of Γ, which is equal to $P_\Gamma(k)$, while the number of k-colourings of $\Gamma\backslash a$ such that x and y have the same colour does not change when we fuse the two vertices x and y, so it is the same as in Γ/a. Thus $P_{\Gamma\backslash a}(k) = P_\Gamma(k) + P_{\Gamma/a}(k)$. If there is a p-edge ($p \geqslant 2$) between x and y, then $P_\Gamma(k) = P_{\Gamma\backslash a}(k)$ and $P_{\Gamma/a}(k) = 0$ since Γ/a possesses a loop. ∎

From this particularly simple theorem, we will deduce that P_Γ can be considered in some sense as a polynomial; more precisely, as a polynomial function of the integer variable k, which essentially comes to the same thing since $\mathbb{Z}$ is infinite (over a finite ring, there exist non-zero polynomials which are identically zero as polynomial functions).

Corollary 7.6.2. *P_Γ is a polynomial function.*

Proof. Let us show that there exists a (necessarily unique) polynomial $F_\Gamma \in \mathbb{Z}[x]$ such that $F_\Gamma(k) = P_\Gamma(k)$ for all $k \geqslant 1$. We use induction on $m = |E|$.

If $m = 0$, then $P_\Gamma(k) = k^n$, where $n = |V|$, so P_Γ is the polynomial function x^n.

$m \curvearrowright m + 1$: Let Γ have $m + 1$ edges. The function $F_\Gamma = P_{\Gamma\backslash a} - P_{\Gamma/a}$ is a polynomial function by the induction hypothesis, but by Theorem 7.6.1 it satisfies $F_\Gamma(k) = P_\Gamma(k)$ for every integer k, so P_Γ is itself a polynomial function. ∎

P_Γ is called the *chromatic polynomial* of Γ, and we have

$$P_\Gamma(x) = P_{\Gamma\backslash a}(x) - P_{\Gamma/a}(x), \quad \forall a \in E. \tag{7.1}$$

Proposition 7.6.3. *Let $\Gamma = (V; E, N)$ be a multigraph whose connected components are denoted $(\Gamma_i)_{1 \leqslant i \leqslant c}$. Then*

$$P_\Gamma = \prod_{1 \leqslant i \leqslant c} P_{\Gamma_i}.$$

Proof. Indeed, the connected components have independent colourings. ■

Theorem 7.6.4. *Let $\Gamma = (V; E, N)$ be a multigraph having $n \geqslant 1$ vertices, m edges and t connected components. Then its chromatic polynomial is*

$$P_\Gamma(x) = x^n + \sum_{1 \leqslant i \leqslant n} (-1)^i a_i x^{n-i} \in \mathbb{Z}[x],$$

where $a_1 = m$, $a_i \geqslant 1$, for $2 \leqslant i \leqslant n - t$, and $a_i = 0$ for $n - t < i \leqslant n$.

Proof. We use induction on m.

If $m = 0$, then clearly $P_\Gamma(x) = x^n$ (we could also use Proposition 7.6.3).

If $m = 1$, denote by a the unique edge. Then $P_{\Gamma \backslash a}(x) = x^n$, $P_{\Gamma / a}(x) = x^{n-1}$, so by (7.1), we obtain $P_\Gamma(x) = x^n - x^{n-1}$, which satisfies the desired properties.

$m \curvearrowright m + 1$: let Γ have $m + 1$ edges. By the induction hypothesis,

$$P_{\Gamma \backslash a}(x) = x^n + \sum_{1 \leqslant i \leqslant n} (-1)^i b_i x^{n-i}$$

with $b_1 = m$, $b_i \geqslant 1$ for $2 \leqslant i \leqslant n - t'$, and $b_i = 0$ for $n - t' < i \leqslant n$ (where t' is the number of connected components of $\Gamma \backslash a$; we have $t' \geqslant t$). As $m \geqslant 1$, we have $n \geqslant 2$, and again by the induction hypothesis,

$$P_{\Gamma / a}(x) = x^{n-1} + \sum_{1 \leqslant i \leqslant n-1} (-1)^i c_i x^{n-1-i},$$

with $c_1 = m + 1 - 1 = m$, $c_i \geqslant 1$ for $2 \leqslant i \leqslant n - t$, and $c_i = 0$ for $n - 1 - t < i < n - 1$. By shifting the summation variable ($j = i + 1$), we can write

$$P_{\Gamma / a}(x) = x^{n-1} + \sum_{2 \leqslant j \leqslant n} (-1)^{j-1} c_{j-1} x^{n-j}.$$

By (7.1), we then obtain

$$P_\Gamma(x) = x^n + (-b_1 - 1)x^{n-1} + \sum_{2 \leqslant i \leqslant n} [(-1)^i b_i + (-1)^i c_{i-1}] x^{n-i};$$

the coefficients a_i of P_Γ are thus $a_1 = b_1 + 1 = m + 1$, $a_2 = b_2 + c_1 = b_2 + m \geqslant 1$, then $a_i = b_i + c_{i-1}$;

- for $3 \leqslant i \leqslant n - t$, $a_i \geqslant 1$ (since $c_{i-1} \geqslant 1$);
- for $n - t < i \leqslant n$: $c_{i-1} = 0$ and $b_i = 0$ (since $i > n - t'$), so $a_i = 0$. ■

If Γ is connected, then $t = 1$, and Theorem 7.6.4 leads to the following corollary.

Corollary 7.6.5. *$\Gamma = (V; E, N)$ is a connected multi-graph if and only if* only *the constant term of its chromatic polynomial is zero.*

Proposition 7.6.6. *Let $\Gamma = (V; E, N)$ be a multigraph with $n \geqslant 1$ vertices. Then Γ is a tree if and only if $P_\Gamma(x) = x(x-1)^{n-1}$.*

Proof. We use induction on n.

If $n = 1$, $P_\Gamma(x) = x$.

$n-1 \curvearrowright n$: Let Γ be a graph with n vertices. If Γ is a tree, then by Proposition 2.2.5, it admits at least two vertices of degree 1. Let x be such a vertex, and let $a = \{x, y\}$ be an edge incident to this vertex. The graph $\Gamma\backslash a$ admits two connected components: one that is reduced to a single vertex, whose chromatic polynomial is x, and the other that is a tree with $n-1$ vertices (namely Γ/a), whose chromatic polynomial is, by the induction hypothesis, equal to $x(x-1)^{n-2}$. Therefore $P_{\Gamma\backslash a}(x) = xP_{\Gamma/a}(x)$, so we have

$$P_\Gamma(x) = P_{\Gamma\backslash a}(x) - P_{\Gamma/a}(x) = (x-1)P_{\Gamma/a}(x) = x(x-1)^{n-1}.$$

Conversely, let $\Gamma = (V; E, N)$ have chromatic polynomial

$$P_\Gamma(x) = x(x-1)^{n-1} = x^n - (n-1)x^{n-1} + \cdots + (-1)^{n-1}x.$$

Then Theorem 7.6.4 shows that $|E| = n - 1 = |V| - 1$, and Corollary 7.6.5 shows that Γ is connected. By Theorem 2.2.1, it follows that Γ is a tree. ∎

Exercise 7.15. Compute the chromatic polynomials of

(i) graphs with 2 edges;

(ii) connected graphs with 3 edges;

(iii) the complete graph K_n.

Exercise 7.16. True or false?

$$\Gamma \simeq \Gamma' \iff P_\Gamma = P_{\Gamma'}.$$

Chapter 8

Matching and factorisation

Consider a set of boys, and a set of girls of which each is acquainted with a specific subset of the set of boys. Under what conditions on these sets and subsets can each girl marry a boy of her acquaintance?

Alternatively, consider a business in which k different specific tasks are necessary to construct a given object; the business has t employees, each of whom is competent in at least one of the tasks. Under what conditions is it possible to distribute all the tasks in such a way that every employee is set a task for which he or she is competent?

Problems such as these, generally called *marriage problems*, were given a complete solution by P. Hall in 1935. The key concept used by Hall is that of a *matching* in a graph; a matching is nothing other than a set of graph edges that are pairwise non-incident. Determining a maximum matching is a problem that arises frequently in combinatorics, engineering, optimisation and other domains. Both of the above problems can be modelled as finding maximum matchings in a bipartite graph.

More generally, we will define the k-factorisation of a graph, which generalises the notion of a matching. Other related questions, such as finding the maximum cardinality of an independent set or finding vertex covers or edge covers (defined below) are also related to matching problems. This chapter is devoted to the study of this type of question. *All graphs are assumed to be of finite order* and usually *simple without isolated vertices.*

8.1 Definitions and first properties

Let $\Gamma = (V; E, N)$ be a graph. Recall that a subset of vertices S of V is an *independent set* or a *stable set* if no two vertices of S are adjacent. A stable set S is said to be *a maximum stable set* if for every stable set S', we have $|S'| \leqslant |S|$. The cardinality of a maximum stable set is called the *stability number* of Γ and denoted $\alpha(\Gamma)$.

Remark 8.1.1. A maximum stable set is maximal among all stable sets in the sense that if $S \subset S'$ with S' stable, then $S' = S$. However, a maximal stable set is not necessarily maximum; it does not necessarily have maximum cardinality.

In contrast, a subset T of V is said to be *transversal* if every edge of Γ is incident to a vertex in T, i.e. $\forall a \in E, \varepsilon(a) \cap T \neq \emptyset$. A *minimum transversal set* is a transversal set of minimum cardinality; this minimum cardinality is denoted $\beta(\Gamma)$ and called the *transversality number* of the graph Γ.

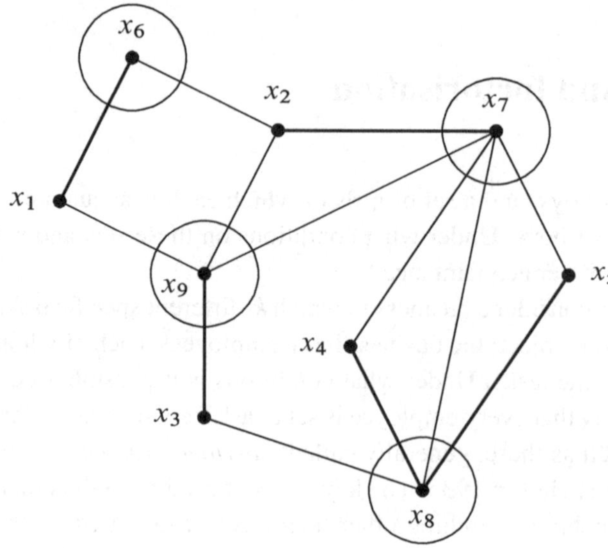

Figure 8.1. Stability, transversality, matching and covering in a graph. The set $\{x_1, x_2, x_3, x_4, x_5\}$ is a maximum stable set: $\alpha(\Gamma) = 5$. The circled vertices form a minimum transversal set: $\beta(\Gamma) = 4$. The edges $\{x_1, x_9\}, \{x_2, x_6\}, \{x_3, x_8\}, \{x_5, x_7\}$ form a maximum matching: $\alpha'(\Gamma) = 4$. The boldface edges form a minimum edge cover: $\beta'(\Gamma) = 5$.

Notions analogous to these can be given for edges instead of vertices.

A *matching* C in a graph Γ is a stable set of edges, i.e. a set such that no two edges of C are incident: $\forall a, a' \in C, \varepsilon(a) \cap \varepsilon(a') = \emptyset$. The matching C is *maximum* if for every matching C' of Γ, we have $|C'| \leqslant |C|$ (as a consequence, such a matching is maximal for the inclusion). The cardinality of a maximum matching is denoted $\alpha'(\Gamma)$.

Conversely, a subset of edges $R \subset E$ is said to be an *edge cover* if every vertex of Γ is incident to one of the edges of R, i.e. $V \subset \varepsilon(R)$. The edge cover $R \subset E$ is *minimum* if for every other edge cover R', we have $|R| \leqslant |R'|$. The cardinality of a minimum edge cover is denoted $\beta'(\Gamma)$. A matching that is also an edge cover is called a *perfect matching*.

Remark 8.1.2. This notion only makes sense if Γ has no isolated vertex.

Remark 8.1.3. It is clear that if Γ_s is the simplified graph of Γ, then

$$\alpha(\Gamma_s) = \alpha(\Gamma), \quad \beta(\Gamma_s) = \beta(\Gamma),$$
$$\alpha'(\Gamma_s) = \alpha'(\Gamma), \quad \beta'(\Gamma_s) = \beta'(\Gamma).$$

For this reason, in the sequel, all graphs will be assumed *simple*.

Figure 8.1 illustrates the different notions introduced above.

Summary and notation.

- Stable/independent set (non-adjacent vertices) S, max $= \alpha$;
- Matching set (non-incident edges) C, max $= \alpha'$;
- Transversal set (every edge is incident) T, min $= \beta$;
- Edge cover (every vertex is incident) R, min $= \beta'$.

As we will see, interesting relations exist between the numbers $\alpha(\Gamma), \alpha'(\Gamma), \beta(\Gamma)$ and $\beta'(\Gamma)$; they are known as Gallai's relations.

Theorem 8.1.4 (Gallai, 1959). *Let $\Gamma = (V; E)$ be a simple graph. Then we have*

$$\alpha(\Gamma) + \beta(\Gamma) = |V|. \tag{8.1}$$

If moreover Γ has no isolated vertex, then

$$\alpha'(\Gamma) + \beta'(\Gamma) = |V|. \tag{8.2}$$

Proof. Let $S \subset V$. Then S stable $\Leftrightarrow$ $V \backslash S$ transversal. Indeed,

- if S is stable, every edge a is incident to $V \backslash S$, since otherwise we would have $\varepsilon(a) \subset S$;
- if $V \backslash S$ is transversal, then for all $x, y \in S$, x and y are not adjacent, since otherwise the edge $e = (\{x, y\}, n)$ would not be incident to $V \backslash S$.

This proves (8.1) for a stable set S of maximum cardinality. Let us prove (8.2).

Let C be a matching of maximum cardinality $\alpha'(\Gamma)$, and let $\varepsilon(C)$ denote as usual the set of extremities of the edges in C. Then $S = V \backslash \varepsilon(C)$ is a stable set, since if an edge a had both extremities in S, then $C \cup \{a\}$ would be a matching. Choose an incident edge for each vertex of S (which is possible since the graph has no isolated vertex); these edges are all distinct and form a set A such that $|A| = |S|$. The subset $C \cup A$ of E is an edge cover of cardinality $|C| + |S| = |C| + |V| - |\varepsilon(C)| = |C| + |V| - 2|C| = |V| - |C| = |V| - \alpha'(\Gamma)$. Thus we have $\beta'(\Gamma) \leqslant |V| - \alpha'(\Gamma)$.

Now let R be an edge cover of cardinality $\beta'(\Gamma)$, and let $\Gamma(R)$ denote the partial graph determined by the edges of R (it is a partial graph since Γ has no isolated vertex). Then for all $a \in R$, $R \backslash \{a\}$ is no longer an edge cover. This implies that each edge of R has one extremity of degree 1 in the partial graph $\Gamma(R)$, so the connected components of $\Gamma(R)$ are stars K_{1,t_i}, $i = 1, \ldots, l$. For a star $K_{1,t}$, we have the relation $|V(K_{1,t})| = |E(K_{1,t})| + 1$ between the numbers of its vertices and edges. Take an edge e_i for each connected component K_{1,t_i}, $1 \leqslant i \leqslant l$, of $\Gamma(R)$; the set $C = \{e_1, \ldots, e_l\}$ is then a matching, and we have

$$|V| = \sum_{i=1}^{l} |V(K_{1,t_i})| = l + \sum_{i=1}^{l} |E(K_{1,t_i})| = |C| + |R| = |C| + \beta'(\Gamma).$$

Since $|C| \leqslant \alpha'(\Gamma)$, we finally obtain $|V| \leqslant \alpha'(\Gamma) + \beta'(\Gamma)$, which concludes the proof. ∎

Exercise 8.1. (i) Prove that in general, if Γ has no isolated vertex, then

$$\alpha'(\Gamma) \leqslant \frac{|V|}{2} \leqslant \beta'(\Gamma).$$

(ii) Show that Γ admits a perfect matching if and only if

$$\alpha'(\Gamma) = \frac{|V|}{2} = \beta'(\Gamma).$$

We note in passing that $|V|$ is even whenever Γ admits a perfect matching.

An *alternating chain* relative to a matching C is a chain

$$(x_1, a_1, x_2, a_2, \dots, x_{n-1}, a_{n-1}, x_n)$$

satisfying

- $x_1 \in V \setminus \varepsilon(C)$;
- $a_1, a_3, \dots, a_{2k-1}, \dots \notin C$;
- $a_2, a_4, \dots, a_{2k}, \dots \in C$.

When the graphs are bipartite, we have the following result, which completes Gallai's theorem (Theorem 8.1.4).

Theorem 8.1.5 (König, 1931). *Let $\Gamma = (V; E)$ be a simple bipartite graph. Then*

$$\alpha'(\Gamma) = \beta(\Gamma). \tag{8.3}$$

If moreover Γ has no isolated vertex, then

$$\alpha(\Gamma) = \beta'(\Gamma).$$

Proof. We may assume that Γ is connected. By Gallai's theorem (Theorem 8.1.4), it suffices to prove (8.3). Write the bipartite graph Γ in the form $\Gamma = (V_1 \sqcup V_2; E)$, and let C be a maximum matching $|C| = \alpha'(\Gamma)$. If $|E| = 1$, we are done; otherwise we define $T \subset V_1 \sqcup V_2$ as follows: for each $a \in C$, $a = \{v_1, v_2\}$, $v_1 \in V_1$, $v_2 \in V_2$, we put either v_2 or v_1 in T according to whether there exists (or not) an alternating chain $(x_1, a_1, x_2, a_2, \dots, x_{n-1}, a_{n-1}, x_n)$ ending at $x_n = v_2$. Thus, by construction, $|T| = |C| = \alpha'(\Gamma)$.

Let us show that T *is a transversal set*, namely that every edge is incident to T. Let $a = \{w_1, w_2\}$, $w_1 \in V_1$, $w_2 \in V_2$, be an edge.

(1) If $a \in C$, then a is incident to T since by construction w_1 or w_2 is in T.

(2) If $a \notin C$, we distinguish two cases:

(a) either $w_1 \notin \varepsilon(C)$; then (w_1, a, w_2) is an alternating chain, so $w_2 \in T$;

(b) or $w_1 \in \varepsilon(C)$; then there exists $a' = \{w_1, v_2\} \in C$:

- if $w_1 \in T$, we are done;
- otherwise $w_1 \notin T$; then by construction $v_2 \in T$: there exists an alternating chain

$$(x_1, a_1, x_2, a_2, x_3, a_3, x_4, \ldots, x_{2n-1}, a_{2n-1}, v_2);$$

hence $(x_1, a_1, x_2, a_2, x_3, a_3, x_4, \ldots, x_{2n-1}, a_{2n-1}, v_2, a', w_1, a, w_2)$ is another alternating chain.

Suppose that $w_2 \notin T$ and consider

$$C' = (C \backslash \{a_2, a_4, \ldots, a_{2n-2}, a'\}) \cup \{a_1, \ldots, a_{2n-1}, a\}.$$

Then $|C'| = |C| + 1$. Let us show that C' is a matching. When we pass from C to C', the extremities of the removed edges are $x_2, x_3, \ldots, x_{2n-1}, v_2, w_1$, while those of the added edges are $x_1, x_2, x_3, \ldots, x_{2n-1}, v_2, w_1, w_2$. As $x_1 \notin \varepsilon(C)$ and $a = \{w_1, w_2\} \notin C$ with $w_1 \in \varepsilon(C)$, we have $w_2 \notin \varepsilon(C)$; thus C' is a matching of cardinality greater than that of C, which is a contradiction. Therefore $w_2 \in T$.

Then T is a minimum transversal set, since an arbitrary transversal set must be incident to every edge in the matching C, so must contain at least $|C|$ vertices (i.e. $\alpha'(\Gamma) \leqslant \beta(\Gamma)$). Thus $|T| = \beta(\Gamma)$ and $\alpha'(\Gamma) = \beta(\Gamma)$. ■

Remark 8.1.6. The proof above is effective: we can construct a maximum matching starting from an arbitrary matching C as follows. If there exists an alternating chain

$$(x_1, a_1, x_2, a_2, \ldots, a_{2n-1}, x_{2n})$$

ending at $x_{2n} \in V_2 \backslash \varepsilon(C)$ (it is an *increasing* chain), then

$$C' = (C \backslash \{a_2, a_4, \ldots, a_{2i}, \ldots, a_{2n-2}\}) \cup \{a_1, a_3, \ldots, a_{2i+1}, \ldots, a_{2n-1}\}$$

is a matching satisfying $|C'| = |C| + 1$.

Exercise 8.2. Let $\Gamma = (V_1 \sqcup V_2; E)$ be a bipartite graph. Show that $\alpha'(\Gamma) = \beta(\Gamma)$ as follows: add a vertex s to the graph adjacent to all the vertices of V_1, and another vertex p adjacent to all the vertices of V_2; this gives a new graph Γ' containing Γ.

(i) Prove that to each edge of Γ there corresponds an elementary chain from s to p, and vice versa.

(ii) Prove that there are as many edges in a maximum matching of Γ as there are in a maximum set of vertex-disjoint elementary chains from s to p in Γ'.

(iii) Prove that a minimum transversal set of Γ is a minimum vertex-separating set of Γ'.

(iv) Conclude using Menger's theorem (Theorem 4.5.3).

Exercise 8.3. Is the graph shown in Figure 8.1 bipartite?

Proposition 8.1.7. *Let $\Gamma = (V; E)$ be a simple graph with no isolated vertex. Then*

$$\frac{|V|}{1+\Delta(\Gamma)} \leqslant \alpha'(\Gamma) \leqslant \frac{|V|}{2}.$$

Furthermore, these bounds are optimal.

Proof. With no loss of generality, we may assume that the graph is connected (otherwise we deal with the connected components one by one). The upper bound of $\alpha'(\Gamma)$ was given in Exercise 8.1; the bound is attained when the graph admits a perfect matching.

For the lower bound of $\alpha'(\Gamma)$, we use induction on the number of edges $m = |E|$. If $m = 1$, then $|V| = 2$, $\Delta(\Gamma) = 1$ and $\alpha'(\Gamma) = 1$; if $m = 2$, then $|V| = 3$ and $\Delta(\Gamma) = 2$ and $\alpha'(\Gamma) = 1$, so we have $\frac{|V|}{1+\Delta(\Gamma)} = 1 = \alpha'(\Gamma)$.

Assume that the assertion holds up to m ($m \geqslant 2$), and let $\Gamma = (V; E)$ be a graph with $m + 1$ edges. There are two cases.

(i) If Γ possesses a cycle, let a be an edge on the cycle; then $\Gamma\backslash a$ is connected and

$$\alpha'(\Gamma) \geqslant \alpha'(\Gamma\backslash a) \geqslant \frac{|V|}{1+\Delta(\Gamma\backslash a)} \geqslant \frac{|V|}{1+\Delta(\Gamma)}.$$

(ii) If Γ possesses no cycle, then Γ is a tree:

- if $\Gamma = K_{1,|V|-1}$ is a star, then $\Delta(\Gamma) = |V| - 1$ and $\alpha'(\Gamma) = 1 = \frac{|V|}{1+\Delta(\Gamma)}$; in this case the bound is attained;
- if Γ is not a star, it contains an edge a such that $\Gamma\backslash a$ admits two connected components Γ_1 and Γ_2 neither of which is reduced to an isolated vertex. Then by the induction hypothesis,

$$\begin{aligned}\alpha'(\Gamma) = \alpha'(\Gamma_1) + \alpha'(\Gamma_2) &\geqslant \frac{|V_1|}{1+\Delta(\Gamma_1)} + \frac{|V_2|}{1+\Delta(\Gamma_2)} \\ &\geqslant \frac{|V_1|}{1+\Delta(\Gamma)} + \frac{|V_2|}{1+\Delta(\Gamma)} = \frac{|V|}{1+\Delta(\Gamma)}.\end{aligned}$$ ∎

Exercise 8.4. Prove the following corollary.

Let $\Gamma = (V; E)$ be a graph with no isolated vertex. Then

$$\frac{|V|}{2} \leqslant \beta'(\Gamma) \leqslant \frac{|V|.\Delta(\Gamma)}{1+\Delta(\Gamma)}.$$

Furthermore, these bounds are optimal.

Hint: use Theorem 8.1.5 and Proposition 8.1.7.

8.2 Matchings in bipartite graphs

8.2.1 Hall's theorem

Let $\Gamma = (V; E)$ be a simple graph. Recall (see Section 1.1.1) that $\Gamma(x) = \{y \in V\backslash\{x\} \mid y \text{ adjacent to } x\}$ is the set of all neighbours of x, and that more generally, the *neighbourhood of a subset of vertices* X is defined by

$$\Gamma(X) = \{y \in V\backslash X \mid y \text{ adjacent to at least one vertex } x \in X\}.$$

Hall's theorem, also called the *marriage theorem*, gives a necessary and sufficient condition for a bipartite graph to admit a perfect matching. There are several different proofs of this theorem; we will give three of them. The first one uses König's theorem (Theorem 8.1.5), the second is algorithmic, and the third uses Menger's theorem (Theorem 4.5.3). The statement of Hall's theorem below gives a very general solution to the marriage problem mentioned at the beginning of this chapter, related to the following notion. Let $X \subset V$, and call $C \subset E$ a *matching of* X if

(i) the edges of C are not pairwise incident (i.e. C is a matching);

(ii) every vertex of X is incident to an edge of C.

If $X = V$, a matching of V is simply a perfect matching of Γ.

Theorem 8.2.1 (Hall, 1935). *Let $\Gamma = (V_1 \sqcup V_2; E)$ be a bipartite graph with no isolated vertex. Then*

(i) *Γ admits a matching of V_1 if and only if*

$$\forall X \subset V_1, \quad |\Gamma(X)| \geqslant |X|. \tag{8.4}$$

(ii) *Γ admits a perfect matching (i.e. a matching of V) if and only if*

$$\forall X \subset V \text{ such that } X \subset V_1 \text{ or } X \subset V_2, \quad |\Gamma(X)| \geqslant |X|. \tag{8.5}$$

Proof. (ii) is a consequence of (i), since if Γ admits a perfect matching C, then this same matching is clearly a matching of V_1 and of V_2. By (8.4) applied to V_1 and V_2, we obtain (8.5). Conversely, suppose that Γ satisfies (8.5). Then Γ admits a matching C_1 of V_1 and a matching C_2 of V_2, so we have $|V_1| = |C_1| \leqslant |V_2|$ and $|V_2| = |C_2| \leqslant |V_1|$. Thus $|C_1| = |V_2|$ and every vertex of V_2 is incident to C_1. Thus C_1 is a perfect matching of Γ.

Let us now prove (i). We start by showing that condition (8.4) is necessary. Let C be a matching of V_1 and X a subset of V_1, and let $E(X)$ denote the set of edges of C that are incident to vertices of X. The extremities of edges of $E(X)$ that lie in V_2 form a subset of $\Gamma(X)$ whose cardinality is $|X|$ (since the edges of C are not incident), so $|\Gamma(X)| \geqslant |X|$.

We now show that condition (8.4) is sufficient. For this, we present three different arguments.

First proof. Let $X = X_1 \sqcup X_2$ be a minimum transversal set of Γ with $X_1 \subset V_1$ and $X_2 \subset V_2$. By König's theorem (Theorem 8.1.5), $|X|$ is the cardinality of a maximum matching C. If C is not a matching of V_1, we have $|C| < |V_1|$, so $|X| = |X_1| + |X_2| = |C| < |V_1|$ and $|X_2| < |V_1| - |X_1| = |V_1 \backslash X_1|$. But since $|X|$ is a transversal set, every edge is incident to X_1 or to X_2, so there is no edge between $V_1 \backslash X_1$ and $V_2 \backslash X_2$. From this, we deduce that every edge incident to $V_1 \backslash X_1$ is incident to X_2, which gives $\Gamma(V_1 \backslash X_1) \leqslant |X_2| < |V_1 \backslash X_1|$, so (8.4) is not satisfied.

Second proof. Suppose condition (8.4) is satisfied. Let $\vec{\Gamma}_{s,p}$ be the network associated to Γ (see Section 8.2.2). Let C be a vertex-separating set for s and p (see Section 4.5). Then $\Gamma(V_1 \backslash C) \subseteq V_2 \cap C$. We have $|C| = |V_1 \cap C| + |V_2 \cap C|$, so $|C| \geqslant |V_1 \cap C| + |\Gamma(V_1 \backslash C)|$. By condition (8.4), $|\Gamma(V_1 \backslash C)| \geqslant |V_1 \backslash C|$. Thus $|C| \geqslant |V_1 \cap C| + |V_1 \backslash C| = |V_1|$, but then by Menger's theorem (Theorem 4.5.3), there are $|V_1|$ vertex-disjoint elementary paths from s to p, which gives a matching of cardinality $|V_1|$.

Third proof. This is a constructive proof. We start with an arbitrary matching C in Γ (it can be empty or an isolated edge). Our aim is to enlarge it step by step in order to get a matching of V_1. If $|C| = |V_1|$, then C is a matching of V_1 and the proof is finished.

Otherwise, let $x_0 \in V_1 \backslash \varepsilon(C)$. We will show the following property.

(P) There exist an integer $1 \leqslant t \leqslant \min(|V_1|, |V_2|)$ and two sequences of distinct vertices $x_0, \dots, x_{t-1} \in V_1$, $y_1, \dots, y_t \in V_2$ such that

- (a) $y_t \notin \varepsilon(C)$,
- (b) $\forall 1 \leqslant i \leqslant t, \exists 0 \leqslant f(i) \leqslant i-1$ such that $e_i = \{y_i, x_{f(i)}\} \in E \backslash C$,
- (c) $\forall 1 \leqslant i \leqslant t-1, a_i = \{y_i, x_i\} \in C$.

By (8.4), we have $|\Gamma(x_0)| \geqslant |\{x_0\}| = 1$; hence there exists at least one vertex of V_2 adjacent to x_0.

If $\Gamma(x_0) \not\subset \varepsilon(C)$, then we take $y_1 \in \Gamma(x_0) \backslash \varepsilon(C)$ and set $e_1 = \{y_1, x_0\}$. It follows that property (P) holds with $t = 1$. We obtain that $C' = C \cup \{e_1\}$ is a matching of Γ.

Otherwise, $\Gamma(x_0) \subset \varepsilon(C)$. We take any $y_1 \in \Gamma(x_0)$. There exists $x_1 \in V_1$ such that $a_1 = \{x_1, y_1\} \in C$. By (8.4), we have $|\Gamma(\{x_0, x_1\})| \geqslant |\{x_0, x_1\}| = 2$. Hence there exists $y_2 \in V_2 \backslash \{y_1\}$ connected to x_0 or x_1. If we can find such a y_2 that is not in $\varepsilon(C)$, then property (P) holds with $t = 2$. If not, there exists $x_2 \in V_1 \backslash \{x_0, x_1\}$ such that $a_2 = \{x_2, y_2\} \in C$, and we proceed the same way recursively.

Let $1 \leqslant j \leqslant |V_1| - 1$. Assume that we have constructed $x_0, \dots, x_j \in V_1$ and $y_1, \dots, y_j \in V_2$ such that

- $\forall 1 \leqslant i \leqslant j, \exists 0 \leqslant f(i) \leqslant i - 1$ with $e_i = \{y_i, x_{f(i)}\} \in E \backslash C$,
- $\forall 1 \leqslant i \leqslant j, a_i = \{y_i, x_i\} \in C$.

By the lower bound $|\Gamma(\{x_0, \dots, x_j\})| \geqslant |\{x_0, \dots, x_j\}| = j + 1$, we can pick some $y_{j+1} \in \Gamma(\{x_0, \dots, x_j\}) \backslash \{y_1, \dots, y_j\}$, giving that there is an integer $0 \leqslant f(j+1) \leqslant j$ such that $e_{j+1} = \{y_{j+1}, x_{f(j+1)}\} \in E$. One distinguishes two cases.

- If $y_{j+1} \in \varepsilon(C)$, then there exists $x_{j+1} \in V_1 \backslash \{x_0, \dots, x_j\}$ such that $a_{j+1} = \{y_{j+1}, x_{j+1}\} \in C$, and the required construction is done at step $j + 1$. As $|C| < |V_1|$, note that we cannot have $j = |V_1| - 1$. And we iterate the process.
- If $y_{j+1} \notin \varepsilon(C)$ (which is automatically the case when $j = |V_1| - 1$), property (P) is valid with $t = j + 1$.

We assume that t is the smallest positive integer such that property (P) is satisfied. As f is decreasing, we have $f^r(t) = 0$ for some $r \geqslant 0$. Then

$$(y_t, e_t, x_{f(t)}, a_{f(t)}, y_{f(t)}, e_{f(t)}, x_{f^2(t)}, \dots, x_{f^r(t)} = x_0)$$

is a chain that is increasing in the sense of the remark following König's theorem (Theorem 8.1.5). We remove from C the $r - 1$ edges $a_{f^k(t)}$ for $1 \leqslant k \leqslant r - 1$ and replace them by the r edges $e_{f^k(t)}$ for $0 \leqslant k \leqslant r - 1$; this gives a new matching C' in Γ satisfying $|C'| = |C| + 1$.

Observe that, while this proof is somewhat technical, it has the advantage of being effective and implementable. ∎

Remark 8.2.2. We see that by this theorem, a necessary and sufficient condition to marry all the girls in the marriage problem is that each group of *friends* should know at least as many boys.

We also observe that if a bipartite graph $\Gamma = (V_1 \sqcup V_2; E)$ admits a perfect matching, then $|V_1| = |V_2|$, since a perfect matching is also an edge cover.

Corollary 8.2.3. *Let $\Gamma = (V; E)$ be a regular bipartite graph of degree $k \geqslant 1$. Then Γ admits a perfect matching.*

Proof. By Hall's theorem (Theorem 8.2.1), it is sufficient to prove that (8.5) is satisfied. Let X be a subset of V_2 of cardinality t. Then there exist kt edges incident to X. As each vertex of V_1 is incident to k edges, the kt edges incident to X must be incident to at least t vertices of V_1. Hence $|\Gamma(X)| \geqslant |X|$. The same reasoning works for subsets X of V_1. ∎

8.2.2 Networks associated to bipartite graphs

Let $\Gamma = (V_1 \sqcup V_2; E)$ be a bipartite graph, and associate to it the network defined as follows:

- two vertices s, p are added to Γ;
- the vertex s is adjacent to all the vertices of V_1;
- the vertex p is adjacent to all the vertices of V_2;
- the edges are oriented as follows: the edges of type $a = \{s, x\}$ for $x \in V_1$ become arcs of type $\vec{a} = (s, x)$, and the edges of type $a = \{y, p\}$, $y \in V_2$, become arcs of type $\vec{a} = (y, p)$; edges between V_1 and V_2 are oriented from V_1 to V_2: $a \curvearrowright \vec{a}$;
- the capacity of each arc $\vec{a}$ is fixed to the value $c(\vec{a}) = 1$.

This network, denoted $\vec{\Gamma}_{s,p}$, is called the *network associated to the bipartite graph* Γ.

Proposition 8.2.4. *Let $\Gamma = (V_1 \sqcup V_2; E)$ be a bipartite graph and $\vec{\Gamma}_{s,p}$ its associated network. To every flow f on $\vec{\Gamma}_{s,p}$ with values in $\mathbb{N}$ there corresponds a matching, and vice versa.*

Proof. Let f be a flow with integer values on $\vec{\Gamma}_{s,p}$. By the properties of the capacity, we have $f(\vec{a}) = 0$ or 1 for each $\vec{a}$ of $\vec{\Gamma}_{s,p}$. Let C be the set of edges of Γ corresponding to the arcs of $\vec{\Gamma}_{s,p}$ with flow equal to 1. From the structure of $\vec{\Gamma}_{s,p}$ and the law of conservation of the flow (see Section 4.4), we see that C is indeed a matching: for each $x \in V_1$, there is at most one arc of C whose initial vertex is x, and for each $y \in V_2$, there is at most one arc of C whose terminal vertex is y.

Conversely, let C be a matching in Γ. For each arc $\vec{a}$ of $\vec{\Gamma}_{s,p}$, set

$$f(\vec{a}) = \begin{cases} 1 & \text{if } a \text{ is an edge of } C, \\ 1 & \text{if } \vec{a} \text{ is an arc of type } (s,x) \text{ or } (x,p) \text{ incident to an edge of } C, \\ 0 & \text{otherwise.} \end{cases} \tag{8.6}$$

By construction, f satisfies the capacity constraint condition and conservation of the flow. ∎

Exercise 8.5. Show that a flow defined as in (8.6) is maximum if and only if the corresponding matching is maximum.

8.2.3 Algorithmic remarks

To obtain an algorithm that constructs a maximum matching in a bipartite graph, we can use the Ford–Fulkerson algorithm giving the maximum value of a flow on a network. We first construct a network as indicated in Section 8.2.2. Then, thanks to the Ford–Fulkerson algorithm (see Section 4.4.2), we construct a maximum flow between s and p of cardinality k. By Proposition 8.2.4, the conclusions of Exercise 8.5, and the law of conservation of the flow, we get a maximum matching by taking all the edges a of Γ such that for the corresponding arcs $\vec{a}$ in $\vec{\Gamma}_{s,p}$, we have $f(\vec{a}) = 1$.

8.3 Matchings in arbitrary graphs

Let us now turn to a general result due to W. T. Tutte.

Notation. Let $\Gamma = (V; E)$ be a simple graph. Let $q(\Gamma)$ denote the number of connected components whose cardinality is odd (we will simply refer to *odd components*). If $X \subset V$, let $\Gamma \backslash X$ denote the induced subgraph $\Gamma(V \backslash X)$, so that $q(\Gamma \backslash X)$ is the number of odd components of the induced subgraph on $V \backslash X$.

One useful property of the invariant $q(\Gamma)$ of a graph Γ is presented in the following easy lemma.

Lemma 8.3.1. $q(\Gamma) \equiv |V| \pmod 2$.

Proof. The total number of vertices in a connected component of even cardinality is even; we write it as $2k$. Let C_i, $i = 1, \dots, q(\Gamma)$, be the connected components of odd cardinality. We thus have

$$|V| - 2k = \sum_{i=1}^{q(\Gamma)} |C_i| = q(\Gamma) + \sum_{i=1}^{q(\Gamma)} (|C_i| - 1),$$

so $|V| \equiv q(\Gamma) \pmod 2$. ∎

Observe that if Γ admits a perfect matching C, then we have *Tutte's property*:

$$q(\Gamma \backslash X) \leqslant |X|, \quad \text{for all } X \subset V. \tag{8.7}$$

Indeed, if M is an odd component of $\Gamma \backslash X$, then M contains edges of C, but since $|M|$ is odd, one of them exits from M to a vertex of $V \backslash M$, so of X since C is a matching, which gives (8.7). This fact is illustrated in Figure 8.2.

Remark 8.3.2. Observe that if (8.7) is satisfied, then Γ has no isolated vertex. For $X = \emptyset$, (8.7) gives $q(\Gamma) = 0$.

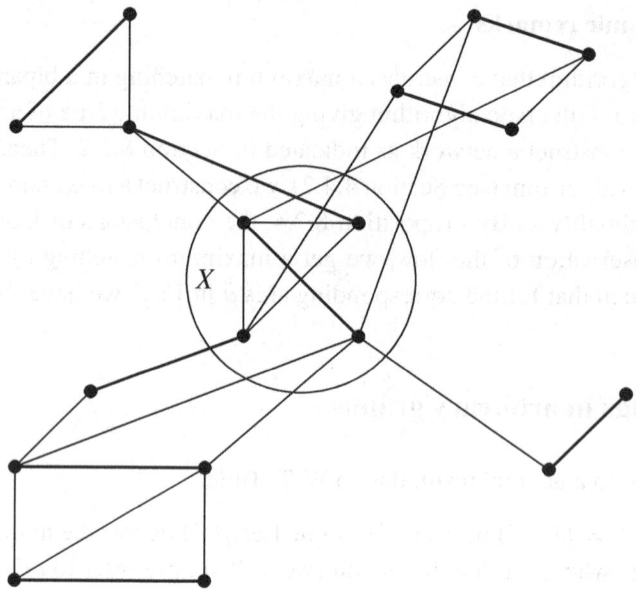

Figure 8.2. Tutte's property. The perfect matching consists of the edges in boldface; X consists of the 4 circled vertices; $\Gamma\backslash X$ admits 4 connected components, two even and two odd; $q(\Gamma\backslash X) = 2 < 4 = |X|$.

Remarkably, this simple necessary condition is actually also sufficient, as the following theorem shows.

Theorem 8.3.3 (Tutte, 1947). *A graph* $\Gamma = (V; E)$ *admits a perfect matching if and only if it satisfies Tutte's condition* (8.7).

Proof. Let us first observe what happens when Γ admits a perfect matching C. Suppose there exists $X_0 \neq \emptyset$ for which we have the equality $q(\Gamma\backslash X_0) = |X_0|$ of (8.7). Let

$$V(\Gamma\backslash X_0) = \bigsqcup_{1\leqslant i\leqslant m} M_i \sqcup \bigsqcup_{1\leqslant j\leqslant p} P_j$$

be the decomposition of $\Gamma\backslash X_0$ into connected components, where $m = |X_0|$ is the number of odd components M_i and p the number of even components P_j. As observed earlier, for each M_i there exists an edge $\{m_i, x_i\} \in C$, $x_i \in X_0$, $m_i \in M_i$, but since $m = |X_0|$, there exists exactly one such edge. Being subgraphs of C, each of the $M_i\backslash m_i$ and the P_j contains a perfect matching. Finally, the edges $\{m_i, x_i\}$, $1 \leqslant i \leqslant m$, form a matching between X_0 and the set $\{M_1, \dots, M_m\}$.

Suppose (8.7) holds. The proof is based on the fact that we can find such an X_0 such that $|q(\Gamma\backslash X_0)| = |X_0|$. For this, we use induction on the number of vertices of Γ. If $|V(\Gamma)| = 0$ there is nothing to prove. If $|V(\Gamma)| = 1$, then by Remark 8.3.2, Γ does not satisfy Tutte's condition (8.7), so the implication is trivially verified.

Now let $n \geqslant 1$, and suppose that the result is proven for all graphs of order $< n$. Let Γ be a graph of order n satisfying (8.7). Observe that (8.7) applied to $X = \emptyset$ implies that $|V|$ is even. Let x be a vertex of Γ. Then $\Gamma - x$ has an odd number of vertices, so it admits at least one odd component, but (8.7) applied to $X = \{x\}$ shows that in fact $\Gamma - x$ possesses exactly one odd component, and thus for $X = \{x\}$ we have $q(\Gamma \setminus X) = |X|$.

We can thus make the following assertion:

$$\text{there exists a } \textit{maximal } X_0 \subset V \text{ such that } q(\Gamma \setminus X_0) = |X_0|. \tag{8.8}$$

Let us write the decomposition of $\Gamma \setminus X_0$ into connected components as above:

$$V(\Gamma \setminus X_0) = \bigsqcup_{1 \leqslant i \leqslant m} M_i \sqcup \bigsqcup_{1 \leqslant j \leqslant p} P_j, \tag{8.9}$$

where $m = |X_0|$ is the number of odd components M_i and p the number of even components P_j of $\Gamma \setminus X_0$.

We proceed in several steps.

(a) Each P_j admits a perfect matching. Let $X \subset P_j$; by (8.9), we have

$$V(\Gamma \setminus (X_0 \sqcup X)) = \bigsqcup_{1 \leqslant i \leqslant m} M_i \sqcup (P_j \setminus X) \sqcup \bigsqcup_{1 \leqslant k \leqslant p, k \neq j} P_k,$$

where we see that the odd components of $\Gamma \setminus (X_0 \sqcup X)$ are the M_i and the odd components of $P_j \setminus X$. Thus

$$q(\Gamma \setminus (X_0 \sqcup X)) = q(\Gamma \setminus X_0) + q(P_j \setminus X),$$

but by (8.7): $q(\Gamma \setminus (X_0 \sqcup X)) \leqslant |X_0 \sqcup X| = |X_0| + |X|$, and since $q(\Gamma \setminus X_0) = |X_0|$, it follows that

$$q(P_j \setminus X) \leqslant |X|;$$

thus P_j satisfies Tutte's condition, so by the induction hypothesis, it admits a perfect matching.

(b) For $x \in M_i$ (an odd component), $\Gamma(M_i \setminus \{x\})$ admits a perfect matching. Suppose this is false. Then by the induction hypothesis, there exists $X \subset M_i \setminus \{x\}$ such that

$$q(M_i \setminus (\{x\} \sqcup X)) > |X|,$$

but by Lemma 8.3.1, $q(M_i \setminus (\{x\} \sqcup X)) \equiv |M_i| - 1 - |X| \equiv |X| \pmod 2$. It follows that

$$q(M_i \setminus (\{x\} \sqcup X)) \geqslant |X| + 2. \tag{8.10}$$

Moreover, $\{x\} \sqcup X \subset M_i$. Remove this set from both sides of (8.9):

$$V(\Gamma\backslash(X_0 \sqcup \{x\} \sqcup X)) = (M_i\backslash(\{x\} \sqcup X)) \sqcup \bigsqcup_{i' \neq i} M_{i'} \sqcup \bigsqcup_j P_j,$$

hence

$$\begin{aligned} q(\Gamma\backslash(X_0 \sqcup \{x\} \sqcup X)) &= q(M_i\backslash(\{x\} \sqcup X)) + q(\Gamma\backslash X_0) - 1 \\ &\geqslant |X| + 2 + q(\Gamma\backslash X_0) - 1 = |X| + 1 + |X_0| \end{aligned}$$

by (8.8) and (8.10), which gives the equality in Tutte's condition (8.7), but this contradicts the maximality of X_0.

(c) Γ contains m disjoint edges of the form $\{m_i, x_i\}, m_i \in M_i, x_i \in X_0, 1 \leqslant i \leqslant m$. To see this, consider the bipartite graph $\Gamma' = (V_1 \sqcup V_2; E')$ where

- $V_1 = \{M_1, \ldots, M_m\}$,
- $V_2 = X_0$,
- we connect M_i to $x \in X_0$ if Γ contains an edge between x and M_i.

Let us show that Γ' admits a matching of V_1. Let $Y \subset V_1$ and set $Z = \Gamma'(Y)$. Then $Z \subset V_2$, and by (8.7) we have $q(\Gamma\backslash Z) \leqslant |Z|$. The elements of Y are still odd components of $\Gamma\backslash Z$ by construction, so $|Y| \leqslant q(\Gamma\backslash Z) \leqslant |Z| = |\Gamma'(Y)|$. Thus Hall's theorem (Theorem 8.2.1) applies, so Γ' admits a matching of V_1.

(d) Let us now create a matching of Γ:

- take the m disjoint edges $\{m_i, x_i\}, m_i \in M_i, x_i \in X_0, 1 \leqslant i \leqslant m$, obtained in (c),
- add the edges of perfect matchings of $M_i\backslash\{x_i\}, 1 \leqslant i \leqslant m$ from step (b),
- finally, take the edges of p perfect matchings of the $P_j, 1 \leqslant j \leqslant p$ of step (a).

This gives a perfect matching of Γ, since each vertex of Γ is clearly incident to an edge of this matching.

In Figure 8.3, we diagram the structure of this proof. ■

Corollary 8.3.4 (Petersen, 1891). *Every cubic graph with no isthmus has a perfect matching.*

Proof. It suffices to check that Tutte's condition (8.7) is satisfied for a cubic graph $\Gamma = (V; E)$ with no isthmus. Let $X \subset V$ and let M be an odd component of $\Gamma(V\backslash X)$. Since Γ is a cubic graph, the sum of degrees (in Γ) of the vertices of M is an odd number. But the sum of degrees of the vertices of M that are incident to the edges contained in M is even (see Exercise 1.1.9). Hence the number of edges between M and X is odd, so there are at least three such edges since the graph has no isthmus. Consequently, the total number of edges between X and $V\backslash X$ is at least equal to $3\,q(\Gamma\backslash X)$. But this number is also at most equal to $3|X|$ since Γ is cubic. Hence $q(\Gamma\backslash X) \leqslant |X|$. ■

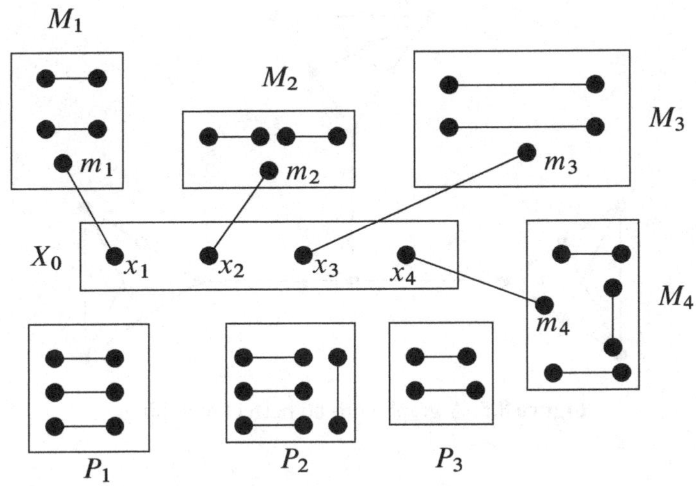

Figure 8.3. Outline of the proof of Tutte's theorem. Here $\Gamma \setminus X_0$ has $|X_0| = 4$ odd components M_i (perfect matchings of $\Gamma(M_i \setminus \{x\})$, step (b)), 3 even components P_j (perfect matchings, step (a)), and edges $\{x_i, m_i\}$ (step (c)).

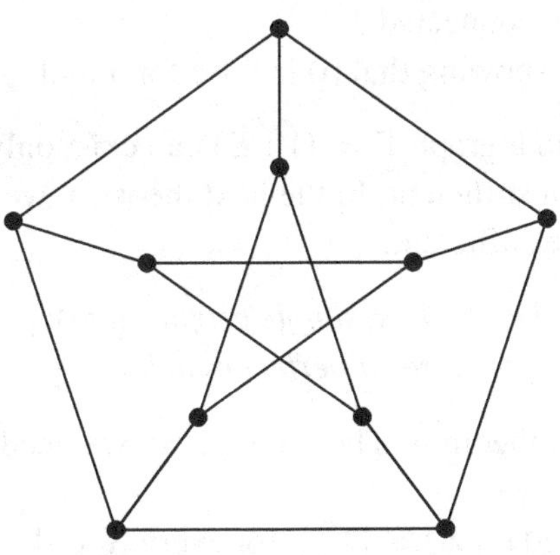

Figure 8.4. Petersen's graph.

Remark 8.3.5. In particular, a 2-edge-connected cubic graph admits a perfect matching.

Exercise 8.6. Petersen's graph (shown in Figure 8.4) is a cubic graph with no isthmus. Find all its perfect matchings.

Exercise 8.7. (i) Find a cubic graph with an isthmus that admits a perfect matching.

(ii) Prove that the graph shown in Figure 8.5 does not admit a perfect matching. Can this construction be generalised?

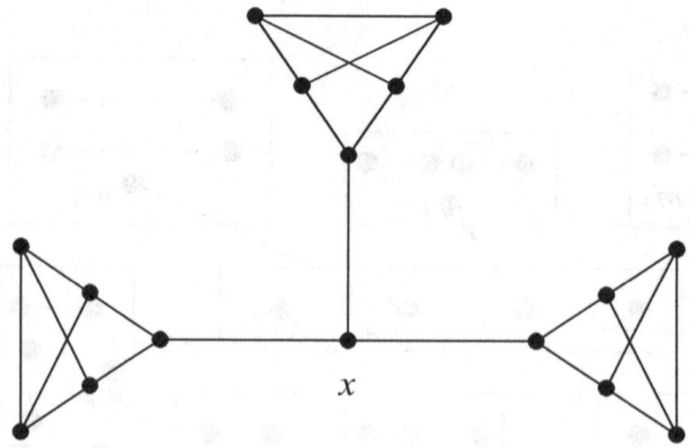

Figure 8.5. A graph with no perfect matching.

Let us give another application of Tutte's theorem (Theorem 8.3.3). We first need the following auxiliary result, whose proof we leave as an exercise.

Exercise 8.8. (i) Prove by contradiction that a simple graph $\Gamma = (V; E)$ satisfying $\delta(\Gamma) \geqslant \frac{|V|}{2}$ is necessarily connected.

(ii) Give an example showing that (i) is false for a multigraph.

A perfect matching in a graph $\Gamma = (V; E)$ can exist only if $|V|$ is even, and this condition is generally not sufficient. In the next theorem, we give a further condition which does turn out to be sufficient.

Theorem 8.3.6. *Let $\Gamma = (V; E)$ be a simple graph with $|V| = 2n$ vertices, of minimal degree $\delta(\Gamma) \geqslant n$. Then Γ possesses a perfect matching.*

Proof. Thanks to Tutte's theorem (Theorem 8.3.3), we need to check Tutte's condition (8.7):

$$q(\Gamma \backslash X) \leqslant |X| \quad \text{for every } X \subset V.$$

(a) If $X = \emptyset$, we have $q(\Gamma \backslash X) = q(\Gamma) = 0$; since $|V|$ is even and by the conclusions of Exercise 8.8, the graph Γ is connected.

(b) Suppose then that X satisfies $1 \leqslant |X| \leqslant 2n$. Let $V(\Gamma \backslash X) = \bigsqcup_{1 \leqslant i \leqslant k} C_i$ denote the decomposition of $\Gamma \backslash X$ into connected components. Each $x \in C_i$ has at most $|X|$ neighbours in X; consequently

$$n \leqslant \delta(\Gamma) \leqslant d_\Gamma(x) = d_{C_i}(x) + |\Gamma(x) \cap X| \leqslant d_{C_i}(x) + |X|.$$

It follows that $d_{C_i}(x) \geqslant n - |X|$, and then

$$|C_i| \geqslant d_{C_i}(x) + 1 \geqslant n - |X| + 1.$$

If $|X| = 1$, then

$$|V| = 2n = 1 + \sum_{1 \leqslant i \leqslant k} |C_i| \geqslant 1 + k(n - |X| + 1) = 1 + kn,$$

so $k \leqslant \frac{2n-1}{n} = 2 - \frac{1}{n} < 2$: $k \leqslant 1$; hence $q(\Gamma \backslash X) \leqslant k \leqslant 1 = |X|$.

If $2 \leqslant |X| \leqslant n$, then

$$|V| = 2n = |X| + \sum_{1 \leqslant i \leqslant k} |C_i| \geqslant |X| + k(n - |X| + 1),$$

which gives $k \leqslant \frac{2n-|X|}{n-|X|+1}$. But $\frac{2n-|X|}{n-|X|+1} \leqslant |X|$ is equivalent to

$$|X|^2 - (n+2)|X| + 2n \leqslant 0.$$

Since the polynomial $P(T) = T^2 - (n+2)T + 2n$ is negative between the roots 2 and n, we obtain $q(\Gamma \backslash X) \leqslant k \leqslant |X|$.

If $n + 1 \leqslant |X| \leqslant 2n$, the number k of connected components of $\Gamma \backslash X$ is less than or equal to $|\Gamma \backslash X|$, and this number satisfies $k \leqslant 2n - |X| \leqslant |X|$; a fortiori $q(\Gamma \backslash X) \leqslant k \leqslant |X|$. ■

For example, K_{2n} admits a perfect matching (see also Proposition 8.4.4, and an illustration of the factorisation of K_6 in Section 8.4).

8.4 Factorisation

A regular partial graph of degree k of a graph Γ is called a *factor of degree k* or simply a *k-factor* of Γ. A perfect matching is thus nothing more than a 1-factor.

If a graph possesses a set of edge-disjoint k-factors whose union contains all the edges of the graph, i.e. $E = \bigsqcup E_i$ where E_i is the set of edges of the i-th k-factor, we call this a *k-factorisation* of Γ.

Proposition 8.4.1. *Let $\Gamma = (V; E)$ be a simple bipartite graph. Then*

(i) *if Γ admits a 1-factorisation, it is regular;*

(ii) *conversely, if Γ is regular of degree k, it admits a 1-factorisation consisting of k factors of degree 1.*

Proof. (i) Write $E = \bigsqcup_{1 \leqslant i \leqslant k} E_i$, where E_i is the set of edges of a 1-factor. Each vertex x is incident to exactly one edge of each E_i since 1-factors are perfect matchings, so $d(x) = k$.

(ii) If Γ is regular of degree k, it admits at least one perfect matching (i.e. a 1-factor) by Corollary 8.2.3 of Hall's theorem (Theorem 8.2.1). Remove the edges of this matching. We obtain a partial graph that is bipartite and $(k-1)$-regular, so we can again apply Corollary 8.2.3, and we continue this way until we finally arrive at a regular bipartite graph of degree 1, which corresponds to a 1-factor. ■

Exercise 8.9. Give an example of a regular graph with an even number of vertices which does not admit a 1-factorisation.

Theorem 8.4.2 (Petersen, 1891). *A simple regular graph of degree $2k$ ($k \geqslant 1$) admits a 2-factorisation.*

Proof. Let $\Gamma = (V; E)$ be a regular graph of degree $2k$. With no loss of generality, we may assume that Γ is connected. By Euler's theorem (Theorem 2.4.2), the graph admits an Euler cycle C. Orient the edges of C in such a way that it becomes a circuit. For every vertex x in the oriented graph $\vec{\Gamma} = (V; \vec{E})$, we have $d^-(x) = d^+(x) = k$: there are k entering arcs and k leaving arcs. Starting from this orientation, we construct a bipartite graph $B = (V^+ \sqcup V^-; A)$ as follows:

- each vertex x is replaced by a pair of vertices x^-, x^+, and $V^+ = \{x^+ \mid x \in V\}$, $V^- = \{x^- \mid x \in V\}$;
- each arc $\vec{e} = (x, y) \in \vec{E}$ is replaced by an edge $a = \{x^+, y^-\}$ of A.

The graph B is bipartite by construction, and it is regular of degree k. Applying Proposition 8.4.1, B admits a 1-factorisation with $E = \bigsqcup E_i$. For each i, we see that for all $x \in V$ there exists a unique edge $a \in E_i$ such that $x^+ \in \varepsilon(a)$ and a unique edge $a' \in E_i$ such that $x^- \in \varepsilon(a')$, with $a' \neq a$, in such a way that by *gluing* x^+ and x^-, the degree of x is equal to 2 in the subgraph of Γ determined by E_i. Thus E_i yields a 2-factor, so we have obtained a 2-factorisation of Γ. ■

Theorem 8.4.3. *Let $\Gamma = (V; E)$ be a regular graph of degree $2k$.*

(i) *If $|V|$ is even, then Γ admits a k-factorisation.*

(ii) *Γ admits a $2t$-factor for all t with $1 \leqslant t \leqslant k$.*

Proof. (i) If Γ is $2k$-regular and connected then Γ admits an Euler cycle C (see Euler's theorem (Theorem 2.4.2)). As the degree of each vertex is equal to $2k$, we pass through each vertex exactly k times. Moreover, since $2k|V| = \sum_{x \in V} d(x) = 2|E|$, the number of edges of the graph is $k|V|$, which is even. Starting from an arbitrary vertex x and following the cycle C until it returns to x, we can divide the edges into two subsets E_1 and E_2 by placing the edges alternatively in E_1 or E_2 as we pass them. Then each vertex $x \in V$ is incident to k edges of E_1 and k edges of E_2. Thus the two partial graphs induced by E_1 and E_2 respectively are k-factors of Γ, and $E = E_1 \sqcup E_2$, which yields a k-factorisation of Γ.

(ii) By Petersen's theorem (Theorem 8.4.2), this graph admits a 2-factorisation. By eliminating a first 2-factor from the graph, it becomes $(2k-2)$-regular. Iterating the process $(k-t)$ times, we obtain a regular partial graph of degree $2t$. ∎

Proposition 8.4.4. *The complete graph K_{2n} admits a 1-factorisation.*

Proof. If $n = 1$, this is obvious. Suppose $n \geqslant 2$. The graph K_{2n} has $n(2n-1)$ edges. Identify the vertices of K_{2n} with the integers $0, 1, \ldots, 2n-1$, and arrange the vertices $0, 1, \ldots, 2n-2$ in that order around a regular polygon with $2n-1$ sides. Place the vertex $2n-1$ in the centre of this polygon.

Set

$$F_0 = \{(0, 2n-1); (1, 2n-2); (2, 2n-3); \ldots; (n-1, n)\},$$

and for $1 \leqslant k \leqslant 2n-2$,

$$F_k = \{\{k, 2n-1\}\} \cup \{\{\overline{i+k}, \overline{2n-1-i+k}\} \mid 1 \leqslant i \leqslant n-1\},$$

where $\overline{x}$ denotes the remainder of the Euclidean division of x by $2n-1$. Let us show that F_k has n edges, and that each vertex is the extremity of only one of these edges. The first point is clear. For the second, it suffices to note that:

(i) when i varies between 1 and $n-1$, $\overline{i+k}$ and $\overline{2n-1-i+k}$ each take $n-1$ distinct values not equal to k or to $2n-1$;

(ii) if $\overline{i+k} = \overline{2n-1-i'+k}$ for $1 \leqslant i, i' \leqslant n-1$, then $\overline{i+i'} = 0$, which is never possible since $i + i' < 2n-1$.

F_k is thus a perfect matching of K_{2n}.

We now need to show that the 1-factors F_k constitute a 1-factorisation. Indeed, we note that for $k \neq k'$, we have $F_k \cap F_{k'} = \emptyset$, and we easily see that $\{k', 2n-1\} \notin F_k$ and $\{k, 2n-1\} \notin F_{k'}$. Furthermore, if we had

$$\{\overline{i+k}, \overline{2n-1-i+k}\} = \{\overline{i'+k'}, \overline{2n-1-i'+k'}\},$$

then modulo $2n-1$, we would have the congruences

$$\begin{cases} i+k \equiv i'+k', \\ -i+k \equiv -i'+k' \end{cases} \quad \text{or} \quad \begin{cases} i+k \equiv -i'+k', \\ -i+k \equiv i'+k', \end{cases}$$

so $2k \equiv 2k' \pmod{2n-1}$ and $k = k'$ since $0 \leqslant k, k' \leqslant 2n-2$.

We have thus constructed $2n-1$ disjoint perfect matchings $F_0, \ldots, F_{2n-2}$, containing a total of $n(2n-1)$ edges, which is the full number of edges of K_{2n}. Thus they form a partition of the edges of K_{2n}, so the F_k constitute a 1-factorisation. ∎

Figure 8.6 gives an illustration of this proof.

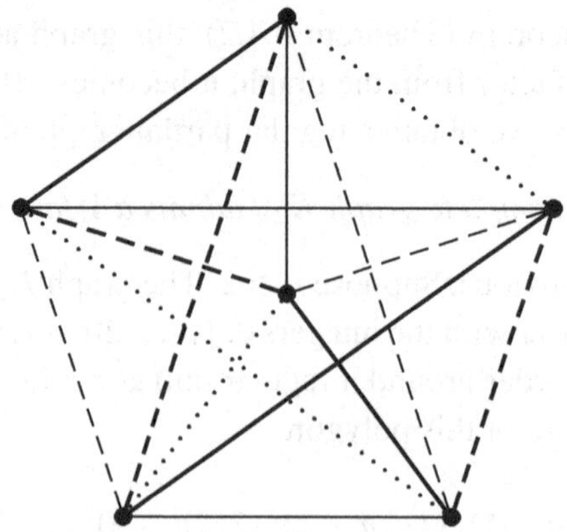

Figure 8.6. A 1-factorisation of K_6.

8.5 Some applications of matchings

Let $\mathcal{F} = (F_1, F_2, F_3, \ldots, F_n)$ be a family of n non-empty subsets of a set X. A *system of representatives of* $\mathcal{F}$ is a set of distinct elements $x_1, x_2, x_3, \ldots, x_n$ such that $x_i \in F_i$ for all $i = 1, 2, 3, \ldots, n$. The question of determining whether a system of representatives exists (and possibly finding one) arises for example when one needs to choose an element in each left class of a group G modulo a subgroup H, or more generally an element in each equivalence class associated to a partition of a set X. In these cases, however, the answer is trivially positive since the left classes, and more generally the subsets in a partition, are all disjoint. The next result, whose proof is based on Hall's theorem (Theorem 8.2.1), gives a necessary and sufficient condition for the existence of a set of representatives even when the subsets are not disjoint.

Proposition 8.5.1. *Let* $\mathcal{F} = (F_1, F_2, F_3, \ldots, F_n)$ *be a family of n non-empty subsets F_i of a finite set X. Then $\mathcal{F}$ admits a system of representatives if and only if for all* $I \subseteq \{1, 2, 3, \ldots, n\}$ *we have*

$$\left|\bigcup_{i \in I} F_i\right| \geqslant |I|. \tag{8.11}$$

Proof. We create a bipartite graph $\Gamma = (V_1 \sqcup V_2; E)$ as follows:

- $V_1 = \mathcal{F}$, $V_2 = X$;
- $e = \{x, F_i\} \in E$ if and only if $x \in F_i$.

The problem can then be rephrased as the question of whether there exists a matching of $V_1 = F$ in Γ. Since the graph Γ has no isolated vertex ($F_i \neq \emptyset$), condition (8.11) is nothing other than condition (8.4) of Hall's theorem (Theorem 8.2.1), which thus yields the desired answer. ∎

The combinatorics of matrices is an interesting application of the notion of matching. The (probabilistic) context is as follows.

A (homogeneous) *Markov chain* (see also § 12.5.1) is an infinite sequence of random variables $X_0, X_1, \ldots,$ taking values in a finite or countable set X (called the state space, in such a way that for all states $x_0, x_1, \ldots, x_{t+1} \in X$, the conditional probability that $X_{t+1} = x_{t+1}$ knowing that $X_i = x_i$ for $i = 0, \ldots, t$ depends only on x_t and x_{t+1}. The transition probabilities (which are independent of t), i.e. the $p_{xy} = \Pr[X_{t+1} = y \mid X_t = x]$ for all $x, y \in X$, are described by a matrix $P = (p_{xy})_{x,y \in X}$ all of whose coefficients are positive or zero, and satisfy $\sum_y p_{xy} = 1$ for all $x \in X$. This matrix is stochastic, where we recall that the definition of a stochastic matrix is as follows: let $A = (a_{ij}) \in \mathcal{M}_{n \times n}(\mathbb{R})$ be a real square matrix of order n, with $0 \leqslant a_{ij} \leqslant 1$; the matrix A is *stochastic* if for all $i = 1, \ldots, n$ we have $\sum_j a_{ij} = 1$. We also say that A is *doubly stochastic* or *bistochastic* if furthermore for all $j = 1, \ldots, n$ we have $\sum_i a_{ij} = 1$.

The simplest doubly stochastic matrices are the *permutation matrices*: these are matrices having exactly one coefficient 1 in each row and in each column, and all the other coefficients are zero. If P is such a matrix, and if $(e_1, \ldots, e_n)$ is a basis of $\mathbb{R}^n$, then P is a representation in this basis of the linear map permuting the e_i according to P.

The following result is due to D. König.

Theorem 8.5.2 (König, 1916). *Let $A = (a_{ij}) \in \mathcal{M}_{n \times n}(\mathbb{R})$ be a real square matrix with $a_{ij} \in \{0, 1\}$ and such that the sum of coefficients of each row (resp. each column) is equal to k for some $k \geqslant 1$. Then A is the sum of k permutation matrices.*

Proof. We associate to A the following bipartite graph $\Gamma = (V_1 \sqcup V_2; E)$:

- V_1 is the set of rows $l_1, l_2, \ldots, l_n$;
- V_2 is the set of columns $c_1, c_2, \ldots, c_n$;
- $\{l_i, c_j\} \in E$ if and only if $a_{ij} = 1$.

Since each row (resp. each column) contains exactly k coefficients equal to 1, the bipartite graph Γ is regular of degree k. Thus by Proposition 8.4.1 (ii), this graph possesses a 1-factorisation based on the partition $E = E_1 \sqcup E_2 \sqcup \cdots \sqcup E_k$. For $1 \leqslant r \leqslant k$, let $P_r = (p_{ij}(r))_{1 \leqslant i,j \leqslant n}$ be the matrix defined as follows:

$$p_{ij}(r) = \begin{cases} 1 & \text{if the edge } \{l_i, c_j\} \in E_r, \\ 0 & \text{otherwise.} \end{cases}$$

As E_r is a 1-factor, there exists only one edge in E_r with extremity l_i (resp. c_j). Thus there only exists one coefficient 1 per row (resp. per column) in the matrix P_r, so P_r is a permutation matrix, and we have $A = P_1 + P_2 + \cdots + P_k$ since the E_r are edge-disjoint. ∎

Recall (see Section 5.2) that if W is a $\mathbb{R}$-vector space, the segment with extremities $x, y \in W$ is parameterised by

$$[\![x, y]\!] = \{z \in W \mid \exists\lambda \text{ such that } z = (1-\lambda)x + \lambda y,\ 0 \leqslant \lambda \leqslant 1\}.$$

A subset A of W is said to be *convex* if

$$x, y \in A \implies [\![x, y]\!] \subset A.$$

The *convex envelope* of A, written $\mathrm{Conv}(A)$, is the smallest convex subset (for inclusion) containing A. Since the intersection of an arbitrary family of convex subsets (of W) remains convex, $\mathrm{Conv}(A)$ can also be defined as the intersection of all convex subsets containing A. The next exercise gives a concrete description.

Exercise 8.10. Let $A \subset W$ be a vector space over $\mathbb{R}$.

(i) Show that the subset

$$B = \Big\{\xi \in W \;\Big|\; \xi = \sum_{1 \leqslant i \leqslant m} \lambda_i a_i,\ a_i \in A,\ 0 \leqslant \lambda_i \leqslant 1, \sum_i \lambda_i = 1\Big\}$$

is a convex subset containing A.

(ii) Show that $B = \mathrm{Conv}(A)$.

Let $W = \mathcal{M}_{n\times n}(\mathbb{R})$ be the $\mathbb{R}$-vector space of $n \times n$ square matrices with real coefficients. Then $\dim_{\mathbb{R}} W = n^2$.

Theorem 8.5.3 (Birkhoff and von Neumann). *The convex envelope of the set $\mathcal{P}$ of permutation matrices in $\mathcal{M}_{n\times n}(\mathbb{R})$ is the set of doubly stochastic matrices.*

Proof. Let $\mathcal{S}$ denote the set of doubly stochastic matrices. We need to prove that $\mathcal{S} = \mathrm{Conv}(\mathcal{P})$.

(i) Let us show that $\mathcal{S}$ is convex. Let $A = (a_{ij})$ and $B = (b_{ij})$ be two doubly stochastic matrices, and let λ be such that $0 \leqslant \lambda \leqslant 1$. Set $C = \lambda A + (1-\lambda)B$. With the obvious notation, we have

$$\begin{aligned}\sum_j c_{ij} &= \sum_j (\lambda a_{ij} + (1-\lambda)b_{ij}) = \lambda \sum_j a_{ij} + (1-\lambda)\sum_j b_{ij}\\ &= \lambda + (1-\lambda) = 1.\end{aligned}$$

The same holds for the $\sum_i c_{ij}$, the sums of column coefficients. Therefore C is doubly stochastic.

(ii) Let us show that $\mathcal{S} \subseteq \mathrm{Conv}(\mathcal{P})$. Let $A = (a_{ij})$ be a doubly stochastic matrix, so that $0 \leqslant a_{ij} \leqslant 1$ and the sums of the coefficients in each row and each column are all equal to 1. As in König's theorem (Theorem 8.5.2), we associate to A a bipartite graph $\Gamma = (V_1 \sqcup V_2; E)$ defined by

- V_1 is the set of rows $l_1, l_2, \dots, l_n$;
- V_2 is the set of columns $c_1, c_2, \dots, c_n$;
- $\{l_i, c_j\} \in E$ if and only if $a_{ij} \neq 0$.

Let us show that this graph satisfies condition (8.5) from Hall's theorem. Let $X \subset V_1$. Then

$$\Gamma(X) = \{c_j \in V_2 \mid \exists l_i \in X : \{l_i, c_j\} \in E\}$$
$$= \{c_j \in V_2 \mid \exists l_i \in X : a_{ij} \neq 0\}.$$

Thus since $\lceil a_{ij} \rceil$ is equal to 1 or 0 according to whether $\{l_i, c_j\}$ lies in E or not, we have

$$|\Gamma(X)| = \sum_{l_i \in X} \sum_{c_j \in V_2} \lceil a_{ij} \rceil,$$

(here, for $x \in \mathbb{R}$, $\lceil x \rceil$ denotes the integer just above x). Consequently,

$$|\Gamma(X)| \geqslant \sum_{l_i \in X} \sum_{c_j \in V_2} a_{ij} = \sum_{l_i \in X} 1 = |X|.$$

We proceed the same way for $Y \in V_2$, and we then obtain $|\Gamma(Y)| \geqslant |Y|$. By Hall's theorem (Theorem 8.2.1), Γ admits a perfect matching C. We now define $P = (p_{ij}) \in \mathcal{M}_{n \times n}(\mathbb{R})$ by

$$p_{ij} = \begin{cases} 1 & \text{if the edge } \{l_i, c_j\} \in C, \\ 0 & \text{otherwise.} \end{cases}$$

Since C is a perfect matching, P is a permutation matrix (for the same reasons as in the proof of König's theorem, Theorem 8.5.2). Set

$$\lambda = \min_{1 \leqslant i,j \leqslant n} \{a_{ij} \mid p_{ij} \neq 0\} \quad \text{and} \quad B = A - \lambda P = (b_{ij}).$$

Clearly $0 \leqslant \lambda \leqslant 1$ and in fact $\lambda \neq 0$ since $p_{ij} \neq 0$ implies $a_{ij} \neq 0$. Furthermore, $b_{ij} \geqslant 0$ since if $p_{ij} \neq 0$, then $p_{ij} = 1$ and $\lambda \leqslant a_{ij}$. Let us compute $\sum_j b_{ij} = \sum_j a_{ij} - \lambda p_{ij} = 1 - \lambda$ and $\sum_i b_{ij} = \sum_i a_{ij} - \lambda p_{ij} = 1 - \lambda$.

If $\lambda = 1$, observe that all the b_{ij} are zero, so $A = P \in \mathcal{P}$.

Otherwise, $0 < \lambda < 1$; the preceding calculations then show that the matrix $A' = \frac{1}{1-\lambda} B$ is doubly stochastic, and moreover

$$A = \lambda P + B = \lambda P + (1 - \lambda) A'.$$

As there exist indices i, j such that $a_{ij} = \lambda$, we have

$$a'_{ij} = \frac{1}{1-\lambda} b_{ij} = \frac{1}{1-\lambda} (a_{ij} - \lambda p_{ij}) = \frac{1}{1-\lambda} (\lambda - \lambda \times 1) = 0$$

so that A' has at least one zero coefficient more than A, and is also doubly stochastic. Thus we have shown that a doubly stochastic matrix A can be written as a *convex linear combination* (i.e. a linear combination with positive coefficients which sum to 1) of a permutation matrix P and a doubly stochastic matrix A' that has one more zero coefficient than A. Iterating the same procedure with A' and its successors as many times as necessary, we end by obtaining $A = \sum_{1 \leqslant k \leqslant m} \mu_k P_k$, where the P_k are permutation matrices and $\sum_k \mu_k = 1$; thus $A \in \mathrm{Conv}(\mathcal{P})$. We can apply this reduction procedure on the number of non-zero coefficients at most $n^2 - n$ times, since the matrix A has at most n^2 non-zero coefficients to start with. But the final matrix P_m is a permutation matrix, so it has n non-zero coefficients. ∎

Remark 8.5.4. The application of this theorem to the case of *regular* matrices A of degree k (i.e. adjacency matrices of regular graphs of degree k) yields a weaker result than König's theorem (Theorem 8.5.2), since we only obtain that A is sum of at most nk permutation matrices.

Let us present another application of Hall's theorem (Theorem 8.2.1). We use the terminology *lines* to indicate both the rows and the columns of an $m \times n$ matrix, so that an $m \times n$ matrix possesses $m + n$ lines.

Theorem 8.5.5 (König and Egerváry, 1931). *Let A be an $m \times n$ matrix whose coefficients are all 0's and 1's. The maximal number of 1's that we can choose in A in such a way that no two of them ever lie on the same line is equal to the minimal number of lines needed to contain all the coefficients 1 of A.*

Proof. The matrix A is associated to the bipartite graph $\Gamma = (V_1 \sqcup V_2; E)$, in which the rows of A are the vertices of V_1 and the columns the vertices of V_2. A matching of Γ is a set of non-incident edges, i.e. a set of 1's on distinct lines. A transversal set of Γ is a set of lines containing all the 1's in A. The result to prove is thus simply the identity $\alpha'(\Gamma) = \beta(\Gamma)$, an equality that was stated and proved in König's theorem (Theorem 8.1.5). ∎

Exercise 8.11. Compare Theorem 8.5.5 and Theorem 8.1.5.

Examples of applications. In a business, producing a certain object requires five distinct operations which must be performed simultaneously on five different machines. The time in minutes for each operation is given in Table 8.1.

One can assign for example operation 1 to machine 3, operation 2 to machine 1, operation 3 to machine 4, operation 4 to machine 2 and operation 5 to the machine 5; in this case the time required to produce the object is equal to $\max\{3, 6, 3, 10, 6\} = 10$ minutes. Is it possible to assign the operations to the machines in such a way that the production process only takes a maximum of five minutes?

	M1	M2	M3	M4	M5
Operation 1	5	6	3	7	5
Operation 2	6	7	2	3	6
Operation 3	3	4	6	3	5
Operation 4	5	10	3	2	9
Operation 5	2	7	8	4	6

Table 8.1

For this, we will create a matrix C whose coefficients are 0's and 1's as follows:

$$c_{ij} = \begin{cases} 1 & \text{if the operation } i \text{ performed on the machine } j \text{ takes} \\ & \text{at most 5 minutes,} \\ 0 & \text{otherwise.} \end{cases}$$

We obtain the following matrix:

$$C = \begin{pmatrix} 1 & 0 & 1 & 0 & \boxed{1} \\ 0 & 0 & \boxed{1} & 1 & 0 \\ 1 & \boxed{1} & 0 & 1 & 1 \\ 1 & 0 & 1 & \boxed{1} & 0 \\ \boxed{1} & 0 & 0 & 1 & 0 \end{pmatrix}.$$

The maximal number of coefficients 1 that we can take on different lines is equal to 5 (see Theorem 8.5.5 of König–Egerváry): the coefficients 1 in the matrix C are distributed in such a way that there is exactly one on each line. Thus the answer to the question is that it is possible to produce the object in a maximum of five minutes.

The minimal number of rows in the König–Egerváry theorem (Theorem 8.5.5) is thus also equal to 5.

Latin squares. A *Latin rectangle* is an $m \times n$ matrix $M = (m_{ij})$ of size with integer coefficients satisfying the following properties:

(i) $1 \leqslant m_{ij} \leqslant n$;

(ii) the coefficients in each row are distinct, as are those in each column.

If $m = n$ we obtain a *Latin square*. Here is an example of a Latin square and of a Latin rectangle:

$$M_1 = \begin{pmatrix} 1 & 2 & 3 & 4 & 5 \\ 2 & 3 & 4 & 5 & 1 \\ 3 & 4 & 5 & 1 & 2 \\ 4 & 5 & 1 & 2 & 3 \\ 5 & 1 & 2 & 3 & 4 \end{pmatrix}, \qquad M_2 = \begin{pmatrix} 4 & 2 & 1 & 3 & 5 \\ 5 & 3 & 2 & 1 & 4 \\ 3 & 4 & 5 & 2 & 1 \end{pmatrix}.$$

Sudokus are also particular examples of 9×9 Latin squares.

The following theorem addresses the question of whether an $m \times n$ Latin rectangle with $m < n$ can be augmented to a Latin square.

Theorem 8.5.6. *Let M be a Latin rectangle of size $m \times n$ with $m < n$. Then M can be augmented to a Latin square by adding on $n - m$ new rows.*

Proof. In fact, it suffices to show that M can be completed to a Latin rectangle of size $(m+1) \times n$.

Let $N \subset \{1, 2, \ldots, n\}$ be the set of integers appearing in the Latin rectangle M. Set $\mathcal{F} = \{F_1, F_2, \ldots, F_n\}$ where for each i, F_i is the set consisting of elements of N which *are not* in the i-th column of M. If we prove that $\mathcal{F}$ admits a system of representatives, i.e. a set of n (distinct) integers each belonging to only one of the F_i, then the result follows by taking the members of the system of representatives $x_1, \ldots, x_n$, in that order, and adding them to the bottom of M as an additional row. If we do this, it is clear that the augmented matrix has the desired properties (i) and (ii).

To prove that a system of representatives exists, we need to check that conditions (8.4) of Hall's theorem (Theorem 8.2.1) are satisfied. Let $X = \{F_{i_1}, F_{i_2}, \ldots, F_{i_k}\}$ be a subset with $k \leqslant n$ elements of $\mathcal{F}$. Then the $(n-m)k$ (not necessarily distinct) integers belonging to one or another of the F_i form a subset of N. Since each element $x \in N$ appears exactly once on each row, x appears exactly m times in M, so it is located on m columns of M; thus it is absent from $n - m$ other columns, meaning it is located in exactly $n - m$ of the subsets F_i. Suppose now that $\bigcup_{1 \leqslant j \leqslant k} F_{i_j}$ contains less than k different elements. Then since $\bigcup_{1 \leqslant j \leqslant k} F_{i_j}$ is the union of k sets of $n - m$ elements, somewhere in this union there is necessarily an element that is repeated more than $n - m$ times. But this contradicts the fact that each element is repeated exactly $n - m$ times in $\bigcup_{1 \leqslant j \leqslant k} F_{i_j}$. Thus we have $|\Gamma(X)| \geqslant |X|$, which proves the existence of a set of representatives. ∎

The following matrix illustrates Theorem 8.5.6, passing from a 3×5 Latin rectangle to a 5×5 Latin square.

$$M_2 \curvearrowright \begin{pmatrix} 4 & 2 & 1 & 3 & 5 \\ 5 & 3 & 2 & 1 & 4 \\ 3 & 4 & 5 & 2 & 1 \\ \hline 2 & 1 & 4 & 5 & 3 \\ 1 & 5 & 3 & 4 & 2 \end{pmatrix}$$

Exercise 8.12. Find two ways to complete the following 2×5 Latin rectangle to a Latin square:

$$\begin{pmatrix} 1 & 2 & 3 & 4 & 5 \\ 5 & 3 & 1 & 2 & 4 \end{pmatrix}.$$

8.6 Generalisation of the notion of factor

We conclude this chapter with a more sophisticated concept: the notion of an f-factor.

Let $\Gamma = (V; E, N)$ be a *loopless* graph *without isolated vertices*, and let $f: V \to \mathbb{N}$ be a map. An *f-factor* of Γ is a partial graph Γ' of Γ such that for every vertex $x \in V$, the degree of x in Γ' is equal to $f(x)$; this implies in particular that $f(x) \leqslant d(x)$ for every vertex x of Γ. This notion generalises the notion of a k-factor, which corresponds to the situation where f is simply the constant function equal to k. One interesting question is knowing under what conditions a graph admits an f-factor, and how to characterise the graphs that have this property.

To do this, we will take the graph $\Gamma = (V; E, N)$ and define a simple auxiliary graph $\Gamma_f = (V_f; E_f)$, called its *blow-up*, as follows:

- assume first that $f(x) \leqslant d(x)$ for all $x \in V$ (otherwise the graph has no f-factor);
- replace each vertex $x \in V$ by a complete bipartite graph $K_{d(x),e(x)} = (D(x) \sqcup F(x); E)$ where $D(x)$ is a set with $d(x)$ elements and $F(x)$ a set with $e(x) = d(x) - f(x)$ elements (if $d(x) = f(x)$ the graph consists of $d(x)$ isolated vertices). Since Γ has no isolated vertex, this construction always makes sense;
- replace each p-edge $a = \{x, y\}$ of Γ by p disjoint edges $\{x_i(a), y_i(a)\}$, $i = 1, \ldots, p$: when a and i vary, the $x_i(a) \in D(x)$ are chosen distinct, as are the $y_i(a) \in D(y)$.

Example 8.6.1. This process of blowing up the graph Γ according to f is illustrated in Figures 8.7, 8.8 and 8.9. Figure 8.7 shows a graph that we will blow up according to the function f defined by $f(x_1) = 3$, $f(x_2) = 1$, $f(x_3) = 2$, $f(x_4) = 3$, $f(x_5) = 1$, $f(x_6) = 2$ (the $f(x_i)$ are indicated in Figure 8.7).

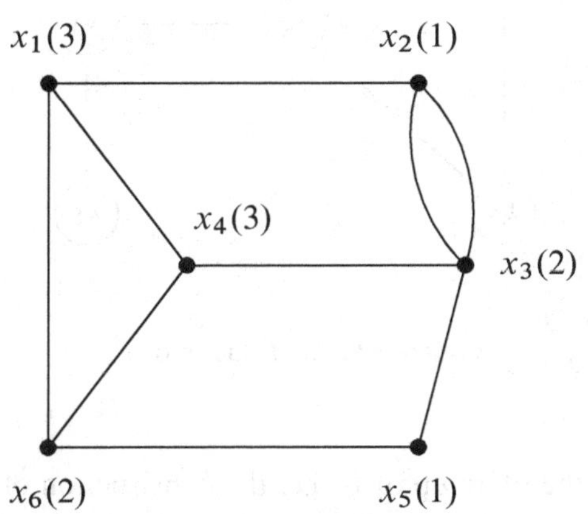

Figure 8.7. A graph Γ to be blown up.

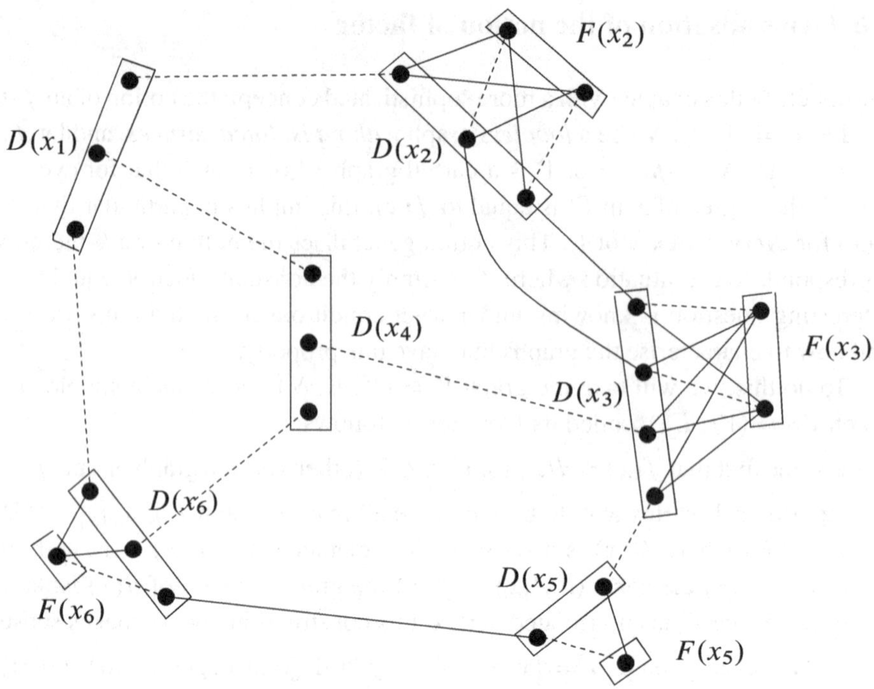

Figure 8.8. The blown-up graph Γ_f.

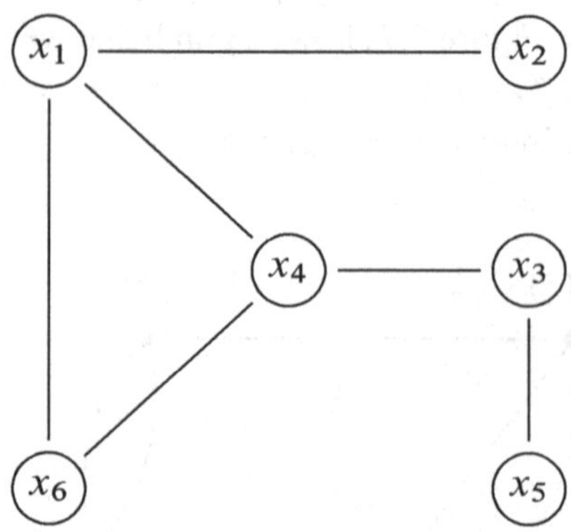

Figure 8.9. An f-factor of Γ.

Figure 8.8 shows the blow-up of Γ, i.e. the construction of the graph Γ_f corresponding to the graph Γ in Figure 8.7; the 1-factor is drawn with dotted lines and is reproduced in Figure 8.9. We can check that it is the unique f-factor of Γ.

Proposition 8.6.2. *A loopless graph* $\Gamma = V; E, N)$ *without isolated vertices admits an* f*-factor if and only if the graph* $\Gamma_f = (V_f; E_f, N_f)$ *constructed as above admits a* 1*-factor.*

Proof. Suppose Γ admits an f-factor. The corresponding edges in Γ_f are incident to $f(x)$ vertices of $D(x)$ for each $x \in V$. This leaves $e(x) = |F(x)|$ free vertices in $D(x)$. Choose $e(x)$ edges in the bipartite graph $K_{d(x),e(x)}$ connecting each of these free vertices of $D(x)$ to as many vertices of $F(x)$. This gives a 1-factor of Γ_f.

Conversely, let C be a 1-factor in Γ_f. Remove from C the edges that are incident to a vertex of $F(x)$. Do this for every vertex x of Γ_f, then fuse each $K_{d(x),e(x)}$ to a single vertex: This yields an f-factor of Γ. ∎

Chapter 9

Graphs and group theory

In this chapter, we explore the relations between graph theory and group theory. Some surprising and elegant results emerge, such as for example the statement that every finite group is isomorphic to the automorphism group of a graph. Group theory and linear algebra give rise to interpretations of many of the combinatorial characteristics of a graph. Groups arise in graph theory mainly in the study of isomorphisms, embeddings and symmetries. Conversely, however, graphs can yield precious hints about the properties of certain groups.

9.1 Permutation groups

Let V be a set. A bijection from V to itself is called a *permutation* of V. The set of all permutations of V forms a group whose multiplication is the composition of maps, called the *symmetric group*. This group is denoted by $\mathfrak{S}_V$. When V is finite, we simply take $V = \{1, \dots, n\}$ as a *model* for V, and write $\mathfrak{S}_n$ for the associated group. Its cardinality is $n!$.

A *permutation group* of V is any subgroup of $\mathfrak{S}_V$. For example, let $\Gamma = (V; E)$ be a simple graph. In Chapter 1 we defined the notion of an automorphism, and we saw that the set of automorphisms of a graph forms a group $\mathrm{Aut}(\Gamma)$. Thus each automorphism of Γ is a permutation of V, and $\mathrm{Aut}(\Gamma)$ is a subgroup of $\mathfrak{S}_V$.

Let $\sigma \in \mathfrak{S}_n$. Cauchy's notation for the permutation σ is

$$\sigma = \begin{pmatrix} 1 & 2 & \dots & n \\ \sigma(1) & \sigma(2) & \dots & \sigma(n) \end{pmatrix}.$$

If $k \geqslant 2$, a permutation σ is a *cyclic permutation* of length k, or a *k-cycle*, if there exists x such that $\sigma^k(x) = x$ and the k elements $x, \sigma(x), \dots, \sigma^{k-1}(x)$ are distinct and are the only elements of $\{1, 2, \dots, n\}$ that are not fixed by σ; in other words, for all $y \in V \setminus \{x, \sigma(x), \dots, \sigma^k(x)\}$, we have $\sigma(y) = y$. For cyclic permutations, we use the notation

$$\sigma = (x\ \sigma(x)\ \dots\ \sigma^{k-1}(x)).$$

This notation is not unique, since we can also write

$$\sigma = (\sigma^j(x)\ \sigma^{j+1}(x)\ \dots\ \sigma^{j+k-1}(x))$$

for every integer j (for example the permutations $(1\,2\,3\,4)$ and $(2\,3\,4\,1)$ are equal).

The *support of a permutation* σ is given by $\mathrm{Supp}(\sigma) = \{x \in V \mid \sigma(x) \neq x\}$; for example the support of the 4-cycle $(7\,2\,4\,1)$ of $\mathfrak{S}_7$ is $\{1, 2, 4, 7\}$. It is easy to see that two cycles with disjoint supports commute, for example in $\mathfrak{S}_7$, we have $(1\,2\,3) \circ (4\,7) = (4\,7) \circ (1\,2\,3)$, where $\circ$ is the composition law for maps. We have the following basic result on permutations.

Theorem 9.1.1. *Let $\sigma \in \mathfrak{S}_n$, $\sigma \neq \mathrm{Id}$. Then σ has a unique decomposition (up to the order of the factors) as a product of disjoint cycles, called the cyclic decomposition of σ.*

Example 9.1.2. The decomposition into disjoint cycles of the permutation

$$\sigma = \begin{pmatrix} 1 & 2 & 3 & 4 & 5 & 6 & 7 & 8 & 9 \\ 4 & 6 & 9 & 7 & 2 & 5 & 8 & 1 & 3 \end{pmatrix} \in \mathfrak{S}_9$$

is given by $\sigma = (1\,4\,7\,8) \circ (2\,6\,5) \circ (3\,9)$.

Decomposition into disjoint cycles allows us to determine the *order* $o(\sigma)$ of a permutation σ, i.e. the smallest integer $k \geqslant 1$ such that $\sigma^k = \mathrm{Id}$. If σ is a k-cycle for $k \geqslant 2$, then its order is equal to k.

Proposition 9.1.3. *Let $\sigma \in \mathfrak{S}_n \backslash \{\mathrm{Id}\}$, and let $\sigma = \sigma_1 \circ \sigma_2 \circ \cdots \circ \sigma_r$ be its cyclic decomposition. Then*

$$o(\sigma) = \mathrm{lcm}_{1 \leqslant i \leqslant r} |\mathrm{Supp}(\sigma_i)|.$$

For example, the order of $(1\,2\,5\,9) \circ (3\,4\,7)$ is equal to $\mathrm{lcm}(4, 3) = 12$.

Exercise 9.1. Let $\sigma = (6\,2\,4) \circ (2\,5\,3) \circ (8\,7\,6) \circ (4\,5) \in \mathfrak{S}_8$. Compute σ^{2022}.
Hint: Start by determining the cyclic decomposition of σ.
Show that there is no element of order 20 in $\mathfrak{S}_8$.

The particular importance of permutation groups in group theory is due to the next, very famous theorem.

Theorem 9.1.4 (Cayley). *Every group G is isomorphic to a subgroup of $\mathfrak{S}_X$ for a non-empty set X (that can be chosen equal to G).*

Proof. We can embed G into $\mathfrak{S}_G$ as follows: to $g \in G$ we associate the permutation $\gamma_g \in \mathfrak{S}_G$ of elements of G given by $\gamma_g(x) = gx$. It is trivial to see that $\gamma_g \in \mathfrak{S}_G$ (i.e. that left multiplication induces a permutation of the elements), that $g \mapsto \gamma_g$ is a group homomorphism, and finally that this homomorphism is injective.

We could of course equally well make use of right translations instead:

$$G \xrightarrow{\delta} \mathfrak{S}_G,$$
$$g \mapsto \delta_g : x \mapsto xg^{-1};$$

we must merely take care to define the right translation by multiplying on the right by g^{-1} and not g so as to ensure that δ is a homomorphism. ∎

9.2 The automorphism group of a graph and of its line-graph

9.2.1 Graph automorphisms and edge-automorphisms

Let $\Gamma = (V; E)$ be a simple graph. An automorphism of Γ is an isomorphism of Γ with itself (see Section 7.3). We can thus characterise automorpshisms of Γ as follows:

$$\mathrm{Aut}(\Gamma) = \{\text{bijective maps } f\colon V \to V \text{ satisfying [AP]}\},$$

where [AP] is the adjacency property

$$\{x, y\} \in E \iff \{f(x), f(y)\} \in E. \qquad \text{[AP]}$$

We have $\mathrm{Aut}(\Gamma) \subset \mathfrak{S}_V$. The study of $\mathrm{Aut}(\Gamma)$ reveals the symmetry properties of Γ. If the group $\mathrm{Aut}(\Gamma)$ is *large*, this indicates that many vertices play the same role within Γ. The complexity of the local structure of Γ, determined by its structure at a given vertex or a grouping of vertices, can be at least partially determined by the structure of $\mathrm{Aut}(\Gamma)$. The automorphism group of the complete graph on the vertices of V coincides with $\mathfrak{S}_V$; the graph with no edges also has automorphism group $\mathfrak{S}_V$. In these two extreme cases, the graph Γ has a maximal amount of symmetry, and this is indicated by the size of its automorphism group.

Note that a graph automorphism preserves the degree at each vertex (although this is not sufficient for a map to be an automorphism), so it preserves the entire degree distribution of the graph. This makes it easy to construct graphs having trivial automorphism group, such as the graph Γ shown in Figure 9.1.

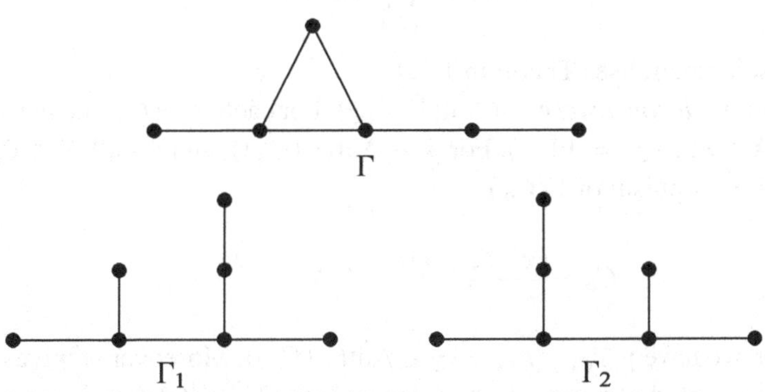

Figure 9.1. $\mathrm{Aut}(\Gamma) = \{\mathrm{Id}\}$; Γ_1 and Γ_2 are isomorphic.

Exercise 9.2. (i) Show that the graph Γ in Figure 9.1 has trivial automorphism group.
(ii) Show that there is exactly one isomorphism between Γ_1 and Γ_2.

If $f \in \mathrm{Aut}(\Gamma)$, then f must send a connected subset to a connected subset (see Exercise 7.8). In fact, we have even more precise information.

We recall some facts: A subset C of $V(\Gamma)$ is called a *connected component* of Γ if C is connected and maximal (i.e. if for all $x \notin C$, the union $C \cup \{x\}$ is not connected). Clearly if C and C' are connected and $C \cap C' \neq \emptyset$, then the union $C \cup C'$ is also connected. It follows that the connected components form a partition of V:

$$V(\Gamma) = \bigsqcup_{C \in \mathcal{C}} C$$

where $\mathcal{C}$ is the set of all the connected components of Γ. For example, a connected component of type $C = \{x\}$ denotes an isolated vertex, while a component $C = \{x, y\}$ denotes a component of type K_2 (isolated edge).

Since an automorphism of Γ permutes the isomorphic connected components, it is natural to regroup them via the equivalence relation $\mathcal{R}$:

$$C \mathcal{R} C' \quad \text{if and only if} \quad \Gamma(C) \simeq \Gamma(C').$$

This gives a partition of the set $\mathcal{C}$ of connected components into t classes $\mathcal{C}_\alpha$, $\alpha \in \{1, 2, \ldots, t\}$. Choose a representative $C_\alpha \in \mathcal{C}_\alpha$ from each equivalence class. Set $\mathbf{C}_\alpha = \bigsqcup_{C \in \mathcal{C}_\alpha} C$; then we have

$$V(\Gamma) = \bigsqcup_{\alpha=1}^{t} \mathbf{C}_\alpha.$$

With notation as above, we have the following proposition.

Proposition 9.2.1.

$$\mathrm{Aut}(\Gamma) \simeq \prod_{\alpha \in A} \mathrm{Aut}(\Gamma(\mathbf{C}_\alpha)).$$

This result generalises Theorem 1.9.4.

It remains to *deconstruct* each $\mathrm{Aut}(\Gamma(\mathbf{C}_\alpha))$. For each $X \simeq C_\alpha$, fix an isomorphism $\gamma_X : C_\alpha \to X$ (with $\gamma_{C_\alpha} = \mathrm{Id}_{C_\alpha}$). For $f \in \mathrm{Aut}(\Gamma(\mathbf{C}_\alpha))$, and for all $X \in \mathcal{C}_\alpha$, consider the *inner* automorphism of $\Gamma(C_\alpha)$

$$C_\alpha \xrightarrow{\gamma_X} X \xrightarrow{f|_X} f(X) \xrightarrow{\gamma_{f(X)}^{-1}} C_\alpha.$$

In particular, we have $\gamma_{f(X)}^{-1} \circ f|_X \circ \gamma_X \in \mathrm{Aut}(\Gamma(C_\alpha))$. Moreover, f gives a permutation of the connected components X isomorphic to C_α, so it determines a permutation σ_f in $\mathfrak{S}_n$, where n is the cardinality of $\mathcal{C}_\alpha$: intuitively, σ_f expresses the permutation of the connected components, and $\gamma_{f(X)}^{-1} \circ f|_X \circ \gamma_X$ represents the effect of f on X, an effect which comes down to the action of a unique automorphism of $\Gamma(C_\alpha)$ via the isomorphisms γ_Y. Thus $\mathrm{Aut}(\Gamma(\mathbf{C}_\alpha))$ has the same order as $(\mathrm{Aut}(\Gamma(C_\alpha)))^n \times \mathfrak{S}_n$. The following result, whose proof is beyond the scope of this book, makes this idea more precise.

Theorem 9.2.2. *Let n be the number of connected components C of the graph Γ that are isomorphic to C_α, i.e. such that $\Gamma(C) \simeq \Gamma(C_\alpha)$. Let $\mathbf{C}_\alpha = \bigsqcup_{\Gamma(C)\simeq\Gamma(C_\alpha)} C$. Then*

(i) $\mathrm{Aut}(\Gamma(\mathbf{C}_\alpha)) \simeq \mathrm{Aut}(\Gamma(C_\alpha))^n \rtimes \mathfrak{S}_n$;

(ii) *in particular* $|\mathrm{Aut}(\Gamma(\mathbf{C}_\alpha))| = |\mathrm{Aut}(\Gamma(C_\alpha))|^n\, n!$.

Let N be an H-module, i.e. there is a map $(h,n) \in H \times N \mapsto h.n \in N$. The *semi-direct product* $N \rtimes H$ is defined by putting the group law $(n,h).(n',h') := (n(h.n'), hh')$ on the set of pairs $N \times H$, which makes the set into a group.

From now on, we will restrict our attention to automorphism groups $\mathrm{Aut}(\Gamma)$ associated to graphs Γ that are *connected and possess at least one edge*. Let $\Gamma = (V; E)$ be (such) a graph. We will define the group of *edge-automorphisms*, which is a dual notion to the group of automorphisms.

An *edge-automorphism* of Γ is a permutation f of E such that

$$\varepsilon(a) \cap \varepsilon(b) \neq \emptyset \iff \varepsilon(f(a)) \cap \varepsilon(f(b)) \neq \emptyset. \qquad [\mathrm{AP}]_{\mathrm{E}}$$

An *edge-isomorphism* is defined analogously. We say that two graphs are *edge-isomorphic* if there exists an edge-isomorphism between them. Let $\mathrm{Isom}_E(\Gamma_1, \Gamma_2)$ be the set of edge-isomorphisms from Γ_1 to Γ_2.

Exercise 9.3. Show that the set of edge-automorphisms of a graph $\Gamma = (V; E)$ forms a group, which can be realised as a subgroup of $\mathfrak{S}_E$.

We denote this group by $\mathrm{Aut}_E(\Gamma)$.

This notion is closely related to that of a line-graph. Let $\Gamma = (V; E)$ be a graph. Recall (see Section 2.4) that the *line-graph* of Γ, denoted $L(\Gamma) = (V'; E')$, is the graph whose vertices are the edges of Γ, $V' = E$, and such that two vertices are adjacent if and only if the corresponding edges in Γ are adjacent. Clearly (exercise), we have

$$\mathrm{Isom}(L(\Gamma_1), L(\Gamma_2)) = \mathrm{Isom}_E(\Gamma_1, \Gamma_2),$$

and in particular,

$$\mathrm{Aut}(L(\Gamma)) = \mathrm{Aut}_E(\Gamma).$$

There is a natural map

$$\begin{aligned} \mathrm{Isom}(\Gamma_1, \Gamma_2) &\xrightarrow{\alpha_{12}} \mathrm{Isom}_E(\Gamma_1, \Gamma_2), \\ f &\longmapsto f|_{E_1}, \end{aligned}$$

where $f|_{E_1}: E_1 \to E_2$ is the bijective map deduced from f by the adjacency property [AP]. In particular, we see that

$$\begin{aligned} \mathrm{Aut}(\Gamma) &\xrightarrow{\alpha_\Gamma} \mathrm{Aut}_E(\Gamma), \\ f &\longmapsto f|_E \end{aligned}$$

is a homomorphism of groups that we will now study.

9.2.2 Ker α_Γ

The homomorphism α_Γ is almost always injective, as the following theorem shows.

Theorem 9.2.3. *Let* $\Gamma = (V; E)$ *be a connected graph. Then*

(i) *if* $|E| = 1$, *then* $\mathrm{Ker}\,\alpha_\Gamma \simeq \mathfrak{S}_2$*;*

(ii) *if* $|E| \geqslant 2$, *then* $\mathrm{Ker}\,\alpha_\Gamma = \{\mathrm{Id}_V\}$.

Proof. If $|E| = 1$, then $\Gamma \simeq K_2$; we easily see that $\mathrm{Aut}(\Gamma) \simeq \mathfrak{S}_2$ and $\mathrm{Aut}_E(\Gamma) = \{\mathrm{Id}_E\}$. Suppose $|E| \geqslant 2$, and let $f \in \mathrm{Ker}\,\alpha_\Gamma$, i.e. $f|_E = \mathrm{Id}_E$. Let $x \in V$. Since Γ is connected and $|E| > 1$, there exists $a = \{x, y\} \in E$. It follows from the condition $f(a) = a$ that $\{f(x), f(y)\} = \{x, y\}$. It remains to show that $f(x) = x$ and $f(y) = y$, otherwise we would have $f(x) = y$ and $f(y) = x$. But since $|E| \geqslant 2$ and Γ is connected, there is at least one other edge $\{x, t\}, t \neq y$, or $\{y, t\}, t \neq x$. By symmetry, we may assume that $\{x, t\} \in E$, so we have $\{f(x), f(t)\} = \{x, t\}$, which then implies that $f(x) \in \{x, t\}$, which is a contradiction. ■

9.2.3 Im α_Γ

An element of $\mathrm{Im}\,\alpha_\Gamma$ is called an *induced edge-automorphism*; the group $\mathrm{Im}\,\alpha_\Gamma$ is denoted $\mathrm{Aut}^{\#}(\Gamma)$.

Theorem 9.2.3 has the following immediate corollary.

Corollary 9.2.4. *Let* $\Gamma = (V; E)$ *be a connected graph. Then* $\mathrm{Aut}(\Gamma) \simeq \mathrm{Aut}^{\#}(\Gamma)$ *if and only if* $|E| \geqslant 2$.

The morphism α_Γ can be surjective, meaning that every edge-automorphism is induced by an automorphism of the graph. Consider the two graphs shown in Figure 9.2:

- $\mathrm{Aut}(K_3) \simeq \mathfrak{S}_3$; it is easy to see that $L(K_3) \simeq K_3$, so $\mathrm{Aut}_E(K_3)$ is isomorphic to $\mathfrak{S}_3$ and α_{K_3} is an isomorphism;
- $\mathrm{Aut}(K_{1,3}) \simeq \mathfrak{S}_3$; $\mathrm{Aut}_E(K_{1,3}) = \mathrm{Aut}(L(K_{1,3}))$ and clearly $L(K_{1,3}) \simeq K_3$, so $\mathrm{Aut}_E(K_{1,3}) \simeq \mathfrak{S}_3$ and $\alpha_{K_{1,3}}$ is an isomorphism.

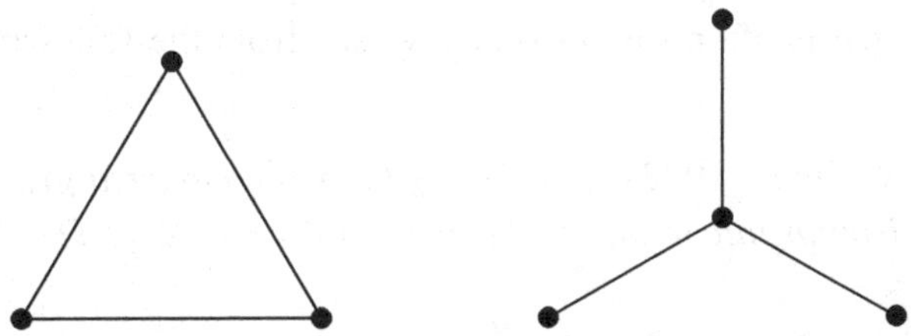

K_3 $K_{1,3}$

Figure 9.2. α_{K_3} and $\alpha_{K_{1,3}}$ are surjective.

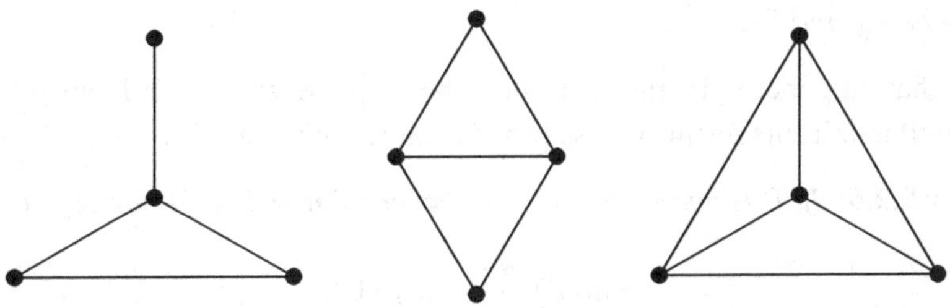

Ω_1 $\Omega_2 = L(\Omega_1)$ $\Omega_3 = K_4$

Figure 9.3. Three graphs Ω_i for which α_{Ω_i} is not surjective.

Note that even though $K_3 \not\simeq K_{1,3}$, we have $L(K_3) \simeq L(K_{1,3})$; the graphs K_3 and $K_{1,3}$ are edge-isomorphic.

In Figure 9.3, we give some examples where the morphism α_Γ is not surjective.

Exercise 9.4. Consider the graphs Ω_1 and Ω_2 in Figure 9.3.

(i) Prove the following facts:

$$\mathrm{Aut}(\Omega_1) \simeq \mathfrak{S}_2, \quad L(\Omega_1) \simeq \Omega_2 \quad \text{and} \quad \mathrm{Aut}_E(\Omega_1) \simeq \mathfrak{S}_2 \times \mathfrak{S}_2.$$

(ii) Show that $\mathrm{Aut}(\Omega_2) \simeq \mathfrak{S}_2 \times \mathfrak{S}_2$. Determine $L(\Omega_2)$ and deduce $\mathrm{Aut}_E(\Omega_2) \simeq \mathfrak{S}_4$.

(iii) Deduce that α_{Ω_i} is not surjective for $i = 1, 2$.

Exercise 9.5. Consider the graph Ω_3 in Figure 9.3.

(i) Show that $\mathrm{Aut}(\Omega_3) \simeq \mathfrak{S}_4$.

(ii) Determine $L(\Omega_3)$ (it is a 4-regular graph with 6 vertices).

(iii) Show that there is a cycle of length 6 in $\mathrm{Aut}\big(L(\Omega_3)\big)$.

(iv) Show that $\mathfrak{S}_4$ contains no element of order 6 (use the results of Section 9.1), and deduce that α_{Ω_3} is not surjective.

Nevertheless, α_Γ is often surjective, as we see from the following more general result.

Theorem 9.2.5 (Whitney, 1932). *Let Γ_1, Γ_2 be two connected graphs with at least two vertices, not isomorphic to any of the five graphs K_3, $K_{1,3}$, Ω_1, Ω_2, or Ω_3. Then the map*

$$\begin{aligned} \operatorname{Isom}(\Gamma_1, \Gamma_2) &\stackrel{\alpha_{12}}{\longrightarrow} \operatorname{Isom}_E(\Gamma_1, \Gamma_2), \\ f &\longmapsto f|_{E_1} \end{aligned}$$

is surjective: every edge-isomorphism between Γ_1 and Γ_2 comes from an isomorphism between Γ_1 and Γ_2.

Note that, in general, $\operatorname{Isom}(\Gamma_1, \Gamma_2)$ can be empty even though $\operatorname{Isom}_E(\Gamma_1, \Gamma_2)$ is not. We will admit this result, whose proof is quite technical [39].

Corollary 9.2.6. *If Γ is connected and not isomorphic to Ω_1, Ω_2 or Ω_3, the morphism*

$$\begin{aligned} \operatorname{Aut}(\Gamma) &\stackrel{\alpha_\Gamma}{\longrightarrow} \operatorname{Aut}_E(\Gamma), \\ f &\longmapsto f|_E \end{aligned}$$

is surjective. If moreover Γ is not isomorphic to K_2, then α_Γ is an isomorphism and

$$\operatorname{Aut}(\Gamma) \simeq \operatorname{Aut}_E(\Gamma) = \operatorname{Aut}^\#(\Gamma).$$

Proof. It suffices to take $\Gamma_1 = \Gamma_2$ in Whitney's theorem, observing that α_Γ is surjective for $\Gamma = K_3$ or $K_{1,3}$ (Figure 9.2). Theorem 9.2.3 then ensures the injectivity. ∎

It is clear that $\Gamma_1 \simeq \Gamma_2 \Rightarrow L(\Gamma_1) \simeq L(\Gamma_2)$. Conversely, we have the following result.

Corollary 9.2.7. *Let Γ_1, Γ_2 be two connected graphs having at least two vertices and not isomorphic to K_3 or $K_{1,3}$. If $L(\Gamma_1) \simeq L(\Gamma_2)$, then $\Gamma_1 \simeq \Gamma_2$.*

Proof. Since we have $\operatorname{Isom}(L(\Gamma_1), L(\Gamma_2)) = \operatorname{Isom}_E(\Gamma_1, \Gamma_2)$, if none of Γ_1 and Γ_2 is $K_3, K_{1,3}, \Omega_1, \Omega_2, \Omega_3$, we get the desired result by Whitney's theorem. Assume that both Γ_1 and Γ_2 are distinct among $\Omega_1, \Omega_2, \Omega_3$. Then, by Exercises 9.4 and 9.5, we infer $L(\Gamma_1) \not\simeq L(\Gamma_2)$, and the result remains valid. Finally, we suppose that $\Gamma_1 = \Omega_i$ and $\Gamma_2 \not\simeq K_3, K_{1,3}, \Omega_1, \Omega_2$ or Ω_3. We shall use the fact that, given a simple loopless graph Γ, if v is a vertex of $L(\Gamma)$, then there exists an edge e of Γ such that $L(\Gamma \backslash e) = L(\Gamma) - v$. Moreover, the degree of v in $L(\Gamma)$ is the number of edges adjacent to e in Γ. We separate the three cases $\Gamma_1 = \Omega_i$, $i = 1, 2, 3$.

(i) Assume $\Gamma_1 = \Omega_1$, and let Γ be a graph such that $L(\Gamma) = \Omega_2$. We remove from $L(\Gamma)$ a vertex v of degree 2 (Figure 9.3). We get the graph K_3, and we infer that there exists an edge e of Γ such that $L(\Gamma \backslash e) = K_3$. This shows that $\Gamma \backslash e$ is K_3 or $K_{1,3}$. In the former case, adjoining e to $\Gamma\backslash e$ gives $\Gamma = \Omega_1$. In the latter case, we adjoin the edge e in such a way that e is adjacent to 2 others edges. We get again $\Gamma = \Omega_1$.

(ii) Assume $\Gamma_1 = \Omega_2$, and let Γ be a graph such that $L(\Gamma) = L(\Omega_2) = K_5\backslash\{a, a'\}$, where a and a' are two non-incident edges of K_5. We remove from $L(\Gamma)$ the unique vertex v of degree 4, and we get the cyclic graph C_4. Hence there exists an edge e of Γ such that $L(\Gamma\backslash e) = C_4$. This implies that $\Gamma\backslash e = C_4$. If e is a pending edge of Γ, then it is adjacent to exactly 2 edges; hence v is of degree 2 in $L(\Gamma)$, a contradiction. We infer that e connects 2 vertices of $\Gamma\backslash e = C_4$, yielding $\Gamma = \Omega_2$.

(iii) Assume $\Gamma_1 = \Omega_3$, and let Γ be a graph such that $L(\Gamma) = L(\Omega_3) = L(K_4)$, which is a 4-regular graph of order 6. We remove from $L(\Gamma)$ an arbitrary vertex v. We can easily see that we get the graph $L(\Gamma) - v = L(\Omega_2)$. By (ii), there exists an edge e adjacent to 4 edges in Γ such that $\Gamma\backslash e = \Omega_2$. The only way to adjoin to Ω_2 an edge adjacent to 4 edges is to connect both vertices of degree 2. This gives the complete graph K_4; hence $\Gamma = \Omega_3$. ■

We now introduce a very useful group. Let $n \geqslant 3$ and let X be the regular polygon with n sides. The rotations that leave X invariant are centred at the centre of the polygon, and their angle lies in the set $\{2k\pi/n,\ 0 \leqslant k \leqslant n-1\}$. The axes of symmetry of X are the lines connecting two diametrically opposite vertices (resp. a vertex and the middle of the opposite side) if n is even (resp. odd). The symmetries of the plane with respect to these n axes of symmetry all preserve X. The $2n$ corresponding isometries of the plane form a group called the *dihedral group*, denoted D_n.

Exercise 9.6. Let $X = \{1, 2, \ldots, n\}$ denote the set of vertices of the regular polygon with n sides placed on a circle in this order, and let $f \in \mathrm{Aut}(X)$. Show that if $f(1)$ is fixed, there are two choices for $f(2)$, and deduce that $\mathrm{Aut}(X)$ has exactly $2n$ elements.

Remark 9.2.8. For $n \geqslant 3$, the group D_n has the structure of a semi-direct product: $D_n \simeq \mathbb{Z}/n\mathbb{Z} \rtimes \mathbb{Z}/2\mathbb{Z}$. This means that the binary operation $\circ$ in this group acts as $(x, i) \circ (y, j) = (x + (-1)^i y, i + j)$, where $x, y \in \mathbb{Z}/n\mathbb{Z}$ and $i, j \in \mathbb{Z}/2\mathbb{Z}$.

Exercise 9.7. Let $\Gamma = (V; E)$ be a graph and $\overline{\Gamma} = (V; P_2(V)\backslash E)$ its complement. Prove the following results:

(i) $\mathrm{Aut}(\Gamma) = \mathrm{Aut}(\overline{\Gamma})$.

(ii) If Γ is connected and $\mathrm{Aut}(\Gamma) = \mathfrak{S}_V$, then Γ is complete.

(iii) $\mathrm{Aut}(\Gamma) = \mathfrak{S}_V$ if and only if Γ or $\overline{\Gamma}$ is complete.

(iv) Let C_n be an elementary cycle with n vertices: $\mathrm{Aut}(C_n)$ contains $2n$ automorphisms; more precisely, $\mathrm{Aut}(C_n) \simeq D_n$ where D_n is the dihedral group of order $2n$.

Exercise 9.8. (i) Determine $\mathrm{Aut}(K_{2,3})$.

(ii) Determine $\mathrm{Aut}(K_{3,3})$.

Remark 9.2.9. The graph $L(K_{n,n})$ is called the *lattice-graph*; it has n^2 vertices and $n^2(n-1)$ edges.

The graph $T(n) = L(K_n)$ is called the *triangular* graph; it has $n(n-1)/2$ vertices and $n(n-1)(n-2)/2$ edges.

9.3 Coloured Cayley graphs

As we have seen, every graph Γ has an associated group, namely its automorphism group $\mathrm{Aut}(\Gamma)$. It is natural to ask the inverse question: can we associate a graph to any given group?

Let G be a *not necessarily finite* group (with unit element 1 and law written multiplicatively), and let S be a non-empty subset of G not containing 1. Let $\langle S \rangle$ denote the subgroup generated by S, which is the intersection of all subgroups containing S, or equivalently, the smallest subgroup of G (for inclusion) containing S. Concretely, $\langle S \rangle$ is the set obtained by taking all products $s_1^{\varepsilon_1} s_2^{\varepsilon_2} \dots s_k^{\varepsilon_k}$ where $\varepsilon_i \in \{-1, 1\}$, $s_i \in S$, and $k \geqslant 0$ (if $k = 0$ the product is taken to be equal to 1). We say that S is a *generating system* of G if $\langle S \rangle = G$.

Let (G, S) be a couple, where G is a group and S a non-empty subset of G such that $1 \notin S$. We can construct a simple digraph associated to (G, S), called the *Cayley digraph* and denoted $\overrightarrow{\mathrm{Cay}}(G, S) = (V, \vec{E})$, as follows:

- the vertices of the graph are the elements of the group: $V = G$;
- for $x, y \in G$, we place an arc (x, y) from x to y if there exists $s \in S$ such that $y = sx$; we say that the arc (x, y) is *coloured* by s (in particular there are no multiple arcs nor loops).

Remark 9.3.1. The use of the word "colouring" here is a mere convenience; it is not the same use of the term "edge colouring" as in Chapter 7.

Lemma 9.3.2. *If x is a vertex of a Cayley digraph, then $d^-(x) \geqslant 1$ and $d^+(x) \geqslant 1$.*

Proof. Let x be a vertex of $\overrightarrow{\mathrm{Cay}}(G, S)$; there is an arc from x to sx and one from $s^{-1}x$ to x ($s \neq 1$). ■

Exercise 9.9. Show that $\langle S \rangle = G$ if and only if the digraph is strongly connected.

Example 9.3.3. Figure 9.4 gives two examples of Cayley digraphs: (A) shows the Cayley digraph of $\overrightarrow{\mathrm{Cay}}(\mathbb{Z}/3\mathbb{Z}, \{\bar{1}, \bar{2}\})$, and (B) shows the digraph $\overrightarrow{\mathrm{Cay}}(\mathbb{Z}/5\mathbb{Z}, \{\bar{1}\})$ (where $\mathbb{Z}/k\mathbb{Z}$ is the cyclic group of order k).

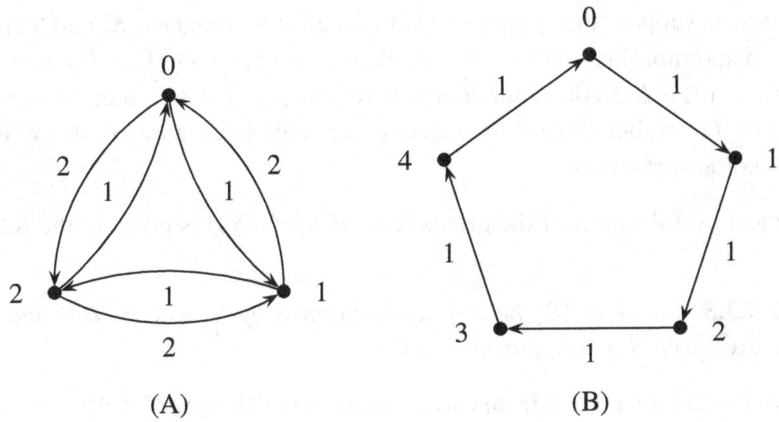

Figure 9.4. Examples of Cayley graphs.

Recall (see Section 1.9) that an automorphism of a simple digraph $\vec{\Gamma} = (V; \vec{E})$ is a bijection $f: V \to V$ satisfying

$$\forall x, y \in V : (x, y) \in \vec{E} \iff (f(x), f(y)) \in \vec{E}. \qquad [\overrightarrow{\text{AP}}]$$

As in the non-oriented case, the set of automorphisms of a digraph forms a group for the composition law, that we denote by $\text{Aut}(\vec{\Gamma})$.

An automorphism f of a Cayley digraph $\overrightarrow{\text{Cay}}(G, S) = (V, \vec{E})$ such that

$$(x, y) \text{ is coloured by } s \implies (f(x), f(y)) \text{ is coloured by } s$$

is called a *colour-preserving automorphism.*

Exercise 9.10. Let $\overrightarrow{\text{Cay}}(G, S) = (V, \vec{E})$ be a Cayley digraph. Show that the set of colour-preserving automorphisms forms a subgroup of $\text{Aut}(\overrightarrow{\text{Cay}}(G, S))$, denoted $\text{Aut}_{\text{col}}(\overrightarrow{\text{Cay}}(G, S))$.

The following lemma gives a characterisation of colour-preserving automorphisms.

Lemma 9.3.4. *Consider a couple (G, S) with $S \neq \emptyset$ and $1 \notin S$, and let $f: V \to V$ be a bijection. Then $f \in \text{Aut}_{\text{col}}(\overrightarrow{\text{Cay}}(G, S))$ if and only if*

$$\forall (x, s) \in G \times S, \quad f(sx) = sf(x). \tag{9.1}$$

Proof. Suppose f is a colour-preserving automorphism. Let $x \in G$ and $s \in S$. Then $a = (x, sx) \in \vec{E}$ so $f(a) = (f(x), f(sx)) \in \vec{E}$; but s colours a, so s also colours $f(a)$: $f(sx) = sf(x)$.

Conversely, suppose that $f(sx) = sf(x)$ for all $x \in G$ and $s \in S$, and let us show that f is an automorphism. If $(x, y) \in \vec{E}$, then $y = sx$, so $f(y) = f(sx) = sf(x)$, and $(f(x), f(y)) \in \vec{E}$. In the other direction, if $(f(x), f(y)) \in \vec{E}$, then $f(y) = sf(x)$ and $f(y) = f(sx)$, but since f is injective, we must have $y = sx$, so $(x, y) \in \vec{E}$. Thus f is colour-preserving. ∎

The most useful aspect of the group $\mathrm{Aut}_{\mathrm{col}}(\overrightarrow{\mathrm{Cay}}(G, S))$ is given in the following theorem.

Theorem 9.3.5. *Let $G = \langle S \rangle$ be a group generated by S, with $S \neq \emptyset$ and $1 \notin S$. Then $\mathrm{Aut}_{\mathrm{col}}(\overrightarrow{\mathrm{Cay}}(G, S))$ is isomorphic to G.*

Proof. We use the injection δ from Cayley's theorem (Theorem 9.1.4):

$$\begin{aligned} G &\xrightarrow{\delta} \mathfrak{S}_G, \\ g &\mapsto \delta_g : x \mapsto xg^{-1}, \end{aligned}$$

and show that $\mathrm{Im}\,\delta = \mathrm{Aut}_{\mathrm{col}}(\overrightarrow{\mathrm{Cay}}(G, S))$.

For $g \in G$, $\delta_g \in \mathrm{Aut}_{\mathrm{col}}(\overrightarrow{\mathrm{Cay}}(G, S))$ since δ_g is bijective and $\delta_g(sx) = sxg^{-1} = s\delta_g(x)$, so we can apply Lemma 9.3.4 to obtain $\mathrm{Im}\,\delta \subset \mathrm{Aut}_{\mathrm{col}}(\overrightarrow{\mathrm{Cay}}(G, S))$.

Let $f \in \mathrm{Aut}_{\mathrm{col}}(\overrightarrow{\mathrm{Cay}}(G, S))$, so f satisfies (9.1):

(a) we apply (9.1) with $x = 1$: $f(s) = sf(1)$,

(b) we apply (9.1) with $x = s^{-1}$: $f(1) = sf(s^{-1})$, so $f(s^{-1}) = s^{-1}f(1)$; more generally, $f(y) = f(ss^{-1}y) = sf(s^{-1}y)$ by (9.1), so $f(s^{-1}y) = s^{-1}f(y)$;

(c) let $x \in G$ and write $x = s_1^{\varepsilon_1} s_2^{\varepsilon_2} \dots s_k^{\varepsilon_k}$ where $\varepsilon_i \in \{-1, 1\}$, $s_i \in S$ and $k \geqslant 0$; by induction on k, we obtain the following by using (a) or (b) according to the sign of ε_i:

$$f(x) = s_1^{\varepsilon_1} f(s_2^{\varepsilon_2} \dots s_k^{\varepsilon_k}) = s_1^{\varepsilon_1} s_2^{\varepsilon_2} \dots s_k^{\varepsilon_k} f(1) = xf(1),$$

which shows that $f = \delta_{f(1)^{-1}} \in \mathrm{Im}\,\delta$. ∎

Remark 9.3.6. This proof works in the same way if the group is infinite uncountable, since every element of the group is still a finite product of elements of S. When G is a finite group, step (b) of the proof is not needed, since $x^{-1} = x^{|G|-1}$, so every element of $\langle S \rangle$ is a finite product of elements of S.

9.4 König's problem

Without a doubt the first ever book on graph theory was published in 1936. Its author was D. König [24]. In this book, he asked the following question, which we frame as an *inverse* problem: *Which groups are isomorphic to the automorphism group of a graph?* This problem was solved in 1938 by R. Frucht, who proved that all finite groups have this property.

If $G = \{1\}$ it suffices to take $\Gamma = K_1$, and if $G = \mathbb{Z}/2\mathbb{Z}$ it suffices to take $\Gamma = K_{1,2}$ or K_2. So we can assume that $|G| \geqslant 3$. Let G be generated by a finite set $S = \{s_1, s_2, s_3, \dots, s_t\}$ not containing 1. The first step in the proof is to construct the Cayley digraph $\overrightarrow{\mathrm{Cay}}(G, S)$. By Theorem 9.3.5, we have

$$\mathrm{Aut}_{\mathrm{col}}(\overrightarrow{\mathrm{Cay}}(G, S)) \simeq G.$$

The second step is to transform this digraph into a graph Γ, in such a way that the automorphism group $\mathrm{Aut}(\Gamma)$ is isomorphic to $\mathrm{Aut}_{\mathrm{col}}(\overrightarrow{\mathrm{Cay}}(G, S))$. Let us describe this idea, which is amazingly simple and pretty.

Construction method. Each arc (x, y) coloured by s_i is replaced by two *garlands*, as illustrated in Figure 9.5.

Precisely, starting from the digraph $\overrightarrow{\mathrm{Cay}}(G, S) = (V, \vec{E})$ where $V = G$, we construct a simple non-oriented graph as follows.

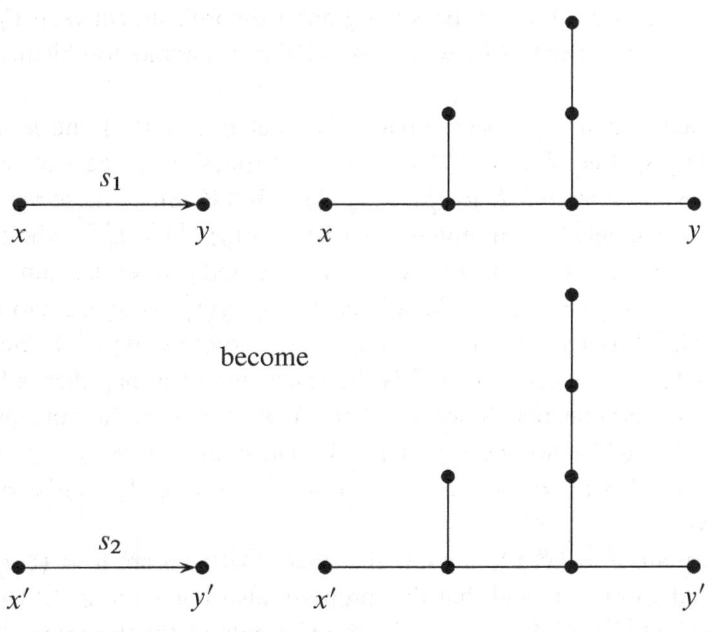

Figure 9.5. Transformation of arcs into garlands. This figure illustrates the transformation of arcs coloured with s_1, s_2: a garland of length 1, plus a garland of length 2 for s_1, of length 3 for s_2, …, of length $i + 1$ for s_i.

For each i, $1 \leqslant i \leqslant t$, we modify all the arcs coloured by s_i: if $e = (x, y)$ is an arc of colour s_i, then

- we replace e by a (simple non-oriented) chain (x, x_e^0, y_e^0, y);
- we add two chains or garlands (Figure 9.6):

$$G_i(x, y) = (x_e^0, x_e^1) \quad \text{and} \quad G_i'(x, y) = (y_e^0, y_e^1, \dots, y_e^{i+1}).$$

Let $\Gamma = (W; A)$ denote the simple graph obtained in this way. Intuitively, the orientation of arcs is *memorised* in Γ by the lengths of the two added garlands: $|G_i| < |G_i'|$. Moreover, an automorphism respects the lengths of the garlands G_i' (the garlands are associated to the colours!) and as there is only one isomorphism between two induced subgraphs $\Gamma_i(x, y) = \Gamma(\{x, x_e^0, x_e^1, y, y_e^0, \dots, y_e^{i+1}\})$ and $\Gamma_i(z, t) = \Gamma(\{z, z_a^0, z_a^1, t, t_a^0, \dots, t_a^{i+1}\})$, the number of automorphisms does not increase.

Define a morphism of groups

$$\varphi\colon \mathrm{Aut}_{\mathrm{col}}(\overrightarrow{\mathrm{Cay}}(G, S)) \to \mathrm{Aut}(\Gamma)$$

as follows: let $f \in \mathrm{Aut}_{\mathrm{col}}(\overrightarrow{\mathrm{Cay}}(G, S))$; then $f\colon V = G \to V$ is bijective; extend f to each $\Gamma_i(x, y)$ by the only possible graph isomorphism between $\Gamma_i(x, y)$ and $\Gamma_i(f(x), f(y))$; this gives $\varphi(f) = \tilde{f} \in \mathrm{Aut}(\Gamma)$; φ is a homomorphism; φ is clearly injective.

The difficult point is the surjectivity of φ. Let $g \in \mathrm{Aut}(\Gamma)$ and let $e = (x, y)$ be coloured by s_i. Let x_e^1 and y_e^{i+1} be the two vertices of degree 1 of two garlands added to e. By Lemma 9.3.2, $g(x_e^1), g(y_e^{i+1}) \in W \setminus V$. Since these vertices are of degree 1, they can only be extremities of garlands: $g(y_e^{i+1}) = t_a^{j+1}$ where $a = (z, t)$ is coloured by s_j. Thus we must have $i = j$, so e and a have the same colour and $g(y_e^i) = t_a^i, \dots, g(y_e^0) = t_a^0$. As x_e^0 is adjacent to y_e^0, $g(x_e^0)$ is adjacent to $g(y_e^0)$, then $g(x_e^1) = z_a^1$; it follows that $g(x) = z$ and $g(y) = t$, so $g(x), g(y) \in V$. Set $f = g|_V$. We have $f(V) \subset V$, where since f is the restriction of a map that is bijective so injective, f is itself injective. Since $g \in \mathrm{Aut}(\Gamma)$, we can apply the same procedure as above. For $f' = g^{-1}|_V$ we have $f'(V) \subset V$. The compositions $f \circ f'$ and $f' \circ f$ are defined, and $f \circ f' = g|_V \circ g^{-1}|_V = \mathrm{Id}_V$, $f' \circ f = g^{-1}|_V \circ g|_V = \mathrm{Id}_V$. Thus f is bijective.

For every arc $e = (x, y)$, we saw that there exists an arc $a = (z, t)$ such that $g(x) = z$ and $g(y) = t$, and that this property also holds for g^{-1}. Thus f is an automorphism of $\overrightarrow{\mathrm{Cay}}(G, S)$ which also preserves colours (by the above). So we have shown that $f \in \mathrm{Aut}_{\mathrm{col}}(\overrightarrow{\mathrm{Cay}}(G, S))$, and that f extends uniquely to Γ: $\varphi(f) = g$. ∎

The above discussion gives the proof of the following theorem.

Theorem 9.4.1 (Frucht, 1938). *Let G be a finite or countable group. Then there exists a graph Γ such that* $\mathrm{Aut}(\Gamma) \simeq G$.

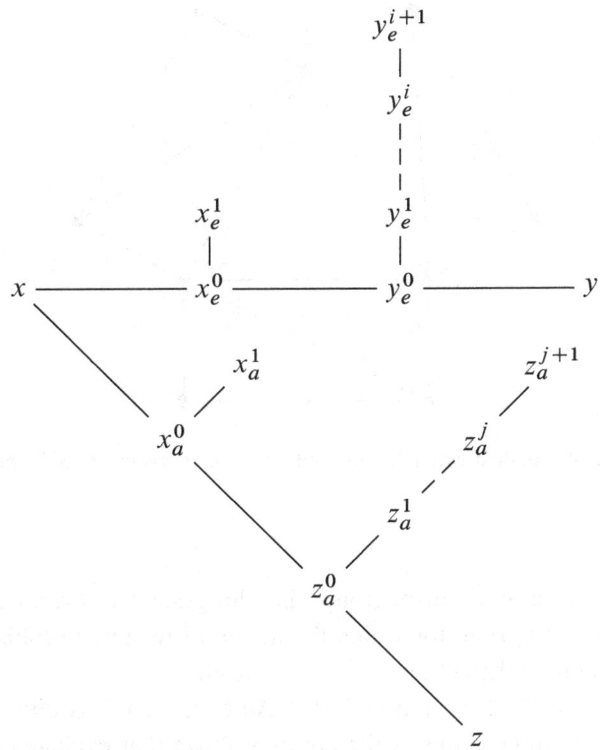

Figure 9.6. Precise numbering of new vertices produced by two arcs $e = (x, y)$ and $a = (x, z)$ coloured by s_i and s_j.

Exercise 9.11. Apply the construction method to the graphs in Figure 9.4. In the case of the graph (A), give the smallest graph whose automorphism group is isomorphic to $\mathbb{Z}/3\mathbb{Z}$.

Exercise 9.12. Give the Cayley digraphs in the following cases:

(1) $G = \mathbb{Z}, S = \{1\}$ (the graph is very simple!)

(2) $G = \mathbb{Z} \times \mathbb{Z}/n\mathbb{Z}, S = \{(1, 0), (0, 1)\} \subset G$.

(3) $G = \mathbb{Q}, S = [0, 1[\cap \mathbb{Q} = \{s_1, s_2, \dots, s_n, \dots\}$.

In seeking to answer König's question, we can required more precision. For example, let G be a subgroup of $\mathfrak{S}_n$. Then G acts on $\{1, \dots, n\}$ (see Section 9.5), and we may ask whether there exists a graph Γ having vertices $1, \dots, n$ in such a way that the action of $\mathrm{Aut}(\Gamma)$ is the same? The answer is not always affirmative. It is if $G = \mathfrak{S}_n$, since $\Gamma = K_n$ works. But consider the cyclic group of order 3 (isomorphic to $\mathbb{Z}/3\mathbb{Z}$) generated by the 3-cycle $(1\,2\,3)$ of $\mathfrak{S}_3$. Figure 9.7 shows the smallest graph

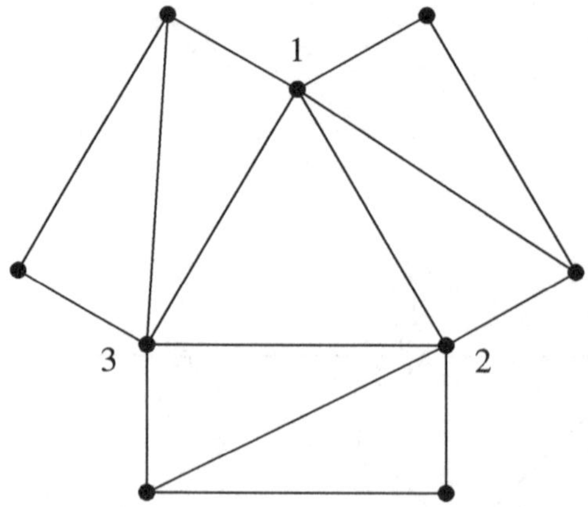

Figure 9.7. Graph of smallest order having automorphism group $\mathbb{Z}/3\mathbb{Z}$, the cyclic group of order 3.

having this group as automorphism group – but this graph has 9 vertices and 15 edges! In fact, there exists no graph for which the action of its automorphism group on its vertices is equivalent to the action of $\mathbb{Z}/3\mathbb{Z}$ on itself.

Also, if $V(\Gamma) = \{1, 2, 3\}$ and $(1\,2\,3) \in \mathrm{Aut}(\Gamma)$, then Γ is either K_3 or $\overline{K_3}$, so $(1\,3)(2) = (1\,3) \in \mathrm{Aut}(\Gamma)$; this is the automorphism that exchanges the vertices 1 and 3 and leaves the vertex 2 invariant, so it is of order 2, but if $\mathrm{Aut}(\Gamma)$ contains an element of order 2, then it cannot be isomorphic to $\mathbb{Z}/3\mathbb{Z}$.

Remark 9.4.2. The method used to prove Frucht's theorem (Theorem 9.4.1) gives a graph with 18 vertices and 18 edges.

However, we can give the following result (see [26]).

Theorem 9.4.3 (Bouwer). *Let G be a finite group acting on a set X. Then there exists a graph Γ such that* $\mathrm{Aut}(\Gamma) \simeq G$ *and*

- $X \subset V(\Gamma)$*;*
- *X is invariant under the action of* $\mathrm{Aut}(\Gamma)$ *(i.e. we have $f(X) \subset X$ for all $f \in$* $\mathrm{Aut}(\Gamma)$*);*
- *the action of* $\mathrm{Aut}(\Gamma)$ *on X is equivalent to the action of G on X.*

Example 9.4.4. Thus for $G = \langle(1\,2\,3)\rangle \subset \mathfrak{S}_3$ acting on $X = \{1, 2, 3\}$, the graph in Figure 9.7 has 9 vertices, X is invariant under the action of $\mathrm{Aut}(\Gamma)$ (exercise) and this action is exactly that of G on X.

9.5 Group actions

A group G (written multiplicatively, with unit element 1) *acts on a set* X if there exists a group morphism

$$\varphi: G \to \mathfrak{S}_X,$$
$$g \mapsto \varphi_g.$$

The morphism φ is called the *action of* G *on* X, and $\mathrm{Ker}(\varphi)$ is the *kernel of the action* of G on X; the action is *faithful* if $\mathrm{Ker}(\varphi) = \{1\}$. We use the notation $\varphi_g(x) = g \cdot x$.

We associate an external law on X to this action:

$$G \times X \to X,$$
$$(g, x) \mapsto g \cdot x$$

satisfying

- for all $g_1, g_2 \in G$ and for all $x \in X$, $g_1 \cdot (g_2 \cdot x) = (g_1 g_2) \cdot x$;
- for all $x \in X$: $1 \cdot x = x$.

When G acts on X, we also say that X is a G*-module* (by analogy with the notion of module over a ring, but with the notable difference that there is no internal law on X).

For example, let H be a subgroup of G. Then G is an H-module whose external law is induced by the translation

$$H \times G \to G,$$
$$(h, x) \mapsto hx.$$

Exercise 9.13. Show that if X is a G-module, it is associated to an action of G on X.

Suppose G_1 acts on X_1 and G_2 acts on X_2. These actions are *equivalent* if there exists an isomorphism $\alpha: G_1 \to G_2$ and a bijection $\beta: X_1 \to X_2$ satisfying

$$\beta(g \cdot x) = \alpha(g) \cdot \beta(x), \quad \forall x \in X_1,\ \forall g \in G_1.$$

A very important example of a group action is the action of *permutation groups* (see Section 9.1), i.e. the action of subgroups G of $\mathfrak{S}_X$: this means that we consider the way in which G acts by permuting the elements (or letters) of X. We already saw an example of this type of action at the beginning of this chapter, with the automorphism group of a graph $\mathrm{Aut}(\Gamma)$ acting on the set $\mathcal{C}$ of connected components of $V(\Gamma)$:

$$\mathrm{Aut}(\Gamma) \times \mathcal{C} \to \mathcal{C},$$
$$(f, C) \mapsto f(C).$$

Let X be a G-module and $x \in X$. The *orbit of x* under the action of G is the set

$$G \cdot x = \{g \cdot x \mid g \in G\},$$

and the *stabiliser of x in G* is

$$\mathrm{Stab}_G(x) = \{g \in G \mid g \cdot x = x\}.$$

The various orbits form a partition of the set X (see Exercise 9.14 below).

Exercise 9.14. (i) Show that the relation x *belongs to the same orbit as* y is an equivalence relation (called *transitivity equivalence*).

(ii) Show that $\mathrm{Stab}_G(x)$ is a subgroup of G.

(iii) Show that if $y = g \cdot x$, then $\mathrm{Stab}_G(y) = g\,\mathrm{Stab}_G(x)g^{-1}$.

We can choose an element x_t in each orbit (i.e. a system of representatives for the equivalence relation); then the orbits $G \cdot x_t$, $t \in T$ satisfy

$$X = \bigsqcup_{t \in T} G \cdot x_t.$$

Thus the set of orbits form a partition of X. In the example given at the beginning of this chapter, the orbits are the isomorphic connected components of a graph, and the above equality translates to the relation (see Section 9.2.1):

$$\mathcal{C} = \bigsqcup_{\alpha} \mathcal{C}_\alpha.$$

An action is *transitive* if there is only a single orbit:

$$\forall x, y \in X,\ \exists g \in G \text{ such that } g \cdot x = y.$$

In this situation, we also say that G *acts transitively* on X.

Let X, Y be two G-modules. A *G-morphism* is a map $f: X \to Y$ satisfying

$$f(g \cdot x) = g \cdot f(x), \quad \forall g \in G,\ \forall x \in X.$$

We write $\mathrm{Hom}_G(X, Y)$ for the set of G-morphisms and $\mathrm{Isom}_G(X, Y)$ for the set of G-isomorphisms from X to Y; the latter are G-morphisms f admitting an inverse G-morphism $f': Y \to X$ such that $f' \circ f = \mathrm{Id}_X$, $f \circ f' = \mathrm{Id}_Y$. It is easy to check that this is equivalent to requiring f to be a bijective G-morphism.

Another very important example of a G-module is the following. Let H be a subgroup of G. The set $xH = \{xh \mid h \in H\}$ is the *left class of x modulo H*. Similarly, we define $Hx = \{hx \mid h \in H\}$ to be the *right class of x modulo H*.

Exercise 9.15. Show that the relation x *and* y *are in the same left (resp. right) class modulo* H is an equivalence relation. Deduce that if $Hx \neq Hy$, then $Hx \cap Hy = \emptyset$.

Let G/H denote the set of left classes xH and $\pi\colon G \mapsto G/H$ the canonical surjection $x \mapsto xH$. The set G/H is not always a group (for this it is necessary and sufficient for H to be a normal subgroup of G, meaning that H satisfies the property that $x^{-1}Hx = H$ for all $x \in G$), but it is always a G-module for the natural action

$$\begin{aligned} G \times G/H &\to G/H, \\ (g, xH) &\mapsto gxH. \end{aligned}$$

It is easy to recall the conventions by observing that G acts *on the left* on the *left classes* modulo H.

Clearly, $\pi \in \mathrm{Hom}_G(G, G/H)$ where $\mathrm{Hom}_G(G_1, G_2)$ denotes the set of morphisms of G-modules of G_1 to G_2.

Another example of G-modules is given by each orbit of a G-module X. Indeed, the orbit $G \cdot x$ of an element $x \in X$ is itself a G-module for the natural left action $g \cdot (h \cdot x) = (gh) \cdot x$, and

$$\begin{aligned} f\colon G &\to G \cdot x, \\ g &\mapsto g \cdot x \end{aligned}$$

is a G-morphism. Thus we have a diagram of G-modules:

$$\begin{array}{ccc} G & \xrightarrow{\ f\ } & G/\mathrm{Stab}_G(x) \\ & \searrow^{\pi} & \downarrow \tilde{f} \\ & & G \cdot x \end{array}$$

Let us show that there exists a unique $\tilde{f} \in \mathrm{Hom}_G(G/\mathrm{Stab}_G(x), G \cdot x)$ such that $\tilde{f} \circ \pi = f$. Let $H = \mathrm{Stab}_G(x)$. The uniqueness of such a morphism is clear). To prove its existence,

- set $\tilde{f}(\pi(g)) = f(g) = g \cdot x$;
- $\tilde{f}$ is well-defined since if $\pi(g) = \pi(g')$, then $gH = g'H$, so $g^{-1}g' \in H$: $(g^{-1}g') \cdot x = x$; $g' \cdot x = g \cdot x$, i.e. $f(g') = f(g)$;
- $\tilde{f}$ is a G-morphism: $\tilde{f}(h \cdot \pi(g)) = \tilde{f}(h \cdot gH) = f(hg) = h \cdot f(g) = h\tilde{f}(gH)$;
- $\tilde{f}$ is a G-isomorphism, since by setting $\theta\colon G \cdot x \to G/H$, $\theta(g \cdot x) = gH$, we easily see that $\theta \in \mathrm{Hom}_G(G \cdot x, G/H)$ and that θ is the inverse of $\tilde{f}$.

We have just proved the following famous theorem.

Theorem 9.5.1 (Orbit-stabiliser theorem). *There exists a* G*-isomorphism between* $G \cdot x$ *and the set* $G/\mathrm{Stab}_G(x)$.

In particular, when the action is transitive (i.e. $X = G \cdot x$) and X is finite, we see that $|X| = |G/\mathrm{Stab}_G(x)|$ must divide the order of G; thus the order of a transitive group of permutations acting on $\{1, 2, \ldots, n\}$ will always be a multiple of n.

9.6 Transitive graphs

Let $\Gamma = (V; E)$ be a graph, and consider the group $\mathrm{Aut}(\Gamma) \subset \mathfrak{S}_V$. We say that Γ is *vertex-transitive* if for every pair of vertices x, y, there exists $f \in \mathrm{Aut}(\Gamma)$ such that $f(x) = y$; in other words, a graph is vertex-transitive if its automorphism group acts transitively on its vertices. We saw also that $\mathrm{Aut}(\Gamma)$ can be considered as a subgroup of $\mathfrak{S}_E$ (on condition that $|E| \geqslant 2$, see Theorem 9.2.3). We say that Γ is *edge-transitive* if for every pair of edges $a = \{x, y\}$, $b = \{z, t\}$, there exists $f \in \mathrm{Aut}(\Gamma)$ such that $\{f(x), f(y)\} = \{z, t\}$, i.e. such that $\alpha_\Gamma(f)(a) = b$. In other words, a graph is edge-transitive if its automorphism group acts transitively on its edges.

Attention: a graph can be vertex-transitive without being edge-transitive, and it can be edge-transitive without being vertex-transitive! Figure 9.8 gives examples of such graphs.

Example 9.6.1. Graph (A) in Figure 9.8 is vertex-transitive but not edge-transitive, since otherwise an automorphism would send the edge a to the edge b. But the smallest cycle containing a is of length 3, and the smallest cycle containing b is of length 4. Graph (B) in Figure 9.8 is edge-transitive but not vertex-transitive, since no automorphism can send a vertex of degree 1 to a vertex of degree 2.

In a vertex-transitive graph, all the vertices basically play the same role, so such a graph is necessarily regular. In an edge-transitive graph, it is the edges that play the same role.

A graph that is both edge-transitive and vertex-transitive is said to be *symmetric*.

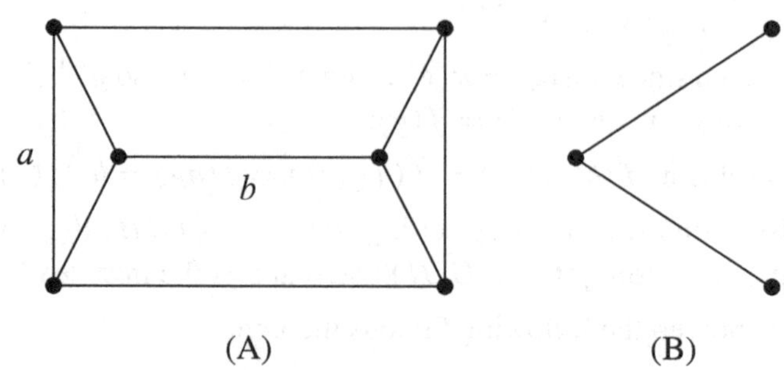

Figure 9.8. Vertex-transitivity and edge-transitivity.

Exercise 9.16. Give a necessary and sufficient condition for the graph $K_{m,n}$ to be symmetric.

The following theorem yields a description of edge-transitive graphs which may not be symmetric.

Theorem 9.6.2. *Let $\Gamma = (V; E)$ be an edge-transitive graph with no isolated vertex. If Γ is not vertex-transitive, then the action of* $\mathrm{Aut}(\Gamma)$ *on V has exactly two orbits, and these two orbits form a partition of V such that $\Gamma = (V; E)$ is bipartite.*

Proof. Observe first that an edge-transitive graph with no isolated vertices that is not vertex-transitive must admit at least 3 vertices. Let $\{x, y\}$ be an edge of Γ, set $G = \mathrm{Aut}(\Gamma)$, and consider the two orbits $V_1 = G \cdot x$, $V_2 = G \cdot y$. Let us show that $V = V_1 \sqcup V_2$. Let $z \neq x, y$; since z is not isolated, there exists an edge $e = \{z, w\} \in E$; by transitivity there exists an automorphism $\sigma \in G$ such that $\sigma(\{x, y\}) = \{z, w\}$, so $\sigma(x) = z$ or $\sigma(y) = z$, and z is in the orbit of x or of y. Note that $V_1 \neq V_2$ (otherwise Γ would be vertex-transitive). It remains to show that the elements of V_1 (resp. V_2) are not adjacent. If there existed vertices $u, v \in V_1$ such that $\{u, v\} \in E$, there would exist $\sigma \in G$ such that $\sigma(\{x, y\}) = \{u, v\}$, so we would have $\sigma(x) = u$ and $\sigma(y) = v$ (or the inverse), and then $v \in G \cdot y = V_2$, so $v \in V_1 \cap V_2$. But this is not possible since distinct orbits are disjoint. ■

Remark 9.6.3. When Γ is vertex-transitive, it follows from the comment after the orbit-stabiliser theorem (Theorem 9.5.1) that $|V|$ divides $|\mathrm{Aut}(\Gamma)|$. If a connected graph Γ is edge-transitive and is not one of the graphs K_2, Ω_1, Ω_2 or Ω_3 of Corollary 9.2.6 of Whitney's theorem (see Section 9.2.3), we find that $|E|$ divides $|\mathrm{Aut}(\Gamma)|$. This is illustrated by graph (A) in Figure 9.8, which has 9 edges, but whose automorphism group is of cardinality 12. Thus (A) is not edge-transitive, as already observed earlier.

Exercise 9.17. Let $\Gamma = (V; E)$ be a regular graph of odd degree. Show that the number of vertices is even. Furthermore, if Γ is vertex-transitive, show that the cardinality of $\mathrm{Aut}(\Gamma)$ is even.

Let $\vec{\Gamma} = (V; \vec{E})$ be a simple digraph. We say that $\vec{\Gamma}$ is *arc-transitive* if for every pair of arcs (x, y) and (z, w), there exists $f \in \mathrm{Aut}(\vec{\Gamma})$ such that $f((x, y)) = (z, w)$, i.e. such that $f(x) = z$ and $f(y) = w$. Let $\Gamma^+(x) = \{y \in V \mid (x, y) \in \vec{E}\}$.

Proposition 9.6.4. *Let $\vec{\Gamma} = (V; \vec{E})$ be a simple digraph whose underlying graph is connected, and let G be a subgroup of* $\mathrm{Aut}(\vec{\Gamma})$ *acting transitively on the vertices. Then G also acts transitively on the arcs if and only if* $\mathrm{Stab}_G(x)$ *acts transitively on the vertices of $\Gamma^+(x)$ for all $x \in V$.*

Proof. Suppose G acts transitively on the arcs, and let $u, v \in \Gamma^+(x)$. Then (x, u), $(x, v) \in \vec{E}$. There exists $f \in G$ such that $f((x, u)) = (x, v)$, so $f(x) = x$, $f(u) = v$ and $f \in \mathrm{Stab}_G(x)$. It follows that $\mathrm{Stab}_G(x)$ acts transitively on the vertices of $\Gamma^+(x)$.

Recall that $x \notin \Gamma(x)$, and suppose now that $\mathrm{Stab}_G(x)$ acts transitively on the vertices of $\Gamma^+(x)$ for all x. Take two arcs (y, u), $(z, v) \in \vec{E}$. Since G acts transitively on V, there exist $f, g \in G$ such that $f(y) = x$ and $g(z) = x$. Thus $f((y, u)) = (x, f(u)) \in \vec{E}$ and $g((z, v)) = (x, g(v)) \in \vec{E}$, so $f(u), g(v) \in \Gamma^+(x)$. Furthermore, as $\mathrm{Stab}_G(x)$ acts transitively on the vertices of $\Gamma^+(x)$, there exists $h \in \mathrm{Stab}_G(x)$ such that $h(f(u)) = g(v)$. Thus we have

$$g^{-1}(h(f((y, u)))) = g^{-1}(h((x, f(u)))) = g^{-1}((x, g(v))) = (z, v),$$

and hence G acts transitively on the arcs of Γ. ■

Chapter 10

Spectral theory and applications

In this chapter, we concentrate on the spectrum of a graph Γ: its distribution, its largest eigenvalue, and the one that is next largest in absolute value can yield very delicate information on various invariants attached to Γ.

Throughout this chapter (with the exception of Section 10.2), all graphs are assumed *simple and finite*. We will make frequent use of the basics of linear algebra, and to this end we gave a summary of basic facts in Section 6.5.

10.1 The spectrum of a graph

10.1.1 The characteristic polynomial

The spectral theory of graphs is the study of certain characteristics of graphs, mainly graph invariants such as for example the stability number (i.e. the maximum cardinality of an independent set), the diameter, or the connectivity, via the characteristic polynomial of the graph and the eigenvalues and eigenvectors of the associated adjacency matrix and Laplacian matrix.

In this paragraph, we will study in more detail the adjacency matrix A of a graph $\Gamma = (V; E)$, which was defined in Chapter 1 and further developed in Chapter 6. In particular, we use the existence of a bijection between the set of vertices $V = \{x_1, x_2, x_3, \ldots, x_n\}$ and the canonical basis of $\mathbb{F}_2^n$, which is thus identified with the *space* of vertices of Γ, denoted $C_0(\Gamma)$. In order to exploit the properties of Hermitian matrices, we will take $\mathbb{C}$ as our base field and interpret V as an orthonormal basis of $\mathbb{C}^V \simeq \mathbb{C}^n$ (where $n = |V|$ and $\mathbb{C}^V$ denotes the space of functions defined on V with values in $\mathbb{C}$), without giving an explicit bijection between V and the orthonormal basis $\mathfrak{B} = \{e_1, \ldots, e_n\}$: we simply identify $V = \mathfrak{B}$. By this slight *abuse of language*, we avoid indexing all the vertices, which considerably simplifies the notation in certain computations. Consequently, the adjacency matrix $A = (a_{ij})$ of a graph is interpreted as a *table* of values $A = (a_{vw})_{v,w \in V}$ without taking into account the ordering of the rows and columns as usual in a matrix.

Let $\Gamma = (V; E)$ be a simple graph, and consider V as an orthonormal basis of $\mathbb{C}^n$, where $n = |V|$. Let $A = (a_{vw})_{v,w \in V}$ denotes its adjacency matrix, so that

- $a_{vw} = 1$ if $\{v, w\} \in E$;
- $a_{vw} = 0$ otherwise.

As A is real and symmetric, it is a Hermitian matrix (see Remark 6.5.1). Thus in the orthonormal basis $\mathfrak{B} = V$ of $\mathbb{C}^V \simeq \mathbb{C}^n$, it represents a Hermitian endomorphism $\mathfrak{A} \in \mathcal{H}(\mathbb{C}^n)$. The *characteristic polynomial* of Γ is the characteristic polynomial of the endomorphism $\mathfrak{A}$; we denote it by $\Phi(\Gamma, x) = \det(x\mathrm{Id} - \mathfrak{A})$. It is a monic polynomial of degree n whose coefficients lie in $\mathbb{Z}$. The *eigenvalues* of the graph Γ are by definition the eigenvalues of $\mathfrak{A}$, which are nothing other than the roots in $\mathbb{C}$ of the characteristic polynomial $\Phi(\Gamma, x) \in \mathbb{Z}[x]$. The *spectrum* of the graph $\Gamma = (V; E)$, denoted $\mathrm{Sp}(\Gamma)$, is the list of eigenvalues of the graph, taking account of their multiplicities. Since the eigenvalues are real, we can list them in decreasing size and write the spectrum as

$$\mathrm{Sp}(\Gamma) = \{\lambda_{\max}^{(m_1)} = \lambda_1^{(m_1)} > \lambda_2^{(m_2)} > \cdots > \lambda_s^{(m_s)} = \lambda_{\min}^{(m_s)}\},$$

where $\sum_1^s m_i = |V(\Gamma)| = n$. Observe that $\sum_1^s m_i \lambda_i = \mathrm{tr}(A) = 0$ (the *trace* of A), since Γ has no loops. It will also be useful to write the spectrum in the form

$$\mathrm{Sp}(\Gamma) = (\lambda_1 \geqslant \lambda_2 \geqslant \cdots \geqslant \lambda_n),$$

where each eigenvalue is written out a number of times equal to its multiplicity.

It is usually convenient to simplify notation by identifying A and $\mathfrak{A}$, considering that A is the matrix of $\mathfrak{A}$ in the canonical basis of $\mathbb{C}^n$. We will use the matrix notation Ax when $x \in \mathbb{C}^n$ and where we identify $x = (x_1, \ldots, x_n)$ with the associated column matrix

$$\begin{pmatrix} x_1 \\ \vdots \\ x_n \end{pmatrix}.$$

In particular, we will write the Rayleigh quotient in the form $R(A, x) = \langle Ax, x \rangle = x^* Ax$ when $\|x\| = 1$.

Exercise 10.1. Show that if $z \in \mathbb{Q}$ is a root of $\Phi(\Gamma, x)$, then $z \in \mathbb{Z}$.

Hint: take z in the form $\frac{p}{q}$ with $\mathrm{pgcd}(p, q) = 1$, and write z as a root of the polynomial $\Phi(\Gamma, x)$, then clear the denominators.

Exercise 10.2. Compute the eigenvalues of the graphs K_2, $K_{1,2}$, K_3, L_3, $K_2 + K_1$ and $K_2 + K_2$.

10.1.2 Interlacing of eigenvalues

Let us recall the relations between $\mathfrak{A}$ and its matrix A in the basis $\mathfrak{B} = V$. Let

$$f = \sum_{x \in V} f_x x \in \mathbb{C}^V, \quad f_x \in \mathbb{C};$$

we have $\mathfrak{A}(f) = \sum_{x\in V} f_x \mathfrak{A}(x)$ and $\mathfrak{A}(x) = \sum_{w\in V} a_{wx} w$ (the column of A corresponding to the vertex x consists of components of $\mathfrak{A}(x)$ in the basis V). This gives the following formulas:

$$\mathfrak{A}(x) = \sum_{w\in V} a_{wx} w = \sum_{w\in V \mid \{x,w\}\in E} w, \tag{10.1}$$

$$\mathfrak{A}(f) = \sum_{x\in V} f_x \mathfrak{A}(x) = \sum_{x,w\in V} f_x a_{wx} w = \sum_{x\in V} f_x \sum_{w\in V \mid \{x,w\}\in E} w; \tag{10.2}$$

the component of $\mathfrak{A}(f)$ corresponding to the vertex x is

$$\mathfrak{A}(f)_x = \sum_{w\in V} a_{wx} f_w = \sum_{w\in V \mid \{x,w\}\in E} f_w. \tag{10.3}$$

Furthermore,

$$\forall x, y \in V \quad \langle \mathfrak{A}(y), x\rangle = \Big\langle \sum_{w\in V} a_{wy} w, x \Big\rangle = \sum_{w\in V} a_{wy} \langle w, x\rangle = a_{xy},$$

since $\langle w, x\rangle = 1$ if $w = x$ and 0 otherwise. In particular, we always have $\langle \mathfrak{A}(x), x\rangle = 0$ since the graph is assumed to have no loops.

Let us begin with some relatively simple results.

Proposition 10.1.1. *Let $\Gamma = (V; E)$ be a graph with $E \neq \emptyset$, and denote by $\lambda_{\min}(\Gamma)$ (resp. $\lambda_{\max}(\Gamma)$) the smallest (resp. largest) eigenvalue of Γ. Then*

(i) $\lambda_{\min}(\Gamma) \leqslant -1 < 1 \leqslant \lambda_{\max}(\Gamma)$;

(ii) $\delta(\Gamma) \leqslant \lambda_{\max}(\Gamma) \leqslant \Delta(\Gamma)$;

(iii) *if Γ' is a induced subgraph of Γ, then*

$$\lambda_{\min}(\Gamma) \leqslant \lambda_{\min}(\Gamma') \leqslant \lambda_{\max}(\Gamma') \leqslant \lambda_{\max}(\Gamma);$$

(iv) *if Γ is k-regular, then $\lambda_{\max}(\Gamma) = \Delta(\Gamma) = k$; if moreover Γ is connected, then $\lambda_{\max}(\Gamma)$ has multiplicity 1 and is associated to the eigensubspace $\mathbb{C}(1, 1, \ldots, 1)$.*

Proof. (i) If $\{x, y\}$ is an edge, then

$$\langle \mathfrak{A}(x+y), x+y\rangle = \langle \mathfrak{A}(x), y\rangle + \langle \mathfrak{A}(y), x\rangle = 2,$$
$$\langle x+y, x+y\rangle = \langle x, x\rangle + \langle y, y\rangle = 2,$$

so, by Proposition 6.5.3, $0 < 1 = R(\mathfrak{A}, x+y) \leqslant \lambda_{\max}(\Gamma)$. Similarly, $\langle \mathfrak{A}(x-y), x-y\rangle = -2$ and $\langle x-y, x-y\rangle = 2$, so $\lambda_{\min}(\Gamma) \leqslant R(\mathfrak{A}, x-y) = -1 < 0$.

(ii) Let $\lambda \in \mathbb{R}$ be an eigenvalue of $\mathfrak{A}$, and let $f = \sum_{x\in V} f_x x$ (with coefficients $f_x \in \mathbb{C}$) be an eigenvector associated to λ, so $\mathfrak{A}(f) = \lambda f$. Let $w \in V$ be such that

$|f_w| = \max\{|f_x| \mid x \in V\}$. Since $f \neq 0_{\mathbb{C}^n}$, $f_w \neq 0$ is the largest coordinate of f (by modulus), the component of $\mathfrak{A}(f)$ corresponding to the vertex w is

$$\mathfrak{A}(f)_w = \sum_{y \in V} a_{wy} f_y = \lambda f_w$$

by equation (10.3). Thus

$$\begin{aligned} |\lambda f_w| &= \Big|\sum_{y\in V} a_{wy} f_y\Big| \leqslant \sum_{y\in V} |a_{wy}||f_y| = \sum_{y\in V|\{w,y\}\in E} |f_y| \\ &\leqslant |f_w| \sum_{y\in V|\{w,y\}\in E} 1 = |f_w| d(w) \leqslant |f_w|\Delta(\Gamma), \end{aligned}$$

so simplifying by f_w, we find $|\lambda| \leqslant \Delta$.

Let $\mathbf{1}$ denote the vector $\sum_{x\in V} x$. Observe that $\mathbf{1}$ is nothing other than the vector $(1, 1, 1, \ldots, 1) \in \mathbb{C}^n$, where V is interpreted as the canonical orthonormal basis of $\mathbb{C}^n$. By (10.1), $\mathfrak{A}(\mathbf{1}) = \mathfrak{A}(\sum_{x\in V} x) = \sum_{w,x\in V} a_{wx} w$, so

$$\langle \mathfrak{A}(\mathbf{1}), \mathbf{1}\rangle = \Big\langle \sum_{w,x\in V} a_{wx} w, \mathbf{1}\Big\rangle = \sum_{w,x\in V} a_{wx}\langle w, \mathbf{1}\rangle = \sum_{w,x\in V} a_{wx}$$

and

$$\langle \mathbf{1}, \mathbf{1}\rangle = \Big\langle \sum_{x\in V} x, \sum_{x\in V} x\Big\rangle = \sum_{x\in V} \langle x, x\rangle = |V| = n.$$

Thus, by Proposition 6.5.3,

$$\begin{aligned} \lambda_{\max}(\Gamma) \geqslant R(\mathfrak{A}, \mathbf{1}) &= \frac{\langle \mathfrak{A}(\mathbf{1}), \mathbf{1}\rangle}{\langle \mathbf{1}, \mathbf{1}\rangle} = \frac{1}{n}\sum_{x\in V}\Big(\sum_{w\in V} a_{wx}\Big) = \frac{1}{n}\sum_{x\in V} d(x) \\ &\geqslant \frac{1}{n}\sum_{x\in V} \delta(\Gamma) = \delta(\Gamma). \end{aligned}$$

(iii) Let Γ' be the induced subgraph on $V' \subset V$, and let A' be its adjacency matrix; A' is real and symmetric, so it is Hermitian (see Remark 6.5.1). Thus it corresponds to a Hermitian endomorphism $\mathfrak{A}'$ of $\mathbb{C}^{V'}$. Let $f = \sum_{x\in V'} f_x x$ be an eigenvector of norm 1 written in the orthonormal basis V' and associated to the eigenvalue $\lambda_{\max}(\Gamma')$. Set $g = \sum_{x\in V} g_x x$ where $g_x = f_x$ if $x \in V'$ and $g_x = 0$ if $x \in V\setminus V'$. Then $\langle g, g\rangle = 1$ in $\mathbb{C}^V$ and $\langle \mathfrak{A}(g), g\rangle = \langle \mathfrak{A}'(f), f\rangle$. By Proposition 6.5.3,

$$\lambda_{\max}(\Gamma') = R(\mathfrak{A}', f) = \frac{\langle \mathfrak{A}(g), g\rangle}{\langle g, g\rangle} = R(\mathfrak{A}, g) \leqslant \lambda_{\max}(\Gamma).$$

The other inequality is proved similarly. With the obvious notation, this then gives

$$\lambda_{\min}(\Gamma') = R(\mathfrak{A}', f) = R(\mathfrak{A}, g) \geqslant \lambda_{\min}(\Gamma).$$

(iv) Γ is regular of degree k if and only if for every vertex $x \in V$, we have $\sum_{y \in V} a_{xy} = k$. If this is the case, then

$$\mathfrak{A}(\mathbf{1}) = \sum_{w,x \in V} a_{wx} w = \sum_{w \in V} \Big(\sum_{x \in V} a_{wx}\Big) w = \sum_{w \in V} kw = k\mathbf{1},$$

which shows that k is an eigenvalue of the graph, and since $k = \Delta(\Gamma)$, it follows from (i) that $\lambda_{\max}(\Gamma) = \Delta(\Gamma)$.

Let $f = \sum_{x \in V} f_x x$ be another eigenvector associated to the eigenvalue k. For all x, we have $\mathfrak{A}(f)_x = \sum_{w \in V} a_{wx} f_w = k f_x$. Take $t \in V$ such that $|f_t| = \max\{|f_x| \mid x \in V\}$: then $k f_t = \mathfrak{A}(f)_t = \sum_{x \in V} a_{xt} f_x = \sum_{x \in V \mid \{t,x\} \in E} f_x$, so

$$|k f_t| = \Big| \sum_{x \in V \mid \{t,x\} \in E} f_x \Big| \leqslant \sum_{\{t,x\} \in E} |f_x| \leqslant k |f_t|,$$

or equivalently

$$|k f_t| = \Big| \sum_{x \in V \mid \{t,x\} \in E} f_x \Big| = \sum_{x \in V \mid \{t,x\} \in E} |f_x| = k |f_t|.$$

Since the graph is regular of degree k, the sum has k terms, so these are necessarily equal: $|f_x| = |f_t|$ for all x. By Lemma 10.1.2 below, it also follows that f_x and f_t have the same argument, so $f_x = f_t$ for every neighbour x of t.

If the graph is connected, then any two vertices are connected by a chain whose intermediate vertices are always neighbours of the preceding and following vertices on the chain. By transitivity, it follows that all the f_x, $x \in V$, have a common value f_0. Finally, $f = \sum_{x \in V} f_x x = f_0 \sum_{x \in V} x = f_0 \mathbf{1}$: the eigenspace associated to k is thus of dimension 1. ∎

Lemma 10.1.2. *If $a_1, \dots, a_k \in \mathbb{C}$ satisfy*

$$\Big| \sum_{1 \leqslant i \leqslant k} a_i \Big| = \sum_{1 \leqslant i \leqslant k} |a_i|,$$

then the complex numbers $a_1, \dots, a_k$ all have the same argument.

Proof. The case $k = 2$ correspond to the case of equality in the triangular inequality for the modulus in $\mathbb{C}$, and the result is well-known in this case. The general case can then be obtained by induction on k. ∎

Exercise 10.3. Show that if $V(\Gamma) = \bigsqcup_{i=1}^{r} C_i$ is the connected decomposition of $V(\Gamma)$, then the eigenspace associated to $\lambda = k$ is the subspace $\mathrm{Vect}(\mathbf{1}_{C_i}, i = 1, \dots, r)$ generated by the vectors $\mathbf{1}_{C_i}$. (Here $\mathbf{1}_{C_i}$ is the vector whose components are 1 on C_i and zero everywhere else.)

Exercise 10.4. Let A be the adjacency matrix of the complete graph K_n (which is $(n-1)$-regular). Compute $\mathrm{Ker}(A + I_n)$ and deduce that

$$\mathrm{Sp}(K_n) = \{(n-1)^{(1)}, -1^{(n-1)}\}.$$

Proposition 10.1.3. *Let $C_1, \dots, C_r$ be the connected components of Γ. Then*

$$\Phi(\Gamma, x) = \Phi(C_1, x) \dots \Phi(C_r, x).$$

We complete this part with Cauchy's interlacing theorem, which gives a strong version of part (iii) of Proposition 10.1.1.

Proposition 10.1.4 (Interlacing of spectra). *Let $\Gamma = (V; E)$ be a graph of order n and Γ' the subgraph of order q induced by a subset $V' \subset V$. Consider the spectra of Γ and Γ':*

$$\mathrm{Sp}(\Gamma) = (\lambda_1 \geqslant \lambda_2 \geqslant \dots \geqslant \lambda_n), \quad \mathrm{Sp}(\Gamma') = (\lambda'_1 \geqslant \lambda'_2 \geqslant \dots \geqslant \lambda'_q).$$

Then

$$\lambda_{k+n-q} \leqslant \lambda'_k \leqslant \lambda_k, \quad k = 1, \dots, q.$$

Proposition 10.1.4 is a special case of the following general result. We use the following terminology: given a family $\mathcal{F}$ of vectors of $\mathbb{C}^n$, let $\mathrm{Vect}\{\mathcal{F}\}$ denote the vector subspace generated by $\mathcal{F}$.

Proposition 10.1.5 (Cauchy's Interlacing Theorem). *Let A be a Hermitian matrix of order n and B a principal square submatrix of order q of A. Let $\lambda_1 \geqslant \lambda_2 \geqslant \dots \geqslant \lambda_n$ and $\lambda'_1 \geqslant \lambda'_2 \geqslant \dots \geqslant \lambda'_q$ denote their respective ordered lists of eigenvalues. Then*

$$\lambda_{k+n-q} \leqslant \lambda'_k \leqslant \lambda_k, \quad k = 1, \dots, q.$$

Proof. The matrix A represents a *sesquilinear* or *Hermitian* form[1] φ on $\mathbb{C}^n$ written in the canonical basis $\{e_1, \dots, e_n\}$. Consequently, the matrix B is a restriction of φ to $F = \mathrm{Vect}\{e_{i_1}, \dots, e_{i_q}\}$. Let $\{v_1, \dots, v_q\}$ (resp. $\{u_1, \dots, u_n\}$) be an orthonormal basis of F (resp. $\mathbb{C}^n$) consisting of eigenvectors of B (resp. A) associated to the eigenvalues in decreasing order.

For $1 \leqslant k \leqslant q$, the Grassmann formula[2] ensures the existence of a vector x of norm 1 in $\mathrm{Vect}\{v_1, \dots, v_k\} \cap \mathrm{Vect}\{u_k, \dots, u_n\}$. We thus have $x = \sum_{i=1}^{k} x_i v_i =$

[1] $\varphi\colon (x, y) \mapsto \varphi(x, y) \in \mathbb{C}$ is a Hermitian form if $x \mapsto \varphi(x, y)$ is linear and $\varphi(x, y) = \overline{\varphi(y, x)}$.

[2] If F and G are two finite-dimensional vector subspaces of E, then $\dim(F \cap G) = \dim F + \dim G - \dim(F + G)$.

$\sum_{i=k}^{n} y_i u_i$, with $\sum_{i=1}^{k} |x_i|^2 = \sum_{i=k}^{n} |y_i|^2 = 1$; then

$$\lambda'_k \leqslant \sum_{i=1}^{k} \lambda'_i |x_i|^2 = \langle Bx, x\rangle = \varphi(x,x) = \langle Ax, x\rangle = \sum_{i=k}^{n} \lambda_i |y_i|^2 \leqslant \lambda_k.$$

The second inequality follows from this result by applying it to the eigenvalues of $-A$ and $-B$. We could also consider a vector x' of norm 1 in $\mathrm{Vect}\{v_k, \dots, v_q\} \cap \mathrm{Vect}\{u_1, \dots, u_{k+n-q}\}$ and evaluate the scalar product $\langle Bx', x'\rangle$ as above. ■

The following general result applies to the eigenvalues of a graph, and reveals a symmetry property involving them. Known as the Courant–Fischer theorem, it leads to a second proof of the interlacing theorem.

Theorem 10.1.6 (Min-max theorem). *Let $\mathfrak{A}$ be a Hermitian endomorphism of $\mathbb{C}^n$ and for $x \in \mathbb{C}^n$, let $R(\mathfrak{A}, x)$ be its Rayleigh quotient. Let $\lambda_1 \geqslant \lambda_2 \geqslant \dots \geqslant \lambda_n$ denote the eigenvalues of $\mathfrak{A}$ in decreasing order. Then, for all $k = 1, 2, \dots, n$, we have*

$$\lambda_k = \max_{\substack{F \text{ subspace of } \mathbb{C}^n \\ \dim F = k}} \min_{\substack{x \in F \\ \|x\|=1}} R(\mathfrak{A}, x) = \min_{\substack{F \text{ subspace of } \mathbb{C}^n \\ \dim F = n-k}} \max_{\substack{x \in F \\ \|x\|=1}} R(\mathfrak{A}, x).$$

Proof. Let $\{u_1, u_2, \dots, u_n\}$ be an orthonormal basis of $\mathbb{C}^n$ consisting of eigenvectors of $\mathfrak{A}$ associated to the eigenvalues $\lambda_1, \lambda_2, \dots, \lambda_n$ in that order. Let F be a vector subspace of $\mathbb{C}^n$ of dimension k. Then there exists a vector

$$x \in F \cap \mathrm{Vect}\{u_k, u_{k+1}, \dots, u_n\}$$

such that $\|x\| = 1$. We write $x = \sum_{j=k}^{n} x_j u_j$ and $1 = \|x\|^2 = \sum_{j=k}^{n} |x_j|^2$. Hence

$$R(\mathfrak{A}, x) = \sum_{j=k}^{n} \lambda_j |x_j|^2 \leqslant \lambda_k.$$

Thus

$$\min_{\substack{x \in F \\ \|x\|=1}} R(\mathfrak{A}, x) \leqslant \lambda_k.$$

Now fix $F = \mathrm{Vect}\{u_1, u_2, \dots, u_k\}$. Then $R(\mathfrak{A}, u_k) = \lambda_k$, so

$$\lambda_k = \max_{\substack{F \text{ subspace of } \mathbb{C}^n \\ \dim F = k}} \min_{\substack{x \in F \\ \|x\|=1}} R(\mathfrak{A}, x).$$

The second identity is obtained by considering $-\mathfrak{A}$, taking account of the fact that $\max\min -\{\ \} = -\min\max\{\ \}$ and that the eigenvalues of $-\mathfrak{A}$ are ordered by $-\lambda_n \geqslant -\lambda_{n-1} \geqslant \dots \geqslant -\lambda_1$. ■

Exercise 10.5. Let A and B be as in Proposition 10.1.5. After writing B in the form P^*AP, use the min-max theorem to recover the Cauchy interlacing property.

10.1.3 Spectra of bipartite graphs

Regular bipartite graphs have spectral properties that are simple enough to permit a complete characterisation.

Theorem 10.1.7. *Let $\Gamma = (V; E)$ be a connected k-regular graph. Then*

(a) *The following properties are equivalent:*

  (i) *Γ is bipartite;*

  (ii) *$\mathrm{Sp}(\Gamma)$ is symmetric with respect to 0;*

  (iii) *$\lambda_{\min}(\Gamma) = -k$.*

(b) *If Γ is bipartite with $V = V_1 \sqcup V_2$, then $-k$ is a simple eigenvalue. Furthermore, its associated eigenspace is the one-dimensional space $\mathbb{C}v$, where $v = \mathbf{1}_{V_1} - \mathbf{1}_{V_2}$.*

(c) $\mathrm{Sp}(K_{r,r}) = \{r^{(1)}, 0^{(2r-2)}, -r^{(1)}\}$.

Proof. (a) We use a circular reasoning, as follows.

(i) $\Rightarrow$ (ii): suppose the graph is bipartite, with $V = V_1 \sqcup V_2$. Let $\lambda \in \mathrm{Sp}(\Gamma)$, and let $f = \sum_{x\in V} f_x x$ be an eigenvector, associated to λ, of the Hermitian endomorphism $\mathfrak{A}$ associated to the adjacency matrix A of Γ. Then $\mathfrak{A}(f) = \lambda f = \mathfrak{A}(\sum_{x\in V} f_x x) = \sum_{x,y\in V} f_x a_{yx} y = \sum_{y\in V}(\sum_{x\in V} f_x a_{yx})y$, so by (10.3), we have $\sum_{x\in V} f_x a_{yx} = \sum_{x\in V\,|\,\{x,y\}\in E} f_x = \mathfrak{A}(f)_y = \lambda f_y$ for all $y \in V$

Consider the non-zero vector

$$g = \sum_{y\in V_1} f_y y - \sum_{y\in V_2} f_y y = \sum_{y\in V} g_y y.$$

By (10.3), we have

$$\mathfrak{A}(g) = \mathfrak{A}\Big(\sum_{y\in V} g_y y\Big) = \sum_{x,y\in V} g_y a_{xy} x = \sum_{x\in V}\Big(\sum_{y\in V} g_y a_{xy}\Big)x,$$

which we write as $\mathfrak{A}(g) = \sum_{x\in V} e_x x$ where for all $x \in V$, we set

$$e_x = \sum_{y\in V} g_y a_{xy} = \sum_{y\in V|\{x,y\}\in E} g_y = \mathfrak{A}(g)_x.$$

Fix $x \in V$. If $x \in V_2$, then $g_x = -f_x$ and

$$\begin{aligned}\mathfrak{A}(g)_x &= \sum_{y\in V_1|\{x,y\}\in E} g_y = \sum_{y\in V_1|\{x,y\}\in E} f_y = \sum_{y\in V|\{x,y\}\in E} f_y \\ &= \mathfrak{A}(f)_x = \lambda f_x = -\lambda g_x,\end{aligned}$$

where we have used $g_y = f_y$ for all $y \in V_1$. If $x \in V_1$, then similarly

$$\begin{aligned}\mathfrak{A}(g)_x &= \sum_{y \in V_2 \mid \{x,y\} \in E} g_y = -\sum_{y \in V_2 \mid \{x,y\} \in E} f_y = -\sum_{y \in V \mid \{x,y\} \in E} f_y \\ &= -\mathfrak{A}(f)_x = -\lambda f_x = -\lambda g_x.\end{aligned}$$

It follows that $\mathfrak{A}(g) = -\lambda g$, and since $g \neq 0_{\mathbb{C}^n}$, we infer $-\lambda \in \mathrm{Sp}(\Gamma)$.

(ii) $\Rightarrow$ (iii): follows from Proposition 10.1.1.

(iii) $\Rightarrow$ (i): let f be an eigenvector associated to $-k$, so $\mathfrak{A}(f) = -kf$. Then for all $y \in V$, we have

$$\mathfrak{A}(f)_y = \sum_{x \in V \mid \{x,y\} \in E} f_x = -kf_y. \tag{10.4}$$

Let $w \in V$ be such that $|f_w| \geqslant |f_x|$ for all $x \in V$ ($f_w \neq 0$). It follows that

$$|\mathfrak{A}(f)_w| = k|f_w| = \Big|\sum_{x \in V \mid \{x,w\} \in E} f_x\Big| \leqslant \sum_{x \in V \mid \{x,w\} \in E} |f_x| \leqslant k|f_w|, \tag{10.5}$$

so $\sum_{x \in V \mid \{x,w\} \in E} |f_x| = k|f_w|$; hence $|f_x| = |f_w|$ for all $x \in \Gamma(w)$, but the inequality (10.5) also implies that

$$\Big|\sum_{x \in V \mid \{x,w\} \in E} f_x\Big| = \sum_{x \in V \mid \{x,w\} \in E} |f_x|,$$

and by Lemma 10.1.2, the f_x for $x \in \Gamma(w)$ all have the same argument. Then by (10.4), since the graph is regular of degree k, we have

$$\sum_{x \in V \mid \{x,w\} \in E} f_x = -kf_w,$$

so $kf_x = -kf_w$ and $f_x = -f_w$. Thus we have shown that

$$x \in \Gamma(w) \implies f_x = -f_w.$$

Now, if $x \in \Gamma(w)$, then since $|f_x| = |f_w|$, f_x has maximal modulus, and we can apply the above with x and y in place of w and x to get that for $y \in \Gamma(x)$, we have $f_y = -f_x = f_w$.

As Γ is connected, we see that for all x we have $f_x = \pm f_w$. Set $V_1 = \{x \in V \mid f_x = f_w\}$, $V_2 = \{x \in V \mid f_x = -f_w\}$. This gives a partition of the vertices $V = V_1 \sqcup V_2$. Moreover, if $\{x, y\} \in E$, then $y \in \Gamma(x)$, so $f_y = -f_x = f_w$; thus $y \in V_1$ and $x \in V_2$, so the graph is bipartite.

(b) Let f be a non-zero vector such that $\mathfrak{A}(f) = -kf$, and let w be such that $|f_w| = \max_{x \in V} |f_x|$. Without loss of generality, we may assume that $w \in V_1$. Then, from above, when $x \in \Gamma(w)$ and $y \in \Gamma(x)$, we have $f_y = -f_x = f_w$. Note that, since Γ is bipartite, we have $\Gamma(w) \subset V_2$ and $\Gamma(x) \subset V_1$. By induction, we obtain that if $d(w,x) = i$, then $f_x = (-1)^i f_w$. By connectivity, we infer that $f_y = -f_x = f_w$ for all $(x,y) \in V_2 \times V_1$, and finally, $f = f_w(\mathbf{1}_{V_1} - \mathbf{1}_{V_2})$.

(c) Here $|V_1| = |V_2| = r$. We have $f \in \operatorname{Ker} \mathfrak{A}$ if and only if

$$\sum_{x \in V_1} f_x = \sum_{x \in V_2} f_x = 0;$$

hence $\operatorname{Ker} \mathfrak{A}$ has dimension $2r - 2$, so 0 is eigenvalue with multiplicity $2r - 2$. ■

Exercise 10.6. Show that $\operatorname{Sp}(K_{r,s}) = \{\sqrt{rs}^{(1)}, 0^{(r+s-2)}, -\sqrt{rs}^{(1)}\}$.

10.1.4 Combinatorial properties of the characteristic polynomial

Let $A = (a_{ij}) \in \mathcal{M}_{n \times n}(\mathbb{C})$. For every subset $I = \{t_1 < t_2 < \cdots < t_k\}$ of $\{1, \ldots, n\}$, let A_I be the submatrix $(a_{t_i t_j})_{1 \leqslant i,j \leqslant k}$; $\det(A_I)$ is called a *principal minor of order* k *of* A, where $k = |I|$. In particular, the principal minors of order 1 are the diagonal elements a_{ii}, and the only principal minor of order n is $\det(A)$.

Exercise 10.7. Let $A = (a_{ij}) \in \mathcal{M}_{n \times n}(\mathbb{C})$. Show that there are n principal minors of order 1, $\frac{n(n-1)}{2}$ principal minors of order 2, and more generally $\binom{n}{k}$ principal minors of order k for $1 \leqslant k \leqslant n$.

Recall that the *trace* of a matrix $A = (a_{ij}) \in \mathcal{M}_{n \times n}(\mathbb{K})$, written $\operatorname{tr}(A)$, is the sum of its diagonal elements: $\operatorname{tr}(A) = \sum_{i=1}^n a_{ii}$. Its characteristic polynomial is $\Phi(A,x) = \det(xI_n - A)$ (see Chapter 6), where I_n denotes the identity matrix of order n.

Theorem 10.1.8. *Let* $A = (a_{ij}) \in \mathcal{M}_{n \times n}(\mathbb{C})$.

(i) *Then*

$$\begin{aligned}\Phi(A,x) &= \det(xI_n - A) \\ &= x^n + c_1 x^{n-1} + c_2 x^{n-2} + \cdots + c_k x^{n-k} + \cdots + c_{n-1} x + c_n,\end{aligned}$$

where $c_k = (-1)^k \sum_{|I|=k} \det(A_I)$ *is the sum of principal minors of order* k, $1 \leqslant k \leqslant n$ *(in particular* $c_1 = -\operatorname{tr}(A)$*).*

(ii) *If* $\Phi(A,x)$ *decomposes as the product*

$$\Phi(A,x) = (x - \lambda_1)(x - \lambda_2) \cdots (x - \lambda_n),$$

where the λ_i *are not necessarily distinct, then we have*

$$(-1)^k c_k = \sum_{1 \leqslant i_1 < i_2 < \cdots < i_k \leqslant n} \lambda_{i_1} \ldots \lambda_{i_k}$$

for $k = 1, \ldots, n$.

Proof. Part (i) follows by using the formula for the expansion of the determinant $\det(xI_n - A) = \Phi(A, x)$ (details left to the reader).

(ii) These are Newton's classical formulas giving the relations between the coefficients of a polynomial and its roots. ■

The characteristic polynomial yields information on the graph.

Theorem 10.1.9. *Let $\Gamma = (V; E)$ be a simple graph with adjacency matrix A, associated to the Hermitian endomorphism $\mathfrak{A}$, and let*

$$\Phi(\Gamma, x) = x^n + c_1 x^{n-1} + c_2 x^{n-2} + \cdots + c_{n-1} x + c_n$$

be its characteristic polynomial. Then

(i) $c_1 = 0$*: the sum of eigenvalues (counted with multiplicity) of Γ is zero;*

(ii) c_2 *is the opposite of the number of edges of Γ;*

(iii) c_3 *is the opposite of the double of the number of triangles (or cliques with three vertices) of Γ.*

Proof. By Theorem 10.1.8 (i), we have $c_1 = -\operatorname{tr}(A)$, which is zero since the diagonal terms of A are all zero.

Since the matrix is symmetric and zero along the diagonal, the only non-zero principal minors of order 2 have the form

$$\det\begin{pmatrix} 0 & 1 \\ 1 & 0 \end{pmatrix},$$

so they are all equal to -1. The position of the 1's on the antidiagonal corresponds to an edge of Γ, and conversely, every edge of Γ induces such a minor, but we have $(-1)^2 c_2 = \sum_{|I|=2} \det(A_I)$, so $(-1)^2 c_2 = -$ the number of edges.

The principal minors of order 3 are of the form

$$\det\begin{pmatrix} 0 & a & b \\ a & 0 & c \\ b & c & 0 \end{pmatrix}, \quad a, b, c \in \{0, 1\}.$$

This determinant is equal to $2abc$, so the only type of non-zero minor of order 3 corresponds to the unique choice $a = b = c = 1$; the positions of these 1's around the diagonal corresponds to a triangle (a cycle of length 3) in the graph, and vice versa. But we have $(-1)^3 c_3 = \sum_{|I|=3} \det(A_I)$, therefore $(-1)^3 c_3$ is equal to twice the number of triangles of Γ. ■

Exercise 10.8. Check the assertions of Theorem 10.1.9 on the graphs given in Exercise 10.2.

Exercise 10.9. Consider the matrices

$$A_1 = \begin{pmatrix} 0 & 1 & 0 & 0 & 0 & 0 \\ 1 & 0 & 1 & 1 & 0 & 0 \\ 0 & 1 & 0 & 1 & 1 & 0 \\ 0 & 1 & 1 & 0 & 1 & 0 \\ 0 & 0 & 1 & 1 & 0 & 1 \\ 0 & 0 & 0 & 0 & 1 & 0 \end{pmatrix}, \quad A_2 = \begin{pmatrix} 0 & 1 & 0 & 1 & 0 & 0 \\ 1 & 0 & 0 & 1 & 0 & 0 \\ 0 & 0 & 0 & 1 & 0 & 0 \\ 1 & 1 & 1 & 0 & 1 & 1 \\ 0 & 0 & 0 & 1 & 0 & 1 \\ 0 & 0 & 0 & 1 & 1 & 0 \end{pmatrix}.$$

(i) Give a graphic representation of the graphs Γ_1 and Γ_2 whose adjacency matrices are A_1 and A_2.

(ii) Show that the coefficients c_1, c_2, c_3, c_6 (with the notation of Theorem 10.1.8) are equal for Γ_1 and Γ_2.

(iii) One can also show that c_4 and c_5 are equal for Γ_1 and Γ_2, so $\Phi(\Gamma_1, x) = \Phi(\Gamma_2, x)$, yet Γ_1 and Γ_2 are not isomorphic.

10.1.5 Spectrum, diameter and colouring

The complete graphs K_n have diameter 1, and their spectrum consists of two distinct eigenvalues: $n-1$, with multiplicity 1, and -1 with multiplicity $n-1$. Hence $|\mathrm{Sp}(\Gamma)| = 2 = \mathrm{diam}(\Gamma) + 1$. One can show (exercise) that the only connected graphs such that $|\mathrm{Sp}(\Gamma)| = 2$ are the complete graphs. The following proposition shows that complete graphs minimise the quantity $|\mathrm{Sp}(\Gamma)| - \mathrm{diam}(\Gamma)$.

Proposition 10.1.10. *Let $\Gamma = (V; E)$ be a connected graph such that $E \neq \emptyset$. Then*

$$|\mathrm{Sp}(\Gamma)| \geqslant \mathrm{diam}(\Gamma) + 1.$$

Proof. Set $d = \mathrm{diam}(\Gamma) = d(x, y)$. Let $C = (x = x_0, x_1, \ldots, x_d = y)$ be a chain of minimal length between x and y. The adjacency matrix A of Γ is such that for all $j = 0, \ldots, d-1$, the coefficient of A^j corresponding to the pair of vertices x, y is zero. Moreover, this same coefficient is strictly positive in A^d, due to the existence of the chain C. Consequently,

$$A^d \notin \mathrm{Vect}\{A^j,\ j = 0, \ldots, d-1\},$$

which implies that the *minimal polynomial*[3] of A is of degree at least equal to $d+1$. As A is diagonalisable, the degree of the minimal polynomial is exactly the number of distinct eigenvalues of A. ∎

[3]This is the monic polynomial P of lowest degree such that $P(A) = 0$.

Now we show two bounds involving the chromatic number of Γ.

Proposition 10.1.11. *Let $\Gamma = (V; E)$ be a graph. Then $\chi(\Gamma) \leqslant \lambda_{\max}(\Gamma) + 1$.*

Proof. By Proposition 7.1.6, there exists an induced subgraph Γ' such that $\chi(G) \leqslant \delta(\Gamma') + 1$. But $\delta(\Gamma') \leqslant \lambda_{\max}(\Gamma')$ by Proposition 10.1.1 (ii) applied to Γ'. Then by (iii) of the same proposition, $\lambda_{\max}(\Gamma') \leqslant \lambda_{\max}(\Gamma)$, which yields the desired upper bound of the chromatic number. ∎

Proposition 10.1.12. *Let $\Gamma = (V; E)$ be such that $E \neq \emptyset$. Then*

$$\chi(\Gamma) \geqslant \frac{\lambda_{\max}(\Gamma)}{|\lambda_{\min}(\Gamma)|} + 1.$$

Proof. If $\{V_1, \ldots, V_k\}$ is a k-colouring of Γ, then its adjacency matrix A consists of k identically zero principal minors situated along its diagonal, completed by of blocks of 0's and 1's outside of them. We apply the following lemma.

Lemma 10.1.13. *Let $M = \begin{pmatrix} A & Q \\ Q^* & B \end{pmatrix}$ be a Hermitian matrix where A (resp. B) is a (Hermitian) matrix of order r (resp. s) with $n = r + s$. For an arbitrary Hermitian matrix P, let $\lambda_{\max}(P)$ (resp. $\lambda_{\min}(P)$) denote its largest (resp. smallest) eigenvalue. Then*

$$\lambda_{\max}(M) + \lambda_{\min}(M) \leqslant \lambda_{\max}(A) + \lambda_{\max}(B).$$

Proof. Let $(e_1, e_2, \ldots, e_n)$ denote the canonical basis of $\mathbb{C}^n$. Set $U = \mathrm{Vect}\{e_1, \ldots, e_r\}$ and $V = \mathrm{Vect}\{e_{r+1}, \ldots, e_n\}$. Every vector x of $\mathbb{C}^n$ can be written uniquely in the form $x = x_U + x_V$ with $x_U \in U$ and $x_V \in V$. Finally, set $\lambda_A = \lambda_{\max}(A)$ and $\lambda_B = \lambda_{\max}(B)$.

(i) Intermediate bound: suppose that M is Hermitian positive definite.

Then its eigenvalues are strictly positive. By Cauchy's interlacing theorem (cf. Proposition 10.1.5), we have $\lambda_A, \lambda_B \geqslant \lambda_{\min}(M) > 0$. Let $z = z_U + z_V \in \mathbb{C}^n$. We again identify the n-tuples with their corresponding column matrix. Then

$$0 \leqslant \langle Mz, z\rangle = \langle Mz_U, z_U\rangle + \langle Mz_V, z_V\rangle + \langle Mz_U, z_V\rangle + \langle Mz_V, z_U\rangle.$$

If $z = (z_1, \ldots, z_n)$, then $z_U = (z_1, \ldots, z_r, \underbrace{0, \ldots, 0}_{s \text{ times}})$, so

$$\langle Mz_U, z_U\rangle = (z_1 \ldots z_r) A \begin{pmatrix} z_1 \\ \vdots \\ z_r \end{pmatrix} \leqslant \lambda_A \|z_U\|^2.$$

Similarly $\langle Mz_V, z_V\rangle \leqslant \lambda_B \|z_V\|^2$, so

$$0 \leqslant \langle Mz, z\rangle \leqslant \lambda_A \|z_U\|^2 + \lambda_B \|z_V\|^2 + \langle Mz_U, z_V\rangle + \langle Mz_V, z_U\rangle.$$

Consider the sesquilinear Hermitian form

$$\Psi(x, y) = \lambda_A \langle x_U, y_U \rangle + \lambda_B \langle x_V, y_V \rangle + \langle M x_U, y_V \rangle + \langle M x_V, y_U \rangle.$$

We thus have

$$\Psi(z, z) = \lambda_A \|z_U\|^2 + \lambda_B \|z_V\|^2 + \langle M z_U, z_V \rangle + \langle M z_V, z_U \rangle \geqslant 0.$$

Let $x = x_U + x_V$ be an eigenvector associated to the eigenvalue $\lambda = \lambda_{\max}(M)$ of M: we have $\langle Mx, x \rangle = \lambda \|x\|^2$. Thus

$$\lambda \|x\|^2 \leqslant \lambda_A \|x_U\|^2 + \lambda_B \|x_V\|^2 + \langle M x_U, x_V \rangle + \langle M x_V, x_U \rangle. \tag{10.6}$$

We compute $\Psi(z, z)$ with $z = \lambda_B x_U - \lambda_A x_V$: this gives

$$\lambda_A \|\lambda_B x_U\|^2 + \lambda_B \|-\lambda_A x_V\|^2 - \lambda_B \lambda_A \langle M x_U, x_V \rangle - \lambda_A \lambda_B \langle M x_V, x_U \rangle \geqslant 0.$$

Simplifying by $\lambda_A \lambda_B$ (which is strictly positive), we obtain

$$\lambda_B \|x_U\|^2 + \lambda_A \|x_V\|^2 \geqslant \langle M x_U, x_V \rangle + \langle M x_V, x_U \rangle.$$

Plugging this into (10.6), we obtain

$$\lambda \|x\|^2 \leqslant \lambda_A (\|x_U\|^2 + \|x_V\|^2) + \lambda_B (\|x_V\|^2 + \|x_U\|^2).$$

But $\|x\|^2 = \|x_U\|^2 + \|x_V\|^2$, so

$$\lambda_{\max}(M) = \lambda \leqslant \lambda_A + \lambda_B. \tag{10.7}$$

(ii) General bound: for $\varepsilon > 0$, we can apply (10.7) to

$$M_\varepsilon = M - (\lambda_{\min}(M) - \varepsilon) I_n.$$

Indeed, $\lambda_{\min}(M_\varepsilon) = \varepsilon > 0$, so the matrix M_ε is a positive definite Hermitian matrix. We obtain

$$\lambda_{\max}(M) - \lambda_{\min}(M) + \varepsilon \leqslant (\lambda_A - \lambda_{\min}(M) + \varepsilon) + (\lambda_B - \lambda_{\min}(M) + \varepsilon),$$

hence

$$\lambda_{\max}(M) + \lambda_{\min}(M) \leqslant \lambda_A + \lambda_B + \varepsilon.$$

We then deduce the desired result by letting ε tend to 0. ■

Exercise 10.10. Considering the matrix $M = \left(\begin{smallmatrix} 0 & 1 \\ 1 & 1 \end{smallmatrix}\right)$, show that the upper bound obtained in the lemma is optimal.

Let us return to the proposition. Since $k = \chi(\Gamma) \geqslant 2$, we start by decomposing its adjacency matrix A in the form

$$A = \left(\begin{array}{c|c} 0_m & Q \\ \hline Q^* & B \end{array}\right)$$

where Q is a rectangular block of 0's and 1's, Q^* its transpose, and B a Hermitian matrix of order $n-m$. The lemma yields the inequality $\lambda_{\max}(A) + \lambda_{\min}(A) \leqslant \lambda_{\max}(B)$. Iterating $k-1$ times until the *last* zero principal submatrix and taking account of the fact that $\lambda_{\min}(B) \geqslant \lambda_{\min}(A)$ by Proposition 10.1.1 (iii), we obtain

$$\lambda_{\max}(A) + (k-1)\lambda_{\min}(A) \leqslant 0,$$

which gives the desired result. ∎

10.2 The graph Laplacian

In this paragraph, the graphs may have multiple edges, but they are assumed to be *loopless*.

Let $\Gamma = (V; E, N)$ be such a graph. Recall (Section 6.1) that the degree matrix is the diagonal matrix $D = (d_{xy})_{x,y \in V} \in \mathcal{M}_{n\times n}(\mathbb{C})$, where $n = |V|$, defined by

$$d_{xy} = \begin{cases} d(x) & \text{if } x = y, \\ 0 & \text{otherwise,} \end{cases}$$

and the adjacency matrix $A = (a_{xy})_{x,y\in V}$ is the $n \times n$ symmetric matrix defined by

$$a_{xy} = \text{number of edges between } x \text{ and } y.$$

As D and A are real and symmetric, they correspond to Hermitian endomorphisms $\mathfrak{D}$ and $\mathfrak{A}$ of $\mathbb{C}^n$.

The *Laplacian matrix* of the graph Γ is the symmetric matrix $L = D - A$. It corresponds to the Hermitian endomorphism $\mathfrak{L}(\Gamma)$ of $\mathbb{C}^n$ given by $\mathfrak{L}(\Gamma) = \mathfrak{D} - \mathfrak{A}$; we generally refer to it as the *Laplacian* of Γ.

Recall that a Hermitian operator $u \in \mathcal{H}(\mathbb{C}^n)$ is *positive* if for all $x \in \mathbb{C}^n$, we have $\langle u(x), x\rangle \geqslant 0$, or equivalently, if its eigenvalues are positive.

With the notation of Section 10.1, recall that $f_x = \langle f, x\rangle$. Indeed,

$$\langle f, x\rangle = \Big\langle \sum_{y\in V} f_y y, x\Big\rangle = \sum_{y\in V} f_y \langle y, x\rangle = f_x.$$

Theorem 10.2.1. *The Laplacian $\mathfrak{L}$ of a graph is a positive Hermitian operator.*

Proof. We have $\mathfrak{L} = \mathfrak{D} - \mathfrak{A}$. For $f = \sum_{x \in V} f_x x$, we have

$$\mathfrak{D}(f) = \mathfrak{D}\Big(\sum_{x \in V} f_x x\Big) = \sum_{x \in V} f_x \mathfrak{D}(x) = \sum_{x \in V} f_x d(x) x.$$

On the other hand,

$$\mathfrak{A}(f) = \sum_{x \in V} f_x \mathfrak{A}(x) = \sum_{y,x \in V} f_x a_{yx} y = \sum_{y \in V}\Big(\sum_{x \in V} f_x a_{yx}\Big) y.$$

Thus $\langle \mathfrak{L}(f), f\rangle$ is equal to

$$\begin{aligned}\langle \mathfrak{D}(f), f\rangle - \langle \mathfrak{A}(f), f\rangle &= \sum_{x \in V} f_x d(x) \overline{f_x} - \sum_{y \in V}\Big(\sum_{x \in V} f_x a_{yx}\Big)\overline{f_y} \\ &= \sum_{x \in V} f_x \overline{f_x} d(x) - \sum_{x,y \in V} a_{yx} f_x \overline{f_y} \\ &= \sum_{x,y \in V} a_{yx} f_x \overline{f_x} - \sum_{x,y \in V} a_{yx} f_x \overline{f_y}.\end{aligned}$$

By noting that $d(x) = \sum_{y \in V} a_{yx}$, we obtain

$$\langle \mathfrak{L}(f), f\rangle = \sum_{x,y \in V} a_{xy} f_x(\overline{f_x} - \overline{f_y}) = \sum_{(x,y) \in V \times V \mid \{x,y\} \in E} f_x(\overline{f_x} - \overline{f_y}).$$

But in the last sum, for each edge $\{x, y\}$, we can group the contributions $f_x(\overline{f_x} - \overline{f_y})$ and $f_y(\overline{f_y} - \overline{f_x})$ whose sum factorises as $|f_x - f_y|^2$. Thus

$$\begin{aligned}\sum_{\{x,y\} \in E} f_x(\overline{f_x} - \overline{f_y}) + f_y(\overline{f_y} - \overline{f_x}) &= \sum_{\{x,y\} \in E} (f_x - f_y)(\overline{f_x} - \overline{f_y}) \\ &= \sum_{\{x,y\} \in E} |f_x - f_y|^2 \geqslant 0.\end{aligned}$$

Thus $\langle \mathfrak{L}(f), f\rangle \geqslant 0$. ■

Corollary 10.2.2. *The eigenvalues of the Laplacian $\mathfrak{L}$ are positive.*

Proof. This is a general fact: if $u \in \mathcal{H}$, $u \geqslant 0$, and $\lambda \in \mathrm{Sp}(u)$: $u(f) = \lambda f$; then $0 \leqslant \langle u(f), f\rangle = \lambda \|f\|^2$ so $\lambda \geqslant 0$. ■

Proposition 10.2.3. *Let $\Gamma = (V; E, N)$ be a graph of order n and $\mathfrak{L}$ its Laplacian. Then*

(i) *$\lambda = 0$ is an eigenvalue of $\mathfrak{L}$; its multiplicity is equal to the number k of connected components of Γ.*

(ii) *The rank of* $\mathfrak{L}$ *is* $n-k$.

(iii) *If* $\lambda_1 = 0 \leqslant \lambda_2 \leqslant \cdots \leqslant \lambda_n$ *is the ordered list of eigenvalues of* $\mathfrak{L}$ *(repeated according to their multiplicity), then* $\sum_{1\leqslant i\leqslant n} \lambda_i = 2|E|$.

Proof. (i) Let $V = \bigsqcup_{1\leqslant i\leqslant k} V_i$ denote the connected decomposition of V, and let us show that the eigenspace associated to $\lambda = 0$ is

$$E_0 = \bigoplus_{1\leqslant i\leqslant k} \mathbb{C}\mathbf{1}_i, \quad \text{where } \mathbf{1}_i = \mathbf{1}_{V_i} = \sum_{x\in V_i} x.$$

For all i, we have

$$\begin{aligned}
\mathfrak{L}(\mathbf{1}_i) = \mathfrak{D}(\mathbf{1}_i) - \mathfrak{A}(\mathbf{1}_i) &= \mathfrak{D}\Big(\sum_{x\in V_i} x\Big) - \mathfrak{A}\Big(\sum_{x\in V_i} x\Big)\\
&= \sum_{x\in V_i}\sum_{w\in V} d_{wx} w - \sum_{x\in V_i}\sum_{w\in V} a_{wx} w\\
&= \sum_{x\in V_i}\sum_{w\in V_i} d_{wx} w - \sum_{x\in V_i}\sum_{w\in V_i} a_{wx} w\\
&= \sum_{x\in V_i} d(x)x - \sum_{w\in V_i}\Big(\sum_{x\in V_i} a_{wx}\Big) w\\
&= \sum_{x\in V_i} d(x)x - \sum_{w\in V_i} d(w)w = 0,
\end{aligned}$$

which proves that $\mathfrak{L}(\mathbf{1}_i) = 0$.

Let $f = \sum_{x\in V} f_x x$ be an eigenvector associated to the eigenvalue $\lambda = 0$. We have $\mathfrak{L}(f) = 0_{\mathbb{C}^n}$, and $\mathfrak{D}(f) = \mathfrak{A}(f)$, so

$$\forall x : d(x) f_x = \sum_{y\in V} a_{yx} f_y. \tag{10.8}$$

Fix $w \in V$ such that $|f_w| = \max\{|f_x| \mid x \in V\}$. Then by (10.8)

$$d(w)|f(w)| = \Big|\sum_{y\in V} a_{wy} f_y\Big| \leqslant \sum_{y\in V} a_{wy}|f_y| \leqslant d(w)|f(w)|,$$

so

$$\sum_{y\in V} a_{wy}|f_y| = d(w)|f(w)|.$$

The left-hand sum consists of $d(w)$ numbers $|f_y|$ and the right-hand term consists of $d(w)$ times the number $|f(w)|$, corresponding to the neighbours of w. We conclude that $|f_y| = |f_w|$ for all neighbours y of w, and by Lemma 10.1.2, the complex numbers f_y, $y \in \Gamma(w)$, all have the same argument, so they are equal: $f_y = f_w$ for all y

adjacent to w. Then by transitivity and connectedness, this holds for all $y \in V_{i_1}$, where V_{i_1} denotes the connected component containing w. Set $g^{(1)} = f_w \sum_{x \in V_{i_1}} x = f_w \mathbf{1}_{i_1}$ and $f^{(1)} = f - g^{(1)}$. If $f^{(1)} = 0_{\mathbb{C}^n}$, this shows that $f = g^{(1)} \in \mathbb{C}\mathbf{1}_{i_1} \subset E_0$; otherwise $f^{(1)} = \sum_{x \in V \setminus V_{i_1}} f_x^{(1)} x$ is an eigenvector associated to $\lambda = 0$, which defines a vertex $w_2 \in V_{i_2} \neq V_{i_1}$ that maximises $|f_x^{(1)}|$ for $x \in V$. We proceed for $f^{(1)}$ exactly as we just did for f, and obtain $f_y^{(1)} = f_{w_2}^{(1)}$ for all $y \in V_{i_2}$ ($i_2 \neq i_1$). Then we let

$$g^{(2)} = f_{w_2}^{(1)} \sum_{x \in V_{i_2}} x = f_{w_2}^{(1)} \mathbf{1}_{i_2} \quad \text{and} \quad f^{(2)} = f^{(1)} - g^{(2)}.$$

If $f^{(2)} = 0_{\mathbb{C}^n}$, we obtain $f = g^{(1)} + g^{(2)} \in \mathbb{C}\mathbf{1}_{i_1} + \mathbb{C}\mathbf{1}_{i_2}$. Otherwise $f^{(2)}$ is an eigenvector associated to $\lambda = 0$ to which we can apply the procedure; we continue this construction until we obtain, for some $j \leqslant k$, distinct connected components $V_{i_1}, \dots, V_{i_j}$ of Γ and vertices $w_1 \in V_{i_1}, \dots, w_j \in V_{i_j}$, and vectors $f^{(p)}$ for $p = 1, \dots, j$ obtained recursively by the formula $f^{(p)} = f^{(p-1)} - f_{w_p}^{(p-1)} \mathbf{1}_{i_p}$, $f^{(0)} = f$, $w_1 = w$, and satisfying the *stopping condition* $f^{(j)} = 0_{\mathbb{C}^n}$.

Since at each step we visit a new connected component V_{i_j}, the procedure eventually stops; hence

$$f = f_{w_1}^{(0)} \mathbf{1}_{i_1} + f_{w_2}^{(1)} \mathbf{1}_{i_2} + \cdots + f_{w_j}^{(j-1)} \mathbf{1}_{i_j} \in \bigoplus_{1 \leqslant p \leqslant j} \mathbb{C}\mathbf{1}_{i_p} \subset E_0.$$

(ii) This follows immediately from (i) since $\dim V = \dim \operatorname{Im} \mathfrak{L} + \dim \operatorname{Ker} \mathfrak{L}$ (see § 6.5.1).

(iii) It follows from $\sum_{1 \leqslant i \leqslant n} \lambda_i = \operatorname{tr}(L) = \sum_{x \in V} d(x) = 2|E|$. ■

When the graph is simple and regular, the eigenvalues of the Laplacian are entirely determined by the eigenvalues of the adjacency matrix.

Proposition 10.2.4. *Let $\Gamma = (V; E)$ be a simple regular graph of degree k. If the adjacency matrix has eigenvalues $\lambda_1, \lambda_2, \dots, \lambda_n$, then the Laplacian matrix $\mathfrak{L}$ has eigenvalues $k - \lambda_1, k - \lambda_2, \dots, k - \lambda_n$.*

Proof. This follows from the fact that if f is an eigenvector of $\mathfrak{A}$ associated to the eigenvalue λ, then $\mathfrak{L} = \Delta(\Gamma)\mathrm{Id} - \mathfrak{A} = k\mathrm{Id} - \mathfrak{A}$, so $\mathfrak{L}(f) = (k\mathrm{Id} - \mathfrak{A})(f) = k\,\mathrm{Id}(f) - \mathfrak{A}(f) = kf - \lambda f = (k - \lambda) f$. ■

We recover the fact that $k = \Delta(\Gamma)$ is an eigenvalue of the adjacency matrix A (since 0 is an eigenvalue of L), and the fact that the eigenvalues of the Laplacian $\mathfrak{L}$ are positive by Proposition 10.1.1.

Proposition 10.2.5. *Every eigenvalue λ of the Laplacian $\mathfrak{L}$ of Γ satisfies $0 \leqslant \lambda \leqslant 2\Delta(\Gamma)$.*

Proof. We know that the eigenvalues of $\mathfrak{L}$ are positive. We have

$$\mathfrak{L}(f) = \sum_{x \in V} d(x) f_x x - \sum_{x \in V} f_x \sum_{y \in V} a_{yx} y.$$

Let f be an eigenvector associated to λ, and let w be such that $|f_w| \geqslant |f_x|$ for all $x \in V$; $g = \frac{f}{|f_w|}$ is then an eigenvector for $\mathfrak{L}$ associated to the eigenvalue λ and we have $\mathfrak{L}(g) = \lambda g$. This means that

$$\sum_{x \in V} g_x d(x) x - \sum_{x \in V} \Big(\sum_{y \in V} a_{xy} g_y \Big) x = \lambda \sum_{x \in V} g_x x,$$

so for each component x, we have

$$d(x) g_x - \sum_{y \in V} a_{yx} g_y = \lambda g_x.$$

In particular, when $x = w$,

$$d(w) g_w - \sum_{y} a_{yw} g_y = \lambda g_w,$$

so we find that

$$\begin{aligned} |\lambda g_w| = \lambda &= \Big| d(w) g_w - \sum_{y} a_{wy} g_y \Big| \\ &\leqslant |d(w) g_w| + \Big| \sum_{y} a_{wy} g_y \Big| \leqslant d(w) + \sum_{y} |a_{wy}| |g_y| \\ &\leqslant d(w) + \sum_{y \mid \{w,y\} \in E} 1 = d(w) + d(w) \leqslant 2\Delta(\Gamma). \end{aligned}$$

■

We can even do a little better.

Proposition 10.2.6. *The eigenvalues of the Laplacian matrix $\mathfrak{L}$ of a graph are contained between* 0 *and* $\max\{d(x) + d(y) \mid \{x, y\} \in E\}$.

Proof. Exercise. ■

Let $a = ([x, y], n)$ be an edge of a loopless multigraph $\Gamma = (V; E, N)$. The multigraph $\Gamma \backslash a$ obtained by deleting the edge a is also obviously loopless, however if there is a p-edge between x and y, the contraction of a gives rise to a $(p-1)$-loop in Γ/a. We shall write $\Gamma/\!/a$ for the loopless graph obtained from Γ/a by suppression of all loops. This process successively produces a $(p-2)$-loop, ..., a 2-loop, a loop, and finally no loop, so $\Gamma/\!/a$ is a loopless multigraph. This operation is called a *strong contraction of the edge a in the multigraph Γ*.

We know how to find a spanning tree of Γ, though such a tree is obviously not unique. Our next goal is to count the number of spanning trees of Γ. We will see that the Laplacian is a very useful tool in solving this problem. Let $\tau(\Gamma)$ denote the number of spanning trees of an arbitrary multigraph with or without loops. If Γ is not connected, then clearly $\tau(\Gamma) = 0$. As a convention, if Γ is reduced to a single vertex, then we set $\tau(\Gamma) = 1$.

Proposition 10.2.7. *Let $\Gamma = (V; E, N)$ be a loopless multigraph and let a be an edge of Γ. Then*

$$\tau(\Gamma) = \tau(\Gamma \backslash a) + \tau(\Gamma /\!\!/ a). \tag{10.9}$$

Proof. Every spanning tree T of Γ not containing a is also a spanning tree of $\Gamma \backslash a$ (observe that $\Gamma \backslash a$ is connected since it admits T as spanning tree). Consequently, $\tau(\Gamma \backslash a)$ is the number of spanning trees of Γ that do not contain the edge a. It is easy to check that there exists a bijection between the set of spanning trees of Γ containing a and the set of spanning trees of $\Gamma /\!\!/ a$. Thus the number of spanning trees containing a is equal to $\tau(\Gamma /\!\!/ a)$, which gives the result. ∎

Let $A = (a_{ij}) \in \mathcal{M}_{n \times n}(\mathbb{C})$. For every ordered subset $I \subset \{1, \dots, n\}$, let A_I denote the submatrix $(a_{ij})_{i,j \in I}$. Then $\det(A_I)$ is called a *principal minor* of order $k = |I|$ of A. If $I = \{1, \dots, i-1, i+1, \dots, n\}$, let $A_{\hat{i}}$ denote the matrix obtained by removing the i-th row and the i-th column from A; the principal minor $\det(A_{\hat{i}})$ of order $n-1$ is called the minor with respect to a_{ii}. Similarly, if $\mathfrak{L}$ denotes the Laplacian of a loopless graph $\Gamma = (V; E, N)$ with n vertices, and L denotes its matrix in the canonical basis formed by the vertices, then for each vertex x, $\det(L_{\hat{x}})$ is a principal minor of order $n-1$ of the matrix $L = (\ell_{x,y})_{x,y \in V}$, that we call the *minor relative to* $\ell_{x,x}$. In this situation, when $L = (\ell_{x,y})_{x,y \in V}$ is the matrix associated to a graph Laplacian, the relative minors do not depend on the order of the vertices defining the basis in which the matrix L is written.

Convention. When $A \in \mathcal{M}_{1 \times 1}(\mathbb{C})$, we set $\det(A_{\hat{1}}) = 1$.

Theorem 10.2.8 (Kirchhoff, *matrix-tree theorem*). *Let $\Gamma = (V; E, N)$ be a loopless multigraph with Laplacian matrix $L = (\ell_{x,y})_{x,y \in V}$. Then for every vertex x, the minor relative to $\ell_{x,x}$ is equal to the number of spanning trees of Γ:*

$$\det(L_{\hat{x}}) = \tau(\Gamma).$$

Proof. We prove the result by induction on $m = |E|$.

For $m = 0$, Γ consists of $n \geqslant 1$ isolated vertices. There are two cases:

(a) if $n = 1$, then $L = (0) \in \mathcal{M}_{1 \times 1}(\mathbb{C})$; since $\det(L_{\hat{x}}) = 1$ by convention and $\tau(\Gamma) = 1$, the result holds;

(b) if $n \geqslant 2$, then $L = 0_n \in \mathcal{M}_{n\times n}(\mathbb{C})$ is the zero matrix, so $\det(L_{\hat{v}}) = 0$ for all $v \in V$. Since Γ is not connected, we also have $\tau(\Gamma) = 0$, which gives the result.

Fix $m \geqslant 0$ and assume the assertion holds for every graph with m edges. Let Γ be a graph with $m + 1$ edges. We distinguish again two cases.

(a) Then if there exists an isolated vertex $x \in V$, we let it be the first element of the basis V in which the matrix L is written as

$$L = \left(\begin{array}{c|c} 0 & 0 \\ \hline 0 & M \end{array}\right),$$

so that $\det(L_{\hat{x}}) = \det(M)$, but M is nothing other that the Laplacian matrix of $\Gamma' = (V\backslash\{x\}; E)$ for which we know that $0 \in \mathrm{Sp}(\Gamma')$ (see Proposition 10.2.3). Consequently, there exists a non-zero vector $u \in \mathbb{C}^n$ such that $Mu = 0$, but then we have $\det(L_{\hat{x}}) = \det(M) = 0$. Clearly, for $y \neq x$, we have $\det(L_{\hat{y}}) = 0$ since the matrix $L_{\hat{y}}$ has a zero row. Since Γ is not connected, we infer $\tau(\Gamma) = 0$ as wanted.

(b) If there is no isolated vertex, then choose $x \in V$; there exists $a \in E$ incident to x. Let y denote the other extremity of this edge. To write the matrix $L = L(\Gamma)$, we take x and y as the first two elements of the basis V, so that

$$L(\Gamma) = \left(\begin{array}{cc|c} d(x) & -r & \\ & & A \\ -r & d(y) & \\ \hline \multicolumn{2}{c|}{{}^tA} & M \end{array}\right),$$

where $r \geqslant 1$ is the multiplicity of the multi-edge between x and y. We also have

$$L(\Gamma\backslash a) = \left(\begin{array}{cc|c} d(x)-1 & -r+1 & \\ & & A \\ -r+1 & d(y)-1 & \\ \hline \multicolumn{2}{c|}{{}^tA} & M \end{array}\right).$$

Then, taking z_a (the vertex formed by the contraction of the edge a) as the first element of the set of vertices of $\Gamma/\!/a$, we have

$$L(\Gamma/\!/a) = \left(\begin{array}{c|c} d(z_a) & B \\ \hline {}^tB & M \end{array}\right),$$

with the property that $d(z_a) = d(x) + d(y) - r$; we could also note that B is the sum of both rows constituting the matrix A. We then show

$$\det(L(\Gamma)_{\hat{x}}) - \det(L(\Gamma\backslash a)_{\hat{x}}) = \det\left(\begin{array}{c|c} 1 & 0 \\ \hline 0 & M \end{array}\right) = \det(M). \tag{10.10}$$

Indeed, $L(\Gamma)_{\hat{x}}$ and $L(\Gamma\backslash a)_{\hat{x}}$ differ only by the coefficient situated in their first row and first column: $d(y) \neq d(y) - 1$. Thus, expanding the two determinants with respect to their first column, we obtain for a certain k,

$$\det(L(\Gamma)_{\hat{x}}) = d(y)\det(M) + k,$$
$$\det(L(\Gamma\backslash a)_{\hat{x}}) = (d(y) - 1)\det(M) + k,$$

which gives the desired formula (10.10).

Furthermore, we have $\det(L(\Gamma/\!/a)_{\hat{u}_a}) = \det(M)$ and by the induction hypothesis, $\det(L(\Gamma\backslash a)_{\hat{x}}) = \tau(\Gamma\backslash a)$, hence $\det(L(\Gamma/\!/a)_{\hat{u}_a}) = \tau(\Gamma/\!/a)$. Using (10.10), we find $\det(L(\Gamma)_{\hat{x}}) = \tau(\Gamma\backslash a) + \tau(\Gamma/\!/a)$, which is equal to $\tau(\Gamma)$ by Proposition 10.2.7. ■

Corollary 10.2.9. *Let*

$$\Lambda(\Gamma, x) = x^n + a_1 x^{n-1} + \cdots + a_{n-1}x + a_0 \in \mathbb{Z}[x]$$

be the characteristic polynomial of the Laplacian $\mathfrak{L}$, and let $0 = \lambda_1 \leqslant \lambda_2 \leqslant \ldots \leqslant \lambda_n$ denote the eigenvalues of $\mathfrak{L}$, repeated according to their multiplicity. Then

$$\tau(\Gamma) = \frac{\lambda_2 \ldots \lambda_n}{n}.$$

Proof. Indeed, Theorem 10.1.8 (ii) asserts that

$$(-1)^{n-1}a_{n-1} = \sum_{1 \leqslant i_1 < i_2 < \cdots < i_{n-1} \leqslant n-1} \lambda_{i_1} \ldots \lambda_{i_{n-1}},$$

since the sum is taken over all subsets with $n-1$ elements of $\{1, \ldots, n\}$. As $\lambda_1 = 0$, we have $(-1)^{n-1}a_{n-1} = \lambda_2 \ldots \lambda_n$, and the other products contain λ_1 as a factor. Then (i) of the same theorem shows that $(-1)^{n-1}a_{n-1}$ is the sum of the principal minors of size $n-1$, i.e. $(-1)^{n-1}a_{n-1} = \sum_{x \in V} \det(L_x) = n\tau(\Gamma)$ by Theorem 10.2.8. ■

Remark 10.2.10. The second eigenvalue (in increasing order) λ_2 of the Laplacian contains important information on the graph: for example one can see that here $\lambda_2 = 0 \Leftrightarrow \Gamma$ is not connected.

Exercise 10.11. (i) Show that $\tau(K_n) = (n-1)(n-2)$, $n \geqslant 1$.

Hint: proceed by induction on n.

(ii) Deduce the values of the minors of the Laplacian $\mathfrak{L}(K_n)$ of K_n.

10.3 The spectra of regular graphs

10.3.1 The spectral gap

Let us summarise some of the properties we saw in the preceding paragraphs. Let Γ be a k-regular graph. Then

- $\lambda_{\max}(\Gamma) = k$ is a simple eigenvalue if Γ is connected with the associated eigenspace equal to $\mathbb{C}\mathbf{1}$, where $\mathbf{1} = (1, \dots, 1)$,
- $\lambda_{\min}(\Gamma) \geqslant -k$,
- $\lambda_{\min}(\Gamma) = -k$ if and only if Γ is bipartite; the spectrum is then symmetric with respect to the origin.

This means that when Γ is connected, the spectrum can be written

$$\mathrm{Sp}(\Gamma) = (k > \lambda_2 \geqslant \cdots \geqslant \lambda_n).$$

We also have precise relations between the respective spectra of Γ, its line-graph $L(\Gamma)$ and its complement $\overline{\Gamma}$.

Proposition 10.3.1. *Let Γ be a (not necessarily connected) k-regular graph with n vertices and m edges. Then*

(i) *$L(\Gamma)$ is $2(k-1)$-regular and*

$$\Phi(L(\Gamma), x) = (x+2)^{m-n} \prod_{i=1}^{n} (x - \lambda_i - k + 2);$$

thus $\mathrm{Sp}(L(\Gamma))$ consists of -2 with multiplicity $m-n$, and $\lambda_i + k - 2$ for $1 \leqslant i \leqslant n$;

(ii) *$\overline{\Gamma}$ is $(n-k-1)$-regular and*

$$\Phi(\overline{\Gamma}, x) = (x - n + k + 1) \prod_{i=2}^{n} (x + \lambda_i + 1);$$

thus $\mathrm{Sp}(\overline{\Gamma})$ consists of $n-k-1$ and $-\lambda_i - 1$ for $2 \leqslant i \leqslant n$.

Proof. (i) As Γ is k-regular, we have $|E| = m = kn/2 \geqslant n$; we see that $L(\Gamma)$ is $2(k-1)$-regular. Let $B(\Gamma)$ be the incidence matrix of $\overline{\Gamma}$, namely the $n \times m$ rectangular matrix in which the rows (resp. columns) are indexed by the vertices (resp. edges) of Γ. The scalar product of two different rows of $B(\Gamma)$ is 0 if the rows are indexed by two non-adjacent vertices and 1 if they are adjacent. The scalar product of a row with itself is k, since Γ is k-regular. Thus we have

$$B(\Gamma)^t B(\Gamma) = kI_n + A(\Gamma)$$

where $A(\Gamma)$ is the adjacency matrix of Γ. Taking the scalar product of two columns, we obtain 2 if the columns are the same, 1 if the two corresponding edges are incident, and 0 if the two edges are not incident. Hence

$${}^t B(\Gamma) B(\Gamma) = 2I_m + A(L(\Gamma)).$$

A classical result asserts that $(-x)^n\Phi(MN,x) = (-x)^m\Phi(NM,x)$ if M and N are two rectangular matrices of dimensions $m \times n$ and $n \times m$ respectively. To see this, consider the products UV and VU of two square matrices

$$U = \begin{pmatrix} -xI_m & -M \\ N & I_n \end{pmatrix}, \quad V = \begin{pmatrix} I_m & M \\ 0_{n,m} & -xI_n \end{pmatrix},$$

of order $(m+n)$, where we write $0_{m,n}$ for the identically zero $m \times n$ rectangular matrix. The products are given by

$$UV = \begin{pmatrix} -xI_m & 0_{n,m} \\ N & NM - xI_n \end{pmatrix}, \quad VU = \begin{pmatrix} MN - xI_m & 0_{m,n} \\ -xN & -xI_n \end{pmatrix}.$$

Since the determinant is a homomorphism, $\det(UV) = \det(U) \times \det(V)$, so both products UV, VU have the same determinant, which gives the desired identity

$$(-x)^m\Phi(NM,x) = (-x)^n\Phi(MN,x).$$

We apply it with $N = B(\Gamma)$ and $M = {}^tB(\Gamma)$, and obtain

$$\begin{aligned}(-x-2)^n\Phi(A(L(\Gamma)),x) &= (-x-2)^n\Phi({}^tB(\Gamma)B(\Gamma),x+2) \\ &= (-x-2)^m\Phi(B(\Gamma){}^tB(\Gamma),x+2) \\ &= (-x-2)^m\Phi(A(\Gamma),x+2-k).\end{aligned}$$

But $m \geqslant n$ so

$$\begin{aligned}\mathrm{Sp}(L(\Gamma)) = \{&-2 \text{ with multiplicity } m-n\} \\ &\cup \{\lambda + k - 2 \mid \lambda \in \mathrm{Sp}(\Gamma), \text{ with the same multiplicities}\}\end{aligned}$$

(the multiplicity of the eigenvalue -2 is increased if $\lambda = -k$). The possibility $m < n$ cannot occur except for $k = 0$ or 1: in the first case, the graph is the zero graph whose characteristic polynomial is X^n and the line-graph is empty; if $k = 1$, then n is even, the graph is bipartite, its characteristic polynomial is $(X^2-1)^{n/2}$ and its line-graph is the graph with no edge of order $n/2$.

(ii) We have $A(\overline{\Gamma}) = J_n - A(\Gamma) - I_n$, where J_n is the square matrix of order n all of whose coefficients are equal to 1, and $A(\Gamma)$ and $A(\overline{\Gamma})$ are the adjacency matrices of Γ and $\overline{\Gamma}$ respectively. It follows that Γ and $\overline{\Gamma}$ have the same eigenvectors. Furthermore, if v is an eigenvector associated to λ_i, $i \geqslant 2$, then v is orthogonal to the n-vector $(1, 1, \dots, 1)$ (still identified with a column vector). Hence $J_n v = 0$ and $A(\overline{\Gamma})v = -(\lambda+1)v$. It follows that the eigenvalues of $\overline{\Gamma}$ are : $n-k-1$, of multiplicity 1, and $-\lambda - 1$ for each of other eigenvalues $\lambda \in \mathrm{Sp}(\Gamma)$ counted with multiplicity.

■

Exercise 10.12. (i) Determine $\overline{C_4}$ and compute its spectrum directly. Check the result with Proposition 10.3.1.

(ii) Show that $\overline{K_4} = K_2 + K_2 + K_2$ and compute its spectrum (use Proposition 10.1.3).

(iii) Use Proposition 10.3.1 to determine first $\mathrm{Sp}(L(K_4))$, then $\mathrm{Sp}(\overline{L(K_4)})$, and check (i).

Let us now define another quantity related to a connected k-regular graph Γ, namely the absolute value of its *second* eigenvalue. Set

$$\lambda^{\#}(\Gamma) := \max_{\substack{\lambda \in \mathrm{Sp}(\Gamma) \\ |\lambda| \neq k}} |\lambda|.$$

Thus $\lambda^{\#}(\Gamma)$ is the largest absolute value of any eigenvalue of Γ apart from k and $-k$. This quantity arises in numerous estimations, as we will see below.

The *spectral gap* of Γ is defined as

$$g(\Gamma) := k - \lambda^{\#}(\Gamma) \in [0, k].$$

By Exercise 10.4, we have $g(K_n) = n - 1 - |-1| = n - 2$, and by Theorem 10.1.7, $g(K_{r,r}) = r - |0| = r$: in these two special cases, the spectral gap almost reaches its maximum, namely the degree of regularity (equal to $k = n - 1$ and $k = r$ respectively).

Remark 10.3.2. In the case of k-regular graphs, we observed the simple relation

$$\mathrm{Sp}(\mathcal{L}(\Gamma)) = k - \mathrm{Sp}(\Gamma)$$

connecting the spectrum of Γ to that of its Laplacian. Thus if $\lambda^{\#}(\Gamma) \in \mathrm{Sp}(\Gamma)$, then the spectral gap of Γ is *the second smallest eigenvalue* of $\mathcal{L}(\Gamma)$.

10.3.2 Expander graphs

Let $\Gamma = (V; E)$ be a connected graph of order n, and let $W \subset V$ be a non-empty subset of its vertices. Recall that the neighbourhood $\Gamma(W)$ is the set of vertices of $V \backslash W$ that are adjacent to at least one vertex of W.

The *boundary of* W in E is defined by

$$\partial_E W = \{e = \{v, w\} \in E \mid v \in V \backslash W,\, w \in W\}.$$

The *isoperimetric number* of Γ (also called the *Cheeger constant*) is given by

$$h(\Gamma) = \min_{\substack{W \subset V \\ 1 \leqslant |W| \leqslant n/2}} \frac{|\partial_E W|}{|W|}.$$

Let $0 < \epsilon \leqslant 1$. We say that Γ is an *ϵ-expander* if

$$|\Gamma(W)| \geqslant \epsilon |W|, \quad \text{for all } W \subset V \text{ such that } 1 \leqslant |W| \leqslant n/2.$$

The largest possible ϵ is equal to $\epsilon(\Gamma) = \max_{W \subset V, 1 \leqslant |W| \leqslant n/2} \frac{|\Gamma(W)|}{|W|}$: it is called the *expansion coefficient* of Γ.

Observe that in a connected graph, we have $2/n \leqslant \epsilon(\Gamma) \leqslant h(\Gamma)$.

Exercise 10.13. Prove that K_n and $K_{r,r}$ are 1-expanders. Check that the Petersen graph (Figure 10.1) is also a 1-expander (it suffices to consider the $W \subset V$ of cardinality 4 or 5).

Exercise 10.14. Prove that

$$\epsilon(\Gamma) \geqslant \frac{h(\Gamma)}{\Delta(\Gamma)}$$

for a connected graph $\Gamma = (V; E)$.

We will first consider a bound for the distance between $e(S, T)$, the number of edges between S and T, and its average value, which will then allow us to deduce a lower bound for the Cheeger constant $h(\Gamma)$ when Γ is regular. If two vertices v, w are chosen at random in a k-regular graph Γ, then the probability that there is an edge between them is equal to k/n. Thus, given two arbitrary subsets of vertices of Γ, the total *expected* number of edges between S and T is equal to $k|S||T|/n$. The following theorem gives bounds on the difference between the effective value $e(S, T)$ and the expected value $k|S||T|/n$.

Theorem 10.3.3. *Let $\Gamma = (V; E)$ be a k-regular non-bipartite graph with n vertices, and let S, T be two subsets of V. Then*

$$\left| e(S, T) - \frac{k|S||T|}{n} \right| \leqslant \lambda^{\#}(\Gamma) \sqrt{|S||T|}.$$

In the case where $\lambda^{\#}(\Gamma)$ is *small*, this result shows that Γ behaves like a *quasi-random* graph. The exact opposite situation is that of a bipartite graph, in which the edges are the opposite of randomly distributed ($-k$ is also in the spectrum).

Proof. Define s and t by $|S| = sn$ and $|T| = tn$. Let $V = \{1, 2, \dots, n\}$ denote the vertices of Γ, and define the vector $v = (v_1, v_2, \dots, v_n)$ by setting $v_j = 1 - s$ if $j \in S$ and $-s$ otherwise. Then clearly $\sum_{j=1}^{n} v_j = 0$, so v is in the subspace H orthogonal to $\mathbb{C}(1, 1, \dots, 1)$.

Let $B := A(\Gamma)|_H$ be the restriction to H. The eigenvalue $\lambda_{\max}(B)$ has absolute value $\lambda^{\#}(\Gamma)$ since Γ is not bipartite.

By the *spectral theorem* (Theorem 6.5.2), there exists an orthonormal basis consisting of eigenvectors of B. Let $\lambda_2 \geqslant \cdots \geqslant \lambda_n$ denote the $n-1$ eigenvalues of B (considered with their multiplicities).

Since v is orthogonal to $(1,1,\dots,1)$, we have $v = e_2 + e_3 + \cdots + e_n$, where e_i lies in the eigenspace associated to λ_i. By the orthogonality of the family $\{e_j,\ j = 2,\dots,n\}$, we deduce that

$$\|Av\|^2 = \|Bv\|^2 = \sum_{j=2}^{n} \lambda_j^2 \|u_j\|^2 \leqslant \lambda^{\#}(\Gamma)^2 \sum_{j=2}^{n} \|u_j\|^2 = \lambda^{\#}(\Gamma)^2 \|v\|^2.$$

We have $\|v\|^2 = sn(1-s)^2 + (n-sn)s^2 = ns(1-s)$. It remains to estimate $\|Av\|^2$. The i-th coordinate of Bv is equal to $(1-s)\deg(i) - s(k-\deg(i)) = \deg(i) - sk$, where we write $1,2,\dots,n$ for the vertices of Γ for convenience, and let $\deg(i)$ denote the number of neighbours of i in S. Then

$$\|Av\|^2 = \sum_{i=1}^{n} (\deg(i) - sk)^2 \leqslant \lambda^{\#}(\Gamma)^2 ns(1-s).$$

Applying Cauchy's inequality

$$\Big(\sum_{i=1}^{N} |a_i|\Big)^2 \leqslant N \sum_{i=1}^{N} |a_i|^2,$$

we obtain

$$\begin{aligned}\sum_{i\in T} |\deg(i) - sk| &\leqslant \sqrt{tn}\Big(\sum_{i\in T} (|\deg(i) - sk|)^2\Big)^{1/2} \\ &\leqslant \sqrt{tn}\Big(\sum_{i=1}^{n} (|\deg(i) - sk|)^2\Big)^{1/2},\end{aligned}$$

so

$$\sum_{i\in T} |\deg(i) - sk| \leqslant \sqrt{tn}\lambda^{\#}(\Gamma)\sqrt{ns(1-s)} \leqslant \lambda^{\#}(\Gamma)\sqrt{|S||T|}.$$

Finally, by the triangle inequality, we have

$$\begin{aligned}\Big|e(S,T) - \frac{k|S||T|}{n}\Big| &= |e(S,T) - kstn| = \Big|\sum_{i\in T} (\deg(i) - sk)\Big| \\ &\leqslant \sum_{i\in T} |\deg(i) - sk| \leqslant \lambda^{\#}(\Gamma)\sqrt{|S||T|},\end{aligned}$$

which gives the desired result. ∎

We can deduce from this the lower bound

$$e(S,T) \geqslant \Big(k - \frac{\lambda^{\#}(\Gamma)n}{\sqrt{|S||T|}}\Big)\cdot\frac{|S||T|}{n},$$

which can be improved if the sets S and T in $V(\Gamma)$ are complementary.

Proposition 10.3.4. *Assume that Γ is not a bipartite graph, and that $V(\Gamma) = S \sqcup T$ is a partition of the vertices of Γ. Then*

$$e(S,T) \geqslant g(\Gamma)\cdot\frac{|S||T|}{n}.$$

Proof. We keep the notation of Theorem 10.3.3. We have $s + t = 1$. Let

$$u = (u_1, u_2, \dots, u_n)$$

be a vector considered as a column vector by the identification of $\mathbb{K}^n$ with $\mathbb{K}^{n,1}$. Then

$${}^t u(kI_n - A)u = \sum_{ij\in E}(u_i - u_j)^2.$$

Define the vector v by its coordinates, by setting $v_i = -s$ if $i \in T$ and $v_i = t$ if $i \in S$. We have $\sum_{i=1}^n v_i = 0$, so v is orthogonal to the eigensubspace H of A associated to the eigenvalue k and generated by $(1,1,\dots,1)$ considered as a column vector. Clearly the eigensubspaces of A and of $kI_n - A$ coincide: $k - \lambda^{\#}(\Gamma)$ is the second smallest eigenvalue of $kI_n - A$, which is thus equal to $\lambda_{\min}((kI_n - A)|_H)$. By Proposition 6.5.3, we have

$$\begin{aligned}{}^t v(kI_n - A)v \geqslant (k - \lambda^{\#}(\Gamma))\|v\|^2 &= (k - \lambda^{\#}(G))(s^2tn + t^2sn)\\ &= (k - \lambda^{\#}(\Gamma))stn\end{aligned}$$

since $s + t = 1$. It remains to calculate $\sum_{ij\in E}(v_i - v_j)^2$: we have

$$\sum_{ij\in E}(v_i - v_j)^2 = \sum_{\substack{i\in S\\ j\in T}}(s+t)^2 = e(S,T)(s+t)^2 = e(S,T).$$

As $g(\Gamma) = k - \lambda^{\#}(\Gamma)$, the result follows. ∎

We deduce from this the lower bounds

$$h(\Gamma) \geqslant \frac{g(\Gamma)}{2}, \quad \epsilon(\Gamma) \geqslant \frac{g(\Gamma)}{2k}$$

for every k-regular graph.

10.4 Ramanujan graphs

In this part, we will concentrate on a certain type of graph that has been the object of a great deal of study [29].

10.4.1 An example: The Petersen graph

One can prove (see Theorem 10.4.7) that for every k-regular connected graph Γ with $\mathrm{diam}(\Gamma) > 3$, we have

$$\lambda^{\#}(\Gamma) \geqslant 2\sqrt{k-1}\Big(1 - \frac{2}{\mathrm{diam}(\Gamma) - 2}\Big).$$

This suggests that for large diameters we have $\lambda^{\#}(\Gamma) \approx 2\sqrt{k-1}$. A k-regular graph Γ is called a *Ramanujan graph* if

$$\lambda^{\#}(\Gamma) \leqslant 2\sqrt{k-1}.$$

For example, $\lambda^{\#}(K_n) = 1$ and $\lambda^{\#}(K_{r,r}) = 0$, so these graphs are Ramanujan graphs. Another example is the Petersen graph shown in Figure 10.1.

Proposition 10.4.1. *The Petersen graph P is a Ramanujan graph.*

Proof. P is connected and 3-regular. Its adjacency matrix is as follows:

$$\begin{pmatrix}
0 & 1 & 0 & 0 & 1 & 1 & 0 & 0 & 0 & 0\\
1 & 0 & 1 & 0 & 0 & 0 & 1 & 0 & 0 & 0\\
0 & 1 & 0 & 1 & 0 & 0 & 0 & 1 & 0 & 0\\
0 & 0 & 1 & 0 & 1 & 0 & 0 & 0 & 1 & 0\\
1 & 0 & 0 & 1 & 0 & 0 & 0 & 0 & 0 & 1\\
1 & 0 & 0 & 0 & 0 & 0 & 0 & 1 & 1 & 0\\
0 & 1 & 0 & 0 & 0 & 0 & 0 & 0 & 1 & 1\\
0 & 0 & 1 & 0 & 0 & 1 & 0 & 0 & 0 & 1\\
0 & 0 & 0 & 1 & 0 & 1 & 1 & 0 & 0 & 0\\
0 & 0 & 0 & 0 & 1 & 0 & 1 & 1 & 0 & 0
\end{pmatrix}$$

It is tedious but possible to compute the characteristic polynomial of P directly, however we give an alternative computation just after Proposition 10.4.2 below: we find that

$$\Phi(P, x) = (x-3)(x+2)^4(x-1)^5.$$

Thus $\mathrm{Sp}(P) = \{3^{(1)}, 1^{(5)}, -2^{(4)}\}$ and $\lambda^{\#}(P) = 2$. As P is 3-regular and $2 \leqslant 2\sqrt{3-1}$, this shows that P is a Ramanujan graph. ∎

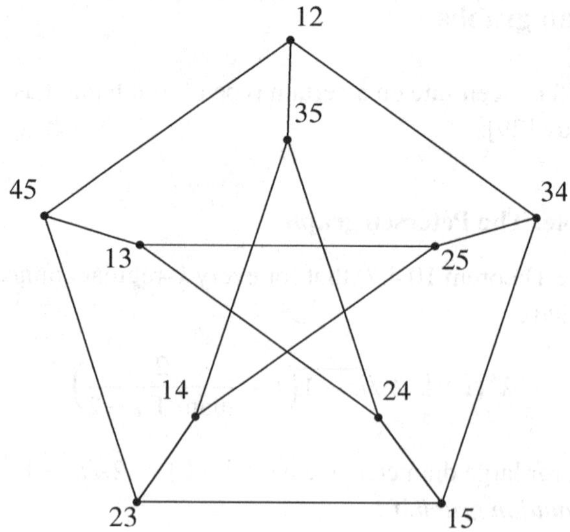

Figure 10.1. The Petersen graph.

It is more instructive (and amusing) to note that the Petersen graph arises in the following way.

Proposition 10.4.2. *The Petersen graph is the complement of the line-graph of the complete graph K_5.*

Proof. Let $1 \leqslant i \leqslant 5$ denote the vertices of K_5 and ij for $1 \leqslant i < j \leqslant 5$ the 10 vertices of $\overline{L(K_5)}$. Observe that $\overline{L(K_5)}$ contains the two 5-cycles $12, 34, 15, 23, 45, 12$ and $35, 14, 25, 13, 24, 35$. Furthermore, these cycles are connected by the 5 edges

$$\{12, 35\}, \{34, 25\}, \{15, 24\}, \{23, 14\}, \{45, 13\}.$$

There is no other edge in $\overline{L(K_5)}$, since $L(K_5)$ is 6-regular when $\overline{L(K_5)}$ is itself 3-regular. It is then easy to give an explicit isomorphism between P and $\overline{L(K_5)}$ using Figure 10.1. We attribute the labels $12, 34, 15, 23, 45$ to the vertices of the outer cycle, and the labels $35, 25, 24, 14, 13$ to the vertices of the inner pentagram. ∎

We can use this and apply Proposition 10.3.1 to compute the characteristic polynomial of the Petersen graph by computing that of the line graph of K_5. First, since K_5 is 4-regular of order 5 with 10 edges, we have

$$\Phi(K_5, x) = (x-4)(x+1)^4, \quad \mathrm{Sp}(K_5) = \{4^{(1)}, -1^{(4)}\}.$$

Then $L(K_5)$ is 6-regular of order 10 with 30 edges and

$$\Phi(L(K_5), x) = (x+2)^5(x-6)(x-1)^4, \quad \mathrm{Sp}(L(K_5)) = \{6^{(1)}, 1^{(4)}, -2^{(5)}\}.$$

Finally, $\overline{L(K_5)}$ is 3-regular of order 10 with 15 edges, and

$$\Phi(\overline{L(K_5)}, x) = (x-3)(x-1)^5(x+2)^4,$$

which is the characteristic polynomial of the Petersen graph given in the proof of Proposition 10.4.1.

10.4.2 Upper bound for the diameter

Let $\Gamma = (V; E)$ be a connected k-regular graph of order n. Recall that the diameter of Γ is given by

$$\operatorname{diam}(\Gamma) = \max_{v,w \in V} d(v,w),$$

where $d(v,w)$ is the distance from v to w, i.e. the minimal chain length between v and w. Let A be the adjacency matrix of Γ. For convenience, we write A^q_{rs} for the (r,s)-coefficient of the matrix A^q. We will make use of the following remark: if q is an integer such that $A^q_{rs} > 0$ for all r and s, then $\operatorname{diam}(\Gamma) \leqslant q$.

Let

$$\operatorname{Sp}(\Gamma) = (\lambda_1 = k > \lambda_2 \geqslant \cdots \geqslant \lambda_n)$$

be the spectrum of Γ and $\mathcal{U} = (U_1, U_2, \ldots, U_n)$ an orthonormal basis of $\mathbb{C}^n$ consisting of eigenvectors of A: $AU_m = \lambda_m U_m$. We naturally fix

$$U_1 = \frac{1}{\sqrt{n}}(1, 1, \ldots, 1).$$

Then $A = UDU^*$, where $D = \operatorname{Diag}(\lambda_1, \ldots, \lambda_n)$ is the diagonal matrix with the eigenvalues of A along the diagonal and U is an orthogonal matrix, namely the transition matrix from the canonical basis $(e_1, \ldots, e_n)$ of $\mathbb{C}^n$ (n-tuples are still identified with column vectors when performing matrix multiplication) to the basis $\mathcal{U}$, which satisfies $U^* = U^{-1} = {}^tU$. Thus, $A^q = UD^qU^*$ and the (r,s) coefficient of A^q is given by

$$A^q_{rs} = \langle e_r, A^q e_s\rangle = \langle U^* e_r, D^q U^* e_s\rangle.$$

But U^*e_r is the vector $U'_r = (U_{1r}, \ldots, U_{nr})$, where U_{mr} denotes the r-th component of U_m in the canonical basis. Therefore

$$A^q_{rs} = \langle U'_r, D^q U'_s\rangle = \sum_{m=1}^{n} \lambda^q_m U_{mr} U_{ms}.$$

As $\lambda_1 = k$ and $U_{1r} = 1/\sqrt{n}$ for all r, the triangular inequality followed by the Cauchy inequality gives

$$A^q_{rs} \geqslant \frac{k^q}{n} - \sum_{m=2}^{n} \lambda^q_m U_{mr} U_{ms} > \frac{k^q}{n} - \lambda^{\#}(\Gamma)^q \Big(\sum_{m=2}^{n} U^2_{mr}\Big)^{1/2} \Big(\sum_{m=2}^{n} U^2_{ms}\Big)^{1/2}.$$

This inequality is strict, since the rows of the invertible matrix U are naturally independent. Hence

$$A^q_{rs} > \frac{k^q}{n} - \lambda^{\#}(\Gamma)^q\Big(1 - \frac{1}{n}\Big).$$

It follows that $A^q_{rs} > 0$ if

$$q \geqslant \frac{\log(n-1)}{\log k - \log \lambda^{\#}(\Gamma)}$$

as long as Γ is not a bipartite graph.

We have proved the following result, due to Fan Chung.

Proposition 10.4.3. *The diameter of a k-regular connected non-bipartite graph Γ satisfies*

$$\operatorname{diam}(\Gamma) < \frac{\log(n-1)}{\log k - \log \lambda^{\#}(\Gamma)} + 1.$$

In the case where Γ is a bipartite graph, we must treat the eigenvalue $-k$ differently.

Proposition 10.4.4. *The diameter of a k-regular connected bipartite graph Γ satisfies*

$$\operatorname{diam}(\Gamma) < \frac{\log(n/2-1)}{\log k - \log \lambda^{\#}(\Gamma)} + 2.$$

Proof. Left as an exercise for the reader. ■

10.4.3 Lower bounds for $\lambda^{\#}(\Gamma)$

From Proposition 10.4.3, we deduce in the non-bipartite case that

$$\lambda^{\#}(\Gamma) \geqslant k \exp\Big(-\frac{\log(n-1)}{\operatorname{diam}(\Gamma)-1}\Big),$$

which yields a *good* lower bound for $\lambda^{\#}(\Gamma)$ in the case where the diameter is *large*. If we assume that Γ is a Ramanujan graph, this gives

$$2\sqrt{k-1} \geqslant k \exp\Big(-\frac{\log(n-1)}{\operatorname{diam}(\Gamma)-1}\Big),$$

hence

$$\operatorname{diam}(\Gamma) \leqslant \frac{\log(n-1)}{\log k - \log 2} + 1.$$

The adjacency matrix A of Γ satisfies

$$\operatorname{tr}({}^tAA) = kn$$

where tr is the trace function. Indeed, the (j, j)-th coefficient of ${}^tAA$ is equal to $\sum_{\ell=1}^n A_{j\ell}^2 = \sum_{\ell=1}^n A_{j\ell} = k$, taking account of the fact that $A_{j\ell} = 0$ or 1. Also we have $\mathrm{Sp}(A^2) = \{\lambda_i^2,\ 1 \leqslant i \leqslant n\}$, so since Γ is assumed bipartite, we have

$$\mathrm{tr}(A^2) = \sum_{i=1}^n \lambda_i^2 \leqslant k^2 + (n-1)\lambda^\#(\Gamma)^2.$$

As A is symmetric, it follows that

$$kn \leqslant k^2 + (n-1)\lambda^\#(\Gamma)^2.$$

This proves the following result.

Proposition 10.4.5. *Let Γ be a k-regular connected non-bipartite graph of order n. Then*

$$\lambda^\#(\Gamma) \geqslant \sqrt{\frac{k(n-k)}{n-1}}.$$

It follows that if $k = o(n)$ as n tends to infinity, then $\lambda^\#(\Gamma) \geqslant (1 + o(1))\sqrt{k}$.

The analogous result for bipartite graphs is the following (proof left as an exercise).

Proposition 10.4.6. *Let Γ be a k-regular connected bipartite graph of order n. Then*

$$\lambda^\#(\Gamma) \geqslant \sqrt{\frac{k(n-2k)}{n-2}}.$$

The last lower bound we will consider is due to Nilli (1991), alias Noga Alon. It yields a very fine bound which illustrates the concept of a Ramanujan graph by showing that the condition $\lambda^\#(\Gamma) \leqslant 2\sqrt{k-1}$ required for Γ to be a Ramanujan graph should be considered as a natural barrier for *generic* k-regular graphs as the number of vertices n tends to infinity (cf. Theorem 10.4.8).

Theorem 10.4.7 (Nilli, 1991). *Let Γ be a k-regular connected graph of diameter at least four. Set*

$$\delta := \left\lfloor \frac{\mathrm{diam}(\Gamma) - 2}{2} \right\rfloor.$$

Then

$$\lambda^\#(\Gamma) \geqslant 2\sqrt{k-1} - \frac{2\sqrt{k-1} - 1}{\delta} \geqslant 2\sqrt{k-1}\Big(1 - \frac{2}{\mathrm{diam}(\Gamma) - 2}\Big).$$

Proof. We start by recalling an identity involving the Laplacian of the graph (see the proof of Theorem 10.2.1):

$${}^tu(kI_n - A)u = \sum_{ij \in E} (u_i - u_j)^2,$$

for every vector $u = (u_1, \ldots, u_n) \in \mathbb{C}^n$. By hypothesis, there exist two vertices g, h such that $d(g,h) \geqslant 2\delta + 2$. We set

$$G_q = \{j \in V \mid d(g,j) = q\}, \quad H_q = \{j \in V \mid d(h,j) = q\},$$

and observe that the G_q and the H_q are pairwise disjoint for $0 \leqslant q \leqslant \delta$. Let

$$G = \bigcup_{q=0}^{\delta} G_q, \quad H = \bigcup_{q=0}^{\delta} H_q.$$

The neighbourhood of G in Γ is disjoint from H and vice versa: indeed if $i \in \Gamma(G)$ and $j \in H$, then $d(g,i) \leqslant \delta + 1$ and $d(h,j) \leqslant \delta$, so $d(i,j) \geqslant d(g,h) - d(g,i) - d(h,j) \geqslant 1$.

Our goal is to construct a vector $f = (f_1, \ldots, f_n)$ whose components are indexed by the vertices i, and which will be orthogonal to the vector $\mathbf{1} = (1, \ldots, 1)$ and realise a *small* value for

$$\frac{\sum_{ij \in E} (f_i - f_j)^2}{\sum_{i \in V} f_i^2}.$$

For this, consider f as a function on V that is constant on each G_q and each H_q. Let $a = f_g$ and $b = f_h$ be non-zero real numbers, and for each $i \in G_q$ (resp. H_q), $q \geqslant 1$, set

$$f_i = a\kappa^{q-1} \quad (\text{resp. } b\kappa^{q-1}),$$

where $\kappa = (k-1)^{-1/2}$. For the other i, we simply fix $f_i = 0$. The constraint that f must be orthogonal to $\mathbf{1}$ is then expressed as

$$a + a\sum_{q=1}^{\delta} |G_q|\kappa^{q-1} + b + b\sum_{q=1}^{\delta} |H_q|\kappa^{q-1} = 0,$$

and this condition is fulfilled for appropriate values of a and b. The square of the norm of the vector f is equal to

$$\|f\|^2 = \|f_G\|^2 + \|f_H\|^2$$

with

$$\|f_G\|^2 = a^2 + a^2 \sum_{q=1}^{\delta} |G_q|\kappa^{2q-2}, \quad \|f_H\|^2 = b^2 + b^2 \sum_{q=1}^{\delta} |H_q|\kappa^{2q-2}.$$

It remains to estimate the *standard deviation* $D := \sum_{ij \in E} (f_i - f_j)^2$.

Firstly, if $i \in G_q$, $1 \leqslant q \leqslant \delta$, then i admits at most $k-1$ neighbours in G_{q+1} since it has at least one in G_{q-1}. This also holds for $i \in H_q$.

Secondly, we can *blow up* D into two complementary parts $D = D_G + D_H$, where

$$D_G = \sum_{\substack{ij \in E \\ i \text{ or } j \in G}} (f_i - f_j)^2, \quad D_H = \sum_{\substack{ij \in E \\ i \text{ or } j \in H}} (f_i - f_j)^2.$$

We obtain

$$\begin{aligned} D_G &\leqslant a^2 \sum_{q=1}^{\delta-1} |G_q|(k-1)(\kappa^{q-1} - \kappa^q)^2 + a^2 |G_\delta|(k-1)\kappa^{2\delta-2} \\ &\leqslant a^2 (1-\kappa)^2 (k-1) \sum_{q=1}^{\delta-1} |G_q| \kappa^{2q-2} + a^2 |G_\delta|(k-1)\kappa^{2\delta-2} \\ &\leqslant a^2 (\sqrt{k-1} - 1)^2 \sum_{q=1}^{\delta-1} |G_q| \kappa^{2q-2} + a^2 |G_\delta|(k-1)\kappa^{2\delta-2} \\ &\leqslant a^2 (\sqrt{k-1} - 1)^2 \sum_{q=1}^{\delta} |G_q| \kappa^{2q-2} + a^2 |G_\delta|(2\sqrt{k-1} - 1)\kappa^{2\delta-2}. \end{aligned}$$

The first term above is

$$(\sqrt{k-1} - 1)^2 \langle f_G, f_G - a^2 \rangle$$

and the second is less than or equal to

$$a^2 (2\sqrt{k-1} - 1) \frac{1}{\delta} \sum_{q=1}^{\delta} |G_q| \kappa^{2q-2} = \frac{2\sqrt{k-1} - 1}{\delta} \langle f_G, f_G - a^2 \rangle.$$

We conclude that

$$\begin{aligned} D_G &\leqslant \Big((\sqrt{k-1} - 1)^2 + \frac{2\sqrt{k-1} - 1}{\delta} \Big) \langle f_G, f_G - a^2 \rangle \\ &< \Big((\sqrt{k-1} - 1)^2 + \frac{2\sqrt{k-1} - 1}{\delta} \Big) \| f_G \|^2. \end{aligned}$$

We obtain a similar bound for D_H by the same method. Then

$$\frac{D}{\|f\|^2} < (\sqrt{k-1} - 1)^2 + \frac{2\sqrt{k-1} - 1}{\delta}.$$

This shows that the Laplacian of Γ (see Section 10.2) has a non-zero eigenvalue that is less than or equal to the right-hand side of the above inequality. Since the eigenvalues of the Laplacian are given by $k - \lambda$ for λ in the spectrum of Γ, we conclude that

$$\lambda^{\#}(\Gamma) > k - \Big((\sqrt{k-1} - 1)^2 + \frac{2\sqrt{k-1} - 1}{\delta} \Big),$$

which implies the desired result. ■

Fix an integer $k \geqslant 2$, and consider the family $\mathfrak{R}_k$ of all simple non-oriented k-regular graphs. When $\Gamma \in \mathfrak{R}_k$ has n vertices, its diameter $\mathrm{diam}(\Gamma)$ clearly satisfies

$$\mathrm{diam}(\Gamma) \geqslant \frac{\log n}{\log k},$$

and it tends to infinity with n. The following statement is known as the Alon–Boppana theorem.

Theorem 10.4.8. *Let $k \geqslant 2$ be an integer. Then*

$$\liminf_{\substack{\Gamma \in \mathfrak{R}_k \\ |V(\Gamma)| \to \infty}} \lambda^{\#}(\Gamma) \geqslant 2\sqrt{k-1}.$$

Proof. The result follows directly from the remark just before the statement together with Theorem 10.4.7. ∎

Remark 10.4.9. Another, somewhat less explicit way to bound $\lambda^{\#}(\Gamma)$ by default when Γ is a k-regular graph is based on a combinatorial interpretation of the coefficients of the powers of the adjacency matrix $A = A(\Gamma)$ of Γ. Observe that after diagonalising A, raising the matrix A to the power $2t$ for every $t \geqslant 0$, we have

$$\mathrm{tr}(A^{2t}) \leqslant k^{2t} + (n-1)\lambda^{\#}(\Gamma)^{2t}.$$

Each diagonal coefficient A^{2t}_{vv} of A^{2t} counts the number of closed chains of length $2t$ starting and ending at v. Therefore the trace of A^{2t} is the total number of such chains, counted with their multiplicity. The number $N(v,t)$ of chains of length $2t$ starting and ending at $v \in V(\Gamma)$ is at least

$$\binom{2t}{t}\frac{1}{t+1}(k-1)^t. \tag{10.11}$$

The quantity $N(v,t)$ is related to the Catalan numbers $c_t = \frac{1}{t+1}\binom{2t}{t}$ in the following way. We can restrict ourselves to the chains $e_1, e_2, \dots, e_{2t}$ (sequences of $2t$ edges) from v to itself with the constraint that each edge $e = \{u, w\}$ must be traversed in each direction (u to w and w to u) the same number of times. At each vertex w, we allow an *ant* to crawl backwards along an edge e (or a series of edges) that it just advanced along, or to crawl forwards along any one of the $k-1$ other edges incident to w, as long as that edge is not forbidden by the constraint. More precisely, there are exactly t vertices at which the ant moves backwards along edges that were already traversed, and exactly t vertices where it moves forward along one of the $k-1$ other edges. Moving forward (resp. backward) is coded by $+1$ (resp. -1).

An *admissible chain* is given by the following data (a) and (b).

(a) A sequence $(a_1, \dots, a_{2t})$ of length $2t$ consisting of $+1$'s and -1's such that

- all partial sums $a_1 + \cdots + a_i \geqslant 0$ are positive for $i = 1, \ldots, 2t$,
- the complete sum of all terms is zero: $a_1 + \cdots + a_{2t} = 0$.

(b) To each index i such that $a_i = +1$ (corresponding to moving forward), exactly one of the $k-1$ possible edges (excluding the edge that was just traversed, which would correspond to moving backward).

There are c_t admissible sequences $(a_1, \ldots, a_{2t})$ and $(k-1)^t$ choices of possible edges for the t components a_i equal to $+1$. The number of such chains is given by the quantity defined in (10.11), which is thus a lower bound of $N(v,t)$.

Then

$$(n-1)\lambda^{\#}(\Gamma)^{2t} \geqslant n\binom{2t}{t}\frac{1}{t+1}(k-1)^t - k^{2t}.$$

From the lower bound $\binom{2t}{t} \geqslant \frac{4^t}{t}$, we deduce that

$$(n-1)\lambda^{\#}(\Gamma)^{2t} \geqslant \frac{n}{t(t+1)}(2\sqrt{k-1})^{2t} - k^{2t}.$$

Choose t in such a way that k^t is small compared to n/t^2 as n tends to infinity (with k fixed):

$$t = \left\lfloor \frac{\log n}{2\log k} \right\rfloor.$$

We finally obtain

$$\lambda^{\#}(\Gamma) \geqslant 2\sqrt{k-1}\Big(1 - \frac{\log t}{t}\Big).$$

10.5 Non-oriented Cayley graphs

10.5.1 Circulant graphs

Recall (Section 9.3) that if G is a group (written multiplicatively, with unit element 1) and $S \subset G$ a non-empty subset not containing 1, we define $\overrightarrow{\mathrm{Cay}}(G,S) = (V; \vec{E})$ with $V = G$ and $(x,y) \in \vec{E}$ if and only if $yx^{-1} \in S$ (i.e. $y = sx$).

When S is symmetric (i.e. $S^{-1} = S$), we can forget the orientation since $y = sx \Leftrightarrow x = s^{-1}y$, and simply define a so-called oriented Cayley graph $\mathrm{Cay}(G,S) = (V; E)$ where $V = G$ and $\{x,y\} \in E$ if and only if $x^{-1}y \in S$. In this case, we have the following result.

Proposition 10.5.1. *Let $\Gamma = \mathrm{Cay}(G,S)$ be a (non-oriented) Cayley graph. Then*

(i) $\mathrm{Cay}(G,S)$ *is connected if and only if $G = \langle S \rangle$ (S generates G).*

(ii) $\mathrm{Cay}(G,S)$ *is a simple $|S|$-regular graph.*

Proof. (i) Assume $\mathrm{Cay}(G, S)$ is connected. Then for all $x \in G$, there exists a chain in $\mathrm{Cay}(G, S)$ between 1 and x; there are elements $s_1, s_2, \dots, s_k \in S$ giving the chain: $1, s_1, s_1 s_2, \dots, s_1 s_2 \cdots s_k = x$, so $x \in \langle S \rangle$. Conversely, for all $x \in G$, there exists $s_1, s_2, \dots, s_k$ in S such that $x = s_1 s_2 \cdots s_k$. This induces a chain

$$1, s_1, s_1 s_2, \dots, s_1 s_2 \cdots s_k$$

between 1 and x in $\mathrm{Cay}(G, S)$, which is thus connected.

(ii) For each vertex $x \in G$, every element of type sx with $s \in S$, is a neighbour of x, and these are the only vertices adjacent to x. As $1 \notin S$, the result follows. ∎

The simplest example is the cyclic group $G = \mathbb{Z}/n\mathbb{Z}$. We say that $\mathrm{Cay}(\mathbb{Z}/n\mathbb{Z}, S)$ is a *circulant graph*; for $S = \{\pm 1\}$, we obtain

$$\mathrm{Cay}(\mathbb{Z}/n\mathbb{Z}, \{\pm 1\}) = C_n.$$

The graph K_n is the circulant graph $\mathrm{Cay}(G, S)$ where $G = \mathbb{Z}/n\mathbb{Z}$ and $S = G \backslash \{0\}$. The complete bipartite graph $K_{r,r}$ is the circulant graph $\mathrm{Cay}(G, S)$ where $G = \mathbb{Z}/2r\mathbb{Z}$ and $S = G \backslash 2G$.

Let p be an odd prime number and set $r = (p-1)/2$. We can also realise $K_{r,r} = \mathrm{Cay}((\mathbb{Z}/p\mathbb{Z})^{\times}, S)$ by taking S to be the set of non-squares in the cyclic multiplicative group $(\mathbb{Z}/p\mathbb{Z})^{\times}$ (to see this, use the fact that the product of 2 non-squares is a square).

Another family of circulant graphs is given by the Paley graphs. For example, $\Gamma = \mathrm{Cay}(\mathbb{Z}/5\mathbb{Z}, S_2)$ where $S_2 = \{1, 4\} = -S_2$. This graph is 2-regular. To compute its spectrum, we use the 5 *characters*:

$$\chi_j \colon \mathbb{Z}/5\mathbb{Z} \to \mu_5$$

defined by $\chi_j(1) = \zeta^j, 0 \leqslant j \leqslant 4$, where $\mu_5 = \{1, \zeta, \zeta^2, \zeta^3, \zeta^4\}$ denotes the multiplicative group of 5-th roots of unity in $\mathbb{C}$, with $\zeta = \exp\left(\frac{2i\pi}{5}\right)$.

For each χ, the vector $(\chi(0), \chi(1), \chi(2), \chi(3), \chi(4)) \in \mathbb{C}^5$ is an eigenvector of the adjacency matrix of $\mathrm{Cay}(\mathbb{Z}/5\mathbb{Z}, S_2)$ associated to the eigenvalue $\lambda_\chi = \sum_{s \in S_2} \chi(s) = \chi(1) + \chi(4)$. A simple computation gives $\lambda_0 = 2$, $\lambda_1 = \lambda_4 = \zeta + \zeta^4 = 2\cos\frac{2\pi}{5} = \frac{-1+\sqrt{5}}{2}$, $\lambda_2 = \lambda_3 = \zeta^2 + \zeta^3 = 2\cos\frac{4\pi}{5} = \frac{-1-\sqrt{5}}{2}$. Thus, the spectrum is

$$\mathrm{Sp}(\mathrm{Cay}(\mathbb{Z}/5\mathbb{Z}, S_2)) = \left\{2^{(1)}, \frac{-1+\sqrt{5}}{2}^{(2)}, \frac{-1-\sqrt{5}}{2}^{(2)}\right\},$$

so $\lambda^{\#}(\Gamma) = \frac{1+\sqrt{5}}{2}$: the golden ratio! Clearly $\frac{1+\sqrt{5}}{2} < 2\sqrt{k-1} = 2$, so Γ is a Ramanujan graph. Observe that $\mathrm{Cay}(\mathbb{Z}/5\mathbb{Z}, S_2) = C_5$.

For $p = 13$, we take $S_2 = \{\pm 1, \pm 3, \pm 4\}$ (the set of squares modulo 13). The graph $\mathrm{Cay}(\mathbb{Z}/13\mathbb{Z}, S_2)$ is thus 6-regular; there are 13 characters, given by $\chi_j(1) = \zeta^j$,

$0 \leqslant j \leqslant 12$, where here $\zeta = \exp(i\theta)$ with $\theta = 2\pi/13$. We find $\lambda_j = 2(\cos j\theta + \cos 3j\theta + \cos 4j\theta)$; $\lambda_0 = 6$ is a simple eigenvalue, the others are double. Thus $\mathrm{Cay}(\mathbb{Z}/13\mathbb{Z}, S_2)$ has spectrum

$$\{6^{(1)}, 2(\cos j\theta + \cos 3j\theta + \cos 4j\theta)^{(2)},\ j = 1, \dots, 6\}.$$

We now call upon the well-known evaluation of quadratic Gauss sums, which are the following identities, valid for all a such that $p \nmid a$:

$$|G(a,p)| = \left|\sum_{x=1}^{p} \exp\Big(\frac{2i\pi a x^2}{p}\Big)\right| = \sqrt{p}.$$

Indeed, $|G(a,p)|^2$ can be written

$$\begin{aligned}\sum_{x,y=1}^{p} \exp\Big(\frac{2i\pi a(x^2-y^2)}{p}\Big) &= \sum_{x,y=1}^{p} \exp\Big(\frac{2i\pi a(x-y)(x+y)}{p}\Big)\\ &= \sum_{y,z=1}^{p} \exp\Big(\frac{2i\pi a z(2y+z)}{p}\Big)\\ &= \sum_{z=1}^{p} \exp\Big(\frac{2i\pi a z^2}{p}\Big) \sum_{y=1}^{p} \exp\Big(\frac{4i\pi a z y}{p}\Big),\end{aligned}$$

which is equal to p since the inner sum over y is equal to p or 0 according to whether $z = p$ or $z \neq p$. Note that every non-zero square x is realised twice since $x = y^2 = (-y)^2$, so

$$\sum_{x \in S_2} \zeta^{jx} = \frac{G(j,13) - 1}{2};$$

hence for $j \neq 0$, we obtain $|\lambda_j| \leqslant \frac{\sqrt{13}+1}{2} \leqslant 2\sqrt{6-1}$. Once again, it follows that Γ is a Ramanujan graph.

This scenario can be generalised to all primes $p \equiv 1 \bmod 4$ by taking

$$S_2 = \{x^2 \mid x \in (\mathbb{Z}/p\mathbb{Z})\backslash\{0\}\}.$$

We then have $S_2 = -S_2$ since -1 is a square; moreover $|S_2| = \frac{p-1}{2}$. Since

$$\frac{\sqrt{p}+1}{2} \leqslant 2\sqrt{\frac{p-1}{2} - 1},$$

the same procedure as above for $p = 13$ allows us to show that $\mathrm{Cay}(\mathbb{Z}/p\mathbb{Z}, S_2)$ is a $(\frac{p-1}{2})$-regular Ramanujan graph.

The characters mentioned above also allow us to determine the spectrum of $C_n = \mathrm{Cay}(\mathbb{Z}/n\mathbb{Z}, \{\pm 1\})$. Consider the characters

$$\chi_j : \mathbb{Z}/n\mathbb{Z} \to \mu_n$$

defined by $\chi_j(1) = \zeta^j, 0 \leqslant j \leqslant n-1$, where $\mu_n = \{\zeta^j, 0 \leqslant j \leqslant n-1\}$ is the multiplicative group of n-th roots of unity in $\mathbb{C}$ and $\zeta = \exp(\frac{2i\pi}{n})$.

For each χ_j, the vector $(\chi_j(0), \chi_j(1), \ldots, \chi_j(n-1)) \in \mathbb{C}^n$ is an eigenvector of the adjacency matrix of C_n, associated to the eigenvalue $\lambda_j = \chi_j(1) + \chi_j(-1) = \zeta^j + \zeta^{-j}$. Note that $\lambda_0 = 2$ is a simple eigenvalue while the others are double, unless n is even, in which case $\lambda_{n/2} = -2$ is simple.

Exercise 10.15. Show that

$$\mathrm{Sp}(C_{10}) = \{2^{(1)}, \Phi^{(2)}, \Phi - 1^{(2)}, -(\Phi - 1)^{(2)}, -\Phi^{(2)}, -2^{(1)}\},$$

where $\Phi = \frac{1+\sqrt{5}}{2}$ is the golden ratio.

Proposition 10.5.2. *Let $n \geqslant 3$, and let $\Gamma = \mathrm{Cay}(\mathbb{Z}/n\mathbb{Z}, S)$ be a circulant graph such that S contains a generator of the cyclic group $\mathbb{Z}/n\mathbb{Z}$. Then Γ is Hamiltonian.*

Proof. As S contains a generator s of $\mathbb{Z}/n\mathbb{Z}$, Γ contains the Hamiltonian cycle $0, s, 2s, \ldots, (n-1)s, 0$. ∎

As we will see below, this result still holds under the weaker assumption that $\mathrm{Cay}(\mathbb{Z}/n\mathbb{Z}, S)$ is connected, i.e. that S is a set of elements such that $\langle S \rangle = \mathbb{Z}/n\mathbb{Z}$. First we give the following exercise.

Exercise 10.16. Show that for $n = 6$, $S = \{2, 3\}$ the circulant graph $\mathrm{Cay}(\mathbb{Z}/6\mathbb{Z}, S)$ is Hamiltonian.

10.5.2 Group actions on Cayley graphs

We will prove a certain characterisation of Cayley graphs involving the structure of their automorphism groups. We begin with a special case.

Proposition 10.5.3. *Let $\Gamma = \mathrm{Cay}(\mathbb{Z}/n\mathbb{Z}, S)$ be a circulant graph. Then* $\mathrm{Aut}(\Gamma)$ *contains a cyclic subgroup of order n.*

Proof. $\sigma: v \mapsto v + 1 \in \mathrm{Aut}(\mathbb{Z}/n\mathbb{Z})$ is an automorphism of Γ since if $\{v, w\}$ is an edge, we have $w - v \in S$. But $\sigma(w) - \sigma(v) = w - v$, so $\{\sigma(v), \sigma(w)\}$ is also an edge; consequently σ is of order n (details left to the reader). ∎

Recall (see Section 9.5) that a group G *acts on* X if there exists a morphism $\sigma: G \to \mathfrak{S}_X$, $\sigma(g)(v) = g \cdot v$.The action is *transitive* if it has a single orbit, and G

acts *freely* (or without fixed points) if the stabilisers are trivial, i.e. if

$$g \cdot v = g' \cdot v \implies g = g'.$$

Proposition 10.5.4. *Let* $\Gamma = \mathrm{Cay}(G, S)$ *be a Cayley graph. Then* $\mathrm{Aut}(\Gamma)$ *contains a subgroup isomorphic to* G *which acts freely and transitively on* $V(\Gamma) = G$.

Proof. Define $\sigma: G \to \mathrm{Aut}(\Gamma) \subset \mathfrak{S}_G$ by $\sigma(g): v \mapsto g \cdot v$. Then $\sigma(g) \in \mathrm{Aut}(\Gamma)$ since $\{v, w\}$ is an edge of Γ if and only if $w = s \cdot v$, which is equivalent to the fact that $\{g \cdot v, g \cdot w\}$ is an edge. This action is clearly free and transitive. ∎

Conversely, we have the following.

Theorem 10.5.5 (Sabidussi). *Let* $\Gamma = (V; E)$ *be a graph, and suppose that* $\mathrm{Aut}(\Gamma)$ *admits a subgroup* G *acting freely and transitively on* V. *Then there exists* $S \subset G$ *such that* $\Gamma \simeq \mathrm{Cay}(G, S)$.

Proof. Since the action is free, for $\sigma, \tau \in G$ we have $\sigma(v) = \tau(v) \Rightarrow \sigma = \tau$. Fix a vertex v_0 of V, and let $\sigma: V \to G$ be defined by $\sigma(v) = \sigma_v$, where σ_v is the unique element of G such that $\sigma_v(v_0) = v$.

- σ is well-defined: σ_v exists since the action is transitive, σ_v is unique since the action is free.
- σ is injective since the action is free.
- σ is surjective: let $\tau \in G$, then $\tau = \sigma_v$ where $v = \tau(v_0)$ since $\sigma_v(v_0) = v = \tau(v_0)$; since the action is free, we have $\sigma_v = \tau$.

Set $S := \sigma(\Gamma(v_0))$. Then

(a) $S = S^{-1}$: let $\sigma_v \in S$, i.e. $\{v, v_0\} \in E$. We have $\sigma_v^{-1} \in G$, so we can write $\sigma_v^{-1} = \sigma_w$ for some $w \in V$. As $G \subset \mathrm{Aut}(\Gamma)$, we have $\{\sigma_w(v), \sigma_w(v_0)\} \in E$; but $\sigma_w(v) = v_0$ and $\sigma_w(v_0) = w$, so $\{v_0, w\} \in E$ and $\sigma_w = \sigma_v^{-1} \in S$.

(b) σ is a graph isomorphism between Γ and $\mathrm{Cay}(G, S)$:

(i) let $\{v, w\} \in E$ and set $x = \sigma_v^{-1}(w)$; then $\sigma_v^{-1}(v) = v_0$ and $\sigma_v^{-1}(w) = x$, so $\{v_0, x\} \in E$ and $\sigma_x \in S$. Furthermore,

$$\sigma_x(v_0) = x = \sigma_v^{-1}(w) = \sigma_v^{-1}(\sigma_w(v_0)) = \sigma_v^{-1} \circ \sigma_w(v_0).$$

Since the action is free, we conclude that $\sigma_x = \sigma_v^{-1} \circ \sigma_w$; it follows that $\{\sigma_v, \sigma_w\} \in E(\mathrm{Cay}(G, S))$.

(ii) Conversely, if $\sigma_v^{-1} \circ \sigma_w \in S$, we have $\sigma_v^{-1} \circ \sigma_w =: \sigma_x$ where $\{v_0, x\} \in E$. Since $G \subset \mathrm{Aut}(\Gamma)$ it follows that $\{\sigma_v(v_0) = v, \sigma_v(x)\} \in E$. But $\sigma_v(x) = \sigma_v \circ \sigma_x(v_0) = \sigma_w(v_0) = w$, so $\{v, w\} \in E$.

This concludes the proof of the theorem. ∎

10.5.3 The case of the Petersen graph

The goal of this paragraph is to show that the Petersen graph P is not a Cayley graph.

We first determine the automorphisms of P.

Proposition 10.5.6. *The automorphism group* $\mathrm{Aut}(P)$ *of the Petersen graph* P *is the symmetric group* $\mathfrak{S}_5$.

Proof. By Proposition 10.4.2, $\mathrm{Aut}(P) = \mathrm{Aut}(\overline{L(K_5)}) = \mathrm{Aut}(L(K_5))$ since a graph and its complement have the same automorphism group. Let $\sigma \in \mathrm{Aut}(K_5)$ be an automorphism of K_5. Set

$$\sigma^*: \begin{cases} L(K_5) \to L(K_5), \\ \{x, y\} \mapsto \{\sigma(x), \sigma(y)\}. \end{cases}$$

This definition is coherent since via σ, every edge $\{x, y\}$ in K_5 induces an edge $\{\sigma(x), \sigma(y)\}$ in K_5. It remains to show that σ^* is an automorphism of $L(K_5)$. If two edges $\{x, y\}$ and $\{x, z\}$ are incident in K_5, i.e. if the vertices $\{x, y\}$ and $\{x, z\}$ in $L(K_5)$ are adjacent, this also holds for their images $\{\sigma(x), \sigma(y)\}$ and $\{\sigma(x), \sigma(z)\}$. Thus we have an injection

$$\varphi: \begin{cases} \mathrm{Aut}(K_5) \to \mathrm{Aut}(L(K_5)), \\ \sigma \mapsto \sigma^*. \end{cases}$$

If $\sigma, \sigma' \in \mathrm{Aut}(K_5)$, then for every edge $\{x, y\}$ of K_5 we have

$$\sigma \circ \sigma'(\{x, y\}) = \{\sigma \circ \sigma'(x), \sigma \circ \sigma'(y)\} = \sigma^*(\{\sigma'(x), \sigma'(y)\}) = \sigma^* \circ \sigma'^*(\{x, y\}).$$

Thus φ is an injective morphism.

Conversely, let $\sigma^* \in \mathrm{Aut}(L(K_5))$. The four edges $\{x, y\}, \{x, z\}, \{x, u\}, \{x, v\}$ induce a complete subgraph K_4 in $L(K_5)$, and σ^* maps this 4-clique onto a 4-clique of $L(K_5)$. There exists a vertex in K_5, denoted $\sigma^{**}(x)$, that is common to the 4 vertices of this 4-clique of $L(K_5)$. The map sending σ^* to σ^{**} is an injection of $\mathrm{Aut}(L(K_5))$ to $\mathrm{Aut}(K_5)$. Note that this map sends $\mathrm{Id}_{L(K_5)}$ to Id_{K_5}. This allows us to assert that $\mathrm{Aut}(L(K_5)) \simeq \mathrm{Aut}(K_5) \simeq \mathfrak{S}_5$, concluding the proof. ∎

Proposition 10.5.7. *The Petersen graph* P *is not a Cayley graph.*

Proof. Assume that $P = \mathrm{Cay}(G, S)$ for a certain finite group G (written multiplicatively) and a symmetric subset S of $G \backslash \{1_G\}$. By Proposition 10.5.4, G acts freely and transitively on

$$V(P) := \{0, 1, 2, 3, 4, 5, 6, 7, 8, 9\}.$$

Then $|G| = 10$. Up to isomorphism, the only groups of order 10 are the cyclic group $\mathbb{Z}/10\mathbb{Z}$ and the dihedral group D_5.

First case: $G = \mathbb{Z}/10\mathbb{Z}$. By Proposition 10.5.3, Aut(Cay(G, S)) should contain an element of order 10. But this contradicts the fact that $\mathrm{Aut}(P) = \mathfrak{S}_5$, since this group only contains elements of orders $1, 2, 3, 4$ and 5.

Second case: $G = D_5 = \langle \sigma, \tau \rangle$, with $\sigma^5 = \tau^2 = 1, \sigma\tau = \tau\sigma^{-1}$. As σ is of order 5, it is either a 5-cycle or the product of two disjoint 5-cycles. If σ is the cyclic permutation (0 1 2 3 4), set $A = \{0, 1, 2, 3, 4\}$. There is at least one edge between A and $A^c = \{5, 6, 7, 8, 9\}$; let us say it has extremities $x \in A$ and $y \in A^c$. Then we have 5 edges $\{\sigma^i(x), \sigma^i(y) = y\}$, $0 \leqslant i \leqslant 4$, so y is of degree at least 5 – but this is impossible since P is 3-regular. Thus $\sigma = (0\ 1\ 2\ 3\ 4)(5\ 6\ 7\ 8\ 9)$. Let us show that $\tau(A) = A^c = \{5, 6, 7, 8, 9\}$. Assume first that $\tau(A) \subset A$. Then since $\sigma(A) = A$, we have $D_5(A) = A$, so D_5 does not act transitively on $V(P) = A \sqcup A^c$, which is a contradiction. Thus $\tau(A) \not\subset A$, and therefore there exists $a \in A$ such that $\tau(a) \in A^c$. But then, for $0 \leqslant i \leqslant 4$, we have $\tau\sigma^i(a) = \sigma^{-i}\tau(a) \in \sigma^{-i}(A^c) = A^c$, so $\tau(A) = A^c$.

If P were bipartite, by Theorem 10.1.7, -3 would be in its spectrum since P is connected and 3-regular, which is contradicting the fact that $\mathrm{Sp}(P) = \{3^{(1)}, 1^{(5)}, -2^{(4)}\}$ (see Propositions 10.4.1 and 10.4.2). We infer that there exists at least one edge e of P between two given vertices of A. By symmetry, we can restrict ourselves to the two cases $e = \{0, 1\}$ and $e = \{0, 2\}$.

If $\{0, 1\} \in E(P)$, then $\sigma \in S$, so there is a cycle C_5 in $\Gamma(A)$, and no other edge (otherwise there would be a triangle in P); this fact can be observed from the definition of P (see page 252) but also shown by developing the characteristic polynomial of P, by noticing that the coefficient of x^7 is zero and finally by using Theorem 10.1.9 (iii). But since the vertex 0 is of degree 3, there exists $i \in \{0, 1, 2, 3, 4\}$ such that $\{0, \tau(i)\} \in E(P)$. Thus there is an edge between $\sigma(0) = 1$ and $\sigma\tau(i) = \tau(i-1)$ since $\sigma \in S$ and the action is transitive. Since these two vertices are neighbours, the cycle $C_4 = (0, 1, \tau(i-1), \tau(i), 0)$ then lies in P. But this is impossible since P contains no C_4 by its definition (see page 252).

If $\{0, 2\} \in E(P)$, we have a C_5 in $\Gamma(A)$ and no other edge for the same reason ($\sigma^2 \in S$), and $\{0, \tau(i)\} \in E(P)$ as above, giving an edge between $\sigma^2(0) = 2$ and $\sigma^2\tau(i) = \tau\sigma^{-2}(i)$, so the $C_4 = (0, 2, \tau\sigma^{-2}(i), \tau(i), 0)$, which is impossible as above. This concludes the proof. ∎

10.5.4 Hamiltonian cycles in Cayley graphs

In the early 1970s, Lovász made the conjecture that every non-oriented connected vertex-transitive graph is Hamiltonian, with the exception of five known counterexamples among which K_2 and the Petersen graph P. This conjecture is still wide open, but the Cayley graphs form a subfamily of more regular vertex-transitive graphs, and restricted to these, the conjecture might be more tractable.

Recall that a Hamiltonian cycle of a graph Γ is a cycle that passes once and only once through each vertex of Γ. The main result of this paragraph is the following statement.

Theorem 10.5.8. *Let G be an abelian group of order $\geqslant 3$, and $S \subset G\backslash\{0\}$ a symmetric subset. If $\langle S\rangle = G$, then* $\mathrm{Cay}(G, S)$ *is a Hamiltonian graph.*

Proof. The proof is constructive and consists in producing an elementary cycle that we extend step by step, taking account of each element of S. We can assume that $S = \{\pm s_1, \dots, \pm s_p\}$ is minimal, in the sense that $\langle S\backslash\{\pm s_i\}\rangle \neq G$ for all $s_i \in S$; we can also assume that their orders satisfy $o(s_1) \geqslant o(s_2) \geqslant \cdots \geqslant o(s_p) \geqslant 2$.

Set $G_k = \langle s_1, \dots, s_k\rangle$, $1 \leqslant k \leqslant p$. Then $G_1 \subsetneq G_2 \subsetneq \cdots \subsetneq G_p = G$. Set $S_k = \{\pm s_1, \dots, \pm s_k\}$. By induction, we construct a Hamiltonian cycle of $\mathrm{Cay}(G_k, S_k)$.

Case 1: $o(s_1) \geqslant 3$. We start with the cycle

$$C_1 : 0 \to s_1 \to 2s_1 \to \cdots \to r_1 s_1 \to 0,$$

where $r_1 = o(s_1) - 1$: C_1 is a Hamiltonian cycle of $\mathrm{Cay}(G_1, S_1)$.

$k \rightsquigarrow k+1$: let $r = |G_k| - 1$ and let

$$C_k : 0 \xrightarrow{+t_1} x_1 \longrightarrow \cdots \longrightarrow x_{r-1} \xrightarrow{+t_{r-1}} x_r \xrightarrow{+t_r} 0$$

be a Hamiltonian cycle of $\mathrm{Cay}(G_k, S_k)$: $x_i \in G_k$, $t_i \in S_k$.

Let $s = s_{k+1}$ and $G_k^* = G_k\backslash\{0\}$. Let $u = \min\{j \geqslant 1 \mid js \in G_k\}$. Note that $2 \leqslant u \leqslant o(s)$ since S is minimal. Let C_{k+1} be the following cycle of $\mathrm{Cay}(G_{k+1}, S_{k+1})$ (shown here when u is odd):

$$\begin{array}{ccccccccc}
0 & \xrightarrow{+t_1} & x_1 & \xrightarrow{+t_2} & x_2 & \xrightarrow{+t_3} \cdots \xrightarrow{+t_{r-1}} & & & x_r \in G_k \\
\uparrow & & & & & & & & \big\downarrow{\scriptstyle +s} \\
\vdots & & x_1 + s & \xleftarrow{-t_2} & x_2 + s & \xleftarrow{-t_3} \cdots \xleftarrow{-t_{r-1}} & & & x_r + s \in G_k^* + s \\
\vdots & & \big\downarrow{\scriptstyle +s} & & & & & & \\
\vdots & & x_1 + 2s & \xrightarrow{+t_2} & x_2 + s & \xrightarrow{+t_3} \cdots \xrightarrow{+t_{r-1}} & & & x_r + 2s \in G_k^* + 2s \\
\vdots & & & & & & & & \big\downarrow{\scriptstyle +s} \\
\vdots & & \vdots & & & & & & \vdots \\
\vdots & & \big\downarrow{\scriptstyle +s} & & & & & & \\
\vdots & & x_1 + (u-1)s & \xrightarrow{+t_2} & \cdots & & \cdots \xrightarrow{+t_{r-1}} & & x_r + (u-1)s \in G_k^* - s \\
\uparrow & & & & & & & & \big\downarrow{\scriptstyle +t_r} \\
s & \xleftarrow{-s} & 2s & \xleftarrow{-s} & \cdots & & \cdots \xleftarrow{-s} & & (u-1)s \in \langle s\rangle^*
\end{array}$$

When u is even, the elements of $G_k^* - s$ are traversed from $x_r + (u-1)s$ to $x_1 + (u-1)s$, which is connected to $(u-1)s$ by the edge $-t_1$; the cycle then closes as above.

The vertices of C_{k+1} are all distinct since all $x \in G_{k+1}$ can be written in a unique way as $x = \lambda s + y, 0 \leqslant \lambda < u, y \in G_k$. Indeed, otherwise we would have $\lambda s + y = \lambda' s + y'$ with $y, y' \in G_k$ and $0 \leqslant \lambda, \lambda' < u$, then $|\lambda - \lambda'| s \in G_k$, and finally $\lambda - \lambda' = 0$ by the definition of u. Thus we have a Hamiltonian cycle C_{k+1} in $\mathrm{Cay}(G_{k+1}, S_{k+1})$. The construction ends once we have exhausted all the s_i of S.

Case 2: $o(s) = 2$ for all $s \in S$ (so $p \geqslant 2$). Here we start the construction with a Hamiltonian cycle of $\mathrm{Cay}(G_2, S_2)$:

$$C_2 \colon 0 \to s_1 \to s_1 + s_2 \to s_2 \to 0,$$

but it then continues exactly as in the preceding case. ∎

It has been conjectured by Parsons and others that the above theorem holds for all connected Cayley graphs.

Conjecture 10.5.9. *Let G be a finite group of order $n \geqslant 3$. Then every connected Cayley graph* $\mathrm{Cay}(G, S)$ *is a Hamiltonian graph.*

We already showed that the Petersen graph P is not Hamiltonian by computing its chromatic index (see the remark following Proposition 7.2.4). As a corollary to the preceding theorem, we now deduce that P is not a Cayley graph of the type $\mathrm{Cay}(G, S)$ with G abelian, since otherwise P would be Hamiltonian.

To show that P contains no Hamiltonian cycle, we could also use the information on the interlacing of spectra of induced subgraphs of a connected graph (see Proposition 10.1.5). Suppose that P contains a cycle C_{10}; then $L(C_{10})$ is the induced subgraph of $L(P)$ by Proposition 2.4.17. We will show a defect of interlacing between the respective spectra of $L(P)$ and $L(C_{10})$. We saw in Proposition 10.4.1 that

$$\mathrm{Sp}(P) = (3, 1, 1, 1, 1, 1, -2, -2, -2, -2).$$

We have $L(C_{10}) = C_{10}$ since C_{10} is an elementary cycle, so by Exercise 10.15, we have

$$\mathrm{Sp}(L(C_{10})) = (2, \Phi, \Phi, \Phi - 1, \Phi - 1, 1 - \Phi, 1 - \Phi, -\Phi, -\Phi, -2)$$

where Φ is the golden ratio. Applying Proposition 10.3.1 to the graph P, we find that

$$\mathrm{Sp}(L(P)) = (4, 2, 2, 2, 2, 2, -1, -1, -1, -1, -2, -2, -2, -2, -2).$$

Finally, apply Proposition 10.1.5: then $n = 15$, $q = 10$ and we choose $k = 7$, which gives a contradiction since $1 - \Phi > -1$.

Chapter 11

Random graphs

Graphs that have several hundred thousand vertices and edges arise more and more frequently nowadays, for example in modelling social media, the internet, etc. Finding the answer to problems such as determining whether a graph is Hamiltonian or computing its edge-connectedness appears very difficult on these extremely *large* graphs. All the results and notions introduced in the preceding chapters remain valid in the context of large graphs; indeed, we saw earlier that some of these questions can be difficult, even for graphs having few vertices or edges, although in practice a small number of vertices does make it possible to attack problems with brute force. However, classical deterministic techniques and algorithms are essentially impossible to use for graphs with *large parameters*.

Instead of seeking for exact answers, the technique we now introduce consists in restricting ourselves to seeking *statistical properties* of given families of graphs. In this spirit, we devote this chapter to the study of random models. These models depend on the nature of the question asked, and also on the characteristics of the graphs in each family. More precisely, for a given family of graphs, the associated random model will depend on the invariants common to all graphs in the family.

The first task is to select the most relevant data types, then to measure whether certain properties are satisfied in terms of local parameters, such as for example:

(i) the degrees of vertices are small with respect to the number of vertices; these are *low density* graphs: $\max_{x\in V} d(x) \ll |V|$;

(ii) the distance between two randomly chosen vertices is small; these are *small world* graphs;

(iii) the distribution of degrees follows a given probability distribution;

and so forth.

In this chapter, we introduce some specific models of random graphs, and concentrate our attention on their *generic* properties. A brief presentation of the basic elements of probability theory is given at the end of the chapter.

11.1 Models of random graphs

A *random graph* is an object obtained thanks to a random process, which is typically a random sampling in a family of graphs characterised by certain given combinatorial invariants shared by all graphs in the family; examples of such invariants could be the order, the distribution of degrees, the number of edges, or the degree of regularity. Random graph theory is thus situated at the interface of combinatorics, probability theory and statistics.

The probability theory required in this context is fairly elementary, mainly based on intuitive considerations. It contributes a new point of view to the objects under consideration (graphs), which become *random* objects, with a distribution based on a probability distribution appropriate for the problem under consideration. We call these objects *models*, with all the implicit underlying assumptions associated to the term. This approach can also be viewed as an attempt to give an *average* interpretation of certain combinatorial questions concerning graphs, such as for example: what proportion of graphs of order n with m edges admit a triangle?

We will study two historical models, introduced in the 1950s by Erdős and Renyi [12] and by Gilbert [16]. Here is the construction procedure for a random graph model. We first fix a sample space $\mathscr{G}$, assumed finite, containing the type of graph we wish to study. Then we attribute a weight, which is an elementary probability, to each graph in $\mathscr{G}$. The most natural, most intuitive choice is a straightforward *uniform distribution*:

$$\mathbf{P}(\{\Gamma\}) = \frac{1}{|\mathscr{G}|} \quad \text{for all } \Gamma \in \mathscr{G}.$$

For every event (i.e. subset) $A \subset \mathscr{G}$, we thus have $\mathbf{P}(A) = \frac{|A|}{|\mathscr{G}|}$. This defines a *probability measure*

$$\mathbf{P}\colon P(\mathscr{G}) \to [0,1]$$

since we have $\mathbf{P}(A \cup B) = \mathbf{P}(A) + \mathbf{P}(B)$ if $A \cap B = \emptyset$ and $\mathbf{P}(\mathscr{G}) = 1$. This probability model is called the *uniform model* or *uniform graph* on $\mathscr{G}$.

A random variable on the probability space $\mathscr{G}$ is nothing other than a map

$$X\colon \mathscr{G} \to \mathbb{R}.$$

As a first example, we consider the uniform graph on $\mathscr{G}(n)$, where $\mathscr{G}(n)$ is the sample space of graphs with n vertices denoted $[n] = \{1, 2, \dots, n\}$: we then have

$$\mathbf{P}(\{\Gamma\}) = \frac{1}{2^N},$$

where $N := \binom{n}{2}$. However, this model is too vast and too generic to display interesting statistical properties.

The *Erdős–Renyi model* is the uniform model on $\mathscr{G}(n,m)$, the set of graphs Γ (up to isomorphism) of order n ($V(\Gamma) = [n]$) with m edges where $m \leqslant N = \binom{n}{2}$. Here we have $|\mathscr{G}(n,m)| = \binom{N}{m}$, and the model is a *uniform random graph*. It is usually just denoted by $\mathscr{G}(n,m)$, but in fact it is the probability space $(\mathscr{G}(n,m), P(\mathscr{G}(n,m)), \mathbf{P})$ equipped with the uniform distribution. We can consider random variables on this space, for example the random variable $\mathfrak{D}_i$ defined by $\mathfrak{D}_i(\Gamma) = \deg_\Gamma(i)$ for $i \in [n]$, and seek for the probability distribution of the variable $\mathfrak{D}_i$, which is determined by the probabilities $\mathbf{P}(\mathfrak{D}_i = k)$, $k \in \mathbb{N}$.

The *Gilbert model* is vaster, and concerns all graphs of fixed order n: it is based on a real parameter $p \in [0, 1]$ which induces a distribution on the graphs as a function of their number of edges, that is dependent on p. We equip the sample space of graphs of order n with the following probability distribution: if Γ has m edges, then

$$\mathbf{P}(\{\Gamma\}) = p^m (1-p)^{N-m}.$$

Since we have

$$\sum_{m=1}^{N} \binom{N}{m} p^m (1-p)^{N-m} = 1$$

by Newton's binomial formula, $\mathbf{P}$ is a probability distribution depending on the parameter p; the random model is thus the probability space

$$\mathcal{G}(n; p) := (\mathcal{G}(n), P(\mathcal{G}(n)), \mathbf{P})$$

(usually denoted simply $\mathcal{G}(n; p)$).

Let us show that this construction comes down to giving a probability p to the existence of an edge ij between any two vertices $i, j \in [n]$. Let $P_2([n])$ denote the set of pairs $e = \{i, j\} \subset [n]$. For each element $e \in P_2([n])$, let ξ_e be the Bernoulli random variable defined by $\xi_e(\Gamma) = 1$ if e is an edge of Γ and 0 otherwise. Then we have the disjoint union

$$\{\xi_e = 1\} = \bigcup_{m=0}^{N} \{\Gamma \in \mathcal{G}(n) \mid |E(\Gamma)| = m \text{ and } e \in E(\Gamma)\}.$$

The number of graphs of order n having $m \geqslant 1$ edges one of which is ij is $\binom{N-1}{m-1}$. Hence

$$\begin{aligned}
\mathbf{P}(\xi_e = 1) &= \sum_{m=1}^{N} \binom{N-1}{m-1} p^m (1-p)^{N-m} \\
&= \sum_{m=0}^{N-1} \binom{N-1}{m} p^{m+1} (1-p)^{N-m} \\
&= p \sum_{m=0}^{N-1} \binom{N-1}{m} p^m (1-p)^{N-m} \\
&= p,
\end{aligned}$$

again by Newton's formula.

It remains to prove the independence of the ξ_e's. We proceed by induction. Let A, B be disjoint subsets of $P_2([n])$, and set

$$\mu_{A,B} = \mathbf{P}(\forall e \in A,\ \xi_e = 1 \text{ and } \forall e \in B,\ \xi_e = 0).$$

If $|A| + |B| = N$, then $\mu_{A,B}$ is the probability of the elementary event $\{\Gamma\}$, where Γ is the graph such that $E(\Gamma) = A$. Therefore

$$\mu_{A,B} = p^{|A|}(1-p)^{N-|A|} = \prod_{e\in A} \mathbf{P}(\xi_e = 1) \times \prod_{e\in B} \mathbf{P}(\xi_e = 0),$$

which initialises the descending induction.

If $|A| + |B| < N$, we fix $e' \notin A \cup B$, and we obtain

$$\mu_{A,B} = \mu_{A\cup\{e'\},B} + \mu_{A,B\cup\{e'\}}.$$

Applying the induction hypothesis to each of the two terms on the right-hand side, we deduce that

$$\begin{aligned}\mu_{A,B} = &\prod_{e\in A\cup\{e'\}} \mathbf{P}(\xi_e = 1) \times \prod_{e\in B} \mathbf{P}(\xi_e = 0)\\ &+ \prod_{e\in A} \mathbf{P}(\xi_e = 1) \times \prod_{e\in B\cup\{e'\}} \mathbf{P}(\xi_e = 0),\end{aligned}$$

which concludes the proof, taking $\mathbf{P}(\xi_{e'} = 1) + \mathbf{P}(\xi_{e'} = 0) = 1$ into account.

In general, we fix a probability space $(\Omega, P(\Omega), \mu)$, on which we consider independent Bernoulli random variables ξ_e, for $e \in P_2([n])$:

$$\mu(\xi_e = 1) = p.$$

For every subset A of $P_2([n])$ of cardinality $|A| = m$, let Γ be the event

$$\Gamma = \bigcap_{e\in A}\{\xi_e = 1\} \cap \bigcap_{e\notin A}\{\xi_e = 0\}.$$

Then

$$\begin{aligned}\mu(\Gamma) &= \prod_{e\in A} \mathbf{P}(\xi_e = 1) \times \prod_{e\notin A} \mathbf{P}(\xi_e = 0)\\ &= \underbrace{ppp\cdots p}_{m \text{ times}}\,\underbrace{(1-p)(1-p)(1-p)\cdots(1-p)}_{N-m \text{ times}}\\ &= p^m(1-p)^{N-m},\end{aligned}$$

thanks to the independence of the ξ_e's.

From the above, we see that for each $\Gamma \in \mathscr{G}(n)$, $|E(\Gamma)|$ denotes the value taken at Γ by the random variable $\mathfrak{M}\colon \mathscr{G}(n) \to \mathbb{N}$ which counts the edges of Γ:

$$\mathfrak{M}(\Gamma) = \sum_{e \in P_2([n])} \xi_e(\Gamma).$$

It is a sum of N independent Bernoulli random variables all having the same parameter p, so we see that $\mathfrak{M}$ follows a binomial distribution with parameters N and p:

$$\mathbf{P}(\mathfrak{M} = m) = \binom{N}{m} p^m (1-p)^{N-m}.$$

The graph model $\mathscr{G}(n; p)$ is also called the *binomial random graph.*

There are many other random variables concerning other classical invariants of graphs, such as

- $\mathfrak{D}_i(\Gamma)$: the degree of a vertex i in Γ,
- $\mathfrak{X}(\Gamma)$: the chromatic number of Γ,
- $\mathfrak{O}(\Gamma)$: the maximal order of a clique of Γ,
- $\mathfrak{A}(\Gamma)$: the maximum cardinality of an independent set of Γ,
- $\mathfrak{T}(\Gamma)$, $\mathfrak{C}_k(\Gamma)$: number of triangles, number of k-cycles in Γ.

Certain properties only make sense or become clear when the number of vertices n is very large. One major task thus consists in studying the behaviour of random models, above all Gilbert's random model, as n tends to infinity. It will be convenient to make the parameter $p = p_n$ of the model $\mathscr{G}(n; p)$ depend on n, for example by asymptotically controlling the product np_n as n tends to infinity.

For example, let $\mathfrak{T}_n$ be the random variable that counts the triangles of Γ in the model $\mathscr{G}(n; p_n)$. Clearly we have

$$\mathfrak{T}_n = \sum_{\substack{\{i,j,k\} \subset [n] \\ i,j,k \text{ distinct}}} \xi_{\{i,j\}} \xi_{\{j,k\}} \xi_{\{k,i\}},$$

hence by linearity of the expected value and independence of the ξ_e, we have

$$\mathbf{E}(\mathfrak{T}_n) = \binom{n}{3} p^3.$$

If we choose $p_n \sim c/n$ for a fixed $c > 0$, we see that $\mathbf{E}(\mathfrak{T}_n) \to c^3/6$ when $n \to +\infty$. In the case where c is *small*, we deduce that

$$\mathbf{P}(\mathfrak{T}_n \geqslant 1) \leqslant \mathbf{E}(\mathfrak{T}_n) \sim \frac{c^3}{6} \quad \text{as } n \to +\infty,$$

so the random graph asymptotically has a high probability of having no triangles.

The study of the asymptotic behaviour of $\mathfrak{T}_n$ in this model when c is *large* necessitates the estimation of its variance, which requires more care. Let $\mathcal{E}$ be the 3-subsets of $[n]$, and for $E = \{i, j, k\} \in \mathcal{E}$, set $\xi_E = \xi_{\{i,j\}}\xi_{\{j,k\}}\xi_{\{k,i\}}$. Then

$$\mathfrak{T}_n^2 = \sum_{E_1, E_2 \in \mathcal{E}} \xi_{E_1}\xi_{E_2}.$$

We have $\mathbf{E}(\xi_{E_1}\xi_{E_2}) = p^{6-t}$ if $|E_1 \cap E_2| = t$, $t = 0, 1, 2, 3$. Hence

$$\mathbf{E}(\mathfrak{T}_n^2) = \binom{n}{3}\binom{n-3}{3}p^6 + 3\binom{n}{3}\binom{n-3}{2}p^5 + 3\binom{n}{3}\binom{n-3}{1}p^4 + \binom{n}{3}p^3$$

which leads to

$$\mathbf{V}(\mathfrak{T}_n) = \binom{n}{3}p^3(1-p)\Big(\frac{3n^2p^2}{2} + \frac{3np(2-5p)}{2} + (1 - 8p + 10p^2)\Big).$$

When $p = p_n \sim c/n$ when $n \to +\infty$, we have

$$\mathbf{V}(\mathfrak{T}_n) \sim \frac{c^3}{6} \times \Big(\frac{3c^2}{2} + 3c + 1\Big).$$

Hence by the Bienaymé–Chebyshev inequality, we have

$$\mathbf{P}(\mathfrak{T}_n = 0) \leqslant \mathbf{P}\big(|\mathfrak{T}_n - \mathbf{E}(\mathfrak{T}_n)| \geqslant \mathbf{E}(\mathfrak{T}_n)\big) \leqslant \frac{\mathbf{V}(\mathfrak{T}_n)}{\mathbf{E}(\mathfrak{T}_n)^2} \sim \frac{3(2 + 6c + 3c^2)}{c^3}$$

as $n \to +\infty$. Thus the greater the value of c, the higher the probability of having a triangle in the random graph, which is exactly the opposite of what happens when c is *small*.

Thus the *asymptotic order* $p_n \asymp 1/n$, i.e. $1/n \ll p_n \ll 1/n$, appears to correspond to a level of equilibrium for the variable $\mathfrak{T}_n$ between two obviously contrary behaviours: it reveals kind of *phase transition*, a notion that we will study further below in Section 11.5.

11.2 Statistical properties of graphs

Recall that $\mathcal{G}(n, m) = \{\Gamma \in \mathcal{G}(n) \mid E(\Gamma) = m\}$. The following proposition shows that for all $0 < p < 1$, the uniform random model on m edges corresponds to the random binomial model depending on the event $\mathfrak{M} = m$.

Proposition 11.2.1. *Let* $0 < p < 1$. *Then*

$$\mathbf{P}(\Gamma \mid \mathfrak{M} = m) = \begin{cases} \frac{1}{\binom{N}{m}} & \text{if } \Gamma \in \mathcal{G}(n, m), \\ 0 & \text{otherwise.} \end{cases}$$

Writing

$$\mathbf{P}_m(A) = \mathbf{P}(A \mid \mathfrak{M} = m) \quad \textit{for all } A \subset \mathcal{G}(n),$$

the probability space $(\mathcal{G}(n), P(\mathcal{G}(n)), \mathbf{P}_m)$ *coincides with the model* $\mathcal{G}(n, m)$.

Proof. Let $\Gamma \in \mathcal{G}(n, m)$ and $0 < p < 1$. Then

$$\mathbf{P}_m(\{\Gamma\}) = \frac{\mathbf{P}(\{\Gamma\})}{\mathbf{P}(\mathfrak{M} = m)} = \frac{p^m(1-p)^{N-m}}{\binom{N}{m} p^m (1-p)^{N-m}} = \frac{1}{\binom{N}{m}}.$$

If $\Gamma \notin \mathcal{G}(n, m)$, then $\Gamma \notin \mathfrak{M}^{-1}(\{m\})$ so $\mathbf{P}_m(\{\Gamma\}) = 0$. ■

Corollary 11.2.2. *Let* $0 < p < 1$. *The probability space* $\mathcal{G}(n; p)$ *with binomial distribution is the disjoint union of* $N + 1$ *uniform subspaces:*

$$\mathcal{G}(n; p) = \bigsqcup_{m=0}^{N} \mathcal{G}(n, m).$$

Proposition 11.2.3. *The expected value and the variance of the random variable* $\mathfrak{M}$ *are given by* $\mathbf{E}(\mathfrak{M}) = Np$ *and* $\mathbf{V}(\mathfrak{M}) = Np(1-p)$.

Proof. $\mathfrak{M}$ follows a binomial distribution with parameters N and p, which gives the result. ■

Proposition 11.2.4. *Let* $i \in [n]$, *and let* $\mathfrak{D}_i : \mathcal{G}(n) \to \mathbb{N}$ *be the random variable indicating the degree of the vertex* i. *Then*

$$\mathbf{P}(\mathfrak{D}_i = d) = \binom{n-1}{d} p^d (1-p)^{n-1-d}.$$

$\mathfrak{D}_i$ *follows a binomial distribution with parameters* $n - 1$ *and* d. *In particular,*

$$\mathbf{E}(\mathfrak{D}_i) = (n-1)p, \quad \mathbf{V}(\mathfrak{D}_i) = (n-1)p(1-p).$$

Proof. We have

$$\mathfrak{D}_i = \sum_{\substack{j \in [n] \\ j \neq i}} \xi_{\{i,j\}}.$$

Thus $\mathfrak{D}_i$ is the sum of $n - 1$ independent Bernoulli random variables, which gives the result. We can also reason combinatorially: if i has exactly d neighbours, then apart from i itself, there are $n - 1 - d$ other vertices that are not neighbours of i. Thus there are $\binom{n-1}{d}$ sets of different possible neighbours for i, and each of them occurs with a probability of $p^d(1-p)^{n-1-d}$; this gives another proof of the result. ■

As a corollary, we obtain that under the hypothesis $\lim_{n\to\infty} np_n = \lambda$, the random variable $\mathfrak{D}_i$ has asymptotic Poisson behaviour (see Section 11.8.4):

$$\mathbf{P}(\mathfrak{D}_i = d) \sim e^{-\lambda}\frac{\lambda^d}{d!} \quad \text{when } n \to +\infty.$$

Exercise 11.1. Prove that

$$\mathbf{P}(\mathfrak{D}_i = 0,\, i = 1,\dots,n) = \sqrt{\prod_{i=1}^{n} \mathbf{P}(\mathfrak{D}_i = 0)}.$$

11.3 Properties asymptotically satisfied by almost all graphs

When $\mathcal{A}$ is a certain property satisfied by a graph Γ, for example being connected, being a tree, etc., we sometimes express this by writing $\Gamma \in \mathcal{A}$. In a fixed random model, for example the Gilbert model $\mathcal{G}(n; p)$, a property $\mathcal{A}$ is simply a subset of $\mathcal{G}(n)$. Thus, evaluating $\mathcal{A}$ consists in computing $\mathbf{P}(\mathcal{A})$, or at least estimating it asymptotically as $n \to +\infty$.

We will say that *asymptotically almost all graphs* $\Gamma \in \mathcal{G}(n)$ *satisfy the property* $\mathcal{A}$ if

$$\lim_{n\to+\infty} \mathbf{P}(\mathcal{A}) = 1.$$

Proposition 11.3.1. *Let* $0 < p < 1$. *Then asymptotically almost all graphs in the model* $\mathcal{G}(n; p)$ *have a chain of length* 2 *between every pair of vertices* i, j.

Proof. Let $k \in [n]\backslash\{i, j\}$. Then the probability that $\{i, k\}$ or $\{j, k\}$ is not an edge is equal to $1 - p^2$. Thus

$$\mathbf{P}\Big(\bigcap_{\substack{k\in[n]\\ k\neq i,j}} \{\xi_{\{i,k\}}\xi_{\{j,k\}} = 0\}\Big) = \prod_{\substack{k\in[n]\\ k\neq i,j}} \mathbf{P}(\xi_{\{i,k\}}\xi_{\{j,k\}} = 0) = (1 - p^2)^{n-2}.$$

It follows that

$$\mathbf{P}(\exists\{i, j\} \subset [n] \text{ such that } \forall k \in [n]\backslash\{i, j\},\ \xi_{\{i,k\}}\xi_{\{j,k\}} = 0) \leqslant \binom{n}{2}(1 - p^2)^{n-2},$$

a quantity that tends to 0 when $n \to +\infty$.

Note here that it suffices to take $p = p_n$ *very large* compared to $(2\log n/n)^{1/2}$ in order for asymptotically almost all graphs in the model $\mathcal{G}(n; p_n)$ to satisfy the stated property. ■

Corollary 11.3.2. *Let* $0 < p < 1$. *Then asymptotically almost all graphs in the model* $\mathcal{G}(n; p)$ *are connected.*

Proof. This follows directly from Proposition 11.3.1. ■

We saw earlier that asymptotically almost all the graphs in the model $\mathscr{G}(n; p)$ possess a triangle when $0 < p < 1$ is fixed. We can easily see that this property should be extendable to the cycles C_k, $k \geqslant 3$, and hence to the cliques K_k, $k \geqslant 3$.

Below we consider the general question of the existence in probability of a subgraph isomorphic to a given graph Γ_0. Observe that when n tends to infinity, the order of Γ_0 does not change: the result does not imply the existence in probability of an elementary cycle of length n for every graph.

Theorem 11.3.3. *Let $0 < p < 1$ and Γ_0 an arbitrary graph. Then asymptotically almost all graphs Γ in the model $\mathscr{G}(n; p)$ admit a subgraph isomorphic to Γ_0.*

Proof. Let Γ_0 be of order n_0. Taking n sufficiently large, we may assume that $n_0 \leqslant n$ and $V_0 = V(\Gamma_0) \subset [n]$. Consider the induced subgraph $\Gamma(V_0)$ of the random graph Γ. We need to measure the probability that $\Gamma(V_0)$ is isomorphic to Γ_0. More precisely, we extract $\lfloor \frac{n}{n_0} \rfloor$ disjoint subsets V_t of size n_0 from $[n]$; we also fix an order on the subsets $\{i, j\} \in P_2(V_t)$. The probability that there exists an isomorphism between $\Gamma(V_t)$ and Γ_0, compatible with the orders between the edges, is equal to

$$p_0 = p^{m_0}(1-p)^{N_0 - m_0} \in]0, 1[, \quad \text{where } N_0 = \binom{n_0}{2} \text{ and } m_0 = |E(\Gamma_0)|.$$

It follows that the probability that none of the $\Gamma(V_t)$ is isomorphic to Γ_0 is at most equal to

$$(1 - p_0)^{\lfloor \frac{n}{n_0} \rfloor},$$

which tends to 0 when $n \to +\infty$ since $0 < p_0 < 1$. ■

Lemma 11.3.4. *Suppose $0 < p < 1$. Let r, s, t be positive integers. Then asymptotically almost all graphs Γ in the model $\mathscr{G}(n; p)$ satisfy the following property: for every pair of disjoint subsets $R, S \subset [n]$ of respective cardinalities r and s, there exists a subset $T \subset [n]$ of cardinality t such that $\{x, z\} \in E(\Gamma)$ for all $(x, z) \in R \times T$ and $\{y, z\} \notin E(\Gamma)$ for all $(y, z) \in S \times T$.*

Proof. The probability that an arbitrary subset T of cardinality t in $[n]$, disjoint from both R and S, has the property of complete adjacency with R and complete non-adjacency with S (i.e. $d(x, z) = 1$ and $d(y, z) \geqslant 2$ for all $x \in R$, $y \in S$ and $z \in T$) is equal to

$$p' = p^{rt}(1-p)^{st}.$$

From the complement $\overline{R \cup S}$ of $R \cup S$ in $[n]$, we can extract $\lfloor \frac{n-r-s}{t} \rfloor$ disjoint subsets T_i of cardinality t. The probability that none of the T_i satisfies the desired result is equal to

$$(1 - p')^{\lfloor \frac{n-r-s}{t} \rfloor},$$

which tends to 0 when $n \to +\infty$ since $0 < p' < 1$. ■

Corollary 11.3.5. *Let $0 < p < 1$ and $k \in \mathbb{N}$. Then asymptotically almost all the graphs in the model $\mathcal{G}(n; p)$ are k-connected.*

Proof. It follows from Lemma 11.3.4 with $r = 2$, $s = 0$ and $t = k$. ■

11.4 Main structural properties of random graphs

Naturally, we would like to study the asymptotic behaviour of the main invariants of graph theory.

Proposition 11.4.1. *Let $0 < p < 1$ and n, k be two integers such that $n \geqslant k \geqslant 2$. Recall that $\mathfrak{A}\colon \mathcal{G}(n) \to \mathbb{N}$ is the random variable indicating the maximum cardinality of an independent set in the model $\mathcal{G}(n; p)$. Then*

$$\mathbf{P}(\mathfrak{A} \geqslant k) \leqslant \binom{n}{k}(1-p)^{\binom{k}{2}}.$$

Proof. Let $S \subseteq [n]$, $|S| = k$. The probability that S is an independent set equals $(1-p)^{\binom{k}{2}}$. But there are $\binom{n}{k}$ possibilities for such a set, which gives the result. ■

Proposition 11.4.2. *Let $0 < p < 1$ and n, k be two integers such that $n \geqslant k \geqslant 2$. Recall that $\mathfrak{O}\colon \mathcal{G}(n) \to \mathbb{N}$ is the random variable indicating the maximum order of a clique in the model $\mathcal{G}(n; p)$. Then*

$$\mathbf{P}(\mathfrak{O} \geqslant k) \leqslant \binom{n}{k} p^{\binom{k}{2}}.$$

Proof. For every graph Γ, we have $\omega(\Gamma) = \alpha(\overline{\Gamma})$. It follows that if Γ is in $\mathcal{G}(n; p)$, then $\overline{\Gamma}$ is in $\mathcal{G}(n; 1-p)$. Thus by Proposition 11.4.1, we have

$$\mathbf{P}(\mathfrak{O} \geqslant k) \leqslant \binom{n}{k}(1-(1-p))^{\binom{k}{2}}$$

and the result follows. ■

Exercise 11.2. Prove that if $p = p_n = o(n^{2/3})$, then $\mathbf{P}(\mathfrak{O} \geqslant 4) = o(1)$ as $n \to +\infty$.

One can prove by induction the following estimate of binomial coefficients.

Lemma 11.4.3. *Let $1 \leqslant k \leqslant n$ be two integers. We have the inequality*

$$\binom{n}{k} \leqslant \frac{e^k n^k}{k^k}.$$

We obtain

$$\binom{n}{k}(1-p)^{\binom{k}{2}} \leqslant \exp\Big(k\Big(1 - \log k + \log n + \frac{(k-1)\log(1-p)}{2}\Big)\Big);$$

we then find that for all $\varepsilon > 0$ and every integer $k \geqslant \frac{(2+\varepsilon)\log n}{-\log(1-p)}$,

$$\mathbf{P}(\mathfrak{A} \geqslant k) = \exp\Big(-\frac{\varepsilon k \log n}{2} + O(1)\Big) = o(1)$$

as $n \to +\infty$. This gives the following corollary to Proposition 11.4.1.

Corollary 11.4.4. *Let $0 < p < 1$, $\varepsilon > 0$ and let n, k be two integers such that $k \geqslant \frac{(2+\varepsilon)\log n}{-\log(1-p)}$. Then in the model $\mathcal{G}(n; p)$ we have*

$$\mathbf{P}(\mathfrak{A} \geqslant k) = o(1) \ \textit{as} \ n \to +\infty.$$

Intuitively, if $0 < p < 1$ the random graph should contain a rather large independent set, which should yield a result symmetric to that of the proposition above. Indeed, we have the following.

Proposition 11.4.5. *Let $0 < p < 1$, $\varepsilon > 0$ and let n, k be two integers such that $k \leqslant \frac{(2-\varepsilon)\log n}{-\log(1-p)}$. Then in the model $\mathcal{G}(n; p)$ we have*

$$\mathbf{P}(\mathfrak{A} \leqslant k) = o(1) \ \textit{as} \ n \to +\infty.$$

Proof. For each k-subset S of $[n]$, let η_S be the Bernoulli random variable which is equal to 1 if S is an independent set in the random graph Γ. Let E_S denote the event: *S is an independent set.* Let

$$\mathfrak{S} = \sum_{\substack{S \subset [n] \\ |S| = k}} \eta_S$$

be the random variable which counts the stable k-subsets of Γ. We have

$$\mathbf{E}(\mathfrak{S}) = \sum_{\substack{S \subset [n] \\ |S| = k}} \mathbf{P}(E_S) = \binom{n}{k}(1-p)^{\binom{k}{2}}$$

and

$$\mathbf{V}(\mathfrak{S}) = \sum_{\substack{S, T \subset [n] \\ |S| = |T| = k}} (\mathbf{P}(E_S \cap E_T) - \mathbf{P}(E_S)\mathbf{P}(E_T)).$$

If $|S \cap T| = 0$ or 1, then $\Gamma(S)$ and $\Gamma(T)$ have no common edge, so E_S and E_T are independent events; thus we have

$$\mathbf{V}(\mathfrak{S}) \leqslant \sum_{\substack{S,T\subset[n]\\|S|=|T|=k\\2\leqslant|S\cap T|\leqslant k}} \mathbf{P}(E_S \cap E_T).$$

Let R_n denote the right-hand side. Then

$$R_n = \sum_{\substack{S\subset[n]\\|S|=k}} \mathbf{P}(E_S) \sum_{j=2}^{k} \sum_{\substack{T\subset[n]\\|T|=k\\|S\cap T|=j}} \mathbf{P}(E_T \mid E_S).$$

When S, T are k-subsets of $[n]$ such that $|S \cap T| = j \geqslant 2$, we have

$$\mathbf{P}(E_T \mid E_S) = (1-p)^{\binom{k}{2}-\binom{j}{2}},$$

hence

$$\begin{aligned} R_n - \mathbf{E}(\mathfrak{S}) &= \mathbf{E}(\mathfrak{S}) \times \sum_{j=2}^{k-1} \binom{k}{j}\binom{n-k}{k-j}(1-p)^{\binom{k}{2}-\binom{j}{2}} \\ &= \mathbf{E}(\mathfrak{S})^2 \times \sum_{j=2}^{k-1} F_k(j), \end{aligned}$$

with

$$F_k(j) = \frac{\binom{k}{j}\binom{n-k}{k-j}}{\binom{n}{k}}(1-p)^{-\binom{j}{2}} \leqslant \frac{k^{2j}}{(1-p)^{j^2/2} n^j},$$

where this inequality holds for all $2 \leqslant j \leqslant k-1$ (although checking this is somewhat tedious). By setting $\theta = -\log(1-p)$, we obtain

$$\log F_k(j) \leqslant j\Big(2\log k - \log n + \frac{j\theta}{2}\Big).$$

The right-hand term is a second degree polynomial in j, so it reaches its maximum on $[2, k-1]$ at $j = 2$ or $j = k-1$:

$$\begin{aligned} \log F_k(2) &\leqslant 4\log k - 2\log n + \theta, \\ \log F_k(k-1) &\leqslant (k-1)\Big(2\log k - \log n + \frac{(k-1)\theta}{2}\Big). \end{aligned}$$

If we assume that $k \leqslant \frac{(2-\varepsilon)\log n}{\theta}$, then the bound for the term $F_k(2)$ is dominant, so

$$\frac{R_n - \mathbf{E}(\mathfrak{S})}{\mathbf{E}(\mathfrak{S})^2} \leqslant k\exp(4\log k - 2\log n + \theta) \leqslant n^{-2+o(1)},$$

hence $R_n = o(\mathbf{E}(\mathfrak{S})^2)$ when $n \to +\infty$. The Bienaymé–Chebyshev inequality then gives

$$\mathbf{P}(\mathfrak{S} = 0) \leqslant \mathbf{P}\big(|\mathfrak{S} - \mathbf{E}(\mathfrak{S})| \geqslant \mathbf{E}(\mathfrak{S})\big) \leqslant \frac{\mathbf{V}(\mathfrak{S})}{\mathbf{E}(\mathfrak{S})^2} = o(1)$$

as $n \to +\infty$. ■

As a corollary to these propositions, we obtain that asymptotically almost all the graphs Γ in the model $\mathcal{G}(n; p)$ satisfy

$$\alpha(\Gamma) \sim \frac{2\log n}{-\log(1-p)}.$$

The notation of this asymptotic equivalence is slightly abusive: in fact we have simply proved that for all $\varepsilon > 0$,

$$\mathbf{P}\Big(\Big|\frac{-\log(1-p)}{2\log n} \times \mathfrak{A} - 1\Big| \leqslant \varepsilon\Big) \xrightarrow[n\to+\infty]{} 1.$$

Since $\omega(\Gamma) = \alpha(\overline{\Gamma})$, we also find (with the same slight abuse) that

$$\omega(\Gamma) \sim \frac{2\log n}{-\log p}$$

asymptotically for almost all graphs Γ of the model $\mathcal{G}(n; p)$.

Proposition 11.4.6. *Let $0 < p < 1$ and let n, k be two integers such that $n \geqslant k \geqslant 2$. Let $\mathfrak{C}_k : \mathcal{G}(n) \to \mathbb{N}$ be the random variable indicating the number of k-cycles in the model $\mathcal{G}(n; p)$. Then*

$$\mathbf{E}(\mathfrak{C}_k) = \frac{n(n-1)(n-2)\dots(n-k+1)}{2k} p^k.$$

Proof. Let $\mathcal{C}_k$ be the set of all k-cycles of K_n (the complete graph on $[n]$), and for each $C \in \mathcal{C}_k$, let $\zeta_C : \mathcal{G}(n) \to \{0, 1\}$ be the Bernoulli random variable that indicates whether C is a subgraph of the random graph Γ. Then $\mathfrak{C}_k = \sum_{C\in\mathcal{C}_k} \zeta_C$, so by linearity of the expected value we have

$$\mathbf{E}(\mathfrak{C}_k) = \sum_{C\in\mathcal{C}_k} \mathbf{E}(\zeta_C) = \sum_{C\in\mathcal{C}_k} \mathbf{P}(\zeta_C = 1).$$

A k-cycle on $[n]$ corresponds to a k-tuple $(i_1, i_2, \dots, i_k)$ of distinct vertices in $[n]$. As $2k$ different k-tuples represent the same k-cycle, we deduce that the cardinality of $\mathcal{C}_k$ is equal to $\frac{n!}{2k\times(n-k)!}$. But $\mathbf{P}(\zeta_C = 1) = p^k$, which gives the result. ■

Proposition 11.4.7. *Let $0 < p < 1$ and $\varepsilon > 0$, and let $\mathfrak{X}\colon \mathcal{G}(n) \to \mathbb{N}$ be the random variable indicating the chromatic number of the random graph. Then*

$$\lim_{n\to+\infty} \mathbf{P}\Big(\mathfrak{X} > \frac{-\log(1-p)}{2+\varepsilon} \times \frac{n}{\log n}\Big) = 1.$$

Proof. By Proposition 7.1.2, we have

$$\mathbf{P}\Big(\mathfrak{X} > \frac{n}{t}\Big) \geqslant \mathbf{P}(\mathfrak{A} < t) = 1 - \mathbf{P}(\mathfrak{A} \geqslant t).$$

Thus we can take advantage of Proposition 11.4.1. Set $\theta = -\log(1-p)$. Then since $\mathfrak{A}$ only takes integer values, we have

$$\begin{aligned} \log \mathbf{P}(\mathfrak{A} \geqslant t) = \log \mathbf{P}(\mathfrak{A} \geqslant \lceil t\rceil) &\leqslant \lceil t\rceil \log n - \frac{\theta\lceil t\rceil^2}{2} + \frac{\theta\lceil t\rceil}{2} \\ &\leqslant \lceil t\rceil\Big(\log n - \frac{\theta\lceil t\rceil}{2} + \frac{\theta}{2}\Big) \\ &\leqslant \lceil t\rceil\Big(\log n - \frac{\theta t}{2} + \frac{\theta}{2}\Big). \end{aligned}$$

Let $t = t(n) = \frac{(2+\varepsilon)\log n}{\theta}$. We then have

$$\log \mathbf{P}(\mathfrak{A} \geqslant t(n)) \leqslant \lceil t(n)\rceil\Big(-\frac{\varepsilon \log n}{2} + \frac{\theta}{2}\Big).$$

When $n \to +\infty$, the right-hand side tends to $-\infty$. We conclude that

$$\lim_{n\to+\infty} \mathbf{P}\Big(\mathfrak{X} > \frac{n}{t(n)}\Big) = 1,$$

which yields the result. ∎

We complete this result by stating Bollobás' theorem.

Theorem 11.4.8 (Bollobás, 1988). *Let $0 < p < 1$. Then asymptotically almost all the graphs Γ in the model $\mathcal{G}(n; p)$ satisfy*

$$\chi(\Gamma) \sim \frac{-\log(1-p)}{2} \times \frac{n}{\log n}.$$

Proof. Let $\varepsilon > 0$ and

$$A_{n,p}(\varepsilon) = \Big\lfloor \frac{(2-\varepsilon)\log n}{-\log(1-p)} \Big\rfloor.$$

Set $q = n/(\log n)^2$ and let S be a q-subset of $[n]$. Then the induced graph $\Gamma(S)$ has model $\mathcal{G}(q; p)$.

By Proposition 11.4.5, given that $\log q \sim \log n$, the probability π of having $\alpha(\Gamma(S)) \leqslant A_{n,p}(\varepsilon)$ is $o(1)$. It is useful here to employ the much more precise estimation of the terms $F_k(j)$ to which the proof actually leads: since

$$\sum_{j=2}^{k-1} F_k(2) = O(q^{-2+o(1)}), \quad \text{where } k = A_{n,p}(\varepsilon),$$

we deduce the upper bound $\pi = O(e^{-q^{2-o(1)}})$ from Janson's inequality (recalled at the end of this chapter).

As there are $\binom{n}{q} \leqslant 2^n = O(e^{q^{1+o(1)}})$ q-subsets in $[n]$, we find that the probability of having $\alpha(\Gamma(S)) \leqslant A_{n,p}(\varepsilon)$ for at least one q-subset S is

$$O(e^{-q^{2-o(1)}+q^{1+o(1)}}) = o(1).$$

Let Γ be a graph from the model $\mathscr{G}(n; p)$ all of whose q-subsets S satisfy $\alpha(\Gamma(S)) > A_{n,p}(\varepsilon)$. Start by taking a q-subset S of $[n]$. Then $\Gamma(S)$ admits an independent set of size k greater than $A_{n,p}(\varepsilon)$. We proceed similarly with the remaining $n - k$ vertices, and we continue as long as at least q vertices remain. We thus obtain $\lfloor (n-q)/k \rfloor$ independent sets that we can colour with as many different colours, while the at most q remaining vertices are coloured with other colours, all distinct. Consequently,

$$\chi(\Gamma) \leqslant \frac{n}{k} + q \leqslant \frac{-\log(1-p)}{2-\varepsilon} \times \frac{n}{\log n} + \frac{n}{(\log n)^2},$$

i.e.

$$\lim_{n\to+\infty} \mathbf{P}\Big(\mathscr{X} < \frac{-\log(1-p)}{2-2\varepsilon} \times \frac{n}{\log n}\Big) = 1,$$

which together with Proposition 11.4.7 yields the desired result. ■

11.5 Phase transition

We studied earlier the behaviour of the model $\mathscr{G}(n; p)$ for fixed p while $n \to +\infty$. We also considered several situations in which the parameter p could depend asymptotically on n. When p varies between 0 and 1, the corresponding random graph passes from an *amorphous* state (with no edge) to its complementary state (the complete graph). Between these two extremes, it passes through several phases that we can now attempt to identify, essentially by

- tracking the evolution of certain invariants of the random graph in the model $\mathscr{G}(n; p)$ as p varies;
- highlighting transition phenomena in the structure of the random graph $\mathscr{G}(n; p)$ related to the existence of critical values p_c of the parameter p.

11.5.1 Bernoulli random variable

We thus make the assumption that the probability measure varies as a function of the number of vertices: $p = p_n$. Let $\mathcal{A}$ be a property of the graphs in the model $\mathcal{G}(n; p_n)$. A function

$$f: \mathbb{N}^* \to \mathbb{N}^*$$

is a *threshold function* for the property $\mathcal{A}$ if

$$\lim_{n \to +\infty} \mathbf{P}(\mathcal{A} \text{ is satisfied}) = \begin{cases} 1 & \text{if } f(n) = o(p_n), \\ 0 & \text{if } p_n = o(f(n)). \end{cases}$$

When there exists a threshold function for the property $\mathcal{A}$, we say that a *phase transition* occurs at this threshold. We will first give an exercise and then several examples illustrating this.

Exercise 11.3. Show that a threshold function of the property *the random graph possesses at least one edge* is given by $f(n) = 1/n^2$.

Proposition 11.5.1. *Let $\mathcal{C}$ be the property:* the random graph possesses at least one cycle. *Then a threshold function is given by $f(n) = 1/n$.*

Proof. Set $\mathfrak{C} = \sum_{k=3}^{n} \mathfrak{C}_k$. By Proposition 11.4.6, for $0 < np < 1$ we have

$$\mathbf{E}(\mathfrak{C}) = \sum_{k=3}^{n} \frac{n(n-1)(n-2)\dots(n-k+1)}{2k} p^k \leqslant \sum_{k=3}^{n} (np)^k \leqslant \frac{(np)^3}{1-np}.$$

If $np = o(1)$, then $\lim_{n \to +\infty} \mathbf{E}(\mathfrak{C}) = 0$. By Markov's inequality,

$$\mathbf{P}(\mathfrak{C} \geqslant 1) \leqslant \mathbf{E}(\mathfrak{C}) \xrightarrow[n \to +\infty]{} 0.$$

If $np \sim c > 1$ when $\to +\infty$, then

$$\lim_{n \to +\infty} \mathbf{E}(\mathfrak{C}) = \sum_{k \geqslant 3} \frac{c^k}{2k} = +\infty.$$

We can thus expect the random graph to admit a cycle.

We saw in Section 11.1 that when np_n becomes large, almost all the graphs in the model $\mathcal{G}(n; p_n)$ possess a triangle, therefore a cycle. We will now prove a stronger result, which expresses the fact that for all $\varepsilon > 0$, if $np_n \geqslant 2 + \varepsilon$, then asymptotically almost all graphs admit a cycle. Set

$$p_n = \frac{2+\varepsilon}{n}.$$

To show that a graph has a cycle, it suffices to show that it is not a forest; thus the number of edges is at least equal to n. By Proposition 11.2.3, we have

$$\mathbf{E}(\mathfrak{M}) = \frac{(2+\varepsilon)(n-1)}{2} = n-1+\frac{\varepsilon(n-1)}{2},$$
$$\mathbf{V}(\mathfrak{M}) \leqslant \frac{(2+\varepsilon)(n-1)}{2}.$$

If $m \leqslant n-1$, then $m - \mathbf{E}(\mathfrak{M}) \leqslant -\varepsilon(n-1)/2$, so $|m - \mathbf{E}(\mathfrak{M})| \geqslant \varepsilon(n-1)/2$. Hence by the Bienaymé–Chebyshev inequality, we obtain

$$\mathbf{P}(\mathfrak{M} \leqslant n-1) \leqslant \mathbf{P}\Big(|\mathfrak{M} - \mathbf{E}(\mathfrak{M})| \geqslant \frac{\varepsilon(n-1)}{2}\Big) \leqslant \frac{2(2+\varepsilon)}{\varepsilon^2(n-1)},$$

which tends to 0 as $n \to +\infty$. This shows that asymptotically almost all graphs in the model $\mathcal{G}(n; p_n)$ have at least n edges, and therefore have a cycle. We conclude that $f(n) = 1/n$ is a threshold function for the property $\mathcal{C}$. ■

Let us now consider the issue of connectedness in the model $\mathcal{G}(n; p)$. We use the property according to which a graph Γ is connected if and only if it possesses a spanning tree, i.e. a tree that connects all the vertices of Γ. We start with the following classical result due to Cayley, which has several different proofs.

Lemma 11.5.2. *The number of spanning trees T_n on n vertices is equal to n^{n-2}.*

Proof. We proceed by a double count. Let ν denote the number of $(n+1)$-tuples

$$(A, v, e_1, \dots, e_{n-1})$$

where A is a spanning tree, v a vertex of A and $E(A) = \{e_1, \dots, e_{n-1}\}$. If we consider that v has the status of a root, we obtain a rooted and thus oriented tree, with its arcs e_i ordered by a fixed order: the rooted tree is thus counted $(n-1)!$ times.

One the one hand, there are T_n possible choices for the spanning tree A, n possible choices for the root v, and $(n-1)!$ possible orders on the arcs; thus in total we have $\nu = n! \times T_n$.

On the other hand, we will estimate ν by counting the possible constructions of a rooted spanning tree on the set of vertices $[n]$, starting from a graph with no edge:

- at time 0, we have a forest consisting of n trees with only one vertex;
- at time $k \leqslant n-1$, we just connected two trees by an oriented edge $e = (i, j)$ in such a way that the end j is the root of the *destination* tree, which does not contain the other endpoint i of e; there are n possible choices for the root i and $n-k$ possible choices for the destination tree containing j;
- at time $n-1$, we clearly have a rooted spanning tree since at each step, the number of trees of the forest decreases by one.

Altogether, this shows that $\nu = \prod_{k=1}^{n-1} n(n-k) = n^{n-1} \times (n-1)!$. By setting the two counting results equal, we conclude that $T_n = n^{n-2}$. ■

Proposition 11.5.3. *A threshold function for the property:* the random graph is connected *is given by* $f(n) = \log n/n$.

Proof. Let $1 \leqslant k \leqslant n$ and let $S \in \mathrm{P}_k([n])$ be a k-subset of $[n]$. Then the induced subgraph $\Gamma(S)$ is connected if and only if it contains a spanning tree, i.e. a tree that connects all the vertices of $\Gamma(S)$. For every spanning tree A on S, let η_A denote the Bernoulli random variable that indicates whether A is a subgraph of $\Gamma(S)$. Let ν_S denote the random variable that counts the spanning trees of $\Gamma(S)$. Then

$$\nu_S = \sum_A \eta_A$$

where the sum runs over all spanning trees A on S. As

$$\mathbf{P}(\eta_A = 1) = p^{k-1}(1-p)^{\binom{k}{2}-(k-1)},$$

and using Lemma 11.5.2, we obtain

$$\mathbf{E}(\nu_S) = k^{k-2} p^{k-1}(1-p)^{\binom{k}{2}-(k-1)} \leqslant k^k p^{k-1}.$$

Let μ_S be the Bernoulli random variable that indicates whether S is not connected to its complement $\overline{S} = [n]\backslash S$. Due to the independence of $\xi_{\{i,j\}}$ [1], the variables ν_S and μ_S are independent:

$$\mathbf{E}(\nu_S \mu_S) = \mathbf{E}(\nu_S)\mathbf{E}(\mu_S)$$

and

$$\mathbf{E}(\mu_S) = \mathbf{P}(\mu_S = 1) = (1-p)^{k(n-k)}.$$

Thus, summing over all the k-subsets S not connected to their complement $\overline{S}$, and over $k = 2, \ldots, \lfloor n/2 \rfloor$, then applying Markov's inequality, we find

$$\mathbf{P}\left(\sum_{k=2}^{\lfloor n/2 \rfloor} \sum_{S \in P_k([n])} \nu_S \mu_S = 0\right) \leqslant \sum_{k=2}^{\lfloor n/2 \rfloor} \binom{n}{k} (1-p)^{k(n-k)} k^k p^{k-1}.$$

By the inequality from Lemma 11.4.3, using $(1-p) \leqslant e^{-p}$, the bound above is less than or equal to

$$\frac{1}{p} \sum_{k=2}^{\lfloor n/2 \rfloor} (enp)^k e^{-kp(n-k)} \leqslant \frac{1}{p} \sum_{k=2}^{\lfloor n/2 \rfloor} (enp)^k e^{-knp/2}.$$

---

[1] Recall that $\xi_{\{i,j\}}$ is the Bernoulli variable that indicates whether $\{i,j\}$ is an edge.

If $np \geqslant (1+\varepsilon)\log n$, we find that the probability that there is no connected component of cardinality between 2 and $\lfloor n/2 \rfloor$ in the random graph is

$$\ll \frac{\log n}{n^{\varepsilon}}.$$

Now consider the number $X = \sum_{j=1}^{n} \zeta_j$ of isolated vertices, where ζ_j is the Bernoulli variable which detects whether j is an isolated vertex. Then

$$\mathbf{E}(X) = n(1-p)^{n-1} \leqslant e^{\log n - np} \leqslant n^{-\varepsilon}$$

when $p \geqslant (1+\varepsilon)\log n/n$, so that from Markov's inequality, we deduce that

$$\mathbf{P}(X \geqslant 1) \leqslant \mathbf{E}(X) \sim n^{-\varepsilon}, \quad n \to +\infty.$$

Conversely, if $p \leqslant (1-\varepsilon)\log n/n$, then $\mathbf{E}(X) \underset{n\to+\infty}{\sim} n^{\varepsilon}$ and

$$\begin{aligned}
\mathbf{V}(X) &= \sum_{i,j\in[n]} \mathbf{P}(\zeta_i = \zeta_j = 1) - \mathbf{P}(\zeta_i = 1)\mathbf{P}(\zeta_j = 1) \\
&= n\big((1-p)^{n-1} - (1-p)^{2n-2}\big) \\
&\quad + n(n-1)\big((1-p)^{2n-3} - (1-p)^{2n-2}\big) \\
&= n(1-p)^{n-1} + n^2 p(1-p)^{2n-3} = \mathbf{E}(X) + O\Big(\frac{\log n}{n^{1-2\varepsilon}}\Big).
\end{aligned}$$

By Proposition 11.8.2, we conclude that the random graph possesses isolated vertices, and thus is not connected. We have shown that the function $\log n/n$ is a threshold function for connectedness, and in particular for the absence of isolated vertices. ■

11.5.2 Some examples of threshold functions

In Table 11.1, we collect some threshold functions in the framework of Gilbert's model $\mathcal{G}(n; p)$. Naturally, one can seek to determine threshold functions for the exact same properties in any other framework, for example in the model $\mathcal{G}(n, m)$.

11.6 Summary of the steps of the evolution of $\mathcal{G}(n; p)$

We saw that the threshold functions play an important role concerning the configuration of the graph. Examples show that there exists a phase transition between random graphs whose largest connected component does not depend on n and those whose largest connected component is of size $O(n)$. A connected component is *giant* if its size (in number of vertices) is $O(n)$. Let us summarise the evolution of the random graph of the model $\mathcal{G}(n; p)$ as a function of p.

Property	Threshold	
Possessing at least one edge	$\frac{1}{n^2}$	cf. Exercise 11.3
Containing a K_4	$\frac{1}{n^{\frac{2}{3}}}$	cf. Exercise 11.2
Containing a tree with n' vertices	$\frac{1}{n^{\frac{n'}{n'-1}}}$	
Containing a cycle	$\frac{1}{n}$	cf. Proposition 11.5.1
Being connected (or being without isolated vertex)	$\frac{\log n}{n}$	cf. Proposition 11.5.3
Having diameter $\geqslant 2$	$\sqrt{\frac{2\log n}{n}}$	cf. Proposition 11.3.1

Table 11.1

- $p = o(1/n)$: the random graph is a forest;
- $p = c/n, 0 < c < 1$: cycles of different lengths appear, the connected components are either trees or pseudo-trees (connected graphs containing only a single cycle). The size of the largest connected component is $O(\log n)$, but most vertices are in connected components of size $O(1)$.
- $p = 1/n$: the size of the largest connected component is $O(n^{2/3})$;
- $p = c/n, c > 1$: the graph contains a single giant connected component; the others are all *small* and most of them are trees;
- $p = c \log n/n, c \geqslant 1$: if $c > 1$ the random graph becomes connected, if $c = 1$ it has a single giant connected component and apart from that, only isolated vertices;
- $p = t(n)\log n/n$ with $t(n) \to +\infty$ as $n \to +\infty$: then the graph is connected and all its vertices asymptotically have the same degree.

Determining the threshold function of a given property can be complicated; indeed, one must first determine whether such a threshold even exists. The next theorem gives a partial answer.

A property $\mathcal{A}$ of graphs is said to be *monotone* if it is stable under addition of edges, i.e. if Γ satisfies $\mathcal{A}$, then $\Gamma + xy$ still satisfies $\mathcal{A}$ for any pair of non-adjacent vertices x, y of Γ. For instance, being connected and being Hamiltonian are monotone properties, whereas being k-colourable or being planar are not.

Proposition 11.6.1. *Let $\mathcal{A}$ be a monotone property of graphs, and let $0 \leqslant p_1 \leqslant p_2 \leqslant 1$ be two real numbers. For $i = 1, 2$, let $\mathbf{P}_i$ denote the probability measure in the model $\mathcal{G}(n; p_i)$. Then $\mathbf{P}_1(\mathcal{A}) \leqslant \mathbf{P}_2(\mathcal{A})$.*

Proof. We express p_2 as a function of p_1:

$$p_1 + (1 - p_1)p_3 = p_2 \text{ for } p_3 \in [0, 1].$$

Since the parameter p_3 is fixed by p_1, p_2, it defines a binomial model $\mathcal{G}(n; p_3)$. Let $\Gamma_3 \in \mathcal{G}(n; p_3)$ be the random graph in this model, and let $\Gamma_1 \in \mathcal{G}(n; p_1)$. Set $\Gamma_2 = \Gamma_1 \cup \Gamma_3$. Let i, j be a pair of vertices of $[n]$. Then $\{i, j\}$ is an edge of Γ_2 if it is an edge of Γ_1 or an edge of Γ_3. Thus, by Poincaré's formula,

$$\begin{aligned}
\mathbf{P}(\{i, j\} \in \Gamma_2) &= \mathbf{P}(\{\{i, j\} \in \Gamma_1\} \cup \{\{i, j\} \in \Gamma_3\}) \\
&= \mathbf{P}(\{i, j\} \in \Gamma_1) + \mathbf{P}(\{i, j\} \in \Gamma_3) \\
&\quad - \mathbf{P}(\{\{i, j\} \in \Gamma_1\} \cap \{\{i, j\} \in \Gamma_3\}) \\
&= p_1 + p_3 - p_1 p_3 = p_2.
\end{aligned}$$

Therefore Γ_2 is in the model $\mathcal{G}(n; p_2)$, since its edges are all selected independently with probability p_2. Consequently, if $\mathcal{A}$ is monotone, so if Γ_1 satisfies $\mathcal{A}$, then Γ_2 also satisfies $\mathcal{A}$, which proves the result. ■

Theorem 11.6.2 (Bollobás–Thomason, 1987). *Every monotone property $\mathcal{A}$ of graphs admits a threshold function.*

Proof. For $p \in [0, 1]$, let $\mathbf{P}_p$ be the probability measure on $\mathcal{G}(n; p)$. Fix $p_0 \in]0, 1[$ such that $\mathbf{P}_{p_0}(\mathcal{A}) = 1/2$. The existence of p_0 follows from the fact that $p \mapsto \mathbf{P}_p(\mathcal{A})$ is a polynomial function of p that varies between 0 and 1 when p varies between 0 and 1, so the intermediate value theorem applies. Note that in general, p_0 depends on $\mathcal{A}$ and n: $p_0 = p_0(\mathcal{A}, n)$.

Let us show that $n \mapsto p_0(\mathcal{A}, n)$ is a threshold function for the property $\mathcal{A}$.

Let $\varepsilon > 0$ and let q be an integer such that $(1 - \varepsilon)^q < 1/2$. Let $0 < p < p_0/q$ and let Γ_i, $i = 1, \dots, q$, be independently chosen graphs in the model $\mathcal{G}(n; p)$. Set $\Gamma = \bigcup_{i=0}^{q} \Gamma_i$. Then Γ is clearly in the model $\mathcal{G}(n; 1 - (1 - p)^q)$. Furthermore, we have $1 - (1 - p)^q \leqslant qp \leqslant p_0$, so because the property $\mathcal{A}$ is monotone, we have

$$\mathbf{P}_{1-(1-p)^q}(\mathcal{A}) \leqslant \mathbf{P}_{p_0}(\mathcal{A}) = \frac{1}{2}.$$

Letting $\overline{\mathcal{A}}$ denote the complementary property of $\mathcal{A}$. Then

$$\frac{1}{2} \leqslant \mathbf{P}_{1-(1-p)^q}(\overline{\mathcal{A}}) = \left(\mathbf{P}_p(\overline{\mathcal{A}})\right)^q = (1 - \mathbf{P}_p(\mathcal{A}))^q.$$

Hence by the definition of ε, we conclude that $\mathbf{P}_p(\mathcal{A}) < \varepsilon$.

To obtain the second inequality, we proceed almost symmetrically. Let Γ be the union of q independent copies of the random graph in the model $\mathcal{G}(n; p_0)$. It is clear that Γ is in the model $\mathcal{G}(n; 1-(1-p_0)^q)$: q is chosen in such a way that $2^{-q} < \varepsilon$. We thus have

$$\mathbf{P}_{1-(1-p_0)^q}(\overline{\mathcal{A}}) = \mathbf{P}_{p_0}(\overline{\mathcal{A}})^q = \frac{1}{2^q},$$

so $\mathbf{P}_{1-(1-p_0)^q}(\mathcal{A}) > 1-\varepsilon$. ∎

11.7 Some remarks

The two models studied in this chapter are both *homogeneous*, which means that the vertices all play an identical role, so they are all equivalent. This does not always correspond to *reality*, but it is the case if we impose strong constraints on the graphs under consideration: they must

- be bipartite,
- have a small diameter,
- have a high *clustering* coefficient[2],
- have a distribution of degrees defined by a certain probability distribution.

Nevertheless, these still allow us to study fundamental aspects of the asymptotic behaviour of graphs. They can be also easily be generalised: for each pair $i, j \in [n]$ of vertices, fix

$$\mathbf{P}(\{i, j\} \text{ is an edge}) = q_{i,j} \in [0, 1],$$

where the matrix $Q = (q_{i,j})$ is symmetric. We obtain a new model of random graph $\mathcal{G}(n; Q)$: the probability of the elementary event $\{\Gamma\}$ where $\Gamma = ([n]; E)$ is given by

$$\mathbf{P}(\{\Gamma\}) = \prod_{\substack{i<j \\ \{i,j\}\in E}} q_{i,j} \times \prod_{\substack{i<j \\ \{i,j\}\notin E}} (1-q_{i,j}).$$

Another example is a *geometric random graph*, whose construction consists in placing n nodes (vertices) in the square $[0, 1]^2 \subset \mathbb{R}^2$ (or any other metric space) according to a fixed probability distribution, and then connecting pairs of nodes whenever the distance between them is less than or equal to a given constant $\delta > 0$. This graph models *spatial networks*. The fundamental question for these graphs is whether they are connected, particularly when they model communication networks.

[2] The *clustering* coefficient of a graph is equal to 3 times the number of triangles divided by the total number of triplets (connected subgraphs with 3 vertices).

The structure of these graphs bears an amazing resemblance to the structure of social networks...

The Barabási–Albert model has two parameters: the number n_0 of initial nodes, and an integer $1 \leqslant k \leqslant n_0$. For $i = n_0 + 1, \ldots, n$, we add at time $i - 1$ the new node i, and connect it to exactly k nodes chosen randomly as follows: the node u is chosen with the probability $p_u = \frac{d_{i-1}(u)}{\sum_v d_{i-1}(v)}$, where $d_{i-1}(u)$ denotes the degree of the vertex u at time $i - 1$ and the sum over v runs over all the vertices of the network at time $i - 1$. This describes a dynamic process of *preferential attachment* which again resembles the formation of social networks in which new members tend to prefer forming relations with the most popular of the older members. One can show that this model is *scale-free*, which means that the degrees of the vertices at time n follow a power law that is invariant over time: for $d \geqslant k$, the proportion of vertices of degree d is equal to $2k(k+1)/d(d+1)(d+2)$ as $n \to +\infty$.

Finally, we give an example of a bipartite random graph: it consists in randomly choosing a fixed number m of edges between two sets of vertices $\{v_1, \ldots, v_r\}$ and $\{v_{r+1}, \ldots, v_{r+s}\}$ (with $n = r + s$), or else choosing each edge with a probability $p \in [0, 1]$.

There is no particular limit to the construction of models of random graphs: some are easier to study than others, but in general the important thing is to take care to construct the model as a function of the precise nature of the situation under consideration, such as

- various real-life networks such as the web, public transportation, social networks, or telecommunications networks;
- data mining and classification;
- modelling the brain;
- many other applications arising from computer science, biology, mathematics, physics and engineering.

11.8 Background in probability theory

11.8.1 Probability spaces and probability measures

Let Ω be a finite set. The elements of Ω are called *elementary events*. The set Ω can be interpreted as the space of all possible outcomes of a certain random test: it is called the *event space*. A sequence of n random tests from Ω is called a *sample*.

The idea is to realise certain *experiments* on random objects with the goal of determining a probability that measures how likely it is that a given property will hold. For example, one may randomly choose simple graphs with n distinct numbered vertices and m edges, and probabilistically measure the property of edge-connectedness.

An *event* A is a subset of Ω, i.e. a collection of elementary events. Let $P(\Omega)$ denote the set of subsets of Ω. The set-theoretic operations in $P(\Omega)$ possess natural probabilistic meanings. We use the following notation:

- for $A, B \subset \Omega$, the event $A \cap B$ means *the events A and B both simultaneously occur*; the event $A \cup B$ means *at least one of two events A and B occurs*;
- for $A, B \subset \Omega$, the event $A \backslash B$ means *the event A occurs but the event B does not*;
- for every $A \subset \Omega$, $\overline{A} = \Omega \backslash A$ denotes *the complementary event* of A (A does not occur);
- the empty set is also an event.

The sample space Ω together with the set of all its subsets $A \subset \Omega$ is called a *measurable space*. The general framework of probability theory does not oblige us to be able to *measure* all the subsets of Ω: it only requires us to rely on a *σ-algebra*, a notion that is not indispensable in the context of random graphs: here, the σ-algebra is $P(\Omega)$.

A *probability measure* is a map $\mathbf{P}\colon P(\Omega) \to [0, 1]$ satisfying

(i) $\mathbf{P}(\Omega) = 1$;

(ii) σ-additivity: for a subset $I \subset \mathbb{N}$ and for every family $\{A_i \mid i \in I\} \subset P(\Omega)$ such that $A_i \cap A_j = \emptyset$ if $i \neq j \in I$, we have

$$\mathbf{P}\Big(\bigcup_{i \in I} A_i\Big) = \sum_{i \in I} \mathbf{P}(A_i).$$

A measurable space equipped with a probability measure, i.e. a triple $(\Omega, P(\Omega), \mathbf{P})$, is called a *probability space*.

Example 11.8.1. Let $\Omega = \{\omega_i \mid 1 \leqslant i \leqslant n\}$ be a set of cardinality n, and suppose that for all $i \in \{1, 2, \ldots, n\}$ we have

$$\mathbf{P}(\{\omega_i\}) = \frac{1}{n}.$$

Then $(\Omega, P(\Omega), \mathbf{P})$ is called the uniform probability space.

Using the above axioms, we can easily prove that:

(i) $\mathbf{P}(\emptyset) = 0$;

(ii) For $A, B \subset \Omega$, if $A \subseteq B$, then $\mathbf{P}(A) \leqslant \mathbf{P}(B)$;

(iii) if $A \subset \Omega$, then $\mathbf{P}(\overline{A}) = 1 - \mathbf{P}(A)$;

(iv) for every finite family $\{A_i \subset \Omega,\ i \in I\}$ we have

$$\mathbf{P}\Big(\bigcup_{i \in I} A_i\Big) \leqslant \sum_{i \in I} \mathbf{P}(A_i);$$

(v) if $A, B \subset \Omega$, then

$$\mathbf{P}(A \cup B) = \mathbf{P}(A) + \mathbf{P}(B) - \mathbf{P}(A \cap B);$$

(vi) if $A, B \subset \Omega$, then

$$\mathbf{P}(A) = \mathbf{P}(A \cap B) + \mathbf{P}(A \cap \overline{B});$$

(vii) if $A, B_1, \ldots, B_m \subset \Omega$ are such that $i \neq j \Rightarrow B_i \cap B_j = \emptyset$ and $B_1 \cup \cdots \cup B_m = \Omega$, then

$$\mathbf{P}(A) = \sum_{i=1}^{m} \mathbf{P}(A \cap B_i).$$

11.8.2 Independence and conditional probabilities

Let $(\Omega, P(\Omega), \mathbf{P})$ be a probability space and let $A, B \in P(\Omega)$ be two events. The *probability of A given B* is then defined by

$$\mathbf{P}(A \mid B) = \frac{\mathbf{P}(A \cap B)}{\mathbf{P}(B)}$$

as long as $\mathbf{P}(B) \neq 0$. Thus we have the identity

$$\mathbf{P}(A \mid B)\mathbf{P}(B) = \mathbf{P}(B \mid A)\mathbf{P}(A).$$

From (vii) above, we obtain the *law of total probability*: if the A_i, $i = 1, \ldots, m$ are events forming a partition of Ω, then

$$\mathbf{P}(B) = \sum_{i=1}^{m} \mathbf{P}(B \mid A_i)\mathbf{P}(A_i).$$

In particular, for any $A \in \Omega$, we have

$$\mathbf{P}(B) = \mathbf{P}(B \mid A)\mathbf{P}(A) + \mathbf{P}(B \mid \overline{A})\mathbf{P}(\overline{A}),$$

which gives the famous formula known as Bayes' theorem:

$$\mathbf{P}(A \mid B) = \frac{\mathbf{P}(B \mid A)\mathbf{P}(A)}{\mathbf{P}(B \mid A)\mathbf{P}(A) + \mathbf{P}(B \mid \overline{A})\mathbf{P}(\overline{A})}.$$

Two events are said to be *independent* if

$$\mathbf{P}(A \mid B) = \mathbf{P}(A),$$

or equivalently,

$$\mathbf{P}(A \cap B) = \mathbf{P}(A) \times \mathbf{P}(B) \quad \text{or} \quad \mathbf{P}(B \mid A) = \mathbf{P}(B).$$

We can generalise this notion to a family of events: the family $\{A_i \in P(\Omega) \mid i \in I\}$ is *independent* if for every finite subset $\{i_1, i_2, \ldots, i_k\}$ of I, we have

$$\mathbf{P}(A_{i_1} \cap A_{i_2} \cap \cdots \cap A_{i_k}) = \mathbf{P}(A_{i_1})\mathbf{P}(A_{i_2}) \cdots \mathbf{P}(A_{i_k}).$$

11.8.3 Finite random variables

Let $(\Omega, P(\Omega), \mathbf{P})$ be a probability space. A *finite random variable* on this space is a map

$$X : \Omega \to \mathbb{R}, \qquad \omega \mapsto X(\omega),$$

which only takes a finite number of values $|X(\Omega)| < +\infty$. Every finite random variable on Ω has a *probability distribution* defined by

$$\mathbf{P}(X = x) \quad \text{for each } x \in X(\Omega),$$

where the event $\{X = x\}$ corresponds to

$$\{X = x\} = X^{-1}(\{x\}) = \{\omega \in \Omega \mid X(w) = x\}.$$

As $\Omega = \bigcup_{x \in X(\Omega)} \{X = x\}$, we have $\sum_{x \in X(\Omega)} \mathbf{P}(X = x) = 1$.

Let X be a finite random variable: $X(\Omega) = \{x_1, x_2, \ldots, x_n\}$. Let

$$p_i = \mathbf{P}(X = x_i), \quad i = 1, \ldots, n.$$

The *expected value* of X is defined by

$$\mathbf{E}(X) = \sum_{i=1}^{n} p_i x_i.$$

It is the expected value or mean value of the map X. We can easily show that the *expected value is linear*:

$$\mathbf{E}(tX + Y) = t\mathbf{E}(X) + \mathbf{E}(Y)$$

for every pair X, Y of finite random variables and every $t \in \mathbb{R}$.

The *variance* of X is given by

$$\mathbf{V}(X) = \sum_{i=1}^{n} p_i (x_i - \mathbf{E}(X))^2 = \mathbf{E}((X - \mathbf{E}(X))^2) = \mathbf{E}(X^2) - \mathbf{E}(X)^2.$$

The *standard deviation* of X is defined by $\sigma(X) = \sqrt{\mathbf{V}(X)}$.

We say that two random variables X and Y are *independent* if the events $\{X = x\}$ and $\{Y = y\}$ are independent for all $x \in X(\Omega)$ and all $y \in Y(\Omega)$. If this is the case, then let us show that

$$\mathbf{E}(XY) = \mathbf{E}(X)\mathbf{E}(Y) \quad \text{and} \quad \mathbf{V}(X+Y) = \mathbf{V}(X) + \mathbf{V}(Y).$$

Let $\mathbf{Cov}(X, Y) = \mathbf{E}(XY) - \mathbf{E}(X)\mathbf{E}(Y)$ denote the *covariance* of the pair of random variables X, Y. Then

$$\mathbf{E}((X+Y)^2) - \mathbf{E}(X+Y)^2 = \mathbf{V}(X) + \mathbf{V}(Y) + 2\mathbf{Cov}(X,Y).$$

But we also have

$$\mathbf{P}(XY = z) = \sum_{\substack{x \in X(\Omega), y \in Y(\Omega) \\ xy = z}} \mathbf{P}(X = x \text{ and } Y = y).$$

By the independence of X and Y, we see that

$$\mathbf{P}(X = x \text{ and } Y = y) = \mathbf{P}(X = x)\mathbf{P}(Y = y),$$

which proves the desired identities.

Two random variables X and Y *are not correlated* if

$$\mathbf{Cov}(X,Y) = 0.$$

11.8.4 Some examples of probability distributions

The discrete uniform distribution. This distribution is given by

$$\forall i \in \{1, 2, 3, \dots, n\}, \quad \mathbf{P}(X = x_i) = \frac{1}{n}.$$

We have

$$\mathbf{E}(X) = \frac{n+1}{2} \quad \text{and} \quad \mathbf{V}(X) = \frac{n^2-1}{12}.$$

It is the distribution according to which all random phenomena occur with the same probability.

Bernoulli distribution. This distribution corresponds to a random experiment having exactly two possible outcomes 0 and 1. It depends on a parameter $p \in [0, 1]$. We have

$$\mathbf{P}(X = 1) = p \quad \text{and} \quad \mathbf{P}(X = 0) = 1 - p.$$

The game of heads or tails can be modelled by the Bernoulli distribution, with the parameter $p = 1/2$ if the coin is well-balanced. We have

$$\mathbf{E}(X) = p \quad \text{and} \quad \mathbf{V}(X) = p(1-p)$$

Binomial distribution. This distribution depends on two parameters $n \in \mathbb{N}$ and $p \in [0, 1]$. It is the distribution that determines the number of successes obtained after n independent Bernoulli experiments with parameter $p \in [0, 1]$. It is obtained by taking the sum of n independent Bernoulli random variables. Its distribution is given by

$$\mathbf{P}(X = k) = \binom{n}{k} p^k (1 - p)^{n-k} \quad \text{for every integer } 0 \leqslant k \leqslant n.$$

Furthermore,

$$\mathbf{E}(X) = np \quad \text{and} \quad \mathbf{V}(X) = np(1 - p).$$

Poisson distribution. This distribution is used to model rare phenomena, such as for example a series of airplane or train accidents. It can also be used to describe the behaviour of the number of events that occur during a fixed time interval. It depends on one real parameter $\lambda > 0$, and its distribution is given by

$$\mathbf{P}(X = k) = \frac{\lambda^k}{k!} \mathrm{e}^{-\lambda} \quad \text{for all } k \in \mathbb{N}.$$

Furthermore,

$$\mathbf{E}(X) = \lambda \quad \text{and} \quad \mathbf{V}(X) = \lambda.$$

When X_n follows a binomial distribution with parameters n and p_n in such a way that $\lim_{n \to +\infty} np_n = \lambda$, then the sequence (X_n) *converges in distribution* to a Poisson distribution with parameter λ: this means that

$$\lim_{n \to +\infty} \mathbf{P}(X_n = k) = \frac{\lambda^k}{k!} \mathrm{e}^{-\lambda}, \quad \forall k \in \mathbb{N}.$$

Geometric distribution. This distribution is used to model the rank of the first success when a given Bernoulli experiment is repeated several times. It depends on a parameter $\theta \in \,]0, 1[$ and is given by

$$\mathbf{P}(X = k) = (1 - \theta)^{k-1} \theta, \quad k \geqslant 1.$$

We have

$$\mathbf{E}(X) = \frac{1}{\theta} \quad \text{and} \quad \mathbf{V}(X) = \frac{1 - \theta}{\theta^2}.$$

11.8.5 Classical inequalities

Probability theory often concentrates on the study of the behaviour of a random variable X, and in particular the *dispersion of its values* with respect to their average $\mathbf{E}(X)$. We will give some inequalities that are quite simple to prove but can also be extraordinarily efficient.

Markov's inequality. Let X be a finite positive random variable and let $a > 0$. Then

$$\mathbf{P}(X \geqslant a) \leqslant \frac{\mathbf{E}(X)}{a}.$$

Indeed,

$$\begin{aligned}\mathbf{E}(X) = \sum_{x \in X(\Omega)} x\mathbf{P}(X = x) &\geqslant \sum_{\substack{x \in X(\Omega)\\ x \geqslant a}} x\mathbf{P}(X = x)\\ &\geqslant a \sum_{\substack{x \in X(\Omega)\\ x \geqslant a}} \mathbf{P}(X = x) = a\mathbf{P}(X \geqslant a).\end{aligned}$$

Bienaymé–Chebyshev inequality. Let X be a finite random variable and let $a > 0$. Then

$$\mathbf{P}(|X - \mathbf{E}(X)| \geqslant a) \leqslant \frac{\mathbf{V}(X)}{a^2}.$$

This inequality follows from applying Markov's inequality to $Y = (X - \mathbf{E}(X))^2$.

We also have the following basic application.

Proposition 11.8.2. *Let X be a random variable depending on the parameter $n \in \mathbb{N}$. If $\mathbf{E}(X) \to +\infty$ and $\mathbf{V}(X) = o(\mathbf{E}(X))^2$ as $n \to \infty$, then*

$$X \underset{n \to +\infty}{\sim} \mathbf{E}(X)$$

and for all $a \geqslant 0$, we have $\mathbf{P}(X \leqslant a) = o(1)$.

One situation that arises frequently corresponds to the sum $X = X_1 + \cdots + X_n$ of n particular Bernoulli random variables: $X_i = \mathbb{1}_{A_i}$, the indicator function of the event $A_i \subset \Omega$, equal to 1 if A_i holds, 0 if not. We have

$$\mathbf{V}(X) = \sum_{1 \leqslant i,j \leqslant n} \mathbf{E}(X_i X_j) - \mathbf{E}(X_i)\mathbf{E}(X_j).$$

Let us write $i \sim j$ if $i \neq j$ but A_i and A_j are not independent. Then

$$\mathbf{V}(X) \leqslant \mathbf{E}(X) + \sum_{i \sim j} \mathbf{P}(A_i \cap A_j).$$

Let Δ denote the second term on the right-hand side of the inequality. By Proposition 11.8.2, we deduce that if $\mathbf{E}(X) \xrightarrow[n \to +\infty]{} +\infty$ and $\Delta = o(\mathbf{E}(X)^2)$, then

$$X \underset{n \to +\infty}{\sim} \mathbf{E}(X).$$

Janson's inequality. When the events A_i, $i = 1, \ldots, I$, are independent, we have the formula

$$\mathbf{P}\left(\bigcap_{i=1}^{I} \overline{A_i}\right) = \prod_{i=1}^{I}(1 - \mathbf{P}(A_i)) \leqslant e^{-\mu}, \quad \text{where} \quad \mu = \sum_{i=1}^{I} \mathbf{P}(A_i).$$

For $X = \mathbb{1}_{A_1} + \mathbb{1}_{A_2} + \cdots + \mathbb{1}_{A_I}$, we obtain $\mathbf{P}(X = 0) \leqslant e^{-\mu}$, which is interesting when μ is large: indeed, we can compare this bound with the weaker bound given by the Bienaymé–Chebyshev inequality: $\mathbf{P}(X = 0) \leqslant \mu^{-1}$.

Even if the A_i are not independent, we can still obtain an inequality that is almost as good as long as we assume that the events A_i are of a particular kind, as explained in the following result.

Proposition 11.8.3 (Janson's inequality). *Let Ω be a finite set from which we choose a random subset R using the procedure $\mathbf{P}(\omega \in R) = p_\omega \in [0, 1]$, where the events $\omega \in R$ are mutually independent. Fix a choice of subsets $A_i \subset \Omega$ for $i = 1, \ldots, I$, and let A_i also denote the corresponding events defined by $A_i \subset R$. Write $i \sim j$ if A_i and A_j are not independent, i.e. if $A_i \cap A_j \neq \emptyset$, and set*

$$\mu = \sum_{i=1}^{I} \mathbf{P}(A_i), \quad \Delta = \sum_{i \sim j} \mathbf{P}(A_i \cap A_j).$$

Then

$$\mathbf{P}\left(\bigcap_{i=1}^{I} \overline{A_i}\right) \leqslant \begin{cases} e^{-\mu + \frac{\Delta}{2}} & \text{if } \Delta \leqslant \mu, \\ e^{-\frac{\mu^2}{2\Delta}} & \text{if } \Delta \geqslant \mu. \end{cases}$$

Applying this result to the situation described just before the proposition, this gives

$$P(X = 0) \leqslant e^{-\mu + \frac{\Delta}{2}} \text{ or } e^{-\frac{\mu^2}{2\Delta}} \quad \text{according to the sign of } \Delta - \mu,$$

which is generally better than $P(X = 0) \leqslant \frac{\mathbf{V}(X)}{\mathbf{E}(X)^2}$.

11.8.6 Asymptotic properties

Knowing the asymptotic behaviour of a sequence of random variables $(Z_n)_{n \geqslant 1}$ is often crucial. We distinguish several notions of convergence of Z_n to a random variable Z:

(i) *convergence in distribution*: $Z_n \xrightarrow[n \to +\infty]{} Z$ in distribution if

$$\lim_{n \to +\infty} \mathbf{P}(Z_n \leqslant t) = \mathbf{P}(Z \leqslant t) \quad \text{for all } t \in \mathbb{R}.$$

(ii) *convergence in probability*: $Z_n \xrightarrow[n\to+\infty]{} X$ in probability if

$$\forall\varepsilon > 0, \qquad \lim_{n\to+\infty} \mathbf{P}(|Z_n - Z| > \varepsilon) = 0.$$

(iii) *convergence in mean*: $Z_n \xrightarrow[n\to+\infty]{} Z$ in mean if

$$\lim_{n\to+\infty} \mathbf{E}(|Z_n - Z|) = 0.$$

(iv) *almost sure convergence*: $Z_n \xrightarrow[n\to+\infty]{} Z$ almost surely if

$$\mathbf{P}\big(\lim_{n\to+\infty} |Z_n - Z| = 0\big) = 1.$$

We note that almost sure convergence implies convergence in probability, which itself implies convergence in distribution (exercise). The following proposition asserts even more.

Proposition 11.8.4. *Let* $(Z_n)_{n\geqslant 1}$ *be a sequence of random variables all of whose expected values are equal. Then*

$$Z_n \xrightarrow[n\to+\infty]{} 0 \textit{ almost surely} \iff \sup_{k\geqslant n}|Z_k| \xrightarrow[n\to+\infty]{} 0 \textit{ in probability.}$$

Proof. If direction: for all $\varepsilon > 0$ and $n \geqslant 1$, set $\mathcal{A}_n(\varepsilon) = \{|Z_n| > \varepsilon\}$. By hypothesis, there exists n_0 such that if $n \geqslant n_0$, then $\mathbf{P}(|Z_n| > \varepsilon) = 0$. Thus $\mathbf{P}(\sup_{n\geqslant n_0}|Z_n| > \varepsilon) = 0$.

Only if direction: observe that

$$\begin{aligned}\limsup_{n\to+\infty} \mathcal{A}_n(\varepsilon) &:= \{\omega \in \Omega \mid \exists \text{ an infinity of } n \text{ such that } |Z_n(w)| > \varepsilon\}\\ &\subset \big\{\omega \in \Omega \mid \sup_{k\geqslant n}|Z_k(\omega)| > \varepsilon\big\},\end{aligned}$$

and the probability of this event tends to 0 as $n \to +\infty$ by hypothesis. Hence

$$\mathbf{P}\big(\limsup_{n\to+\infty} \mathcal{A}_n(\varepsilon)\big) = 0.$$

Consequently, the event

$$\mathcal{A} = \bigcup_{k\geqslant 1} \{\limsup_{n\to+\infty} \mathcal{A}_n(1/k)\}$$

is a countable union of events of probability 0, so $\mathbf{P}(\mathcal{A}) = 0$.

Let $\omega \in \Omega\backslash\mathcal{A}$. Then for all $k \geqslant 1$, $\omega \notin \limsup_{n\to+\infty} \mathcal{A}_n(1/k)$, so $|Z_n(\omega)| \leqslant 1/k$ for only a finite number of n; hence $|Z_n(\omega)| \to 0$ as $n \to +\infty$. ■

Proposition 11.8.5. *Let $(X_n)_{n\geqslant 1}$ be a sequence of random variables all having the same distribution. Set $Z_n = (X_1 + X_2 + \cdots + X_n)/n$. Then*

$$Z_n \xrightarrow[n\to+\infty]{} 0 \text{ in mean} \implies Z_n \xrightarrow[n\to+\infty]{} 0 \text{ almost surely.}$$

Proof. Let $\varepsilon > 0$. For $1 \leqslant n \leqslant N$, set

$$\mathcal{B}_{n,N}(\varepsilon) = \bigl\{\omega \in \Omega \mid \sup_{n\leqslant k\leqslant N} |Z_k(\omega)| > \varepsilon\bigr\}.$$

Let $Y_{n,N}$ denote the random variable equal to the largest $n \leqslant k \leqslant N$ such that $|Z_k| > \varepsilon$, so that we have the disjoint union

$$\mathcal{B}_{n,N}(\varepsilon) = \bigcup_{k=n}^{N} \{Y_{n,N} = k\}.$$

Then

$$\mathbf{P}(\mathcal{B}_{n,N}(\varepsilon)) = \sum_{k=n}^{N} \mathbf{P}(Y_{n,N} = k). \tag{11.1}$$

We have

$$\varepsilon \mathbf{P}(Y_{n,N} = k) < \mathbf{E}(|Z_k| \mathbb{1}_{Y_{n,N}=k}). \tag{11.2}$$

Partition the event $\{Y_{n,N} = k\}$ into

$$\{Y_{n,N} = k\} = \underbrace{\{Y_{n,N} = k \text{ and } Z_k \geqslant 0\}}_{\mathcal{C}^+_{n,N,k}} \cup \underbrace{\{Y_{n,N} = k \text{ and } Z_k < 0\}}_{\mathcal{C}^-_{n,N,k}}.$$

Then

$$\mathbf{E}(|Z_k| \mathbb{1}_{Y_{n,N}=k}) = \mathbf{E}(Z_k \mathbb{1}_{\mathcal{C}^+_{n,N,k}}) - \mathbf{E}(Z_k \mathbb{1}_{\mathcal{C}^-_{n,N,k}}).$$

Observe that each expected value is an average of k expected values of random variables $X_i \mathbb{1}_{\mathcal{C}^\pm_{n,N,k}}$ all having the same distribution. Thus, restricted to the average of the n first expected values, the right-hand side does not change. Thus

$$\begin{aligned}
\mathbf{E}(|Z_k| \mathbb{1}_{Y_{n,N}=k}) &= \mathbf{E}(Z_n \mathbb{1}_{\mathcal{C}^+_{n,N,k}}) - \mathbf{E}(Z_n \mathbb{1}_{\mathcal{C}^-_{n,N,k}}) \\
&\leqslant \mathbf{E}(|Z_n| \mathbb{1}_{\mathcal{C}^+_{n,N,k}}) + \mathbf{E}(|Z_n| \mathbb{1}_{\mathcal{C}^-_{n,N,k}}) \\
&= \mathbf{E}(|Z_n| \mathbb{1}_{Y_{n,N}=k}),
\end{aligned}$$

by gluing the pieces together. Hence by (11.2),

$$\varepsilon \mathbf{P}\Biggl(\bigcup_{k=n}^{N} \{Y_{n,N} = k\}\Biggr) \leqslant \sum_{k=n}^{N} \mathbf{E}(|Z_n| \mathbb{1}_{Y_{n,N}=k}) \leqslant \mathbf{E}(|Z_n|),$$

then by (11.1),

$$\varepsilon \mathbf{P}(\mathcal{B}_{n,N}(\varepsilon)) \leqslant \mathbf{E}(|Z_n|).$$

As this holds for all $N \geqslant n$, we obtain

$$\varepsilon \mathbf{P}\big(\sup_{k \geqslant n} |Z_k|\big) \leqslant \mathbf{E}(|Z_n|),$$

and we conclude by applying Proposition 11.8.4. ∎

Law of large numbers. There exist several versions of the law of large numbers, from *weak* to *strong* versions. We put them all together in the following theorem. We will use the same notation for a constant $\mu \in \mathbb{R}$ and for the constant random variable $\mu\colon \omega \mapsto \mu$. The weak law of large numbers is due to Khintchine, and the strong law to Kolmogorov (1929). The statement of the strong law that we give here is not the most general one: we use the additional hypothesis that the random variables X_n have finite variance. On the other hand, we only assume that they are not correlated, which is a weaker hypothesis than complete independence.

Theorem 11.8.6. *Let $(X_n)_{n \geqslant 1}$ be a sequence of random variables. Set*

$$Z_n = \frac{X_1 + X_2 + \cdots + X_n}{n}, \quad n \geqslant 1.$$

Then

(i) *if the X_n are not correlated and have the same expected value $\mu \in \mathbb{R}$ and variance $\sigma^2 < +\infty$, then $Z_n \xrightarrow[n \to +\infty]{} \mu$ in probability.*

(ii) *If the X_n are not correlated and have the same expected value $\mu \in \mathbb{R}$ and variance $\sigma^2 < +\infty$, then $Z_n \xrightarrow[n \to +\infty]{} \mu$ in mean.*

(iii) *If the X_n are not correlated and have the same probability distribution of expected value $\mu = \mathbf{E}(X_1)$ such that $\mathbf{E}(|X_1|) < +\infty$ and $\mathbf{V}(X_1) < +\infty$, then $Z_n \xrightarrow[n \to +\infty]{} \mu$ almost surely.*

Proof. Part (i) follows from the Bienaymé–Chebyshev inequality. We have $\mathbf{E}(Z_n) = \mu$ and $\mathbf{V}(Z_n) = \sigma^2/n$, so for all $\varepsilon > 0$,

$$\mathbf{P}(|Z_n - \mu| > \varepsilon) \leqslant \frac{\sigma^2}{n\varepsilon^2} \xrightarrow[n \to +\infty]{} 0.$$

Part (ii) uses essentially the same argument. For all $\varepsilon > 0$,

$$\begin{aligned} \mathbf{E}(|Z_n - \mu|) &= \mathbf{E}(|Z_n - \mu| \mathbb{1}_{|Z_n - \mu| \leqslant \varepsilon}) + \mathbf{E}(|Z_n - \mu| \mathbb{1}_{|Z_n - \mu| > \varepsilon}) \\ &\leqslant \varepsilon + \frac{1}{\varepsilon} \mathbf{E}((Z_n - \mu)^2) = \varepsilon + \frac{\sigma^2}{n\varepsilon} \xrightarrow[n \to +\infty]{} \varepsilon. \end{aligned}$$

We conclude by letting ε tend to 0.

Part (iii) follows from part (ii) and Proposition 11.8.5, considering the centred sequence $(X_n - \mu)_{n \geqslant 1}$. ■

The most familiar illustration of this result is the game of *heads or tails* with a well-balanced coin. Suppose that X_n is a Bernoulli random variable which takes the value 1 if the result of the i-th toss is *tails*. Then the frequency of *tails* obtained is almost surely equal to $1/2$.

Central limit theorem. This theorem formulates the convergence in distribution of the sequence of *normalised averages* of random variables to the centred normal distribution $\mathcal{N}(0,1)$. This distribution has *density for the Lebesgue measure* on $\mathbb{R}$ given by

$$d\mu(x) = \frac{1}{2\pi} \exp\Big(-\frac{x^2}{2}\Big)\, dx.$$

In other words, if Z is a random variable with distribution $\mathcal{N}(0,1)$, then

$$\mathbf{P}(Z \leqslant x) = \frac{1}{2\pi} \int_{-\infty}^{x} \exp\Big(-\frac{t^2}{2}\Big)\, dt.$$

Theorem 11.8.7. *Let $(X_n)_{n \geqslant 1}$ be a sequence of independent random variables having the same probability distribution. Suppose that*

$$\mu := \mathbf{E}(X_1) \quad \textit{and} \quad \sigma := \sqrt{\mathbf{V}(X_1)}$$

exist, and set

$$Z_n = \frac{X_1 + X_2 + \cdots + X_n}{n} \quad \textit{for } n \geqslant 1.$$

Then

$$\sqrt{n}\Big(\frac{Z_n - \mu}{\sigma}\Big) \xrightarrow[n \to +\infty]{} \mathcal{N}(0,1)$$

in distribution.

In a particular form, this result has been known since the work of de Moivre and Laplace in the early 1800s. There are more general versions, eliminating the hypothesis that the variables X_n all have the same distribution. The hypothesis of independence can also be weakened if the correlation of the variables X_n is sufficiently controlled. The reader may consult any classical book on probability theory to discover the various statements and their proofs.

Chapter 12

Weighted graphs and their applications

Over the years, graph theory has become not just a field of research in its own right, but a field crucially situated at the interface of many different domains. Apart from irrigating various areas of fundamental and applied mathematics, it plays a significant role in physics, particularly in superconductivity, statistical physics, quantum electrodynamics, nanotechnology and so forth, as well as in more classical subjects such as electrical networks and vibration problems. One reason for this plethora of relatively recent developments is the flexible nature of graphs themselves, which can be viewed as combinatorial objects, or as logical structures via the binary relations, or as algebraic constructions via the associated matrices, or as geometrical or topological spaces. The concept of a graph is remarkably supple, and constitutes a tool that can be integrated easily into all kinds of different theories. In this chapter, we will consider the case of weighted graphs, in the framework of networks and Markov chains.

12.1 Weighted graphs

A *weighted* graph is a graph $\Gamma = (V; E)$ equipped with a map w called *the weight*:

$$\begin{aligned} w\colon V \times V &\to [0, \infty[, \\ (x, y) &\mapsto w(x, y) \end{aligned}$$

satisfying

- $w(x, y) = w(y, x)$
- $w(x, y) > 0$ if and only if $\{x, y\} \in E$.

A weighted graph will be written as a triple $\Gamma = (V; E, w)$, or simply $(V; w)$. Every finite weight, i.e. every weight on a graph satisfying $\sum_{x,y \in V} w(x, y) < +\infty$, induces a map on the subsets of vertices $A \subset V$

$$\mu(A) = \sum_{x \in A} \mu_x \quad \text{with } \mu_x = \sum_{y \in \Gamma(x)} w(x, y)$$

satisfying $\mu(\emptyset) = 0$.

For every subset $I \subset \mathbb{N}$ and every family $(A_i)_{i \in I}$ of pairwise disjoint subsets of V, we have

$$\mu\Big(\bigcup_{i \in I} A_i\Big) = \sum_{i \in I} \mu(A_i).$$

Thus it is a *positive finite measure* on the set of subsets of V.

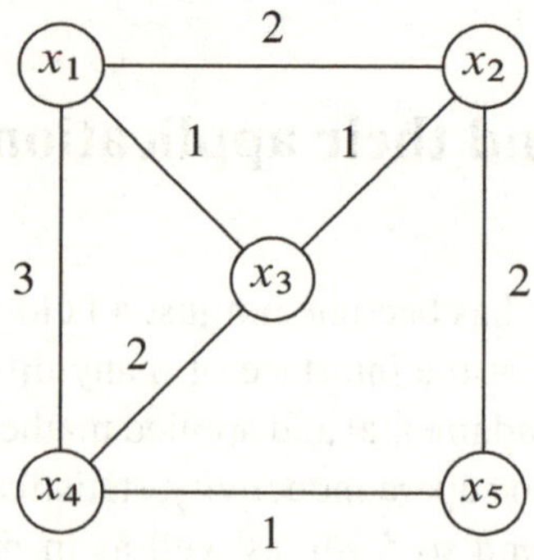

Figure 12.1. Example of a weighted graph with 5 vertices.

We can also define a weight on a digraph $\vec{\Gamma} = (V; \vec{E})$. Such a weight is not required to be symmetric; in a digraph we may allow $w(x, y) \neq w(y, x)$ for some vertices.

12.2 The Laplacian in physics

The Laplacian operator or Laplacian

$$f \mapsto \Delta f = \sum_{i=1}^{n} \frac{\partial^2 f}{\partial x_i^2} \quad (= \nabla^2 f)$$

generalises the notion of second derivative to several variables. Physically, $\Delta f(x_0)$ measures the difference between $f(x_0)$ and the average of f in the neighbourhood of x_0. The Laplacian arises in many different situations. Some examples are the following.

(i) Electromagnetism: $\Delta V = 0$, where V is the potential in the absence of charge.

(ii) Fluid mechanics: $\Delta \Phi = 0$, where Φ is the curl-free (irrotational) incompressible flow.

(iii) Wave equation: $\Delta f = \frac{1}{c^2} \frac{\partial^2 f}{\partial t^2}$, where c is the speed of the wave.

(iv) Poisson equation: for a charged elastic membrane,

$$\begin{cases} \Delta f = h & \text{on } U, \\ f = g & \text{on } \delta U, \end{cases}$$

where h is the charge density, g the boundary condition, and f the elevation of the membrane.

(v) Schrödinger equation: $\mathcal{H}\Psi = i\hbar\frac{\partial \Psi}{\partial t}$, where

$$\mathcal{H} = -\frac{\hbar^2}{2m}\nabla^2 + \mathcal{V}$$

is the Hamiltonian of Schrödinger with $\mathcal{V}$ the operator of multiplication by the potential.

(vi) Heat equation: $\frac{\partial T}{\partial t} = \frac{\kappa}{\rho c}\Delta T$ with thermal conductivity κ, mass density ρ, and specific heat c.

12.3 From physics to graphs: the harmonic Laplacian

12.3.1 The discrete second derivative

Let f be a differentiable function on $X \subset \mathbb{R}$ (or $\mathbb{C}$); then

$$\frac{\mathrm{d}f}{\mathrm{d}x}(x) = \lim_{h\to 0} \frac{f(x+h) - f(x)}{h}.$$

If h is very small, x and $x \pm h$ are *neighbours* and we have

$$\begin{aligned}\frac{\mathrm{d}f}{\mathrm{d}x}(x) &\approx \frac{f(x+h) - f(x)}{h}\\ &\approx \frac{f(x) - f(x-h)}{h}.\end{aligned}$$

The two operators above are called *difference operators* and constitute a good (numerical) approximation of the derivative of f at each point x where the function is differentiable.

Let us continue the computation

$$\begin{aligned}\frac{\mathrm{d}^2 f}{\mathrm{d}x^2}(x) &\approx \frac{\frac{\mathrm{d}f}{\mathrm{d}x}(x+h) - \frac{\mathrm{d}f}{\mathrm{d}x}(x)}{h}\\ &\approx \frac{f(x+h) - 2f(x) + f(x-h)}{h^2}\\ &\approx \frac{2}{h^2}\Big(\frac{f(x+h) + f(x-h)}{2} - f(x)\Big).\end{aligned}$$

This indicates the fact that for fixed h, $-\frac{h^2}{2}\frac{\mathrm{d}^2 f}{\mathrm{d}x^2}(x)$ is approximated by the difference between $f(x)$ and the average value of f at the 2 neighbouring points $x + h$ and $x - h$.

12.3.2 The Laplacian of a weighted graph

Let $\mathcal{F} = \mathbb{C}^V$. Recall that the Laplacian $\mathcal{L}$ of a graph $\Gamma = (V; E)$ is given by the formula

$$\mathcal{L} f(x) = d(x) f(x) - \sum_{y \in \Gamma(x)} f(y)$$

for $f \in \mathcal{F}$, $x \in V$. Observe that x is an isolated vertex if and only if $d(x) = 0$, and in this case $\mathcal{L} f(x) = 0$ for all $f \in \mathcal{F}$. If $\Gamma = (V; E)$ has no isolated vertex, we can write

$$\frac{\mathcal{L} f(x)}{d(x)} = f(x) - \frac{1}{d(x)} \sum_{y \in \Gamma(x)} f(y),$$

which is exactly the difference between the value of f at x and its average value on the neighbours of x. For $f \in \mathcal{F}$ and $x \in V$, we are thus naturally led to define the operator Δ by

$$\Delta f(x) = f(x) - \frac{1}{d(x)} \sum_{y \in \Gamma(x)} f(y).$$

The operator Δ is called the *harmonic Laplacian* of the graph Γ.

More generally, if $\Gamma = (V; E, w)$ is a finite weighted graph with no isolated vertex, we can define the *weighted harmonic Laplacian operator* by

$$\Delta_w f(x) = \frac{1}{\mu_x} \sum_{y \in V} (f(x) - f(y)) w(x, y) = f(x) - \frac{1}{\mu_x} \sum_{y \in V} f(y) w(x, y)$$

for $f \in \mathcal{F}$, $x \in V$.

Clearly Δ_w is a linear operator on $\mathcal{F}$. The operator Δ corresponds to the special case where all the weights of the edges $\{x, y\} \in E$ are equal to 1.

Because there are no isolated vertices, we have $\mu_x = \sum_{y \in V} w(x, y) > 0$ for all x. Thus we can associate to w a scalar product on $\mathcal{F}$, given by

$$\langle f, g \rangle_\mu := \sum_{x \in V} \mu_x f(x) \overline{g(x)}.$$

Recall that if $\mathcal{O}$ is a linear operator on $\mathcal{F}$, the adjoint $\mathcal{O}^*$ of $\mathcal{O}$ is defined by the identity

$$\langle \mathcal{O}^* f, g \rangle_\mu = \langle f, \mathcal{O} g \rangle_\mu$$

for all $f, g \in \mathcal{F}$, and that the operator $\mathcal{O}$ is said to be *Hermitian* if $\mathcal{O} = \mathcal{O}^*$.

Proposition 12.3.1. *Let $\Gamma = (V; E, w)$ be a weighted graph with no isolated vertex.*

(i) Δ_w *is a positive operator, i.e. for all* $f \in \mathcal{F}$ *we have* $\langle \Delta_w f, f \rangle_\mu \geqslant 0$.

(ii) Δ_w *is a Hermitian operator, i.e.* $\Delta_w = \Delta_w^*$.

Proof. (i) For all $f \in \mathcal{F}$, since $w(x,y) = w(y,x)$, we have

$$\sum_{x\in V} \mu_x \Delta_w f(x)\overline{f(x)} = \sum_{x\in V}\sum_{y\in V}(f(x)-f(y))\overline{f(x)}w(x,y)$$
$$= -\sum_{y\in V}\sum_{x\in V}(f(x)-f(y))\overline{f(y)}w(x,y)$$

by exchanging x and y. Taking the half-sum, we have

$$\langle \Delta_w f, f\rangle_\mu = \frac{1}{2}\sum_{x,y\in V}|f(x)-f(y)|^2 w(x,y) \geqslant 0.$$

(ii) For $f, g \in \mathcal{F}$, the difference $\langle \Delta_w f, g\rangle_\mu - \langle f, \Delta_w g\rangle_\mu$ can be written

$$\sum_{x,y\in V}(f(x)-f(y))\overline{g(x)}w(x,y) - \sum_{x,y\in V} f(x)\overline{(g(x)-g(y))}w(x,y)$$
$$= \sum_{x,y\in V} f(x)\overline{g(y)}w(x,y) - \sum_{x,y\in V} f(y)\overline{g(x)}w(x,y) = 0,$$

since these two sums are identical under exchanging x and y. ■

Exercise 12.1. Let $\vec{\Gamma} = (V; \vec{E}, w)$ be a weighted digraph such that $d^+(x) > 0$ for all $x \in V$. For $f \in \mathcal{F}$ and $x \in V$, define

$$\Omega_w f(x) = \frac{1}{\mu_x}\sum_{y\in V}\big(f(x)w(x,y)^{1/2} - f(y)w(y,x)^{1/2}\big)w(x,y)^{1/2}$$
$$= f(x) - \frac{1}{\mu_x}\sum_{y\in V} f(y)w(x,y)^{1/2}w(y,x)^{1/2}.$$

Show that Ω_w is a Hermitian operator for the scalar product $\langle\cdot,\cdot\rangle_\mu$.

In the context of the harmonic Laplacian, we can redefine the operators introduced above in the differential framework. For example, the Schrödinger Hamiltonian on the weighted graph $(V; w)$ can be written

$$\mathcal{H}_w f(x) = \frac{1}{\mu_x}\sum_{y\in\Gamma(x)}(f(x)-f(y))w(x,y) + U(x)f(x)$$
$$= f(x) - \frac{1}{\mu_x}\sum_{y\in\Gamma(x)} f(y)w(x,y) + U(x)f(x),$$

where $U\colon V \to \mathbb{R}$ is a potential function. Thus $\mathcal{H}_w = \Delta_w + \mathcal{U}$, where $\mathcal{U}$ denotes the operator of multiplication by the function U.

12.3.3 The gradient of a function and Green's identities

Let $f \in \mathcal{F} = \mathbb{C}^V$. The *gradient* of f on $\Gamma = (V; E, w)$, denoted by ∇f, is the map defined on $V \times V$ by

$$\nabla f(x, y) = (f(x) - f(y))\mathbb{1}_{\{x,y\} \in E}.$$

The *gradient* ∇ is a linear map from $\mathbb{C}^V$ to $\mathbb{C}^{V \times V}$. It corresponds to a notion of differentiation in several variables in physics.

The weighted Laplacian Δ_w is a function of the gradient:

$$\Delta_w f(x) = \frac{1}{\mu_x} \sum_{y \in V} \nabla f(x, y) w(x, y).$$

Green's formula is an important tool when working with the Laplacian. It generalises the notion of integration by parts to dimensions greater than one. Here is a discrete version:

Theorem 12.3.2 (Green's formula, discrete version). *Let $\Gamma = (V; E, w)$ be a weighted graph and $\emptyset \neq W \subset V$. For all functions $f, g \in \mathcal{F}$, we have*

$$\begin{aligned} 2\langle \Delta_w f_{|W}, g_{|W} \rangle_\mu = &\sum_{x,y \in W} \nabla f(x, y) \overline{\nabla g(x, y)} w(x, y) \\ &+ 2 \sum_{\substack{x \in W \\ y \in V \setminus W}} \nabla f(x, y) \overline{g(x)} w(x, y). \end{aligned}$$

Proof. Expanding $S = \langle \Delta_w f_{|W}, g_{|W} \rangle_\mu = \sum_{x \in W} \mu_x \Delta_w f(x) \overline{g(x)}$, we find that

$$\begin{aligned} S &= \sum_{x \in W} \sum_{y \in V} \nabla f(x, y) \overline{g(x)} w(x, y) \\ &= \sum_{x \in W} \sum_{y \in W} \nabla f(x, y) \overline{g(x)} w(x, y) + \sum_{x \in W} \sum_{y \in V \setminus W} \nabla f(x, y) \overline{g(x)} w(x, y) \\ &= -\sum_{y \in W} \sum_{x \in W} \nabla f(x, y) \overline{g(y)} w(x, y) + \sum_{x \in W} \sum_{y \in V \setminus W} \nabla f(x, y) \overline{g(x)} w(x, y). \end{aligned}$$

We deduce the desired identity by adding the two last equalities. ■

On the space $\{F : V \times V \to \mathbb{C}\} = \mathbb{C}^{V \times V}$, we can define the scalar product

$$\langle\langle F, G \rangle\rangle_w = \sum_{(x,y) \in V \times V} F(x, y) \overline{G(x, y)} w(x, y).$$

We then have the following *transfer principle*.

Corollary 12.3.3. *Let $\Gamma = (V; E, w)$ be a weighted graph. Then for all $f, g \in \mathcal{F}$,*

$$\langle f, \Delta_w g\rangle_\mu = \langle \Delta_w f, g\rangle_\mu = \frac{1}{2}\langle\langle \nabla f, \nabla g\rangle\rangle_w.$$

Proof. It suffices to take $W = V$ in Green's theorem and note that the second term of the right-hand side vanishes since $V \setminus W = \emptyset$. ∎

Exercise 12.2. Let $\vec{\Gamma} = (V; \vec{E}, w)$ be a weighted digraph and $f, g \in \mathcal{F} = \mathbb{C}^V$. Set

$$\Delta_w^+ f(x) = \frac{1}{\mu_x} \sum_{y \in V} (f(x) - f(y)) w(x, y),$$

$$\Delta_w^- f(x) = \frac{1}{\mu_x} \sum_{y \in V} (f(x) - f(y)) w(y, x).$$

Prove the oriented version of Green's theorem:

$$\langle \Delta_w^+ f, g\rangle_\mu + \langle \Delta_w^- f, g\rangle_\mu = \langle\langle \nabla f, \nabla g\rangle\rangle_w.$$

12.3.4 The oriented Laplacian

Let $\vec{\Gamma} = (V; \vec{E}, w)$ be a weighted digraph. We will define a Laplacian operator on $\vec{\Gamma}$ that satisfies the symmetry principle. The *oriented Laplacian* is defined by

$$\vec{\Delta}_w = \Delta_w^+ + \Delta_w^-.$$

Proposition 12.3.4. *Let $\vec{\Gamma} = (V; \vec{E}, w)$ be a weighted digraph. The generalised Laplacian satisfies the two following properties:*

(i) *$\vec{\Delta}_w$ is a linear Hermitian operator acting on $\mathcal{F}$,*

(ii) *$\vec{\Delta}_w$ is a positive operator.*

Proof. (i) Since Δ_w^+ and Δ_w^- are linear, $\vec{\Delta}_w$ is also linear. By Exercise 12.2, we see that $\vec{\Delta}_w$ is Hermitian.

(ii) Also by Exercise 12.2, we see that for all $f \in \mathcal{F}$, we have

$$\begin{aligned}\langle \vec{\Delta}_w f, f\rangle_\mu &= \langle \Delta_w^+ f, f\rangle_\mu + \langle \Delta_w^- f, f\rangle_\mu \\ &= \sum_{x, y \in V} |f(x) - f(y)|^2 w(x, y) \geqslant 0,\end{aligned}$$

which shows that $\vec{\Delta}_w$ is positive. ∎

12.4 Dirichlet's problem

Let $\Gamma = (V; E, w)$ be a finite weighted graph with no isolated vertex, and let $W \subset V$ be a non-empty subset. A function $f : V \to \mathbb{R}$ is *harmonic* on W if for all $x \in W$, we have

$$\Delta_w f(x) = 0.$$

The property of being harmonic means that the average value of the function on the neighbourhood of a point is equal to the value of the function at that point:

$$f(x) = \frac{1}{\mu_x} \sum_{y \in \Gamma(x)} f(y) w(x, y).$$

Dirichlet's problem is a famous problem that turned out to be fundamental in numerous areas of mathematics and physics, arising in electrostatics, heat conduction, elasticity, gravitational theory and more, and stimulating significant advances of mathematical understanding in many different directions. The problem is stated as follows:

For a bounded domain W (i.e. an open connected bounded region) of $\mathbb{R}^n$ and given functions h and g, find a function f on the closure $\overline{W}$ of W that satisfies the following conditions:

$$\begin{aligned} \Delta f(x) &= h(x) \quad \text{for all } x \in W, \\ f(x) &= g(x) \quad \text{for all } x \in \overline{W} \backslash \mathrm{W}. \end{aligned}$$

Under certain very particular hypotheses, this problem admits a unique solution.

One example is the case where f is the potential of an electrostatic field on $W \subset \mathbb{R}^n$. In this situation, it is known that f satisfies the equation

$$\Delta f = h,$$

where $-h$ is the charge density in the interior of W (denoted $\mathring{W}$), while the values of f are entirely determined by the conditions in $\overline{W} \backslash W$.

We now introduce the analogue of Dirichlet's problem for weighted graphs. Let $\Gamma = (V; E, w)$ be a finite weighted graph with no isolated vertex and $W \subset V$ a non-empty subset. Recall that the *closure* of W is the set

$$\overline{W} = \{x \in V \mid \exists x' \in W \text{ such that } \{x, x'\} \in E\},$$

and the *boundary* of W is $\mathrm{Bo}(W) = \overline{W} \backslash W$.

Theorem 12.4.1. *Let $\Gamma = (V; E, w)$ be a weighted graph, $W \subsetneq V$ a non-empty subset and $h: W \to \mathbb{R}$, $g: V \setminus W \to \mathbb{R}$ two functions. Dirichlet's problem for weighted graphs is stated as follows: find the functions $f: V \to \mathbb{R}$ that satisfy the* differential *system*

$$\begin{aligned} \Delta_w f(x) &= h(x) \quad \textit{for all } x \in W, \\ f(x) &= g(x) \quad \textit{for all } x \in V \setminus W. \end{aligned}$$

In this version, Dirichlet's problem admits a unique solution f.

To prove this result, we will use the principle of the *minimum-maximum*, usually called the *min-max principle*.

Lemma 12.4.2. *Let $\Gamma = (V; E, w)$ be a connected weighted graph, $W \subset V$ such that $W \neq \emptyset$, and $f: V \to \mathbb{R}$. If f is harmonic on W reaches its maximum or its minimum on W, then f is constant on $\overline{W}$.*

Proof. Let $M = \max_{x \in V} f(x)$ and $V' = \{x \in V \mid f(x) = M\}$. By hypothesis, there exists $x \in W \cap V'$. As f is harmonic on W, we have $\Delta_w f(x) = 0$, so

$$0 = \frac{1}{\mu_x} \sum_{y \in \Gamma(x)} (f(x) - f(y)) w(x, y) = \frac{1}{\mu_x} \sum_{y \in \Gamma(x)} (M - f(y)) w(x, y),$$

which implies $f(y) = M$ for all $y \in \Gamma(x)$ since $(M - f(y)) \geqslant 0$ and $w(x, y) > 0$. Since Γ is connected, we conclude that $\overline{W} \subseteq V'$. We do the same for the minimum. ■

Proof of Theorem 12.4.1. Let us first show the uniqueness. Let f_1, f_2 be maps from V to $\mathbb{R}$ that are solutions of the problem, and set $f := f_1 - f_2$. Then f is a solution of the system

$$\begin{aligned} \Delta_w f(x) &= 0 \quad \text{for all } x \in W, \\ f(x) &= 0 \quad \text{for all } x \in V \setminus W. \end{aligned}$$

Suppose $f \neq 0$. Then there exists $x \in V$ such that $f(x) \neq 0$, so $x \in W$.

- If $f(x) > 0$, then f attains its maximum M at a value $x_0 \in W$, so it has constant value M on $\overline{W} = W \cup \mathrm{Bo}(W)$ (by Lemma 12.4.2), but this is impossible since $f = 0$ on $V \setminus W$.
- If $f(x) < 0$, we use the analogous principle for the minimum.

Therefore $f(x) = 0$ for all $x \in V$, so $f = 0$ and $f_1 = f_2$.

We now prove the existence of a solution. Let us rewrite the equation $\Delta_w f(x) = h(x)$ for all $x \in W$ as

$$f(x) - \sum_{y \in W} \frac{w(x, y)}{\mu_x} f(y) - \sum_{y \in V \setminus W} \frac{w(x, y)}{\mu_x} f(y) = h(x).$$

Since $f(y) = g(y)$ on $V\backslash W$, we obtain for $x \in W$

$$f(x) - \sum_{y\in W} \frac{w(x,y)}{\mu_x} f(y) = h(x) + \sum_{y\in V\backslash W} \frac{w(x,y)}{\mu_x} g(y).$$

For all $x \in W$, let $h_1(x)$ denote the right-hand side of this equality, $\mathcal{O}_w f(x)$ the left-hand side, and $\mathcal{F}'$ the subspace of $\mathbb{R}^V$ whose elements are the functions f' that vanish on $V\backslash W$. Then $\mathcal{O}_w$ is a linear operator on the space $\mathcal{F}'$: it is the restriction of Δ_w to $\mathcal{F}'$. Consider the functions f' and h'_1 in $\mathcal{F}'$ which coincide with f and h_1 respectively on W. Then the initial Dirichlet problem can be phrased as:

$$\begin{cases} \mathcal{O}_w f' = h'_1, \\ \quad f = g \quad \text{on } V\backslash W. \end{cases}$$

Let $f' \in \mathcal{F}'$ be such that $\mathcal{O}_w f' = 0$. Then for all $x \in W$, we have

$$f'(x) = \sum_{y\in V} \frac{w(x,y)}{\mu_x} f'(y),$$

i.e. f' is harmonic on W. By Lemma 12.4.2, f' must then vanish on $\mathrm{Bo}(W)$, so it reaches its maximum on W. Consequently, $f' \equiv 0$ on V. It follows that $\mathcal{O}_w$ is injective, so bijective since $\mathcal{F}'$ is a $\mathbb{R}$-vector space of finite dimension $|W|$. We then obtain the unique solution f to Dirichlet's problem by setting

$$\begin{cases} f_{|W} = \mathcal{O}_w^{-1} h'_1, \\ f_{|V\backslash W} = g. \end{cases}$$

■

12.5 Random walks in graphs

12.5.1 Markov chains

In this section, we consider graphs with a finite or countable number of vertices. A *random walk* on a (possibly oriented) graph is a random process consisting in moving from vertex to vertex along edges (or arcs), each move from one vertex to one of its neighbours depending on a given probability distribution.

Most propagation-diffusion systems can be modelled by random walks, e.g. the type of propagation that occurs in social or spatial networks. They are also found in automatic learning or classification processes, and in certain physical processes such as heat diffusion or electrical networks.

Let $\Gamma = (V; E, w)$ be a weighted graph. Recall that

$$\mu_x = \sum_{z\in\Gamma(x)} w(x,z)$$

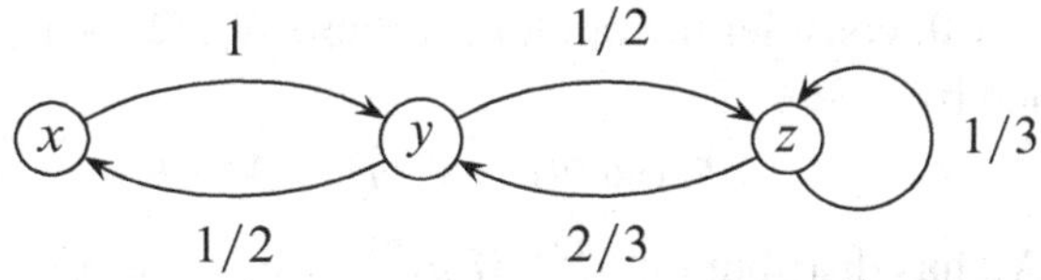

Figure 12.2. Example of a probability graph.

is strictly positive if x is not an isolated vertex of Γ. A *probability graph* is obtained by normalising the weight of the edges coming out of a given vertex as follows: if there is an edge between x and y, we set

$$p_{xy} = \frac{w(x,y)}{\mu_x};$$

otherwise $p_{xy} = 0$. Thus $p_{xy} \in]0,1]$ if and only if $\{x,y\} \in E$, and $\sum_{z\in\Gamma(x)} p_{xz} = 1$.

Insofar as the symmetry $w(x,y) = w(y,x)$ is not necessarily conserved by normalisation (since one may have $p_{xy} \neq p_{yx}$), the notion of a probability graph is even more relevant for digraphs than for graphs. Loops are also permitted in considering normalised weighted (di)graphs.

The vertices of these graphs are called *states* and the normalised weights p_{xy} are the *transition probabilities*.

Exercise 12.3. Construct the probability graph associated to the weighted graph of Figure 12.1.

Every probability graph induces a *Markov chain*, i.e. a discrete Markov process[1]. We first define an appropriate probability space $(\Omega, \mathcal{T}, \mathbf{P})$ as follows:

- the sample space $\Omega = V^{\mathbb{N}}$ is the set of sequences indexed by $0, 1, 2, \dots$ with values in V;
- the event space $\mathcal{T}$ is the σ-algebra generated by all the subsets

$$\{v_0\} \times \{v_1\} \times \cdots \times \{v_k\} \times V \times V \times \cdots, \quad k \geqslant 0, \quad v_i \in V;$$

- the probability measure $\mathbf{P}\colon \mathcal{T} \to [0,1]$ satisfies

$$\mathbf{P}(\{v_0\} \times \{v_1\} \times \cdots \times \{v_k\} \times V \times V \times \cdots) = \pi^{(0)}_{v_0} \times \prod_{i=0}^{k-1} p_{v_i v_{i+1}}$$

where $\pi^{(0)}_v$, $v \in V$, is a given distribution on the vertices of V, i.e. a distribution satisfying $0 \leqslant \pi^{(0)}_v \leqslant 1$ and $\sum_{v\in V} \pi^{(0)}_v = 1$.

[1] This means that the state space is discrete, either finite or countable.

For each $i \geqslant 0$, consider the random variable $X_i\colon \Omega \to V$ with values in the state space V defined by

$$X_i(v_0, v_1, \dots, v_i, \dots) = v_i.$$

Assume that X_0 has distribution $\pi^{(0)}$. If $\pi_x^{(0)} = \mathbf{P}(X_0 = x) > 0$, then

$$\begin{aligned}\mathbf{P}(X_1 = y \mid X_0 = x) &= \frac{\mathbf{P}(X_1 = y \text{ and } X_0 = x)}{\mathbf{P}(X_0 = x)} \\ &= \frac{\mathbf{P}(\{x\} \times \{y\} \times V \times \cdots)}{\mathbf{P}(\{x\} \times V \times \cdots)} = \frac{\pi_x^{(0)} p_{xy}}{\pi_x^{(0)}} = p_{xy}.\end{aligned}$$

Observe that if $\mathbf{P}(X_i = x) > 0$, then for all $x, y \in V$ we also have

$$\mathbf{P}(X_{i+1} = y \mid X_i = x) = p_{xy}.$$

This gives a *stochastic process* denoted $M = (X_i)_{i \geqslant 0}$, which depends on the initial state X_0 and on the transition probabilities p_{xy}. It is also said to be *homogeneous*, in the sense that the transition probabilities are independent of the instant i: if $\mathbf{P}(X_i = x_i) > 0$, then

$$\mathbf{P}(X_{i+1} = y \mid X_j = x_j \ \forall 0 \leqslant j \leqslant i) = p_{x_i y}.$$

The variable X_i is the state of the process (or the chain) at the instant i. The stochastic process M is called a *Markov chain*.

A *random walk* starting from the initial state $x_0 \in V$ on the probability graph is a Markov chain $(X_i)_{i \geqslant 0}$ for which $\mathbf{P}(X_0 = x_0) = 1$, i.e. X_0 is the constant random variable x_0: $X_0(\omega) = x_0$ for all $\omega \in \Omega$. We can extend this notion by assuming that the initial vertex of the walk is defined by a certain probability distribution $\pi^{(0)}$ for the random variable X_0.

The transition probabilities p_{xy}, $x, y \in V$, of the probability graph form a matrix

$$T = (p_{xy})_{x,y \in V}$$

called the *transition matrix* of the random walk: it is independent of the initial state.

When the weighted graph has no isolated vertex, the sums of the coefficients of each row of T are equal to 1: $\sum_{y \in V} p_{xy} = 1$. A matrix T with this property is called *stochastic*.

Exercise 12.4. Show that the transition matrix associated to the probability graph in Figure 12.2 is

$$\begin{pmatrix} 1 & 0 & 0 \\ 1/2 & 0 & 1/2 \\ 0 & 2/3 & 1/3 \end{pmatrix}.$$

For each i, let $\pi^{(i)}$ denote the distribution of the random variable X_i:

$$\pi_x^{(i)} = \mathbf{P}(X_i = x), \quad x \in V.$$

We can also consider $\pi^{(i)}$ as a row vector (which may be infinite if the graph is infinite):

$$\pi^{(i)} = (\pi_x^{(i)})_{x \in V}.$$

Let $i \geqslant 0$ and $x \in V$. Then for all $y \in V$, by the law of total probability, we have

$$\begin{aligned}\pi_y^{(i+1)} &= \sum_{x \in V} \mathbf{P}(X_{i+1} = y \mid X_i = x)\mathbf{P}(X_i = x) \\ &= \sum_{x \in V} \pi_x^{(i)} p_{xy},\end{aligned}$$

which is equivalent to the matrix equality $\pi^{(i+1)} = \pi^{(i)} T$. Then by induction we find that $\pi^{(i+j)} = \pi^{(i)} T^j$, where $T^j = (p_{xy}^{(j)})_{x,y \in V}$ is the j-th power of T. In this way, we recover the fact that the random walk is completely determined by $\pi^{(0)}$ and T, namely by the initial conditions and the transition matrix.

Remark 12.5.1. Conversely, any stochastic matrix T induces an oriented probability graph associated to a random walk with transition matrix T.

Proposition 12.5.2. *Let M be a Markov chain with transition matrix T and initial probability vector $\pi^{(0)}$. Then for all $i \in \mathbb{N}$, we have $\pi^{(i+1)} = \pi^{(i)} T$, $\pi^{(i)} = \pi^{(0)} T^i$ and Chapman-Kolmogorov's equation:*

$$p_{xz}^{(i+j)} = \sum_{y \in V} p_{xy}^{(i)} p_{yz}^{(j)}.$$

The σ-algebra generated by $X_1, \ldots, X_n$, denoted $\sigma(X_1, \ldots, X_n)$, is the space generated by the events $\{X_i = x\}$, $i = 1, \ldots, n$, $x \in V$. A *stopping time* adapted to the Markov chain $(X_i)_{i \geqslant 0}$ is a random variable $\Theta\colon \Omega \to \mathbb{N} \cup \{+\infty\}$ such that for all $n \in \mathbb{N}$,

$$\{\Theta \leqslant n\} \in \sigma(X_1, \ldots, X_n).$$

The *property of strong homogeneity* expresses the fact that we can make the variable n non-deterministic, or more precisely, defined by a stopping time. It is stated as follows.

Proposition 12.5.3. *Let $M = (X_i)_{i \geqslant 0}$ be a Markov chain and Θ a stopping time adapted to M. Let x_0 be a state such that*

$$\mathbf{P}(\Theta < +\infty \text{ and } X_\Theta = x_0) > 0.$$

Then

$$\mathbf{P}(X_{\Theta+i} = x_i \ \forall 1 \leqslant i \leqslant n \mid \Theta < +\infty \text{ and } X_\Theta = x_0)$$
$$= \mathbf{P}(X_i = x_i \ \forall 1 \leqslant i \leqslant n \mid X_0 = x_0) = \prod_{i=0}^{n-1} p_{x_i x_{i+1}}.$$

Proof. Let A_Θ be an event depending on Θ. Set $B_\Theta = \{\Theta < +\infty,\ X_\Theta = x_0\}$ and $B_k = \{\Theta = k,\ X_k = x_0\}$. Conditioning on the event $\Theta < +\infty$ makes it possible to expand

$$\mathbf{P}(A_\Theta \cup B_\Theta) = \sum_{k \geqslant 0} \mathbf{P}(A_k \cup B_k) = \sum_{k \geqslant 0} \mathbf{P}(A_k \mid B_k)\mathbf{P}(B_k).$$

For $A_k = \{X_{k+i} = x_i \ \forall 1 \leqslant i \leqslant n\}$, we obtain

$$\mathbf{P}(A_k \mid B_k) = \mathbf{P}(A_k \mid X_k = x_0) = \mathbf{P}(A_0 \mid X_0 = x_0) = \prod_{i=0}^{n-1} p_{x_i x_{i+1}}.$$

But $\sum_{k \geqslant 0} \mathbf{P}(B_k) = \mathbf{P}(B_\Theta)$, which gives the result by gluing the pieces. ■

Example 12.5.4. The most elementary Markov chain is the one with just two states, whose transition matrix can be written as follows with $0 < p, q < 1$:

$$T = \begin{pmatrix} 1-p & p \\ q & 1-q \end{pmatrix}.$$

The eigenvalues of T are 1 and $1 - p - q \neq 1$. Here is a representation of the associated probability graph.

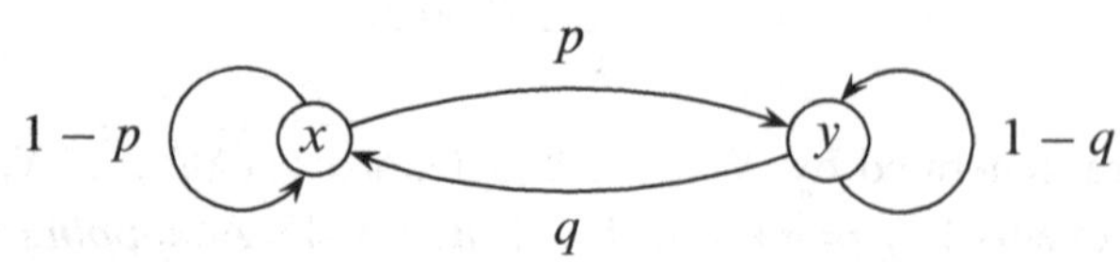

Exercise 12.5. Show that $\lim_{n \to +\infty} T^n = \frac{1}{p+q}\begin{pmatrix} q & p \\ q & p \end{pmatrix}$.

To compute T^n, note that $T^2 - (2 - p - q)T + (1 - p - q)I_2 = 0$ and use Euclidean division of the polynomial X^n by $X^2 - (2 - p - q)X + (1 - p - q)$.

12.5.2 Stationarity and reversibility

When the probability (di)graph is (strongly) connected, a given random walk can be *predictable* in the sense that it visits the different states according to a given probability distribution over a *long period of time*. This type of distribution reveals a principle of invariance that we now define.

Let $M = (X_i)_{i \geqslant 0}$ be a Markov chain with transition matrix $T = (p_{xy})_{x,y \in V}$. A probability distribution $\pi = (\pi_x,\ x \in V)$ is said to be *stationary* or *invariant* for M if $\pi = \pi T$, i.e.

$$\forall y \in V, \quad \pi_y = \sum_{x \in V} \pi_x p_{xy}.$$

It is thus a fixed point of the matrix T.

Stationarity can be interpreted as the fraction of time spent at each state of the Markov chain M, as the time interval becomes arbitrarily long. If we choose the initial distribution to be a probability distribution π that is invariant for M, then by Proposition 12.5.2, X_i has distribution π for all $i \in \mathbb{N}$. We obtain a *dynamic equilibrium*:

$$\forall k \geqslant 0, \quad \pi T^k = \pi.$$

It is natural to ask when a dynamic equilibrium exists. The following proposition gives a sufficient condition.

Proposition 12.5.5. *Let M be a Markov chain with finite state space and transition matrix T. Let $(\pi^{(i)})_{i \in \mathbb{N}}$ denote the sequence of states. If $\lim_{i \to +\infty} \pi_x^{(i)} = \mu_x$ exists for all $x \in V$, then the row vector $\mu = (\mu_x)_{x \in V}$ is a stationary distribution for M.*

Proof. By Proposition 12.5.2, for all $y \in V$ we have

$$\pi_y^{(i+1)} = (\pi^{(i)} T)_y = \sum_{x \in V} \pi_x^{(i)} p_{xy}.$$

Passing to the limit as $i \to +\infty$, we obtain $\mu_y = \sum_{x \in V} \mu_x p_{xy}$ for all $y \in V$, i.e. $\mu = \mu T$. Furthermore, μ is a probability distribution, since $\mu_x \geqslant 0$ for all x (being a limit of positive terms), and

$$\sum_{x \in V} \mu_x = \sum_{x \in V} \lim_{i \to +\infty} \pi_x^{(i)} = \lim_{i \to +\infty} \sum_{x \in V} \pi_x^{(i)} = 1$$

(in which exchanging the sum and the limit is allowed since $|V|$ is finite). ∎

Let $T = (p_{xy})_{x,y \in V}$ be a stochastic matrix on V. We saw that it is then the transition matrix of some Markov chain M. A probability distribution $\pi = (\pi_x)_{x \in V}$ on V is said to be *reversible* for T (or for M) if for all $x, y \in V$,

$$\pi_x p_{xy} = \pi_y p_{yx}.$$

A Markov chain is said to be *reversible* if it admits a reversible distribution π.

Exercise 12.6. Show that a stationary distribution in Example 12.5.4 is given by $(\frac{q}{p+q}, \frac{p}{p+q})$.

Proposition 12.5.6. *If M is a Markov chain with transition matrix $T = (p_{xy})_{x,y \in V}$, then every reversible distribution π of M is stationary.*

Proof. The hypothesis implies that for all $y \in V$, we have

$$\pi_y = \pi_y \sum_{x \in V} p_{yx} = \sum_{x \in V} \pi_y p_{yx} = \sum_{x \in V} \pi_x p_{xy},$$

and this shows that $\pi = \pi T$. ∎

Proposition 12.5.7. *Let $\Gamma = (V; E, w)$ be a weighted graph of finite degree with no isolated vertex. Then the probability distribution defined by*

$$\pi_x = \frac{\sum_{y \in \Gamma(x)} w(x, y)}{\sum_{\substack{y,z \in V \\ y \in \Gamma(z)}} w(z, y)} \quad \textit{for all } x \in V,$$

is a reversible and stationary distribution of the associated Markov chain M.

Proof. Let $x \in V$. Set $\mu_x = \sum_{y \in V} w(x, y)$, $|\mu| = \sum_{x \in V} \mu_x$. Then $\pi_x = \frac{\mu_x}{|\mu|}$. The transition probabilities of M are the numbers p_{xy} equal to $\frac{w(x,y)}{\mu_x}$ or 0 according to whether $\{x, y\}$ is an edge or not. Thus $p_{xy} = 0$ if and only if $p_{yx} = 0$. Thus $\pi_x p_{xy} = \pi_y p_{yx}$ if $p_{xy} = 0$. Otherwise, we have

$$\pi_x p_{xy} = \frac{\mu_x}{|\mu|} \frac{w(x, y)}{\mu_x} = \frac{\mu_y}{|\mu|} \frac{w(y, x)}{\mu_y} = \pi_y p_{yx}.$$ ∎

Remark 12.5.8. Given a Markov chain with state space V and transition matrix $T = (p_{xy})_{x,y}$, one can naturally associate to it the weighted oriented graph $(V; p)$. The induced weight p is not always symmetric. Nonetheless, the fact that a Markov chain on a finite number of states possesses a stationary distribution remains true, as we will see in Theorem 12.5.22.

Proposition 12.5.9. *Every Markov chain with a finite number of states possesses at least one stationary distribution.*

Exercise 12.7. Determine a stationary distribution for the chain associated to the probability graph in Figure 12.2.

12.5.3 Communication classes

Two vertices x, y are said to be *communicating* if they lie in the same connected component of the weighted graph (strongly connected if the graph is oriented): in other words if there exists a chain from x to y (resp. a path from x to y and a path from y to x). The vertices y which communicate with x form an equivalence class called the *communication class* of x.

Two vertices x, y communicate if and only if there exist i_1 and i_2 such that $p_{xy}^{(i_1)} > 0$ and $p_{yx}^{(i_2)} > 0$. The communication classes are thus the strongly connected components of the associated oriented graph. A communication class $\mathcal{C}$ is not always *closed*, in the sense that there can be paths between $x \in \mathcal{C}$ and $y \in V \backslash \mathcal{C}$; however, if so, we cannot return to $\mathcal{C}$ by a path from the class $\mathcal{C}'$ containing y.

A Markov chain is *irreducible* or *ergodic* if it has only a single communication class, i.e. every pair of vertices is communicating.

A Markov chain with transition matrix $T = (p_{xy})_{x,y \in V}$ is *regular* if there exists i such that $p_{xy}^{(i)} > 0$ for all $x, y \in V$. Every regular Markov chain is ergodic.

Exercise 12.8. Show that the Markov chain in Example 12.5.4 is regular, and thus ergodic.

12.5.4 Classification of states

A vertex (or state) $x \in V$ of a Markov chain is said to be *recurrent* if the probability that starting from this state x, the chain will return to it after a finite time is equal to 1. Writing Θ_x for the stopping time corresponding to the time it takes to reach the vertex x starting from the instant 1, then when x is recurrent we obtain

$$\gamma_x := \mathbf{P}(\Theta_x < +\infty \mid X_0 = x) = 1.$$

Let $N_x = \sum_{n \geqslant 0} \mathbb{1}_{X_n = x}$ denote the number of times the vertex x is visited during the random walk. In what follows, we will show that

$$\mathbf{P}(N_x = +\infty \mid X_0 = x) = 1.$$

In a digraph, a recurrent vertex x necessarily belongs to one of its strongly connected components C_x: this component coincides with the connected component of the underlying graph (there is no arc that exits C_x). It is a closed communication class.

A vertex or state x that is not recurrent is said to be *transient*:

$$\gamma_x = \mathbf{P}(\Theta_x < +\infty \mid X_0 = x) < 1.$$

We show below that the random variable N_x follows a geometric distribution with parameter γ_x.

For every event $\mathcal{A}$, set

$$\mathbf{P}_x(\mathcal{A}) = \mathbf{P}(\mathcal{A} \mid X_0 = x).$$

Let $k \geqslant 2$ be an integer. We have

$$\begin{aligned}\mathbf{P}_x(N_x = k \mid \Theta_x < +\infty) &= \mathbf{P}_x\Big(\mathbb{1}_{X_0 = x} + \sum_{n \geqslant \Theta_x} \mathbb{1}_{X_n = x} = k \mid \Theta_x < +\infty\Big) \\ &= \mathbf{P}_x\Big(\sum_{n \geqslant \Theta_x} \mathbb{1}_{X_n = x} = k - 1 \mid \Theta_x < +\infty\Big).\end{aligned}$$

The property of strong homogeneity (cf. Proposition 12.5.3) thus implies that

$$\mathbf{P}_x(N_x = k \mid \Theta_x < +\infty) = \mathbf{P}_x(N_x = k-1),$$

i.e.

$$\mathbf{P}_x(N_x = k) = \mathbf{P}_x(N_x = k-1)\gamma_x.$$

Case 1: if $\gamma_x = 1$ (x is recurrent), then all the terms $\mathbf{P}_x(N_x = k)$ for $k \geqslant 1$ are equal. But their sum satisfies

$$\sum_{k\geqslant 1} \mathbf{P}_x(N_x = k) = \mathbf{P}_x(N_x < +\infty) \leqslant 1.$$

This is a convergent series whose general term is constant: thus we have $\mathbf{P}_x(N_x = k) = 0$ for all $k \geqslant 1$ and $\mathbf{P}(N_x < +\infty) = 0$.

Case 2: if $\gamma_x < 1$ (x is transient), then we have

$$\mathbf{E}_x(N_x) = \sum_{k\geqslant 1} k\gamma_x^{k-1} = \frac{1}{(1-\gamma_x)^2}.$$

We can state this result as follows.

Proposition 12.5.10. *A recurrent state x is visited infinitely often with probability* 1 *in a random walk starting from x.*

We have also the following useful characterisation of recurrent states, which thus also characterises transient states.

Proposition 12.5.11. *A vertex x is recurrent if and only if the series $\sum_{n\geqslant 0} p_{xx}^{(n)}$ diverges.*

Proof. Set

$$\mathbf{E}_x(N_x) = \sum_{k\geqslant 1} k\mathbf{P}_x(N_x = k)$$

in $\mathbb{N} \cup \{+\infty\}$; it is the conditional expected value of N_x given that $X_0 = x$. If x is recurrent, then for all n we have

$$\mathbf{E}_x(N_x) \geqslant \sum_{k\geqslant n} n\mathbf{P}_x(N_x = k) = n\mathbf{P}_x(N_x \geqslant n) = n.$$

Thus $\mathbf{E}_x(N_x) = +\infty$. Furthermore,

$$\mathbf{E}_x(N_x) = \mathbf{E}_x\Big(\sum_{n\geqslant 0} \mathbb{1}_{X_n = x}\Big) = \sum_{n\geqslant 0} \mathbf{E}_x(\mathbb{1}_{X_n = x}),$$

where we may exchange the expected value and the $\sum$ sign thanks to the linearity of the expected value and the fact that the sum has only nonnegative terms. Thus we have

$$\mathbf{E}_x(N_x) = \sum_{n\geqslant 0} \mathbf{P}_x(X_n = x) = \sum_{n\geqslant 0} \mathbf{P}(X_n = x \mid X_0 = x) = \sum_{n\geqslant 0} p_{xx}^{(n)},$$

which proves the only if statement. For the if statement, if x is transient, we saw that $\mathbf{E}_x(N_x) < +\infty$, hence $\sum_{n\geq 0} p_{xx}^{(n)}$ is convergent. ∎

Exercise 12.9. Show that if y is transient then $\lim_{n\to+\infty} p_{xy}^{(n)} = 0$, by showing that the conditional expected value $\mathbf{E}_x(N_y)$ is finite for all x.

Proposition 12.5.12. *Let $x, y \in V$ be two vertices in the same communication class. Then x is recurrent (resp. transient) if and only if y is recurrent (resp. transient).*

Proof. Since x and y are communicating, there exists $i_1, i_2 \in \mathbb{N}$ such that $p_{xy}^{(i_1)} > 0$ and $p_{yx}^{(i_2)} > 0$. Hence

$$\sum_{n\geqslant i_1+i_2} p_{yy}^{(n)} = \sum_{n\geqslant 0} p_{yy}^{(i_2+n+i_1)} \geqslant \sum_{n\geqslant 0} p_{yx}^{(i_2)} p_{xx}^{(n)} p_{xy}^{(i_1)} = p_{yx}^{(i_2)} p_{xy}^{(i_1)} \sum_{n\geqslant 0} p_{xx}^{(n)}.$$

If x is recurrent, then $\sum_{n\geq 0} p_{xx}^{(n)}$ diverges, and the same holds for $\sum_{n\geq 0} p_{yy}^{(n)}$. ∎

Corollary 12.5.13. *The states belonging to the same communication class are of the same type: either all recurrent or all transient.*

This gives a decomposition of the state space.

Proposition 12.5.14. *Let M be a Markov chain. Then*

(i) *Every closed finite communication class of M is recurrent, i.e. all its states are recurrent.*

(ii) *Every communication class of M that is not closed and finite is transient: all its vertices are transient.*

Proof. Left as an exercise. The proposition does not deal with the case of non-closed infinite classes. However, this result does provide a complete classification in the case of a finite state space. ∎

When the state space of a random walk is finite, there exists at least one closed class. Then all its vertices are recurrent, so we obtain the following corollary.

Corollary 12.5.15. *In a finite ergodic Markov chain, all states are recurrent.*

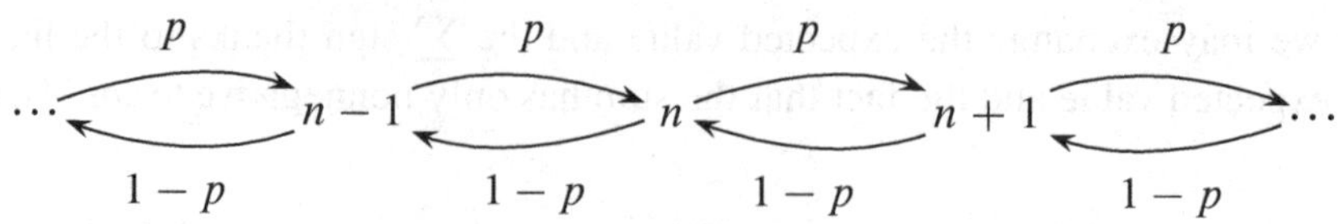

Figure 12.3. Random walk on $\mathbb{Z}$.

Exercise 12.10. Show that the Markov chain associated to the probability graph in Figure 12.2 is ergodic and that its vertices are recurrent.

Remark 12.5.16. Let $M = (X_i)_{i\geqslant 0}$ be a Markov chain on the graph $(V; E)$, with transition matrix $T = (p_{xy})_{x,y\in V}$. We already saw that a recurrent state $x \in V$ satisfies $\sum_{i\geqslant 0} p_{xx}^{(i)} = +\infty$ (cf. Proposition 12.5.11). Suppose that x lies in a closed finite class $\mathcal{C}$. For all $y \in \mathcal{C}$, we have $p_{yx}^{(j)} > 0$ for a $j = j(y) \geqslant 0$. But

$$p_{xx}^{(i+j)} \geqslant p_{xy}^{(i)} p_{yx}^{(j)},$$

so $\lim_{i\to+\infty} p_{xx}^{(i)} = 0$ implies that $\lim_{i\to+\infty} p_{xy}^{(i)} = 0$. This contradicts

$$\sum_{y\in\mathcal{C}} p_{xy}^{(i)} = 1 \quad \text{for all } i \geqslant 0.$$

We conclude from this that when x is recurrent in a closed finite class, we have $p_{xx}^{(i)} \nrightarrow 0$ when $i \to +\infty$.

Example 12.5.17. Consider the random walk on $\mathbb{Z}$ (Figure 12.3). Let $0 < p < 1$; the transition probability from n to $n+1$ is p, and from n to $n-1$ it is $1-p$, so it is ergodic. One can show that $\lim_{i\to+\infty} p_{nn}^{(i)} = 0$ for every state n.

Example 12.5.18. A player is playing heads or tails. She wins or loses \$ 1 according to whether she lands on heads or tails. She has \$ A at the start of the game, while the bank against whom she is playing has \$ B. The game stops when the player or the bank runs out of money. The coin tossed in this game is not necessarily well-balanced: the probability that it falls on heads is equal to some value $p \in \,]0,1[$.

Let $\{0, 1, \dots, A+B\}$ denote the states, namely the sum of money that the player may possess at each instant of the game. This game can be represented by a random walk having 3 communication classes: two closed recursive classes $\{0\}$ and $\{A+B\}$, and one transient class $\{1, 2, \dots, A+B-1\}$. Almost surely, either the player or the bank will be ruined after a finite time. Let $\mathcal{R}$ denote the event "the player ends up ruined", and let p_A denote the probability of the event $\mathcal{R}$. Then if $(X_i)_{i\geqslant 0}$ is the Markov chain induced by this process, we find

$$p_A = \mathbf{P}(\mathcal{R} \mid X_0 = A).$$

Suppose $A, B \geqslant 1$. By the formula of total probability, we see that

$$p_A = \sum_{j=0}^{A+B} \mathbf{P}(\mathcal{R} \mid X_0 = A \text{ and } X_1 = j)\mathbf{P}(X_1 = j \mid X_0 = A).$$

The only non-zero terms in this sum are for $j = A - 1$ and $A + 1$. Hence, by using the properties of Markov chains (homogeneous processes without memory), this gives

$$p_A = p p_{A+1} + (1 - p) p_{A-1},$$

which can be solved for p_A, since it is a linear recursive sequence of order 2. For the recurrent classes, we have $p_0 = 1$ and $p_{A+B} = 0$.

We distinguish 2 cases: $p \neq 1/2$ and $p = 1/2$. If $p \neq 1/2$, setting $\theta = \frac{1-p}{p}$, we find that

$$p_A = \frac{\theta^A - \theta^{A+B}}{1 - \theta^{A+B}}.$$

If $p = 1/2$, then an easy computation gives $p_A = \frac{B}{A+B}$.

12.5.5 Periodicity

The *period of a state* x of a Markov chain with transition matrix $T = (p_{xy})_{x,y \in V}$ is the greatest common divisor (gcd) of the set

$$\{k \in \mathbb{N}^* \mid p_{xx}^{(k)} > 0\}.$$

When this set is empty, we set the period of x to 0. A vertex is said to be *aperiodic* if its period is equal to 1. If all the vertices of the probability graph are aperiodic, the Markov chain is said to be *aperiodic*. Note that the period of x is the gcd of the lengths of all the closed chains (resp. closed paths) passing through x.

Proposition 12.5.19. *If x and y are in the same communication class, then their periods coincide.*

Proof. Since x and y are communicating, there exist integers i_1 and i_2 such that $p_{xy}^{(i_1)} > 0$ and $p_{yx}^{(i_2)} > 0$. Hence

$$p_{yy}^{(i_2+i_1)} = \sum_{z \in V} p_{yz}^{(i_2)} p_{zy}^{(i_1)} \geqslant p_{yx}^{(i_2)} p_{xy}^{(i_1)} > 0.$$

Let r_x be the period of x. There exist multiples $s_1, s_2, \dots, s_m$ of r_x such that $p_{xx}^{(s_j)} > 0$, $1 \leqslant j \leqslant m$, and $r_x = \gcd(s_j,\ j = 1, \dots, m)$. For each j, we have

$$p_{yy}^{(i_2+s_j+i_1)} = \sum_{z,z' \in V} p_{yz}^{(i_2)} p_{zz'}^{(s_j)} p_{z'y}^{(i_1)} \geqslant p_{yx}^{(i_2)} p_{xx}^{(s_j)} p_{xy}^{(i_1)} > 0.$$

The period r_y of y divides $i_2 + i_1$ and $i_2 + s_j + i_1$, so it also divides their difference $s_j = (i_2 + s_j + i_1) - (i_2 + i_1)$. It follows that r_y divides r_x. By symmetry, then r_x also divides r_y, which gives the result. ∎

Proposition 12.5.20. *Let M be a Markov chain with a finite state space and transition matrix $T = (p_{xy})_{x,y \in V}$. We assume that M is ergodic and aperiodic. Then M is regular. More precisely, there exists i_0 such that for all $q \geqslant i_0$ and all pairs x, y of vertices, $p_{xy}^{(q)} > 0$.*

Proof. Let x be a state of the chain. Since M is ergodic, there exists $i = i(x)$ such that $p_{xx}^{(i)} > 0$. But M is aperiodic, so there exists an integer j coprime to i with i such that $p_{xx}^{(j)} > 0$. By Bézout's identity, there exist positive integers u, v such that $ui = 1 + vj$ or $vj = 1 + ui$. We thus obtain an integer $k = k(x)$ such that $p_{xx}^{(k)} > 0$ and $p_{xx}^{(k+1)} > 0$.

Let $k' \geqslant k^2$. Then $k' = k^2 + sk + r$ with $s \geqslant 0$ and $0 \leqslant r \leqslant k - 1$. Thus

$$k' = (k + r - s)k + s(k + 1), \quad k + r - s > 0,\ s \geqslant 0.$$

Hence $p_{xx}^{(k')} > 0$.

As the number of states is finite, there exists an integer i_1 such that $p_{xx}^{(q)} > 0$ for all $x \in V$ and all $q \geqslant i_1$. As M is ergodic, for all $x \neq y$ there exists $i(x, y) \in \mathbb{N}$ such that $p_{xy}^{(i(x,y))} > 0$. Take

$$i_0 = \max_{x,y \in V} i(x, y) + i_1.$$

For $q \geqslant i_0$ and $x, y \in V$, we have $q = i(x, y) + q'$ with $q' = q - i(x, y) \geqslant i_1$, so $p_{xy}^{(q)} \geqslant p_{xx}^{(q')} p_{xy}^{(i(x,y))} > 0$. ∎

Proposition 12.5.21. *Let M be a Markov chain associated to $\Gamma = (V; E, w)$, a non-oriented weighted graph that possesses at least one cycle. Let $T = (p_{xy})_{x,y \in V}$ denote its transition matrix. Then M is aperiodic if and only if Γ is not bipartite.*

Proof. Recall that a graph is bipartite if and only if all its cycles are of even length.

As the graph admits at least one edge $\{x, y\}$, there exists a chain x, y, x of length 2 from x to x. If the graph is bipartite, then

$$\gcd\{i \in \mathbb{N}^* \mid p_{xx}^{(i)} > 0\}$$

is a multiple of 2, so M is not aperiodic.

Conversely, if the graph is not bipartite there exists a cycle of odd length i from x to x, so $p_{xx}^{(i)} > 0$. Then since $\gcd(2, i) = 1$, we deduce that M is aperiodic. ∎

Exercise 12.11. Show that in the random walk on $\mathbb{Z}$ (cf. Example 12.5.17) every state has period 2.

Exercise 12.12. Show that the Markov chain associated to the probability graph in Figure 12.2 is aperiodic and regular.

12.5.6 Memorandum of fundamental properties

Theorem 12.5.22. *Let $M = (X_i)_{i \geqslant 0}$ be a Markov chain on a finite set of states V, $|V| = n$, with transition matrix $T = (p_{xy})_{x,y \in V}$. Then*

(i) *M possesses at least one stationary probability distribution;*

(ii) *if M is ergodic, the stationary probability distribution is unique: we denote it $\mu = (\mu_x,\ x \in V)$;*

(iii) *if M is ergodic and aperiodic, then $\lim_{i \to +\infty} T^i$ is the matrix with every row equal to μ.*
Furthermore, whatever the distribution of the initial state X_0, that of X_i converges to μ (convergence in distribution).

Proof. (i) We first need to find a row vector v with positive coordinates such that $vT = v$, or equivalently, a column vector $w = {}^t v$ such that ${}^t T w = w$. Given that T is stochastic, 1 is an eigenvalue of T associated to the eigenvector $\mathbf{1} = (1, 1, \ldots, 1)$ viewed as a column vector. Thus 1 is also an eigenvalue for the transpose ${}^t T$ since $\det({}^t T - I_n) = \det(T - I_n) = 0$. Now let ${}^t v = {}^t(v_x,\ x \in V)$ be any column vector invariant under ${}^t T$, and define the column vector w by setting $w_x = |v_x|$, $x \in V$. Then for all $y \in V$, by the triangle inequality, we have

$$\delta_y := \sum_{x \in V} w_x p_{xy} - w_y = \sum_{x \in V} |v_x| p_{xy} - |v_y| \geqslant \Big| \sum_{x \in V} v_x p_{xy} \Big| - |v_y| = 0.$$

However,

$$\sum_{y \in V} \delta_y = \sum_{x \in V} w_x \sum_{y \in V} p_{xy} - \sum_{y \in V} w_y = \sum_{x \in V} w_x - \sum_{y \in V} w_y = 0,$$

so $\delta_y = 0$ for each $y \in V$. We conclude that ${}^t T w = w$, so that ${}^t w T = {}^t w$. The row vector ${}^t w$ thus yields a stationary probability distribution for M.

(ii) As the rank of the matrix ${}^t T - I_n$ is equal to that of $T - I_n$, it suffices to show that $\operatorname{Ker}(T - I_n) = \mathbb{R}\mathbf{1}$.

Let $x, y \in V$ be two distinct vertices such that $v_x = \max_{z \in V} v_z$, $v_y = \min_{z \in V} v_z$.

Since M is ergodic, there exists an integer $i \geqslant 1$ such that $p_{xy}^{(i)} > 0$. Furthermore, v is an eigenvector of the stochastic matrix T^i associated to 1. We thus have

$$0 = v_x - v_x = \sum_{z \in V} (v_x - v_z) p_{xz}^{(i)};$$

all terms in this sum are positive or zero, so they are all zero. In particular, $v_x = v_y$: $v \in \mathbb{R}\mathbf{1}$.

(iii) By Proposition 12.5.20 we have $p_{xy}^{(q)} > 0$ for all sufficiently large q and all $x, y \in V$. The eigenvalues of T^q are those of T to the power q. By (ii), 1 is a simple eigenvalue of T^q. Let us temporarily admit (see below) the Perron–Frobenius theorem, which asserts that the other eigenvalues of T, so also of T^q, all have module strictly less than 1. We write T in the form

$$T = Q \begin{pmatrix} 1 & 0 \\ 0 & S \end{pmatrix} Q^{-1}$$

where S is a square matrix of order n and the 0's indicates that we complete this region of the matrix with coefficients 0. Then we have

$$T^i = Q \begin{pmatrix} 1 & 0 \\ 0 & S^i \end{pmatrix} Q^{-1}.$$

The first column of Q can be chosen equal to the vector $\mathbf{1}$: it corresponds to an eigenvector associated to the eigenvalue 1. As $\mu T = \mu$, the first line of Q^{-1} is a multiple $\alpha\mu$ of the invariant distribution μ. But $QQ^{-1} = I_n$, so we must have $\alpha = 1$.

By hypothesis, all the eigenvalues of S in $\mathbb{C}$ have modulus < 1, so S^i tends to the zero matrix as $i \to +\infty$. We conclude that

$$\lim_{i \to +\infty} T^i = Q \begin{pmatrix} 1 & 0 \\ 0 & 0 \end{pmatrix} Q^{-1}$$

is the matrix each row of which is the invariant distribution $\mu = (\mu_x, x \in V)$. ■

Lemma 12.5.23 (Perron–Frobenius theorem). *Let $A = (a_{ij})_{1 \leqslant i,j \leqslant n}$ be a real square matrix of order n all of whose coefficients are strictly positive. Then A admits a simple eigenvalue $\lambda > 0$, and all the other eigenvalues in $\mathbb{C}$ have modulus strictly less than λ.*

Proof. We only need this result in the special case where A is stochastic, so we prove it only in this case. Let $\lambda \in \mathbb{C}$ be an eigenvalue of A different from 1, and let $v = (v_1, \dots, v_n)$ be a (non-zero) eigenvector associated to λ. Let v_i denote one of the coordinates of v of maximum modulus: $|v_i| = \max_{1 \leqslant j \leqslant n} |v_j|$. Then

$$|\lambda v_i| = \Big|\sum_{j=1}^{n} a_{ij} v_j\Big| \leqslant \sum_{j=1}^{n} a_{ij} |v_j| \leqslant |v_i|$$

since $\sum_{j=1}^{n} a_{ij} = 1$. We obtain $|\lambda| \leqslant 1$. In fact, we can even do better.

If there exists at least one j such that $|v_j| < |v_i|$, then since $a_{ij} > 0$,

$$|\lambda v_i| < \sum_{j=1}^{n} a_{ij}|v_i| = |v_i|,$$

hence $|\lambda| < 1$.

Otherwise all the v_j have the same module, in which case we have

$$\Big|\sum_{j=1}^{n} a_{ij}v_j\Big| = \sum_{j=1}^{n}|a_{ij}v_j| = |v_i|,$$

so the complex numbers $a_{ij}v_j$ for $j = 1, \dots, n$ all have the same argument, which implies that the complex numbers v_j have the same argument since $a_{ij} > 0$. We conclude that the v_j's are equal: v is an eigenvector attached to the eigenvalue 1. ∎

Theorem 12.5.24. *Let M be a Markov chain with transition matrix $T = (p_{xy})_{x,y\in V}$. Then*

(i) *if M is reversible (and thus stationary) for the vector $\pi = (\pi_x)_{x\in V}$, then we can associate to it a weighted graph $\Gamma = (V; E, w)$:*
   - *the set of vertices V is the set of states of M;*
   - *$w(x, y) = \pi_x p_{xy}$;*
   - *$\{x, y\} \in E$ if and only if $p_{xy} > 0$.*

(ii) *Conversely, let $\Gamma = (V; E, w)$ be a weighted graph with no isolated vertex such that for all $x \in V$, $\mu_x = \sum_{y\in V} w(x, y) \in]0, +\infty[$. We can associate to it a reversible Markov chain M:*
   - *the set of states of M are the vertices of the graph;*
   - *the transition probabilities are defined by $p_{xy} = \frac{w(x,y)}{\mu_x}$;*
   - *the vector $(\mu_x)_{x\in V}$ is a reversible vector for the transition matrix $T = (p_{xy})_{x,y\in V}$;*
   - *if moreover $|\mu| := \sum_{x\in V} \mu_x < +\infty$, then the following probability distribution is stationary:*

$$(\pi_x)_{x\in V} = \left(\frac{\mu_x}{|\mu|}\right)_{x\in V}.$$

We give the following result without proof.

Theorem 12.5.25. *Let $M = (X_i)_{i\geqslant 0}$ be an ergodic Markov chain. Assume that starting from the state $x \in V$, the time of return to x satisfies $\mathbf{E}_x(\Theta_x) < +\infty$ for every state $x \in V$. Then the unique probability distribution μ stationary for M is given by*

$$\mu_x = \frac{1}{\mathbf{E}_x(\Theta_x)}, \quad x \in V.$$

Note that when V is finite, the condition $\mathbf{E}_x(\Theta_x) < +\infty$ is always satisfied. One can show that this condition is a property of a whole class, so that it suffices for it to be satisfied for just one of the (recurrent) states of the class. Thus, the above result yields an expression of the stationary probability distribution for an ergodic Markov chain on a finite number of states.

The theorem that we now state, known as the *ergodic theorem*, expresses a predictable asymptotic behaviour for ergodic Markov chains satisfying the condition $\mathbf{E}_x(\Theta_x) < +\infty$ for every state $x \in V$.

Theorem 12.5.26. *Suppose that $M = (X_i)_{i \geqslant 0}$ is an ergodic Markov chain such that $\mathbf{E}_x(\Theta_x) < +\infty$ for every state $x \in V$. Let μ denote its unique stationary probability distribution. Let $f : V \to \mathbb{C}$ be a μ-integrable map*[2]*. Then almost surely, i.e. with probability* 1*, we have*

$$\lim_{n \to +\infty} \sum_{k=0}^{n-1} f(X_k) = \sum_{x \in V} \mu_x f(x) = \sum_{x \in V} \frac{f(x)}{\mathbf{E}_x(\Theta_x)}.$$

We can interpret this deep result as a *strong law of large numbers*, for which the usual independence hypotheses are not even necessary! Under the same conditions, we have a *central limit theorem*:

$$\sqrt{n}\left(\frac{1}{n}\sum_{k=0}^{n-1} f(X_k) - \sum_{x \in V} \mu_x f(x)\right) \xrightarrow[n \to +\infty]{} \mathcal{N}(0,1),$$

the reduced centred normal distribution.

12.5.7 Examples of Markov chains

The knight's tour. In chess, the knight moves around the 64-square chessboard according to a particular well-known rule. For each given position i on the chessboard, there is a precise number c_i of positions the knight can reach from there. Consider the graph $\Gamma = (V; E)$ of knight's moves (V is the set of 64 numbered squares, $\{i, j\}$ exists if the knight can get from i to j). Suppose furthermore that all moves starting from a given square i are equiprobable. Then the *knight's trajectory* can be modelled by a Markov chain, whose transition matrix is given by

$$p_{ij} = \begin{cases} 1/c_i & \text{if } \{i, j\} \in E, \\ 0 & \text{otherwise.} \end{cases}$$

[2] That is $\sum_{x \in V} \mu_x |f(x)| < +\infty$.

The vector $c = (c_i)_{1 \leqslant i \leqslant 64}$ is a reversible vector. Set $|c| = \sum_{i=1}^{64} c_i$; then the distribution $\frac{1}{|c|} \cdot c$ is stationary.

The web. The success of the search engine *Google* is based on an algorithm called `PageRank`, which attributes to each web page a weight according to its *importance*.

The network known as the *internet* (or the *World Wide Web*) can be modelled by an oriented graph called the *graph of the web*:

- the vertices are the web pages,
- an arc from x to y represents a link from the page x to the page y; we say that x *links to* y.

The basic principle of the algorithm is as follows: the importance of a page x is measured as a function of several variables: the number of *important* pages that link to x, the number of links to x from other pages, and the *popularity* of those other pages, i.e. the volume of links to them. Thus the whole structure is a weighted oriented graph. Robots browse the web, going from one page to another by clicking on random links on the pages they visit. Thus a robot's trajectory is nothing other than a random walk on the associated probability graph, in which all arcs coming out of a given vertex are equiprobable. This oriented weighted graph $(V; \vec{E}, w)$ is not necessarily strongly connected: if the graph possesses hanging vertices, i.e. pages that do not link to any other page (*hanging pages*), the robot will then swing randomly to another arbitrarily chosen page.

If x is not a hanging vertex, then its exit degree satisfies $d^+(x) \geqslant 1$. Define the *impact factor* of x by setting

$$f_y = \sum_{\substack{x \in V \\ (x,y) \in \vec{E}}} \frac{f_x}{d^+(x)} + \frac{n_{\text{hang}}}{|V|},$$

where n_{hang} is the number of hanging pages of the graph. The impact factor of y is a weighted average of impact factors of all the pages x linking to y, corrected by uniformly taking account of hanging pages. The definition of f above is *implicit*, since f_y depends on the f_x with $(x, y) \in \vec{E}$, which can themselves depend on f_y. An appropriate fixed-point theorem (precisely Theorem 12.5.22 (i)) proves the existence (but not necessarily the uniqueness) of the impact factor function f.

Start by associating probabilities to the graph as follows: the transition probability p_{xy} of passing from the vertex x to the vertex y is given by

$$p_{xy} = \begin{cases} 1/d^+(x) & \text{if } (x, y) \in \vec{E}, \\ 1/|V| & \text{if } x \text{ is hanging}, \\ 0 & \text{otherwise.} \end{cases}$$

The matrix $T=(p_{xy})_{x,y\in V}$ is the transition matrix of the Markov chain $M=(X_i)_{i\geqslant 0}$ associated to the web probability graph:

$$\mathbf{P}(X_{i+1}=y \mid X_i=x)=p_{xy}.$$

Observe that $f=fT$, so that f is a stationary distribution for M.

Google's idea for measuring the (importance) score of a web page is to compute the probability that it will be visited in time t: in other words it measures the visits to the page by considering whether

(i) visitors to the page are numerous;

(ii) the time spent on the page by the visitors is long;

(iii) the other pages linking to this page are numerous.

Given that a *surfer* often leaves a page without clicking on any of its links, the Markov chain M is in practice fairly abstract. Furthermore, it is not always ergodic, since the graph is not necessarily strongly connected. Because of this, Google introduces a correction in the form of a real number $\alpha\in\,]0,1[$ for which $1-\alpha$ corresponds to the probability of leaving a non-hanging page without clicking on any link. Set

$$T^{(\alpha)}=\alpha T+(1-\alpha)\times\frac{1}{n}\times J,$$

where $J={}^t\mathbf{1}\mathbf{1}$ is the square matrix of order n all of whose coefficients are equal to 1. All terms of the stochastic matrix $T^{(\alpha)}$ are strictly positive, so it induces an ergodic Markov chain $M^{(\alpha)}$. Consequently, there exists a unique stationary probability distribution $\mu^{(\alpha)}$, by Theorem 12.5.22 (ii) for $M^{(\alpha)}$. Furthermore, by (iii) of the same theorem, X_i converges in distribution to $\mu^{(\alpha)}$. This stationary distribution yields a criterion measuring the *importance* of each web page, depending on α, namely

$$\mu_x^{(\alpha)}=\frac{1}{\mathbf{E}_x(T_x)},$$

where T_x is the time of the first return to x.

12.5.8 The 2-dimensional Ising model

Before describing the Ising model, let us introduce a statistical method that starts from a probability distribution π on a finite set $K\subset\mathbb{R}^n$ and constructs a Markov chain whose stationary distribution is π.

Metropolis–Hastings algorithm. Let π be a probability distribution on the state space K. We can use π to construct other probability distributions on K. Let $T=$

$(p_{xy})_{x,y\in K}$ be a stochastic matrix on K. Define the matrix $T' = (p'_{xy})_{x,y\in K}$ by

$$p'_{xy} = p_{xy} \min\Big\{1; \frac{\pi_y p_{yx}}{\pi_x p_{xy}}\Big\} \quad \text{for all } x \neq y \in K,$$

$$p'_{xx} = 1 - \sum_{\substack{y\in K\\ y\neq x}} p'_{xy} \quad \text{for all } x \in K.$$

Proposition 12.5.27. *The matrix T' is stochastic and induces a reversible Markov chain which admits π as stationary distribution.*

Proof. For all $x \neq y \in K$, we clearly have $0 \leqslant p'_{xy} \leqslant p_{xy} \leqslant 1$. For all fixed $x \in K$, we have

$$\sum_{\substack{y\in K\\ y\neq x}} p'_{xy} + p'_{xx} = 1.$$

Thus T' is a stochastic matrix.

Let $x \neq y \in K$ be such that $\pi_y p_{yx} \leqslant \pi_x p_{xy}$. Then

$$\pi_x p'_{xy} = \pi_x p_{xy} \frac{\pi_y p_{yx}}{\pi_x p_{xy}} = \pi_y p_{yx} = \pi_y p'_{yx}.$$

Thus every Markov chain M' with transition matrix T' is reversible, and consequently π is a stationary distribution for M'. ∎

The Ising distribution. The spin is a characteristic quantum property of particles in physics, which takes only integer or half-integer values. For example electrons, positrons, neutrinos, and quarks have spin $\frac{1}{2}$, whereas the Higgs boson has spin 0 and the photon has spin 1.

The Ising model is one of the most important models in statistical mechanics. The model that we consider (for example to study a ferromagnetic material) is supported by a *bounded grid* in $\mathbb{Z}^2$. As an example, we can take $\mathcal{E}$ to be a finite subset of $\mathbb{Z}^2$ (imagine that its elements represent electrons). Two particles $u = (u_1, u_2)$ and $v = (v_1, v_2)$ of $\mathcal{E}$ are neighbours if $\|u - v\|^2 = (u_1 - v_1)^2 + (u_2 - v_2)^2 = 1$, which we denote by $u \sim v$; a vertex can thus have between 0 and 4 neighbouring particles. This defines a graph, denoted by $\Gamma = (\mathcal{E}; \mathbf{A})$, where $n = |\mathcal{E}|$ is finite. To each particle $u = (u_1, u_2) \in \mathcal{E}$, we associate its *spin measure*, denoted $\sigma(u)$ and equal to $+1$ or -1. Given a vector $\vec{V}$ of $\mathbb{R}^2$, we are only able to determine the sign of the scalar product $\vec{V} \cdot \vec{S}_u$, where $\vec{S}_u$ is the real spin at $u \in \mathcal{E}$.

These two values encode the state of the particle u, modelled by an element of $\mathcal{E}$ placed in a magnetic field: the spin of u points either up $(+1)$ or down (-1). The state space is given by

$$\mathcal{S} = \{-1, 1\}^{\mathcal{E}} = \{\sigma : \mathcal{E} \to \{-1, 1\}\}.$$

Let $\sigma \in \mathcal{S}$ be a particular configuration of spins in the model, and associate to it a quantity called the energy defined by

$$E_\sigma = -J \sum_{u \sim v \in \mathcal{E}} \sigma(u)\sigma(v),$$

where J is a positive physical constant: in our example it is a constant depending on the ferromagnetic nature of the material under study.

Let T denote the temperature and k_B Boltzmann's constant, and set

$$Z_E = \sum_{\sigma \in \mathcal{S}} \exp\Big(-\frac{E_\sigma}{k_B T}\Big).$$

This is a normalisation constant called the *partition function*. The distribution of each spin is modelled by the frequency of each configuration $\sigma \in \mathcal{S}$:

$$\operatorname{freq}(\sigma) = \frac{1}{Z_E} \exp\Big(-\frac{E_\sigma}{k_B T}\Big).$$

We thus obtain an explicit probability distribution, called the *Ising distribution*. This measure on $\mathcal{S}$ favours low energies, i.e. orientations of spins that are coherent between neighbouring particles.

The full number 2^n of spin configurations is generally too large to be able to study the system exhaustively. The goal is thus simply to *approach* the Ising model by adding a random ingredient. We construct an ergodic Markov chain whose stationary distribution is precisely the Ising distribution. The graph supporting the chain has state space $\mathcal{S}$, the set of spin configurations, and two configurations are neighbours if they differ only by a single spin.

We then proceed so as to respect the rules of statistical mechanics, in particular with the two following hypotheses:

- the spin measure of the particle u is a Bernoulli random variable with values in $\{-1, +1\}$;
- if $u \sim v \in \mathcal{E}$, then $\mathbf{P}(\sigma(u)\sigma(v) = 1) > \mathbf{P}(\sigma(u)\sigma(v) = -1)$; this condition implies that neighbouring particles have a tendency to align their spins.

Take the transition matrix $T = (p_{\sigma\sigma'})_{\sigma,\sigma' \in \mathcal{S}}$ defined by

$$p_{\sigma\sigma'} = \begin{cases} 1/n & \text{if } \sigma(u) \neq \sigma'(u) \text{ for only one } u \in \mathcal{E}, \\ 0 & \text{otherwise.} \end{cases}$$

We now *correct* the transitions using the Metropolis–Hastings algorithm. Starting from a spin configuration σ, we choose a vertex $u \in \mathcal{E}$ at random with probability $\frac{1}{n}$ (discrete uniform distribution on $\mathcal{E}$), and flip its spin: this gives a new configuration σ'. If the energy of the new configuration $E_{\sigma'}$ is less than that of σ, $E_{\sigma'} < E_\sigma$, we accept the transition to σ'. If $E_\sigma \geqslant E_{\sigma'}$, we accept the transition with probability $\exp(-\frac{E_{\sigma'}-E_\sigma}{k_B T})$. Proceeding in this manner, when σ and σ' differ only by a single spin, we obtain the corrected transition probability

$$p'_{\sigma\sigma'} = \frac{1}{n} \min\Big\{1; \exp\Big(-\frac{E_{\sigma'} - E_\sigma}{k_B T}\Big)\Big\}.$$

If σ and σ' differ by at least two spins, we fix $p'_{\sigma\sigma'} = 0$. The value of $p'_{\sigma\sigma}$ is fixed by the fact that the matrix $T' = (p_{\sigma\sigma'})_{\sigma,\sigma' \in \mathcal{S}}$ is a transition matrix.

For σ and σ' differing only by a single spin, we have

$$\frac{\mathrm{freq}(\sigma') p_{\sigma'\sigma}}{\mathrm{freq}(\sigma) p_{\sigma\sigma'}} = \exp\Big(-\frac{E_{\sigma'} - E_\sigma}{k_B T}\Big).$$

Consequently, the matrix T' is exactly the one obtained by the Metropolis–Hastings construction. We thus obtain a reversible Markov chain $(X_i)_{i\geqslant 0}$ admitting the stationary distribution $(\mathrm{freq}(\sigma),\ \sigma \in \mathcal{S})$. One can show that this chain is ergodic and aperiodic; hence by Theorem 12.5.22, the distribution of X_i converges to the Ising distribution, whatever the distribution of X_0.

A physical application of this situation is the magnetisation of the system

$$M_\sigma = \sum_{u \in \mathcal{E}} \sigma(u),$$

considered here as a random variable of $\mathcal{S}$ with values in $\mathbb{Z}$. Considering the sequence of means on the configurations accepted by the algorithm, the strong law of large numbers (cf. Theorem 12.5.26) yields an almost-sure approximation of the magnetisation of the system.

Remark 12.5.28. The version of the Ising model described above concerns a random walk in a situation that has a specific notion of energy attached to the system (depending on the material). Assume that the same material is now influenced by a magnetic field $\vec{B}$, which induces an additional energy on each particle. In this case, there is an additional rule of statistical mechanics that must be respected: $\vec{V}$ being the direction used for the physical measure, we have

$$\mathbf{P}(\sigma(u) = +1) > \mathbf{P}(\sigma(u) = -1).$$

for all $u \in \mathcal{E}$ whenever $\vec{B} \cdot \vec{V} > 0$. This gives rise to the following model, called the *Ising Hamiltonian*:

$$E_\sigma = -J \sum_{u \sim v \in \mathcal{E}} \sigma(u)\sigma(v) - \mu B \sum_{v \in \mathcal{E}} \sigma(u),$$

where $B = \|\vec{B}\|$ and $\mu = \mu_{\vec{B}}$ is a number between -1 and 1. The Metropolis–Hastings algorithm applies as above, and yields a random walk that gives an approximation of the stationary distribution attached to the spins of the system embedded in a magnetic field. The general case corresponds to

$$E_\sigma = -\beta \sum_{u \sim v \in \mathcal{E}} \sigma(u)\sigma(v) - \alpha \sum_{v \in \mathcal{E}} \sigma(u)$$

with two parameters $\alpha \in \mathbb{R}$ and $\beta \in \mathbb{R}_+$.

The simple case is given in Algorithm 12.1.

```
Function Ising(Γ = (ℰ; A): graph, N: integer): vector
  configuration vector: σ, σ';
  begin
    /* fix a configuration */
    Take σ ∈ S;
    Fix β = J/(k_B T);
    Compute the energy E_σ;
    for t from 0 to N − 1 do
      Generate a particle u ∈ ℰ according to a uniform distribution;
      Set σ'(v) = σ(v) if v ≠ u and σ'(u) = −σ(u);
      Compute Δ = E_σ' − E_σ and δ = exp(−βΔ);
      if Δ < 0 then
        σ = σ';                                  /* the transition is accepted */
      end
      else
        Generate u ∈ [0, 1] according to a continuous uniform distribution;
        if u < δ then
          σ = σ';                                /* the transition is accepted */
        end
      end
      return σ;
    end
  end
EndFunction
```

Algorithm 12.1. The Ising distribution via the Metropolis–Hastings algorithm.

12.6 Electrical networks

12.6.1 Definitions

Let $\Gamma = (V; E, c)$ be a non-oriented weighted graph. We introduced several notions from physics:

(i) For every edge $\{x, y\} \in E$, $c(x, y)$ is the *conductance* of the edge $\{x, y\}$ and its inverse $r(x, y) = c(x, y)^{-1}$ is its *resistance*; observe that $r: E \to \mathbb{R}_+$.

(ii) Let a be a fixed vertex and V' a set of vertices such that $a \notin V'$; we define a function $v: V \to \mathbb{R}$ called the *potential function* which satisfies the following relations:

- $v(a) = 1$ and $v(y) = 0$ for all $y \in V'$;
- $i_{x,y} = (v(x) - v(y))c(x, y)$ (*Ohm's law*): $i_{x,y}$ is the *current* between x and y.

The graph $\mathcal{R} = (V; E, c)$ is called an *electrical network*. Set

$$i_x := \sum_{y \in \Gamma(x)} i_{x,y}.$$

The map $i: E \to \mathbb{R}$ satisfies:

- $i_{x,y} = -i_{y,x}$ for all $x \in V \backslash (V' \cup \{a\})$,
- $i_x = 0$ for all $x \in V \backslash (V' \cup \{a\})$ (*Kirchhoff's law*).

The vertex a can be considered as the *entry* or *source* of the network, while the set V' can be considered as the set of vertices *connected to the earth*. Generally, $V' = \{b\}$: b correspond to a certain type of *absorbing* vertex [3] of the Markov chain that we will now investigate.

Proposition 12.6.1. *The potential function $v: V \to \mathbb{R}$ is harmonic on $V \backslash (V' \cup \{a\})$.*

Proof. We have $(v(x) - v(y))c(x, y) = i_{x,y}$, so for all $x \in V \backslash (V' \cup \{a\})$,

$$\sum_{y \in \Gamma(x)} (v(x) - v(y))c(x, y) = \sum_{y \in \Gamma(x)} i_{x,y} = 0.$$

Let $c(x) = \sum_{y \in \Gamma(x)} c(x, y)$ denote the *capacity* of the vertex x. Then

$$\sum_{y \in \Gamma(x)} (v(x) - v(y))c(x, y) = c(x)\Big(v(x) - \sum_{y \in \Gamma(x)} \frac{c(x, y)}{c(x)} v(y)\Big) = 0.$$

As $c(x) \neq 0$ on $V \backslash (V' \cup \{a\})$, we obtain the result. ■

[3] An absorbing vertex is a state for which all incident arcs are entering.

It follows that the map v is the unique solution of the following Dirichlet system:

$$\begin{cases} \Delta_c v(x) = 0 & \text{for all } x \in V \backslash (V' \cup \{a\}), \\ v(x) = 0 & \text{for all } x \in V', \\ v(a) = 1. \end{cases}$$

12.6.2 Electrical networks and random walks

In this paragraph, we will interpret electrical networks in terms of random walks. Let $\mathcal{R} = (V; E, c)$ be a network, and arbitrarily choose two vertices a and b. Let v be a function defined on V. Connect a battery whose difference of potential is equal to one volt between a and b in such a way that $v(b) = 0$ and $v(a) = 1$. This gives an electrical network.

Proposition 12.6.2. *For each vertex $x \in V$, the electric potential $v(x)$ at x is equal to the probability that starting from the state x, the random walk returns to the state a before reaching the state b.*

Proof. By Proposition 12.6.1, the function $v \colon V \to \mathbb{R}$ is harmonic on $V \backslash \{a, b\}$, so $v(x) = \sum_{y \in \Gamma(x)} \frac{c(x,y)}{c(x)} v(y)$ for all $x \neq a, b$.

Let $p(x)$ be the probability that the random walk starting from x reaches a before b. We clearly have $p(a) = 1 = v(a)$ and $p(b) = 0 = v(b)$.

Furthermore, starting from x, the probability of reaching a before b can be written in the following form, taking account of the neighbours of x:

$$p(x) = \sum_{y \in \Gamma(x)} \frac{c(x, y)}{c(x)} p(y). \tag{12.1}$$

But since both functions v and p are harmonic and satisfy the same initial conditions at a and b (the boundaries), the principle of the maximum which implies the uniqueness gives $p = v$, hence the result. ∎

We have just given an interpretation of the potential in terms of probabilities; we can also give a probabilistic interpretation of the current. The current corresponds to positive electrical charges that leave the vertex a and move around randomly in the circuit, where b is grounded, i.e. b is absorbing.

Proposition 12.6.3. *Suppose that a receives a current of one ampere. Then the quantity of current $i_{x,y}$ that goes through an edge $\{x, y\}$ from x to y is equal to the expected value of the algebraic number (i.e. with sign $\pm$) of passages through the edge $\{x, y\}$ from x to y during a random walk starting at a and stopping when b is reached.*

Proof. Let $\{x, y\} \in E$ and let $\mathbf{e}_x$ denote the expected value of the number of passages through x by the current: passing through x means both entering and exiting. Set

$$p_{xy} = \mathbf{P}(X_{t+1} = y \mid X_t = x) = \frac{c(x, y)}{c(x)}.$$

We have $\mathbf{e}_b = 0$ since the walk terminates when b is reached.

For all $x \neq a$, $\mathbf{e}_x = \sum_{y \in \Gamma(x)} \mathbf{e}_y p_{yx}$ since before reaching x, the walk must pass through one of its neighbours. Furthermore,

$$c(x)p_{xy} = c(x)\frac{c(x, y)}{c(x)} = c(x, y) = c(y, x) = c(y)\frac{c(y, x)}{c(y)} = c(y)p_{yx}$$

because the conductance of an edge is *symmetric*. Thus $c(x)p_{xy} = c(y)p_{yx}$, which corresponds to a reversibility property. We thus obtain

$$\frac{\mathbf{e}_x}{c(x)} = \sum_{y \in \Gamma(x)} p_{xy}\frac{\mathbf{e}_y}{c(y)} = \sum_{y \in \Gamma(x)} \frac{c(x, y)}{c(x)}\frac{\mathbf{e}_y}{c(y)}.$$

Therefore the function $x \mapsto \frac{\mathbf{e}_x}{c(x)}$ is harmonic on $V \setminus \{a, b\}$. It has value 0 at b and $\frac{\mathbf{e}_a}{c(a)}$ at a. By the uniqueness principle (cf. Theorem 12.4.1) and (12.1), we conclude that $v = \frac{\mathbf{e}}{c}$. Consequently,

$$i_{x,y} = (v(x) - v(y))c(x, y) = c(x, y)\Big(\frac{\mathbf{e}_x}{c(x)} - \frac{\mathbf{e}_y}{c(y)}\Big) = p_{xy}\mathbf{e}_x - p_{yx}\mathbf{e}_y.$$

This is the expected value of the number of passages from x to y minus the number of passages from y to x. This proves the result. ■

12.6.3 Thomson's principle and Rayleigh's law

Let v be a potential function in the network $\mathcal{R} = (V; E, c)$, and let i be the associated current. The *effective resistance* of the network between a and b is given by

$$r_{\mathtt{eff}} = \frac{v(a)}{i_a}.$$

When $v(a) = 1$, then the *effective conductance* is $c_{\mathtt{eff}} = i_a$. We have

$$\begin{aligned} i_a &= \sum_{x \in \Gamma(a)} (v(a) - v(x))c(a, x) = \sum_{x \in \Gamma(a)} (1 - v(x))\frac{c(a, x)}{c(a)}c(a) \\ &= c(a)\Big(\sum_{x \in \Gamma(a)} \frac{c(a, x)}{c(a)} - \sum_{x \in \Gamma(a)} v(x)\frac{c(a, x)}{c(a)}\Big) = c(a)(1 - p(a)) \end{aligned}$$

by (12.1). We saw that $v(x) = p(x)$ is the probability that the walk starting from x meets a before arriving at b. We can thus interpret the quantity

$$p(a) = \sum_{x \in \Gamma(a)} v(x) \frac{c(a,x)}{c(a)} = \sum_{x \in \Gamma(a)} p_{ax} v(x)$$

as the probability that the walk starting from a returns to a without have reached b at all. Consequently,

$$p_{\mathrm{esc}} = 1 - \sum_{x \in \Gamma(a)} p_{ax} v(x)$$

is the probability that the walk starting from a reaches b before returning to a. In other words, if the vertex b is connected to the earth, then p_{esc} is the probability that the electric charge exits the circuit. We thus have $c_{\mathtt{eff}} = c(a) p_{\mathrm{esc}}$.

We now generalise the notion of a current. Let $\mathcal{R} = (V; E, c)$ be a weighted graph, equipped with resistances that make it into an electrical network. Let $a, b \in V$ be two distinct vertices. A *flux* through $\mathcal{R}$ from a to b is a map $j \colon V \times V \to \mathbb{R}$ satisfying

(i) $j_{x,y} = 0$ if and only if $\{x, y\} \notin E$;

(ii) $j_{y,x} = -j_{x,y}$;

(iii) for all $x \in V \backslash \{a, b\}$, $j_x := \sum_{y \in \Gamma(x)} j_{x,y} = 0$.

Note that $j_b = -j_a$. We say that the flux is *unitary* if moreover $j_a = 1$. If we impose a tension v between a and b in the circuit $\mathcal{R} = (V; E, c)$, the current which circulates in the branches $\{x, y\} \in E$ defines a flux i.

Note the analogy with the notion of flow already seen in Chapter 4. A flux is a function that describes the way in which current behaves on each edge. The orientation can be fixed arbitrarily. The *dissipated energy* in the electrical network $\mathcal{R}$ is defined by

$$E_{\mathcal{R}}(j) = \frac{1}{2} \sum_{x \sim y} r(x,y) j_{x,y}^2,$$

where $x \sim y$ means that $\{x, y\} \in E$: each edge appears twice in the sum.

Lemma 12.6.4. *Let $\mathcal{R} = (V; E, c)$ be an electrical network, $f \colon V \to \mathbb{R}$ a function and j a flux through $\mathcal{R}$ from a to b. Then*

$$(f(a) - f(b)) j_a = \frac{1}{2} \sum_{x \sim y} (f(x) - f(y)) j_{x,y}.$$

Proof. Apply the relations $j_{x,y} = -j_{y,x}$, $j_x = 0$ for all $x \in V \backslash \{a, b\}$ and $j_a = -j_b$. We obtain

$$\sum_{x \sim y} (f(x) - f(y)) j_{x,y} = \sum_{x \in V} \sum_{y \in \Gamma(x)} f(x) j_{x,y} - \sum_{y \in V} \sum_{x \in \Gamma(y)} f(y) j_{x,y}$$

$$= 2\sum_{x\in V} f(x) \sum_{y\in\Gamma(x)} j_{x,y}$$
$$= 2(f(a)j_a + f(b)j_b)$$
$$= 2(f(a) - f(b))j_a,$$

which gives the result. ■

If $f = v$ is a potential function and $j = i$ is the current determined by Ohm's and Kirchhoff's laws, we obtain the well-known principle of conservation of electrical energy: the dissipated energy (Joule heating effect) is equal to the energy supplied, i.e.

$$E_{\mathfrak{R}}(i) = \frac{1}{2}\sum_{x\sim y} r(x,y)i_{x,y}^2 = (v(a) - v(b))i_a.$$

Proposition 12.6.5. *Assume that i is the unitary flux determined by the Ohm's and Kirchhoff's laws, running through an electrical network $\mathfrak{R} = (V; E, c)$ from a to b, on which we have imposed an appropriate difference of potential v. Then $r_{\mathtt{eff}} = E_{\mathfrak{R}}(i)$.*

Proof. We have $i_a = 1$. By Ohm's law, for every edge $\{x, y\} \in E$, we have

$$r(x,y)i_{x,y} = v(x) - v(y).$$

By Lemma 12.6.4, we obtain

$$r_{\mathtt{eff}} = (v(a) - v(b))i_a = \frac{1}{2}\sum_{x\sim y}(v(x) - v(y))i_{x,y} = \frac{1}{2}\sum_{x\sim y} r(x,y)i_{x,y}^2,$$

i.e. precisely $E_{\mathfrak{R}}(i)$. ■

Theorem 12.6.6. *If j is an arbitrary unitary flux running through an electrical network $\mathfrak{R} = (V; E, c)$ from a to b, then $E_{\mathfrak{R}}(j) \geqslant r_{\mathtt{eff}}$.*

Proof. Let i denote the flux of unitary current: $E_{\mathfrak{R}}(i) = r_{\mathtt{eff}}$. The difference $\delta = j - i$ is a circulation (cf. Section 6.3): we have $\delta_a = \delta_b = 0$. This gives

$$E_{\mathfrak{R}}(j) = \frac{1}{2}\sum_{x\sim y} r(x,y)j_{x,y}^2 = \frac{1}{2}\sum_{x\sim y} r(x,y)(i_{x,y} + \delta_{x,y})^2$$
$$= E_{\mathfrak{R}}(i) + E_{\mathfrak{R}}(\delta) + \sum_{x\sim y}(v(x) - v(y))\delta_{x,y}.$$

As no flux either comes in or goes out of a and b, we deduce from Lemma 12.6.4 that $\sum_{x\sim y}(v(x) - v(y))\delta_{x,y} = 0$, so $E_{\mathfrak{R}}(j) = E_{\mathfrak{R}}(i) + E_{\mathfrak{R}}(\delta)$. We conclude by Proposition 12.6.5 together with the fact that $E_{\mathfrak{R}}(\delta) \geqslant 0$. ■

The result of Thomson's theorem (Theorem 12.6.6) is known as *Thomson's principle*[4]. It says that the dissipated energy in an electrical network by the flux of unitary current (determined by Ohm's and Kirchhoff's laws) is a minimum for the dissipated energy in the network by the other unitary fluxes.

Theorem 12.6.7. *Let $(V; E)$ be a graph equipped either with a set of conductances $c(x, y)$, or a different set of conductances $\tilde{c}(x, y)$, satisfying $c(x, y) \geqslant \tilde{c}(x, y)$ for every edge $\{x, y\} \in E$. Then $r_{\mathtt{eff}} \leqslant \tilde{r}_{\mathtt{eff}}$.*

Proof. Let i (resp. j) be the flux of unitary current in $\mathcal{R} = (V; E, c)$ (resp. $\tilde{\mathcal{R}} = (V; E, \tilde{c})$) determined by Ohm's and Kirchhoff's laws. By Proposition 12.6.5 and Thomson's principle, we obtain

$$\begin{aligned}\tilde{r}_{\mathtt{eff}} = E_{\tilde{\mathcal{R}}}(j) = \frac{1}{2}\sum_{x\sim y}\tilde{r}(x, y)j_{x,y}^2 \geqslant \frac{1}{2}\sum_{x\sim y} r(x, y)j_{x,y}^2 = E_{\mathcal{R}}(i) + E_{\mathcal{R}}(j - i)\\ \geqslant E_{\mathcal{R}}(i) = r_{\mathtt{eff}},\end{aligned}$$

which proves the result. ∎

This result is known as *Rayleigh's monotonicity law*. Since

$$c(a)p_{\mathtt{esc}} = \frac{1}{r_{\mathtt{eff}}},$$

the fact of deleting an edge $e = \{x, y\}, x, y \neq a$, in the circuit (letting the conductance $c(x, y)$ tend to 0) increases the effective resistance, so decreases the probability $p_{\mathtt{esc}}$ of reaching b before a. Observe that the capacity $c(a)$ is not modified by deleting the edge e. However, adding an edge does cause it to actually increase.

12.6.4 Random walk in $\mathbb{Z}^2$

Consider the infinite graph whose support is the algebraic network $\mathbb{Z}^2$. The vertices are the pairs of integers $x = (x_1, x_2)$, and the edges are the segments connecting the pairs of points separated by a distance equal to 1. Setting

$$p_{xy} = \begin{cases} 1/4 & \text{if } \|x - y\| = 1,\\ 0 & \text{otherwise,}\end{cases}$$

we obtain an ergodic Markov chain whose associated graph is clearly connected. The unique communication class is either transient or recursive. We call this process *an isotropic random walk* in $\mathbb{Z}^2$. Thanks to the properties of electrical circuits [11], we will show the following result.

[4] William Thomson is better known under the name of Lord Kelvin.

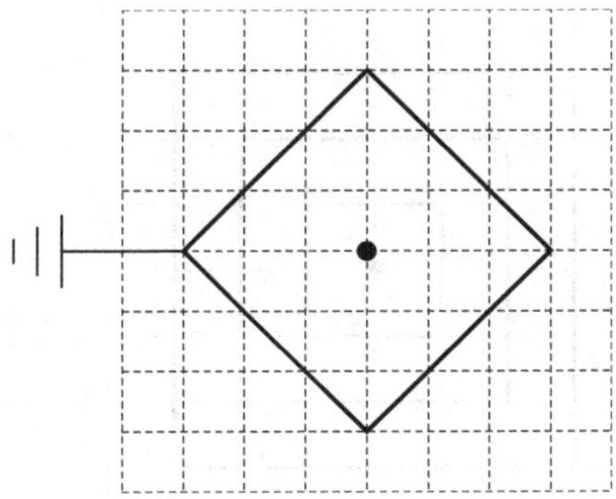

Figure 12.4. The square C_d connected to the earth.

Theorem 12.6.8 (Pólya, 1921). *The isotropic random walk in $\mathbb{Z}^2$ is recursive.*

Proof. Since the walk is ergodic, it suffices to show that the fixed origin O of the walk, for example $O = (0, 0)$, is a recurrent state, or that $p_{\mathtt{esc}} = 0$. Let T_O be the random variable of the first return to O. Then $p_{\mathtt{esc}} = \mathbf{P}(T_O = +\infty \mid X_0 = O)$.

Let dist denote the distance on $\mathbb{Z}^2$ defined by

$$\operatorname{dist}(x, y) = |x_1 - y_1| + |x_2 - y_2|, \quad x = (x_1, x_2),\ y = (y_1, y_2) \in \mathbb{Z}^2.$$

For $d \in \mathbb{N}$, set

$$\begin{aligned} B_d &= \{x = (x_1, x_2) \in \mathbb{Z}^2 : \operatorname{dist}(O, x) \leqslant d\}, \\ C_d &= \{x = (x_1, x_2) \in \mathbb{Z}^2 : \operatorname{dist}(O, x) = d\}. \end{aligned}$$

Fix $d \geqslant 1$. We *truncate* the Markov chain by connecting the vertices of C_d to the earth: this modifies only the transition probabilities of the vertices $x \in C_d$, bringing p_{xx} to 1 and p_{xy} to 0 if $y \neq x$.

We obtain an electrical circuit by imposing a tension of 1 volt at the origin. Assume for simplicity that all the resistors of the circuit are equal to 1 ohm. Then the effective conductance $c_{\mathtt{eff}}^{(d)}$ of the circuit B_d is exactly equal to the intensity $i_O^{(d)}$. But the probability $p_{\mathtt{esc}}^{(d)}$ of coming out is equal to $\frac{c_{\mathtt{eff}}^{(d)}}{c(O)}$, so

$$p_{\mathtt{esc}}^{(d)} = \frac{1}{4 r_{\mathtt{eff}}^{(d)}},$$

writing $r_{\mathtt{eff}}^{(d)}$ for the effective resistance of B_d. We easily see that the monotonicity law applies, and that consequently $p_{\mathtt{esc}}^{(d)}$ is decreasing at d. We deduce that O is a recurrent state in $\mathbb{Z}^2$, i.e. $p_{\mathtt{esc}} = 0$, if and only if

$$\lim_{d \to +\infty} r_{\mathtt{eff}}^{(d)} = +\infty.$$

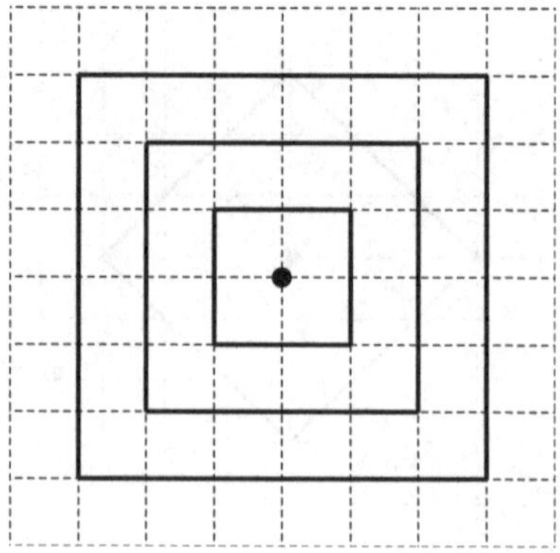

Figure 12.5. The squares Q_d of the electrical network $\mathbb{Z}^2$.

For each $d \geqslant 1$, we bring all the resistors between the vertices of the square $Q_d = \{x \in \mathbb{Z}^2 \mid \max(|x_1|, |x_2|) = d\}$ to 0: the vertices of Q_d are then all identified! See Figure 12.5.

By the monotonicity law, the effective resistance increases to $r'_{\mathtt{eff}}$. The circuit thus obtained corresponds to the states $d = 0, 1, 2, \ldots$ with resistors of 1 ohm connected in parallel between each pair of states d and $d + 1$: there are exactly $8d + 4$ of these. By Ohm's law, these $8d + 4$ parallel resistors are equivalent to a single resistor of $1/(8d + 4)$ ohm. Finally, we obtain a simple circuit made up of a series of resistors: it is equivalent to a simple circuit with a resistance of

$$\sum_{d \geqslant 0} \frac{1}{8d + 4} = +\infty,$$

so $r_{\mathtt{eff}} \geqslant r'_{\mathtt{eff}} = +\infty$, and finally $p_{\mathtt{esc}} = 0$, so the vertex O is recurrent. Then the others are also recurrent, since there is only one communication class. ∎

Exercise 12.13. Show that the isotropic random walk in $\mathbb{Z}$ (i.e. with transition probabilities all equal to $1/2$) is recursive.

12.7 Cheeger's inequality

An *isoperimetric inequality* is an inequality involving the *surface* and the *volume* of a bounded domain in Euclidean space. Such an inequality can be used to address the question: what is the maximum area that can be circumscribed within a given perimeter?

This is an optimisation problem, whose version in the Euclidean plane is easy to describe: in the plane, the question is *how to determine, among all domains bounded by a closed curve of given length, the one with the greatest area*. This problem was already known in Antiquity; a formulation can be found in Virgil's Aeneid, where the queen Elissa (also known as Dido), a Phoenician princess destined for the throne of Tyr, fled after her brother Pygmalion murdered her husband Acerbas in order to seize his wealth. Together with some faithful companions, the queen embarked on a ship and arrived at the coast of what is today Tunisia. Upon receiving a gift of money, the ruler of that region allowed the group to settle as much land as they could cover with an oxhide. Dido had the hide cut into long thin bands, which she arranged in a circle around the top of a hill; thus, given the length of the bands, she arrived at the maximum possible area, which was obviously much bigger than what was originally intended by the ruler. In this way, Dido founded the kingdom of Carthage.

It took until the 19th century, however, to prove that the optimal solution to this problem is really the one guessed by the queen and her companions. The following result was proved by Weierstrass. Let the term *domain* indicate a closed convex subset of the space under consideration.

Theorem 12.7.1. *If A and P denote respectively the area and the perimeter of a bounded domain D in the Euclidean plane, then $P^2 \geqslant 4\pi S$. Equality holds if and only if D is a disk.*

This inequality can be generalised to higher dimensions. For a bounded domain $D \subset \mathbb{R}^3$, we have $S^3 \geqslant 36\pi V^2$, where S and V denote the surface and the volume of D respectively, with equality holding uniquely when D is a closed ball. Generally, for a bounded domain $D \subset \mathbb{R}^n$, let $\mathrm{vol}(\partial D)$ be the surface of D and $\mathrm{vol}(D)$ its volume. Then

$$\mathrm{vol}(\partial D)^n \geqslant nC_n \mathrm{vol}(D)^{n-1},$$

where C_n is the volume of the unit ball of $\mathbb{R}^n$ and ∂D is the *boundary* of D. Equality holds if and only if D is a closed ball of $\mathbb{R}^n$.

Cheeger's inequality was introduced as a invariant in the theory of Riemannian manifolds. It is an *isoperimetric constant* $h(M) \in \mathbb{R}_+$ associated to a compact Riemann manifold M equipped with its Riemannian metric g which gives meaning to the notion of volume. For every domain $D \subset M$ such that $\mathrm{vol}(D) \leqslant \mathrm{vol}(M)/2$, consider the quotient $\frac{\mathrm{vol}(\partial D)}{\mathrm{vol}(D)}$: the Cheeger constant $h(M)$ is the minimum of all such quotients. It is related to the spectrum of the Laplacian of the variety M: the smallest positive eigenvalue of the Laplacian is less than or equal to $\frac{h^2(M)}{4}$.

As we already saw in Section 10.3.2, these considerations are not unrelated to graphs. Let $\Gamma = (V; E, w)$ be a weighted graph. For all $A \subset V$, set

$$\mu(A) = \sum_{x \in A} \mu_x \quad \text{with } \mu_x = \sum_{y \in \Gamma(x)} w(x, y),$$

and for all $F \subset V \times V$ set

$$w(F) = \sum_{(x,y) \in F} w(x, y).$$

Recall that

$$\partial_E A := \{\{x, y\} \in E \mid x \in A, y \in V \setminus A\}$$

is the *boundary of* A in E. Letting $\mathbb{1}$ denote the function identically equal to 1 on $V \times V$, we have

$$\mu(A) = w(A \times V), \quad \mathbb{1}(A \times (V \setminus A)) = |\partial_E A|.$$

We can now generalise the notion of the *Cheeger constant* that we saw in Section 10.3.2 by setting

$$h_w(\Gamma) := \min\Big\{\frac{w(A \times (V \setminus A))}{\mu(A)} \Bigm| A \subset V, 0 < \mu(A) \leqslant \frac{\mu(V)}{2}\Big\}.$$

Observe that

$$\mu(A) + \mu(V \setminus A) = \mu(V) \quad \text{and} \quad w(A \times (V \setminus A)) = \sum_{\substack{x \in A \\ y \in V \setminus A}} w(x, y).$$

Consider the eigenvalues of the Laplacian Δ_w of $(\Gamma; w)$, listed in increasing order with their multiplicities: $0 = \lambda_1 \leqslant \lambda_2 \leqslant \cdots$. The difference $\lambda_2 - \lambda_1 = \lambda_2$ is called the *spectral gap* of the operator Δ_w. We prove the following result.

Proposition 12.7.2. *We have* $\lambda_2 \leqslant 2h_w(\Gamma)$.

Proof. Let $A \subset V$ be the subset realising the minimum in the definition of $h_w(\Gamma)$.

We know that the zero eigenvalue $\lambda_1 = 0$ of the Laplacian Δ_w is associated to the eigenvector $\mathbf{1}$. Let $k \in \mathbb{R}_+$ be a parameter and let $f : V \to \mathbb{R}$ be defined by

$$f(x) = \begin{cases} 1 & \text{if } x \in A, \\ -k & \text{otherwise.} \end{cases}$$

We have

$$\langle f, \mathbf{1} \rangle_\mu = \sum_{x \in V} \mu_x f(x) = \mu(A) - k\mu(V \setminus A).$$

Thus, choosing $k = \mu(A)/\mu(V\backslash A)$, we have $f \perp \mathbf{1}$ and $k \leqslant 1$ since

$$\mu(A) + \mu(V\backslash A) = \mu(V) \geqslant 2\mu(A).$$

By the min-max theorem of Courant–Fischer (cf. Theorem 10.1.6), we have

$$\lambda_2 = \min_{\substack{g \perp \mathbf{1} \\ g \neq 0}} \frac{\langle \Delta_w g, g \rangle_\mu}{\langle g, g \rangle_\mu} \leqslant \frac{\langle \Delta_w f, f \rangle_\mu}{\langle f, f \rangle_\mu}.$$

But

$$\langle f, f \rangle_\mu = \sum_{x \in V} \mu_x f^2(x) = \mu(A) + \mu(V\backslash A)k^2 = (1+k)\mu(A)$$

and

$$\langle \Delta_w f, f \rangle_\mu = \frac{1}{2} \sum_{x,y \in V} (f(y) - f(x))^2 w(x,y) = (1+k)^2 w(A \times (V\backslash A)).$$

Hence

$$\frac{\langle \Delta_w f, f \rangle_\mu}{\langle f, f \rangle_\mu} = \frac{(1+k) w(A \times (V\backslash A))}{\mu(A)} = (1+k) h_w(\Gamma) \leqslant 2 h_w(\Gamma),$$

which gives the desired result. ∎

We can also give an upper bound for $h_w(\Gamma)$.

Theorem 12.7.3 (Cheeger's inequality). *We have*

$$\frac{h_w^2(\Gamma)}{2} \leqslant \lambda_2.$$

Proof. Let $f : V \to \mathbb{R}$ be a function. Set

$$S := \sum_{x,y \in V} |f(x)^2 - f(y)^2| w(x,y).$$

By Cauchy's inequality, we have

$$\begin{aligned} S^2 &\leqslant \Big(\sum_{x,y \in V} (f(x) - f(y))^2 w(x,y) \Big) \Big(\sum_{x,y \in V} (f(x) + f(y))^2 w(x,y) \Big) \\ &=: S_- S_+. \end{aligned}$$

We already saw that $S_- = 2\langle \Delta_w f, f \rangle_\mu$. But also,

$$\begin{aligned} S_+ &= \sum_{x,y \in V} (f(x)^2 + f(y)^2 + 2f(x)f(y)) w(x,y) \\ &\leqslant \sum_{x,y \in V} (2f(x)^2 + 2f(y)^2) w(x,y) = 4 \sum_{x,y \in V} f(x)^2 w(x,y) \end{aligned}$$

by symmetry. Therefore

$$S_+ \leqslant 4\sum_{x\in V} f(x)^2 \sum_{y\in V} w(x,y) = 4\sum_{x\in V} f(x)^2\mu_x = 4\langle f, f\rangle_\mu,$$

so

$$S \leqslant 2\sqrt{2\langle \Delta_w f, f\rangle_\mu \langle f, f\rangle_\mu}. \tag{12.2}$$

We now proceed to find a lower bound for S involving the quantity $h_w(\Gamma)$. Let $\{u_0 = 0 < u_1 < \cdots < u_t\}$ denote the different values taken by f^2, and set

$$A_i := \{x \in V \mid f(x)^2 \geqslant u_i\}, \quad i = 0, \ldots, t.$$

By convention, we set $u_{t+1} = +\infty$ of where $A_{t+1} = \emptyset$. We start by observing that in the sum, we can restrict ourselves to pairs (x, y) such that $f(x)^2 > f(y)^2$, by introducing a factor of 2 to compensate. Next, if (x, y) is such a pair, we have $u_i \leqslant f(x)^2 < u_{i+1}$ and $u_{j-1} \leqslant f(y)^2 < u_j$ for a unique pair of integers (i, j) such that $1 \leqslant j \leqslant i \leqslant t$. We then see that

$$f(x)^2 - f(y)^2 = \sum_{k=j}^{i} (u_k - u_{k-1}),$$

and therefore

$$S = 2\sum_{k=0}^{t} (u_{k+1} - u_k) w(A_{k+1} \times (V \backslash A_{k+1})).$$

By the definition of $h_w(\Gamma)$, we deduce that

$$\begin{aligned} S &\geqslant 2h_w(\Gamma)\sum_{k=0}^{t}(u_{k+1} - u_k)\mu(A_{k+1}) \\ &= 2h_w(\Gamma)\sum_{k=1}^{t} u_k \sum_{x\in A_k\backslash A_{k+1}} \mu_x \\ &= 2h_w(\Gamma)\sum_{x\in V}\mu_x f(x)^2 = 2h_w(\Gamma)\langle f, f\rangle_\mu. \end{aligned}$$

By comparing this lower bound with the upper bound (12.2) of S obtained earlier, we get

$$\frac{h_w(\Gamma)^2}{2} \leqslant \frac{\langle \Delta_w f, f\rangle_\mu}{\langle f, f\rangle_\mu}. \tag{12.3}$$

For f_0, choose a (non-zero) eigenfunction of Δ_w associated to the eigenvalue λ_2. Set $A = \{x \in V \mid f_0(x) > 0\}$. Up to replacing f_0 by $-f_0$, we may assume that $|A| < |V|/2$. Set $f = f_0 \times \mathbb{1}_A$ (where $\mathbb{1}_A(x)$ takes the value 1 if $x \in A$ and 0 otherwise).

For all $x \in A$ and all $y \in V$, we clearly have $f(x) - f(y) \leqslant f_0(x) - f_0(y)$. If $y \in A$, equality holds; if $y \notin A$, then $f_0(y) \leqslant 0$ so $f(x) - f(y) = f_0(x) \leqslant f_0(x) - f_0(y)$. Hence for all $x \in A$, we have

$$\Delta_w f(x) = \frac{1}{\mu_x} \sum_{y \in V} (f(x) - f(y)) w(x, y) \leqslant \Delta_w f_0(x),$$

and thus

$$\begin{aligned} \langle \Delta_w f, f \rangle_w &= \sum_{x,y \in V} (f(x) - f(y))^2 w(x, y) \\ &= 2 \sum_{x,y \in V} (f(x) - f(y)) f(x) w(x, y) \\ &= 2 \sum_{x \in A} \mu_x f_0(x) \Delta_w f(x), \end{aligned}$$

then

$$\begin{aligned} \langle \Delta_w f, f \rangle_w &\leqslant 2 \sum_{x \in A} \mu_x f_0(x) \Delta_w f_0(x) \\ &= \lambda_2 \sum_{x \in A} \mu_x f_0(x)^2 \\ &= \lambda_2 \sum_{x \in V} \mu_x f(x)^2 = \lambda_2 \langle f, f \rangle_w. \end{aligned}$$

Finally, by (12.3), we have

$$\lambda_2 \geqslant \frac{\langle \Delta_w f, f \rangle_w}{\langle f, f \rangle_w} \geqslant \frac{h_w(\Gamma)^2}{2},$$

which concludes the proof of Cheeger's inequality. ∎

In the particular case where $w = \mathbb{1}_E$ (the function defined on $V \times V$ that has value 1 if $\{x, y\} \in E$ and 0 otherwise), we have $\mu_x = d(x)$, the degree of x for each vertex x. The inequality can then be expressed in the following form:

$$\lambda_2 \geqslant \frac{h(\Gamma)^2}{2\Delta(\Gamma)},$$

where $\Delta(\Gamma)$ is the maximal degree in Γ, and $h(\Gamma)$ denotes the classical Cheeger constant introduced in Section 10.3.2, defined by

$$h(\Gamma) = \inf\Big\{ \frac{|\partial_E A|}{|A|} \;\Big|\; A \subset V,\ 0 < |A| \leqslant \frac{|V|}{2} \Big\}.$$

Chapter 13

Other perspectives

In this chapter, we introduce some new algebraic, combinatorial and topological objects which for the most part emerged from graph theory, such as the Tutte polynomials. Even though they are still quite mysterious, these polynomials have a great many applications, particularly in statistical physics. Ramsey theory, also, comes up in applications to computer science and information theory. We will also introduce matroids. Finally, a short section on hypergraphs concludes this volume.

13.1 Tutte polynomials

In Chapter 7, we introduced the chromatic polynomial of a graph $\Gamma = (V; E, N)$, which satisfies the following properties:

- $P_\Gamma(k) = P_{\Gamma\backslash a}(k) - P_{\Gamma/a}(k)$, $a \in E$ (Theorem 7.6.1);
- if the graph Γ has n vertices and no edges then $P_\Gamma(k) = k^n$.

These two assertions allow us to recursively compute the values of the chromatic polynomial $P_\Gamma(k)$ of an arbitrary graph Γ by applying a sequence of contractions and deletions of various edges. Theorem 7.6.1 shows that $P_\Gamma(k)$ is independent of the order of the contractions and deletions used in its construction.

Motivated by the two properties described above, W. T. Tutte, in 1954, defined a 2-variable polynomial that he called the *dichromatic polynomial* [32].

This polynomial, which we will call the *Tutte polynomial*, is an important invariant in graph theory. It has applications to knot theory, statistical physics, computer science, combinatorics, etc. [8].

In this paragraph, all graphs are multigraphs in full generality; we admit *multiple edges and loops*.

Let $\Gamma = (V; E, N)$ be a multigraph and let $a \in E$. Recall the notation $\Gamma\backslash a = (V; E\backslash a, N)$ for the graph obtained by deleting a (there is no ambiguity in keeping the same set of labels N). Now let $a \in E$. The contraction $\Gamma/a = (V/a; E/a, N)$ of Γ was defined in Section 5.5. We see that

$$|E/a| = |E\backslash a| = |E| - 1.$$

There is a bijection $E\backslash a \to E/a$ which associates to every edge $e \neq a$ the corresponding edge e', whose extremities may have been modified by the contraction. Under this bijection, a subset $A \subset E\backslash a$ has the same cardinality as its image (denoted A/a).

For $A \subset E$, the notation A/a thus makes sense in the two following situations:

- if $a \notin A$;
- if $A = E$.

More precisely, let $A \subset E$. For $a = (\{x, y\}, n_a) \notin A$, we denote by $(\Gamma/a)(A)$ the partial graph $(V/a; A/a, N)$ obtained by contracting a in Γ, where

$$\begin{aligned} V/a &= (V \backslash \{x, y\}) \cup \{z_a\}, \\ E/a &= \{e \in E \mid e \text{ not adjacent to } a\} \\ &\qquad \cup \{(\{z_a, t\}, n) \mid (\{x, t\}, n) \in E\} \\ &\qquad \cup \{(\{z_a, t\}, n) \mid (\{y, t\}, n) \in E\} \\ &\qquad \cup \{([z_a, z_a], n) \mid ([x, y], n) \in E \backslash a\}. \end{aligned}$$

Let $\Gamma = (V; E, N)$ be a multigraph with c_Γ connected components. For $A \subset E$, we recall that $\Gamma(A) = (V; A, N)$ denotes the associated partial graph and let $c_{\Gamma(A)}$ be the number of its connected components. We define

- the *rank of A in Γ* by $\mathrm{rank}_\Gamma(A) = |V(\Gamma)| - c_{\Gamma(A)}$;
- the *corank of A in Γ* by $\mathrm{corank}_\Gamma(A) = |A| - \mathrm{rank}_\Gamma(A)$.

When $A = E$, we note that by this definition, the rank of E coincides with the quantity $\rho(\Gamma)$ introduced in Section 6.1.2: $\rho(\Gamma) = \mathrm{rank}_\Gamma(E)$.

The *Tutte polynomial* of a multigraph $\Gamma = (V; E, N)$ is defined by

$$\begin{aligned} T_\Gamma(x, y) &= \sum_{A \subset E} (x-1)^{\mathrm{rank}_\Gamma(E) - \mathrm{rank}_\Gamma(A)} (y-1)^{\mathrm{corank}_\Gamma(A)} \\ &= \sum_{A \subset E} (x-1)^{c_{\Gamma(A)} - c_\Gamma} (y-1)^{\mathrm{corank}_\Gamma(A)} \in \mathbb{Z}[x, y]. \end{aligned}$$

13.1.1 Basic properties

In Theorem 13.1.2 below, we give the relations between the respective Tutte polynomials of the graphs Γ, $\Gamma \backslash a$ and Γ/a; one of them is reminiscent of the property recalled above concerning chromatic polynomials. For this, we will also need the relations between the numbers of connected components of $\Gamma(A)$, $\Gamma(A \cup \{a\})$, $(\Gamma \backslash a)(A)$ and $(\Gamma/a)(A)$.

Proposition 13.1.1. *Let $\Gamma = (V; E, N)$ be a graph, $A \subset E$ and $a \notin A$. Then we have $c_{\Gamma(A)} = c_{\Gamma \backslash a(A)}$. Furthermore,*

(i) *if a is an isthmus, then*
   - $c_{\Gamma(A \cup \{a\})} = c_{\Gamma(A)} - 1 = c_{(\Gamma \backslash a)(A)} - 1$,
   - *in particular* $c_\Gamma = c_{\Gamma \backslash a} - 1$.

(ii) *If a is a loop, then*

- $c_{\Gamma(A\cup\{a\})} = c_{\Gamma(A)} = c_{(\Gamma\backslash a)(A)}$,
- *in particular* $c_\Gamma = c_{\Gamma\backslash a}$.

(iii) *If a is neither an isthmus nor a loop, then*

- $c_\Gamma = c_{\Gamma\backslash a}$,
- *contracting an edge does not change the number of connected components:* $c_{\Gamma(A)} = c_{(\Gamma/a)(A)}$,
- $c_{\Gamma(A\cup\{a\})} = c_{\Gamma(A)} = c_{(\Gamma/a)(A)}$,
- *in particular* $c_\Gamma = c_{\Gamma/a}$.

Proof. (i) The first formula is clear. Let us prove the second. If $\varepsilon(a) = \{x, y\}$ and C_x, C_y are the connected components containing x resp. y in $\Gamma(A \cup \{a\})$, then $C_x = C_y$; these two components become distinct in $(\Gamma\backslash a)(A)$ since a is an isthmus. The other connected components are not modified, hence $c_{\Gamma(A\cup\{a\})} = c_{(\Gamma\backslash a)(A)} - 1$.

(ii) Left as an exercise.

(iii) Only the identity $c_\Gamma = c_{\Gamma\backslash a}$ needs attention, but the deletion of a does not change the connectedness since a is not an isthmus. ■

Theorem 13.1.2. *Let* $\Gamma = (V; E, N)$ *be a multigraph. Then for* $a \in E$ *we have*

$$T_\Gamma(x,y) = \begin{cases} xT_{\Gamma\backslash a}(x,y) & \text{if } a \text{ is an isthmus,} \\ yT_{\Gamma\backslash a}(x,y) & \text{if } a \text{ is a loop,} \\ T_{\Gamma/a}(x,y) + T_{\Gamma\backslash a}(x,y) & \text{otherwise.} \end{cases}$$

Proof. For $A \subset E$, we set

$$\begin{aligned} f_\Gamma(A) &:= (x-1)^{\operatorname{rank}_\Gamma(E) - \operatorname{rank}_\Gamma(A)} (y-1)^{\operatorname{corank}_\Gamma(A)} \\ &= (x-1)^{c_{\Gamma(A)} - c_\Gamma} (y-1)^{\operatorname{corank}_\Gamma(A)}, \end{aligned}$$

so that

$$T_\Gamma(x,y) = \sum_{A\subset E} f_\Gamma(A) = \sum_{A\subset E \mid A \not\ni a} f_\Gamma(A) + \sum_{A\subset E \mid A\ni a} f_\Gamma(A)$$

or

$$T_\Gamma(x,y) = \sum_{A\subset E\mid A\not\ni a} \big(f_\Gamma(A) + f_\Gamma(A\cup\{a\})\big), \tag{13.1}$$

since every B containing a can naturally be written $B = A \cup \{a\}$ with $a \notin A$.

Set $n = |V|$.

(i) If a is an isthmus, then by Proposition 13.1.1 (i), for $A \subset E$ and $a \notin A$, we have

- $\text{rank}_\Gamma(E) = n - c_\Gamma = n - (c_{\Gamma\backslash a} - 1) = \text{rank}_{\Gamma\backslash a}(E\backslash a) + 1$,
- $\text{rank}_\Gamma(A) = n - c_{\Gamma(A)} = \text{rank}_{\Gamma\backslash a}(A)$ since $a \notin A$,
- $\text{corank}_\Gamma(A) = |A| - \text{rank}_\Gamma(A) = |A| - \text{rank}_{\Gamma\backslash a}(A) = \text{corank}_{\Gamma\backslash a}(A)$;

thus

$$\begin{aligned} f_\Gamma(A) &= (x-1)^{\text{rank}_\Gamma(E) - \text{rank}_\Gamma(A)}(y-1)^{\text{corank}_\Gamma(A)} \\ &= (x-1)^{\text{rank}_{\Gamma\backslash a}(E\backslash a) + 1 - \text{rank}_{\Gamma\backslash a}(A)}(y-1)^{\text{corank}_{\Gamma\backslash a}(A)} \\ &= (x-1)\, f_{\Gamma\backslash a}(A). \end{aligned}$$

Furthermore,

- $\text{rank}_\Gamma(A \cup \{a\}) = n - c_{\Gamma(A\cup\{a\})} = n - (c_{(\Gamma\backslash a)(A)} - 1) = \text{rank}_{\Gamma\backslash a}(A) + 1$,
- $\text{rank}_\Gamma(E) = n - c_\Gamma = n - c_{\Gamma\backslash a} + 1 = \text{rank}_{\Gamma\backslash a}(E\backslash a) + 1$,
- $\text{corank}_\Gamma(A \cup \{a\}) = \text{corank}_{\Gamma\backslash a}(A)$;

thus

$$\begin{aligned} f_\Gamma(A \cup \{a\}) &= (x-1)^{\text{rank}_\Gamma(E) - \text{rank}_\Gamma(A\cup\{a\})}(y-1)^{\text{corank}_\Gamma(A\cup\{a\})} \\ &= (x-1)^{\text{rank}_{\Gamma\backslash a}(E\backslash a) + 1 - \text{rank}_{\Gamma\backslash a}(A) - 1}(y-1)^{\text{corank}_{\Gamma\backslash a}(A)} \\ &= f_{\Gamma\backslash a}(A). \end{aligned}$$

It then follows from (13.1) that

$$\begin{aligned} T_\Gamma(x,y) &= \sum_{A \subset E \mid A \not\ni a} \big((x-1)\, f_{\Gamma\backslash a}(A) + f_{\Gamma\backslash a}(A)\big) \\ &= x \sum_{A \subset E \mid A \not\ni a} f_{\Gamma\backslash a}(A) \\ &= x \sum_{B \subset E\backslash a} f_{\Gamma\backslash a}(B) \\ &= x T_{\Gamma\backslash a}(x,y). \end{aligned}$$

(ii) If a is a loop, then (ii) of Proposition 13.1.1 shows that for $A \subset E$ and $a \notin A$, we have

- $\text{rank}_\Gamma(E) = n - c_\Gamma = n - c_{\Gamma\backslash a} = \text{rank}_{\Gamma\backslash a}(E\backslash a)$;
- $\text{rank}_\Gamma(A) = n - c_{\Gamma(A)} = n - c_{(\Gamma\backslash a)(A)} = \text{rank}_{\Gamma\backslash a}(A)$, so
- $\text{corank}_\Gamma(A) = \text{corank}_{\Gamma\backslash a}(A)$;
- $\text{rank}_\Gamma(A \cup \{a\}) = n - c_{\Gamma(A\cup\{a\})} = n - c_{(\Gamma\backslash a)(A)} = \text{rank}_{\Gamma\backslash a}(A)$.

The three first points give $f_\Gamma(A) = f_{\Gamma\backslash a}(A)$, and then together with the fourth point, we find

$$\begin{aligned} f_\Gamma(A \cup \{a\}) &= (x-1)^{\operatorname{rank}_\Gamma(E) - \operatorname{rank}_\Gamma(A\cup\{a\})}(y-1)^{\operatorname{corank}_\Gamma(A\cup\{a\})} \\ &= (x-1)^{\operatorname{rank}_{\Gamma\backslash a}(E\backslash a) - \operatorname{rank}_{\Gamma\backslash a}(A)}(y-1)^{|A\cup\{a\}| - \operatorname{rank}_\Gamma(A\cup\{a\})} \\ &= (x-1)^{\operatorname{rank}_{\Gamma\backslash a}(E\backslash a) - \operatorname{rank}_{\Gamma\backslash a}(A)}(y-1)(y-1)^{|A| - \operatorname{rank}_{\Gamma\backslash a}(A)}. \end{aligned}$$

Thus $f_\Gamma(A \cup \{a\}) = (y-1) f_{\Gamma\backslash a}(A)$; then $T_\Gamma(x,y) = yT_{\Gamma\backslash a}(x,y)$ by (13.1).

(iii) If a is neither an isthmus nor a loop, then by Proposition 13.1.1 (iii), for $A \subset E$ and $a \notin A$, we have

- $c_\Gamma = c_{\Gamma\backslash a}$ since a is not an isthmus;
- $c_{\Gamma(A)} = c_{(\Gamma\backslash a)(A)}$ since a is not an edge of $\Gamma(A)$;
- $\operatorname{rank}_\Gamma(E) = n - c_\Gamma = n - c_{\Gamma\backslash a} = \operatorname{rank}_{\Gamma\backslash a}(E\backslash a)$;
- $\operatorname{corank}_\Gamma(A) = \operatorname{corank}_{\Gamma\backslash a}(A)$.

Thus we see that $f_\Gamma(A) = f_{\Gamma\backslash a}(A)$.

We also have

- $\operatorname{rank}_\Gamma(E) = n - c_\Gamma = n - c_{\Gamma/a} = \operatorname{rank}_{\Gamma/a}(E/a) + 1$ since contracting decreases the number of vertices by one;
- $\operatorname{rank}_\Gamma(A \cup \{a\}) = n - c_{\Gamma(A\cup\{a\})} = n - c_{(\Gamma/a)(A)} = \operatorname{rank}_{\Gamma/a}(A/a) + 1$;
- $\operatorname{corank}_\Gamma(A \cup \{a\}) = |A \cup \{a\}| - \operatorname{rank}_\Gamma(A \cup \{a\}) = 1 + |A| - \operatorname{rank}_{\Gamma/a}(A/a) - 1 = |A/a| - \operatorname{rank}_{\Gamma/a}(A/a) = \operatorname{corank}_{\Gamma/a}(A/a)$;

hence $f_\Gamma(A \cup \{a\}) = f_{\Gamma/a}(A/a)$.

Finally, we then have

$$\begin{aligned} T_\Gamma(x,y) &= \sum_{A\subset E \mid A \not\ni a} \big(f_{\Gamma\backslash a}(A) + f_{\Gamma/a}(A/a)\big) \\ &= \sum_{B\subset E\backslash a} f_{\Gamma\backslash a}(B) + \sum_{B\subset E/a} f_{\Gamma/a}(B), \end{aligned}$$

which gives the desired formula. ■

Proposition 13.1.3. *If a is an isthmus, then $T_\Gamma(x,y) = xT_{\Gamma/a}(x,y)$.*

Proof. Indeed, in this case, if $A \not\ni a$,

$$\begin{aligned} c_{\Gamma(A)} &= c_{(\Gamma/a)(A)} + 1, \\ c_{\Gamma(A\cup\{a\})} &= c_{(\Gamma/a)(A)}. \end{aligned}$$

The first formula comes from the fact that if $\varepsilon(a) = \{x, y\}$ is an isthmus in Γ, then in $\Gamma(A)$, the connected components containing x resp. y are distinct, but in Γ/a

the corresponding connected components coincide. The second formula follows from Proposition 13.1.1 (iii).

Hence we have

- $\mathrm{rank}_\Gamma(E) = \mathrm{rank}_{\Gamma/a}(E/a) + 1$,
- $\mathrm{rank}_\Gamma(A) = \mathrm{rank}_{\Gamma/a}(A/a)$,
- $\mathrm{corank}_\Gamma(A) = \mathrm{corank}_{\Gamma/a}(A/a)$.

These are the same relations that we already saw for $\Gamma\backslash a$. Thus they lead to the same expressions, i.e.

$$f_\Gamma(A) = f_{\Gamma/a}(A/a), \quad f_\Gamma(A \cup \{a\}) = f_{\Gamma/a}(A),$$

and finally $T_\Gamma(x, y) = xT_{\Gamma/a}(x, y)$. ■

Conversely, we can prove the following proposition.

Proposition 13.1.4. *Let $\Gamma = (V; E, N)$ be a multigraph and $U_\Gamma(x, y) \in \mathbb{C}[x, y]$ a polynomial satisfying the following properties:*

$$U_\Gamma(x, y) = \begin{cases} 1 & \text{if } \Gamma \text{ has no edge,} \\ xU_{\Gamma\backslash a}(x, y), & \text{if } a \text{ is an isthmus,} \\ yU_{\Gamma\backslash a}(x, y), & \text{if } a \text{ is a loop,} \\ U_{\Gamma/a}(x, y) + U_{\Gamma\backslash a}(x, y) & \text{otherwise.} \end{cases}$$

Then $U_\Gamma(x, y) = T_\Gamma(x, y)$.

Proof. This result is easily proved by induction on the number of edges m.

If $m = 0$ and $|V| = n$, then $\emptyset$ is the only subset of E: $c_{\Gamma(\emptyset)} = n$, $\mathrm{rank}_\Gamma(\emptyset) = 0$ and $\mathrm{corank}_\Gamma(\emptyset) = 0$. Hence

$$\begin{aligned} 1 = U_\Gamma(x, y) &= (x-1)^0(y-1)^0 \\ &= \sum_{A \subset E} (x-1)^{c_\Gamma - c_{\Gamma(A)}}(y-1)^{\mathrm{corank}_\Gamma(A)} \\ &= T_\Gamma(x, y). \end{aligned}$$

$m - 1 \curvearrowright m$: let $a \in A$.

- If a is an isthmus, then by the induction hypothesis and Theorem 13.1.2 we have

$$U_\Gamma(x, y) = xU_{\Gamma\backslash a}(x, y) = xT_{\Gamma\backslash a}(x, y) = T_\Gamma(x, y);$$

- if a is a loop, idem;
- if a is neither an isthmus nor a loop, the same reasoning holds. ■

Proposition 13.1.5. *Let $\Gamma = (V; E, N)$ be a multigraph.*

(i) *If $\Gamma = \Gamma' \cup \Gamma''$ and there are no edges between Γ' and Γ'', then $T_\Gamma(x, y) = T_{\Gamma'}(x, y)T_{\Gamma''}(x, y)$.*

(ii) *In particular, if $(\Gamma_i)_{i\in I}$ (for a finite set I) denote the connected components of Γ, then*
$$T_\Gamma(x, y) = \prod_{i\in I} T_{\Gamma_i}(x, y).$$

Proof. We prove (i) by writing $T_\Gamma(x, y) = \sum_{A\subset E}(x - 1)^{u_A}(y - 1)^{v_A}$ (with the obvious notation). Every subset A of E can be uniquely written as $A = A' \cup A''$, where $A' \subset E' = E(\Gamma')$ and $A'' \subset E'' = E(\Gamma'')$. Furthermore, the exponents satisfy $u_A = u_{A'} + u_{A''}$, $v_A = v_{A'} + v_{A''}$ (with $u_A = c_\Gamma - c_{\Gamma(A)}$, $v_A = \operatorname{corank}_\Gamma(A)$ and similar notation for A' and A''). Hence clearly $T_\Gamma(x, y) = T_{\Gamma'}(x, y) \cdot T_{\Gamma''}(x, y)$. ■

Now we give some examples of actual computation of the Tutte polynomial. We use the following notation:

- $\overline{K}_n$ denotes the graph with n isolated vertices (the complement of K_n);
- B_m denotes the graph consisting of a single m-loop;
- A_m denotes the graph consisting of a single m-edge;
- C_n denotes the elementary cycle with n vertices.

Then

- $T_{\overline{K}_n}(x, y) = 1$.
- $T_{B_m}(x, y) = y^m$: by induction, $T_{B_{m+1}}(x, y) = yT_{B_m}(x, y)$.
- If Γ is a forest with m edges, then $T_\Gamma(x, y) = x^m$: by induction, as every a is an isthmus, $T_\Gamma(x, y) = xT_{\Gamma\backslash a}(x, y)$. In particular, this gives the Tutte polynomial for a chain with m edges.
- $T_{K_3}(x, y) = x^2 + x + y$: we remove an edge a, giving
$$T_{K_3}(x, y) = T_{K_3/a}(x, y) + T_{K_3\backslash a}(x, y),$$
but $K_3/a = A_2$ and $K_3\backslash a$ is a tree with 2 edges, so $T_{K_3}(x, y) = x + y + x^2$.

Exercise 13.1. Show that

(i) $T_{C_n} = y + x + \cdots + x^{n-1} (n \geqslant 2)$;

(ii) $T_{A_m}(x, y) = x + y + \cdots + y^{m-1} (m \geqslant 2)$;

(iii) $T_{K_4}(x, y) = x^3 + 3x^2 + 2x + 4xy + 2y + 3y^2 + y^3$.
Hint: in this problem, one is led to compute $T_{A_2^2}(x, y)$, where A_2^2 is the connected graph with 3 vertices consisting of 2 double edges. For this, one should find $T_{A_2^2}(x, y) = (T_{A_2}(x, y))^2$.

On these examples, one can check the identities $\deg_x T_\Gamma(x,y) = \mathrm{rank}_\Gamma(E)$ and $\deg_y T_\Gamma(x,y) = \mathrm{corank}_\Gamma(E)$.

Proposition 13.1.4 can be generalised in the following way. With the goal of recovering several polynomials that were introduced earlier, we introduce three further indeterminates α, β, γ apart from the indeterminates x, y.

Theorem 13.1.6 (Tutte, polynomial version). *Let $\Gamma = (V; E, N)$ be a multigraph, and let $U_\Gamma \in \mathbb{C}[x, y, \alpha, \beta, \gamma]$ be a polynomial satisfying the following recursive properties:*

$$U_\Gamma = \begin{cases} \alpha^{|V|} & \text{if } \Gamma \text{ has no edge,} \\ xU_{\Gamma\setminus a} & \text{if } a \in E \text{ is an isthmus,} \\ yU_{\Gamma\setminus a} & \text{if } a \in E \text{ is a loop,} \\ \beta U_{\Gamma\setminus a} + \gamma U_{\Gamma/a} & \text{otherwise.} \end{cases}$$

Then this polynomial can be written

$$U_\Gamma = \alpha^{c_\Gamma} \beta^{\mathrm{corank}_\Gamma} \gamma^{\mathrm{rank}_\Gamma} T_\Gamma\Big(\frac{\alpha x}{\gamma}, \frac{y}{\beta}\Big),$$

where $\mathrm{corank}_\Gamma = \mathrm{corank}_\Gamma(E)$, $\mathrm{rank}_\Gamma = \mathrm{rank}_\Gamma(E)$. *Furthermore,*

$$U_\Gamma \in \mathbb{Z}[x, y, \alpha, \beta, \gamma].$$

Proof. Let us first use induction on the number of edges to prove that there exists at most one polynomial satisfying the desired hypotheses. Suppose that there are two such polynomials, U_Γ and U'_Γ. If $E = \emptyset$, then $U_\Gamma = \alpha^{|V|} = U'_\Gamma$. Assume the property holds for all graphs with $m-1$ edges, and let Γ be a graph with m edges. Then

- if $a \in E$ is an isthmus, then $U_\Gamma = xU_{\Gamma\setminus a} = xU'_{\Gamma\setminus a} = U'_\Gamma$,
- if $a \in E$ is a loop, then $U_\Gamma = yU_{\Gamma\setminus a} = yU'_{\Gamma\setminus a} = U'_\Gamma$,
- if $a \in E$ is neither a loop nor an isthmus, then $U_\Gamma = \beta U_{\Gamma\setminus a} + \gamma U_{\Gamma/a} = \beta U'_{\Gamma\setminus a} + \gamma U'_{\Gamma/a} = U'_\Gamma$.

It thus suffices to prove that the rational function

$$U_\Gamma = \alpha^{c_\Gamma} \beta^{\mathrm{corank}_\Gamma} \gamma^{\mathrm{rank}_\Gamma} T_\Gamma\Big(\frac{\alpha x}{\gamma}, \frac{y}{\beta}\Big)$$

satisfies the four desired properties. Observe first that it is a polynomial with integer coefficients in α, β, γ, x and y, since $\deg_x T_\Gamma(x,y) = \mathrm{rank}_\Gamma$, $\deg_y T_\Gamma(x,y) = \mathrm{corank}_\Gamma$; indeed U_Γ is a sum of terms of the form

$$\alpha^{c_\Gamma} \beta^{\mathrm{corank}_\Gamma} \gamma^{\mathrm{rank}_\Gamma} \Big(\frac{\alpha x - \gamma}{\gamma}\Big)^i \Big(\frac{y-\beta}{\beta}\Big)^j, \qquad \begin{array}{l} 0 \leqslant i \leqslant \mathrm{rank}_\Gamma, \\ 0 \leqslant j \leqslant \mathrm{corank}_\Gamma. \end{array}$$

Secondly, we have the following.

- If Γ has no edge, then $\Gamma = \overline{K}_n$: $c_\Gamma = n$, $\text{rank}_\Gamma = \text{corank}_\Gamma = 0$, so $U_\Gamma = \alpha^{|V|}$.
- If $a \in E$ is an isthmus, then $c_{\Gamma\backslash a} = c_\Gamma + 1$, $\text{rank}_{\Gamma\backslash a} = \text{rank}_\Gamma - 1$, $\text{corank}_{\Gamma\backslash a} = \text{corank}_\Gamma$, so

$$U_\Gamma = \alpha^{c_{\Gamma\backslash a}-1}\beta^{\text{corank}_{\Gamma\backslash a}}\gamma^{\text{rank}_{\Gamma\backslash a}+1}\frac{\alpha x}{\gamma}T_{\Gamma\backslash a}\Big(\frac{\alpha x}{\gamma}, \frac{y}{\beta}\Big) = xU_{\Gamma\backslash a}.$$

- If $a \in E$ is a loop, then $c_{\Gamma\backslash a} = c_\Gamma$, $\text{rank}_{\Gamma\backslash a} = \text{rank}_\Gamma$, $\text{corank}_{\Gamma\backslash a} = \text{corank}_\Gamma - 1$, so

$$U_\Gamma = \alpha^{c_{\Gamma\backslash a}}\beta^{\text{corank}_{\Gamma\backslash a}+1}\gamma^{\text{rank}_{\Gamma\backslash a}}\frac{y}{\beta}T_{\Gamma\backslash a}\Big(\frac{\alpha x}{\gamma}, \frac{y}{\beta}\Big) = yU_{\Gamma\backslash a}.$$

- Otherwise $c_{\Gamma\backslash a} = c_\Gamma$, $\text{rank}_{\Gamma\backslash a} = \text{rank}_\Gamma$, $\text{corank}_{\Gamma\backslash a} = \text{corank}_\Gamma - 1$, $c_{\Gamma/a} = c_\Gamma$, $\text{rank}_{\Gamma/a} = \text{rank}_\Gamma - 1$, $\text{corank}_{\Gamma/a} = \text{corank}_\Gamma$, so

$$\begin{aligned} U_\Gamma &= \alpha^{c_\Gamma}\beta^{\text{corank}_\Gamma}\gamma^{\text{rank}_\Gamma}\Big(T_{\Gamma\backslash a}\Big(\frac{\alpha x}{\gamma}, \frac{y}{\beta}\Big) + T_{\Gamma/a}\Big(\frac{\alpha x}{\gamma}, \frac{y}{\beta}\Big)\Big) \\ &= \alpha^{c_{\Gamma\backslash a}}\beta^{\text{corank}_{\Gamma\backslash a}+1}\gamma^{\text{rank}_{\Gamma\backslash a}}T_{\Gamma\backslash a}\Big(\frac{\alpha x}{\gamma}, \frac{y}{\beta}\Big) \\ &\quad + \alpha^{c_{\Gamma/a}}\beta^{\text{corank}_{\Gamma/a}}\gamma^{\text{rank}_{\Gamma/a}+1}T_{\Gamma/a}\Big(\frac{\alpha x}{\gamma}, \frac{y}{\beta}\Big) \\ &= \beta U_{\Gamma\backslash a} + \gamma U_{\Gamma/a}. \end{aligned}$$

■

Remark 13.1.7. It is easy to see that in the isthmus case of this statement, $xU_{\Gamma\backslash a}$ can be replaced by $\alpha x U_{\Gamma/a}$.

Theorem 13.1.8 (Tutte, specialisation version). *Let $x, y \in \mathbb{C}$ and $\alpha, \beta, \gamma \in \mathbb{C}$, with $\beta, \gamma \neq 0$ fixed. Then for each multigraph $\Gamma = (V; E, N)$, there exists a unique $f(\Gamma) \in \mathbb{C}$ such that*

$$f(\Gamma) = \begin{cases} \alpha^{|V|} & \text{if } \Gamma \text{ has no edge,} \\ xf(\Gamma\backslash a) & \text{if } a \in E \text{ is an isthmus,} \\ yf(\Gamma\backslash a) & \text{if } a \in E \text{ is a loop,} \\ \beta f(\Gamma\backslash a) + \gamma f(\Gamma/a) & \text{otherwise,} \end{cases}$$

and we have $f(\Gamma) = \alpha^{c_\Gamma}\beta^{\text{corank}_\Gamma}\gamma^{\text{rank}_\Gamma}T_\Gamma(\frac{\alpha x}{\gamma}, \frac{y}{\beta})$.

The proof is exactly the same. Indeed, the numerical type of the parameters plays no role whatsoever in the proof.

Remark 13.1.9. As above, in the isthmus case of the statement, $xf(\Gamma\backslash a)$ can be replaced by $\alpha x f(\Gamma/a)$.

The correspondence $\Gamma \mapsto f(\Gamma)$ is an *invariant* under isomorphism, in the sense that $\Gamma_1 \simeq \Gamma_2$ implies $f(\Gamma_1) = f(\Gamma_2)$. Indeed, if Γ_1 and Γ_2 are isomorphic, then we have $c_{\Gamma_1} = c_{\Gamma_2}$, $\mathrm{corank}_{\Gamma_1} = \mathrm{corank}_{\Gamma_2}$ and $T_{\Gamma_1} = T_{\Gamma_2}$.

This theorem allows us to recover some known polynomials; we give some examples in the following.

13.1.2 The Tutte polynomial and chromatic polynomials

We saw in Section 7.6 that

$$P_\Gamma(x) = P_{\Gamma\backslash a}(x) - P_{\Gamma/a}(x).$$

This suggests a possible relationship with the Tutte polynomial.

Proposition 13.1.10. *Let $\Gamma = (V; E, N)$ be a graph such that $\Gamma = \Gamma_1 \cup \Gamma_2$, $\Gamma_1 \cap \Gamma_2 = K_r, r \geqslant 1$ (the complete graph with r vertices). Then*

$$P_\Gamma(x) = \frac{P_{\Gamma_1}(x) P_{\Gamma_2}(x)}{P_{K_r}(x)}.$$

Proof. It suffices to prove the formula for every integer $x = k \geqslant r$. We have

$$P_{\Gamma_1}(k) = P_{K_r}(k)\frac{P_{\Gamma_1}(k)}{P_{K_r}(k)}, \quad P_{\Gamma_2}(k) = P_{K_r}(k)\frac{P_{\Gamma_2}(k)}{P_{K_r}(k)}.$$

The quotient $p_1 := \frac{P_{\Gamma_1}(k)}{P_{K_r}(k)}$ can be viewed (exercise) as the number of ways of extending a colouring of K_r to a colouring of Γ_1; similarly, $p_2 := \frac{P_{\Gamma_2}(k)}{P_{K_r}(k)}$ is the number of ways of extending a colouring of K_r to a colouring of Γ_2. But a colouring of Γ can also be obtained by taking a colouring of K_r and extending it to Γ_1 (in p_1 ways), then to Γ_2 (in p_2 ways), in which case we see that, since the graph differences $\Gamma_1 \backslash K_r$ and $\Gamma_2 \backslash K_r$ have no common edge and no common vertex, the full number of possible extensions is $p_1 p_2$. Thus we have

$$P_\Gamma(k) = p_1 p_2 P_{K_r}(k) = \frac{P_{\Gamma_1}(k)}{P_{K_r}(k)} \frac{P_{\Gamma_2}(k)}{P_{K_r}(k)} P_{K_r}(k),$$

which gives the desired formula. ∎

This result leads to the following proposition.

Proposition 13.1.11. *Let $\Gamma = (V; E, N)$ be a multigraph and $a \in E$ an isthmus. Then*

$$P_\Gamma(x) = \frac{x-1}{x} P_{\Gamma\backslash a}(x).$$

Proof. Set $\varepsilon(a) = \{u, v\}$. We have $\Gamma = \Gamma_1 \cup \Gamma_2$ with $\Gamma_1 = \Gamma$ and $\Gamma_2 = (\{u, v\}; \{a\})$. As $\Gamma_2 = K_2$ and $P_{K_2}(x) = x(x-1)$, Proposition 13.1.10 shows that

$$P_\Gamma(x) = \frac{P_{\Gamma_1}(x) P_{\Gamma_2}(x)}{x(x-1)},$$

but also, $P_{\Gamma_1}(x) = (x-1)P_{\Gamma_1 \backslash a}(x)$ and $P_{\Gamma_2}(x) = (x-1)P_{\Gamma_2 \backslash a}(x)$ since a is an hanging edge in Γ_1 and in Γ_2. Hence

$$\begin{aligned} P_\Gamma(x) &= \frac{(x-1)^2 P_{\Gamma_1 \backslash a}(x) P_{\Gamma_2 \backslash a}(x)}{x(x-1)} \\ &= \frac{x-1}{x} P_{\Gamma_1 \backslash a}(x) P_{\Gamma_2 \backslash a}(x) = \frac{x-1}{x} P_{\Gamma \backslash a}(x) \end{aligned}$$

since $\Gamma_1 = \Gamma$ and $P_{\Gamma_2 \backslash a}(x) = 1$. ∎

Remark 13.1.12. This statement is equivalent to $P_{\Gamma \backslash a}(x) = x P_{\Gamma / a}(x)$ (by applying the induction formula for chromatic polynomials).

Theorem 13.1.13. *Let k be a positive integer, and let $\Gamma = (V; E, N)$ be a multigraph with n vertices and c connected components. Then $P_\Gamma(k) = (-1)^{n-c} k^c T_\Gamma(1-k, 0)$.*

Proof. Using Proposition 13.1.11, we have

$$P_\Gamma(k) = \begin{cases} k^n & \text{if } \Gamma \text{ has no edge,} \\ \frac{k-1}{k} P_{\Gamma \backslash a}(k) & \text{if } a \text{ is an isthmus,} \\ 0 & \text{if } a \text{ is a loop,} \\ P_\Gamma(\Gamma \backslash a)(k) - P_{\Gamma / a}(k) & \text{otherwise,} \end{cases}$$

and then Theorem 13.1.8 with parameters $x = \frac{k-1}{k}$, $y = 0$, $\alpha = k$, $\beta = 1$ and $\gamma = -1$ gives the result (recall that $P_\Gamma(x) = 0$ whenever Γ possesses a loop). ∎

13.1.3 The Tutte polynomial and spanning trees

Formula (10.9) also suggests a link between the Tutte polynomial and $\tau(\Gamma)$.

Recall the definition of a contraction in a loopless multigraph $\Gamma = (V; E, N)$: if $a = ([x, y], n)$ is an edge of Γ, then we let $\Gamma /\!/ a$ denote the (still loopless) multigraph obtained by successively contracting each edge between x and y.

We actually have $\tau(\Gamma /\!/ a) = \tau(\Gamma / a)$ since the only difference between the two graphs $\Gamma /\!/ a$ and Γ / a is the possible presence of loops at the fused vertex z_a in the latter, but this does not change the number of spanning trees. According to this observation, formula (10.10) can be written

$$\tau(\Gamma) = \tau(\Gamma \backslash a) + \tau(\Gamma / a).$$

Theorem 13.1.14. *Let $\Gamma = (V; E, N)$ be a non-empty connected multigraph. Then the number of spanning trees of Γ is equal to $\tau(\Gamma) = T_\Gamma(1, 1)$.*

Proof. We prove this result by induction on the number of edges together with the above results on the Tutte polynomial. If $|E| = \emptyset$, we use in a direct way the definition of the Tutte polynomial given in the preamble of Section 13.1. The graph has only one vertex, so $\tau(\Gamma) = 1 = T_\Gamma(1, 1)$ (with the convention that $0^0 = 1$). Suppose now that the assertion holds for every graph with $m - 1$ edges, $m \geqslant 1$. Then

- if $a \in E$ is a loop, then $\tau(\Gamma) = \tau(\Gamma \backslash a) = T_{\Gamma\backslash a}(1, 1) = \frac{1}{1} T_\Gamma(1, 1) = T_\Gamma(1, 1)$ by Theorem 13.1.2;
- if $a \in E$ is an isthmus, then $\tau(\Gamma) = \tau(\Gamma/a) = T_{\Gamma/a}(1, 1) = \frac{1}{1} T_\Gamma(1, 1) = T_\Gamma(1, 1)$ by Proposition 13.1.3;
- if $a \in E$ is neither a loop nor an isthmus, then by the observation preceding the statement we have $\tau(\Gamma) = \tau(\Gamma \backslash a) + \tau(\Gamma /\!/ a) = T_{\Gamma\backslash a}(1, 1) + T_{\Gamma/\!/a}(1, 1) = T_\Gamma(1, 1)$. ∎

13.1.4 The Tutte polynomial and planarity

In this paragraph, we introduce an elegant result on planarity.

Theorem 13.1.15. *Let $\Gamma = (V; E)$ be a non-empty planar graph and Γ^* its dual. Then $T_\Gamma(x, y) = T_{\Gamma^*}(y, x)$.*

Proof. By the definition of the dual of a graph, there exists a bijection f^* between the set of edges E of Γ and the set of edges E^* of Γ^*. Furthermore, one can easily *see* that a^* is a loop (resp. an isthmus) of Γ^* if and only if a is an isthmus (resp. a loop) of Γ (we leave this result as an exercise). Thus, $\Gamma^* \backslash a^*$ and Γ^*/a^* are the respective duals of Γ/a and of $\Gamma\backslash a$, i.e. $(\Gamma\backslash a)^* \simeq \Gamma^*/a^*$.

We now prove the result, using induction on the number of edges. If $m = 0$, then Γ has one face, so Γ^* is an isolated vertex; thus $T_\Gamma(x, y) = 1 = T_{\Gamma^*}(y, x)$.

$m - 1 \curvearrowright m$: let Γ a graph with m edges and let a be an edge. If a is an isthmus, then by Proposition 13.1.3,

$$T_\Gamma(x, y) = x T_{\Gamma/a}(x, y) = x T_{(\Gamma/a)^*}(y, x) = x T_{\Gamma^*\backslash a^*}(y, x)$$

since a^* is a loop of Γ^*, so $T_{\Gamma^*}(y, x) = x T_{\Gamma^*\backslash a^*}(y, x)$ and $T_\Gamma(x, y) = T_{\Gamma^*}(y, x)$; if a is a loop, the calculation is similar; otherwise

$$\begin{aligned} T_\Gamma(x, y) &= T_{\Gamma/a}(x, y) + T_{\Gamma\backslash a}(x, y) \\ &= T_{\Gamma^*\backslash a^*}(y, x) + T_{\Gamma^*/a^*}(y, x) \\ &= T_{\Gamma^*}(y, x). \end{aligned}$$

∎

13.1.5 Other applications

In statistical mechanics, the cumulative distribution function of the Ising model is given by

$$R_\Gamma(p)\colon v \mapsto (1-p)^{|E|} 2^{c(\Gamma)} v^{r(\Gamma)} T_\Gamma\Big(\frac{v+2}{v}, v+1\Big).$$

This model was first generalised to the Potts model, subsequently as the *random cluster model*, in which, letting $0 < p < 1, q > 0$, the cumulative distribution function is

$$R_\Gamma(q,p)\colon v \mapsto (1-p)^{|E|} q^{c(\Gamma)} v^{r(\Gamma)} T_\Gamma\Big(\frac{v+q}{v}, v+1\Big).$$

Exercise 13.2. Let Γ be graph or multigraph. Show that

(i) $T_\Gamma(2,1)$ is the number of spanning forests;

(ii) $T_\Gamma(1,2)$ is the number of partial connected graphs;

(iii) $T_\Gamma(2,2)$ is the number of partial graphs.

Some examples:

- If we fix an abelian group A and write $q_\Gamma(A)$ for the number of everywhere non-zero A-flows, then $q_\Gamma(A) = (-1)^{\mathrm{corank}_\Gamma} T_\Gamma(0, 1-|A|)$.
- Let $a(\Gamma)$ be the number of acyclic orientations (i.e. the orientations of the graph Γ such that it contains no circuit): then $a(\Gamma) = T_\Gamma(2,0)$.
- Let $s \in V$ and let $a_s(\Gamma)$ denote the number of acyclic orientations in which s is the unique source: then $a_s(\Gamma) = T_\Gamma(1,0)$.
- Random graphs (cf. Chapter 11): fixing n, we consider the complete graph with n vertices $K_n = (V; E)$. The probability that a randomly chosen edge a will be equal to a fixed edge $e \in E$ is uniform with probability $p := \mathbf{P}(a = e) = \frac{1}{N}$, where $N = n(n-1)/2$. Set $q = 1 - p$. If H_0 is an arbitrary fixed subgraph of K_n, the random choice of a subgraph K of K_n then has a probability of

  $$\mathbf{P}(K = H_0) = p^{e(H_0)} q^{N-e(H_0)} \quad (e(H_0) = \text{number of edges of } H_0).$$

  More generally, if we replace K_n by a graph $\Gamma = (V; E)$ and we want to know the rank of a given subgraph K, we have the following relationship with the Tutte polynomials:

  $$\mathbf{P}(\mathrm{rank}_K = \mathrm{rank}_\Gamma) = p^{\mathrm{rank}_\Gamma} q^{\mathrm{corank}\,\Gamma} T_\Gamma\Big(1, \frac{1}{q}\Big)$$

  with $0 < p < 1$ and $q = 1 - p$.
- Other important polynomials beyond the scope of this book are also related to the Tutte polynomials in interesting ways, for example pairing polynomials, and Jones polynomials, which are fundamental invariants in knot theory.

13.2 Ramsey theory

In this paragraph, *all graphs are assumed simple*.

Ramsey theory has many applications in mathematics, information theory, logic, computer science, etc. [31]. It *measures* how much an object is *ordered or disordered*, or rather, how *the order or disorder* appears within a structure. It can be viewed as a generalisation of Dirichlet's pigeonhole principle (1834): if n pigeons must be distributed among m pigeonholes with $n > m$, then at least one pigeonhole must contain at least two pigeons. The pigeonhole principle can be reformulated in the following way: if E and F are two finite sets such that $|E| > |F|$ and f is a map from E to F, then there exists an element of F which admits at least two preimages under f in E. This principle, which seems completely elementary, was used in a masterly manner by C. Siegel in number theory. Ramsey's theorem gives a certain generalisation of this principle, by asking the question: how many elements of a structure must be considered in order to verify that a particular property holds? In the pigeonhole situation, this question would be framed as follows: how many pigeons do we need in order to be certain that at least one pigeonhole is going to contain at least two pigeons?

A famous puzzle asks to prove that in any union of at least six people, either there are at least three who know each other, or there are at least three who do not know each other.

To solve this problem, we first create a graph model in which:

- the vertices are the six people;
- there is an edge between two people if and only if they know each other.

The question then becomes that of either finding either a clique K_3 or an independent set (or anticlique, i.e. a set of vertices in which no two are adjacent) $\overline{K_3}$. Alternatively, we could model the problem by using the completely graph $\Gamma = K_6$, with the vertices again representing the six people, and attributing a colour to each edge between two people

- blue if they know each other;
- red if they do not.

In this situation, solving this problem comes down to proving that there exists a clique K_3 of a single colour.

Remark 13.2.1. Putting colours on the edges of the graph in this problem does not constitute a *colouring* as introduced in Chapter 7, which is why we do not use the term. In particular, here only the edges have colours, and we do not prevent two adjacent edges to have the same colour.

The six-people problem can be generalised to the following question: Given two integers $r, s \geqslant 1$, does there exist an integer $n = n(r, s)$ such that for every way of attributing one of two colours (say blue or red) to each edge of K_n, K_n contains either a subgraph K_r that is entirely blue or a subgraph K_s that is entirely red? The smallest integer n satisfying this property is called the *Ramsey number* and denoted $R(r, s)$. Equivalently, $R(r, s)$ can be considered as the smallest positive integer N such that every graph Γ of order N contains either a clique of order r or an independent set of order s.

The existence of Ramsey numbers is not a priori obvious.

Theorem 13.2.2 (Ramsey, 1930). *Let $r, s \geqslant 1$ be two integers. Then there exists a smallest integer $R(r, s)$ such that for all $n \geqslant R(r, s)$, no matter how we attribute one of two colours (blue or red) to each edge of K_n, then either there exists a completely blue $K_r \subset K_n$ or a completely red $K_s \subset K_n$.*

This result is a consequence of the statement of Erdős–Szekeres' theorem proved below. It is a more precise theorem than Ramsey's original result, but was proved later.

Note that clearly $R(r, s) = R(s, r)$, $r, s \geqslant 1$, since it suffices to exchange the two colours. Also, we have $R(r, 1) = R(1, r) = 1$. We can also show that $R(r, 2) = R(2, r) = r$, as follows. We first observe that $r - 1 < R(r, 2)$, since if we attribute the colour red to every edge of K_{r-1}, there will be neither a red subgraph K_r nor a blue subgraph K_2. Therefore $R(r, 2) \geqslant r$. But if we attribute red to every edge of K_r, then there does exist an entirely red subgraph K_r, whereas if we attribute blue to at least one edge, then there exists a blue K_2. Thus $R(r, 2) = R(2, r) = r$.

Theorem 13.2.3 (Erdős and Szekeres, 1935). *For all $r, s \geqslant 2$, the number $R(r, s)$ exists and we have*

(i) $R(r, s) \leqslant R(r - 1, s) + R(r, s - 1)$;

(ii) $R(r, s) \leqslant \binom{r+s-2}{r-1}$;

(iii) *if $R(r - 1, s)$ and $R(r, s - 1)$ are even numbers, then $R(r, s) \leqslant R(r - 1, s) + R(r, s - 1) - 1$.*

Proof. Let us first show the existence of $R(r, s)$ and inequality (i). For this, we will use induction on $r + s \geqslant 4$.

If $r + s = 4$, we have $r = s = 2$ and $R(2, 2) = 2 \leqslant 1 + 1 = R(1, 2) + R(2, 1)$. Suppose that for $r + s < k$, $(k \geqslant 5)$ the numbers $R(r, s)$ exist and satisfy inequality (i). Let r, s be such that $r + s = k$: the numbers $R(r - 1, s)$ and $R(r, s - 1)$ thus exist. We now show that every graph Γ having $n = R(r - 1, s) + R(r, s - 1)$ vertices contains either a clique K_r or an independent set $\overline{K_s}$. Let x be a vertex. There are two cases:

(a) either $d_\Gamma(x) \geqslant R(r-1,s)$: then the induced subgraph by the vertices adjacent to x has at least $R(r-1,s)$ vertices; by the induction hypothesis, we find that

(1) either it contains a clique K_{r-1}, so by adding the vertex x and the edges from x to this clique, we obtain a clique K_r of Γ;

(2) or it contains an independent set $\overline{K}_s$, so Γ does as well;

(b) or $d_\Gamma(x) < R(r-1,s)$: In this case we have $d_{\overline{\Gamma}}(x) \geqslant R(r,s-1)$ in the complement $\overline{\Gamma}$. We proceed in $\overline{\Gamma}$ of the same manner as above; thus $\overline{\Gamma}$ contains a clique K_s or an independent set $\overline{K}_r$ since $R(r,s-1) = R(s-1,r)$ (exchanging the colours). But then Γ contains either a clique K_r or an independent set $\overline{K}_s$.

We have shown that every graph of order $n = R(r-1,s) + R(r,s-1)$ contains either a clique K_r or an independent set $\overline{K}_s$. This proves that $R(r,s)$ exists and

$$R(r,s) \leqslant R(r-1,s) + R(r,s-1). \tag{13.2}$$

(ii) We use induction on $r+s \geqslant 4$. For $r+s=4$, we have $r=s=2$ and $R(2,2) = 2 \leqslant C_2^1 = 2$. To pass from k to $k+1$, it suffices to apply (13.2), recalling that the binomial coefficients satisfy $\binom{n}{p} = \binom{n-1}{p-1} + \binom{n-1}{p}$; thus

$$R(r,s) \leqslant R(r-1,s) + R(r,s-1) \leqslant \binom{r+s-3}{r-2} + \binom{r+s-3}{r-1} = \binom{r+s-2}{r-1}.$$

(iii) We reason as in (i). Let Γ be a graph with $n = R(r-1,s) + R(r,s-1) - 1$ vertices. By Lemma 1.1.9, we know that the number of vertices of odd degree is even; as Γ has an odd number of vertices, it follows that Γ contains a vertex of even degree. Let x be such a vertex. Then we have $d_\Gamma(x) + d_{\overline{\Gamma}}(x) = R(r-1,s) + R(r,s-1) - 2$, so

(a′) either $d_\Gamma(x) \geqslant R(r-1,s)$ since $d_\Gamma(x) \neq R(r-1,s) - 1$, and we conclude by the same reasoning as in (a);

(b′) or $d_\Gamma(x) < R(r-1,s)$, so $d_\Gamma(x) \leqslant R(r-1,s) - 2$ and in the complement $\overline{\Gamma}$ we have $d_{\overline{\Gamma}}(x) \geqslant R(r,s-1) = R(r-1,s)$, and we conclude by the same reasoning as in (b).

We thus have shown that if $R(r-1,s)$ and $R(r,s-1)$ are even numbers, then every graph of order $n = R(r-1,s) + R(r,s-1) - 1$ contains either a clique K_r or an independent set $\overline{K}_s$. ■

As an illustration, let us use Theorem 13.2.3 (i) to show that $R(3,3) = 6$. We already have $R(3,3) \leqslant R(3-1,3) + R(3,3-1) = R(2,3) + R(3,2) = 6$ since $R(r,2) = R(2,r) = r$. Now, to show that $R(3,3) > 5$, it suffices to find a way to

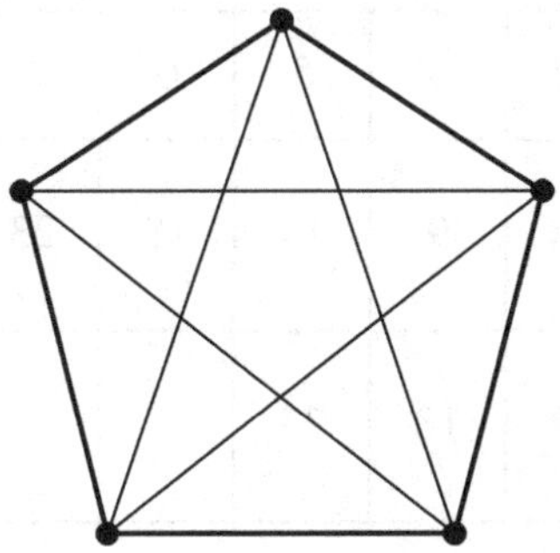

Figure 13.1. A way of attributing two colours (thin and thick lines) to the edges of K_5 such that the result contains neither a *thin* K_3 nor a *thick* K_3.

attribute two colours to K_5 that contains neither a blue clique with 3 vertices nor a red clique with 3 vertices. A way to do this is shown in Figure 13.1. We have thus proved the following theorem which gives the solution to the six-people problem posed above.

Theorem 13.2.4. $R(3,3) = 6$.

By arguing similarly, we can also obtain the following values for $R(3,s)$.

Theorem 13.2.5. $R(3,4) = 9$.

Proof. Applying Theorem 13.2.3 (iii) and Theorem 13.2.4, we find that $R(3,4) \leqslant 9$. To prove the equality, we need to attribute two colours to K_8 in such as way as to contain neither a red K_3 nor a blue K_4: Figure 13.2 shows that this can be done. ■

Theorem 13.2.6. $R(3,5) = 14$.

Proof. Applying Theorem 13.2.3 and Theorem 13.2.5, we obtain $R(3,5) \leqslant 14$. To prove the equality, we need to attribute two colours to K_{13} in such a way as to contain neither a red clique K_3 nor a blue K_5: Figure 13.3 shows such a configuration. ■

In general, computing Ramsey numbers is a very difficult task: if we conjecture that $R(r,s) = n$, we must first show that every graph having n vertices contains either a clique K_r or an independent set $\overline{K_s}$, then show that n is the smallest integer satisfying this property. Some Ramsey numbers are given in Table 13.1.

Another lower bound for $R(r,s)$ is given in the following theorem.

Theorem 13.2.7. $R(r,s) \geqslant (r-1)(s-1) + 1$.

Proof. Set $n = (r-1)(s-1)$. Let us show that there exists a way to attribute two colours to the edges of K_n for which there exists no monochromatic clique K_r or K_s.

r \ s	3	4	5	6	7	8	9
3	6	9	14	18	23	28	36
4	9	18	25				

Table 13.1

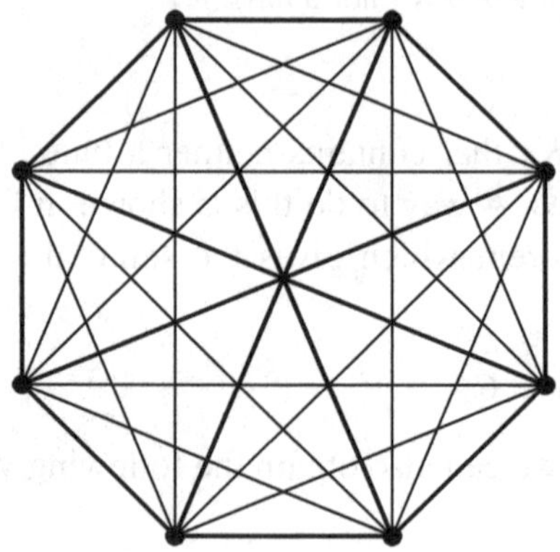

Figure 13.2. A way of attributing two colours (thin and thick lines) to K_8 such that the result contains neither a *thin* K_4 nor a *thick* K_3.

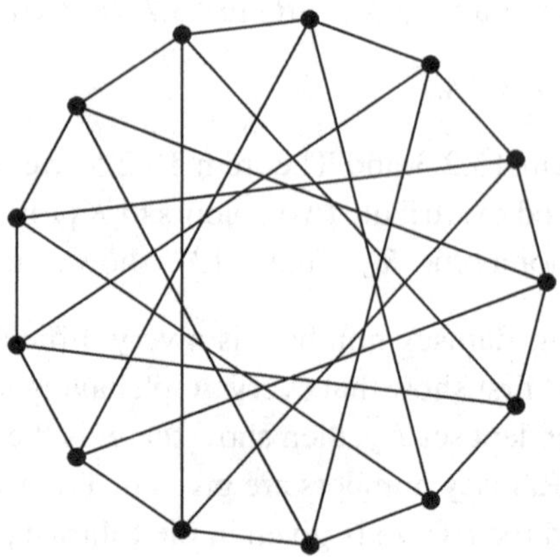

Figure 13.3. An attribution of two colours to K_{13} (in which the shown edges are all of one colour and the missing edges of the other, for better visualisation), that contains neither a (visible) K_3 nor a $\overline{K_5}$.

Order the vertices in an $(r-1)\times(s-1)$ table. If the vertices x and y lie on the same row, colour the edge $\{x, y\}$ blue; otherwise colour it red. Now, take r arbitrary vertices; then necessarily two of them must lie on the same row (by Dirichlet's pigeonhole principle) so K_n cannot contain any red clique K_r. Similarly, if we take s vertices, then two of them must lie in the same column, so K_n cannot contain any blue clique K_s. ■

The diagonal Ramsey numbers $R(r, r)$ have been the subject of a great deal of study.

Theorem 13.2.8 (Erdős, 1947). $R(r, r) > 2^{\frac{r}{2}}$, $r \geqslant 3$.

This theorem was proven by P. Erdős using random graph theory (cf. Chapter 11).

By combining Theorem 13.2.3 with Theorem 13.2.8, we obtain the following bounds.

Theorem 13.2.9. *For $r \geqslant 3$, we have $2^{\frac{r}{2}} < R(r, r) < 2^{2r-2}$.*

Proof. Indeed, by Theorem 13.2.3 (ii), $R(r, r) \leqslant \binom{2r-2}{r-1} < 2^{2r-2}$. ■

It is conjectured that $R(r, r) \simeq 2^{c\times r}$ (c constant), but its proof appears out of reach at present!

To conclude this paragraph, here are some known upper and lower bounds:

- $43 \leqslant R(5, 5) \leqslant 49$.
- $36 \leqslant R(6, 4) \leqslant 41$, $58 \leqslant R(6, 5) \leqslant 87$, $102 \leqslant R(6, 6) \leqslant 165$.
- $49 \leqslant R(7, 4) \leqslant 61$, $80 \leqslant R(7, 5) \leqslant 143$, $113 \leqslant R(7, 6) \leqslant 298$, $205 \leqslant R(7, 7) \leqslant 540$.

13.3 Matroids

In this paragraph *all graphs are assumed simple*.

In Section 2.2, we saw Kruskal's algorithm to find a tree of minimum weight spanning a graph. Let us recall the gist. Let $\Gamma = (V; E)$ be a connected graph and let $\omega\colon E \to \mathbb{R}^+$ be a weight on the edges. Order the edges by increasing value. Create a set A by taking one edge at a time from the list, and adding it to A on the condition that it does not form a cycle with the edges already in A. Continue until the end of the list. By Theorem 2.2.1, the set A forms a forest $\Gamma(A) = (V; A)$ which is of minimum weight in the sense that $\sum_{a\in A}\omega(a)$ is minimum. Furthermore, by Theorem 2.2.7, $\Gamma(A)$ actually forms a single tree T (of minimum weight).

We recall the proof that the spanning tree obtained by the algorithm has minimum weight. Let T_1 be any other spanning tree, and let a_i be the first edge produced by the algorithm (so $a_i \in T$) which is not in T_1. Then $a_0, a_1, \ldots, a_{i-1}$ are edges produced by the algorithm which lie in both T and T_1. If we add a_i to T_1, we obtain a cycle C, but then C must contain an edge a which is not in T (otherwise T would contain a cycle), so we can replace the edge a in T_1 by the edge a_i, and in this way we obtain a new spanning tree T_2 which has more edges in common with T than T_1. We have $\omega(a_i) \leqslant \omega(a)$. Indeed, if we suppose that $\omega(a) < \omega(a_i)$, then the edge a would have been considered by the algorithm before a_i, so it would have been rejected in the construction of T due to forming a cycle with edges in $\{a_0, \dots, a_{i-1}\}$, but all these edges are also in T_1, so there would be a cycle in T_1, which is impossible since T_1 is a tree. Thus $\omega(a_i) \leqslant \omega(a)$, so the total weight satisfies $\omega(T_2) \leqslant \omega(T_1)$. We now repeat the same operation with T_2 to produce a new tree T_3 with even more edges in common with T, and continue in this manner until we obtain T itself. In doing this, we never increase the weight, so $\omega(T) \leqslant \omega(T_1)$, which proves that $\omega(T)$ is minimum among all spanning trees.

If we invert the order of the edges, listing them in decreasing order, the same process yields a spanning tree of maximum weight. This algorithm can be generalised to a class of algorithms called *greedy algorithms*.

Let U be a finite set and $\mathcal{H}$ a family of subsets of U. The pair $(U, \mathcal{H})$ is a *hereditary system* if $\mathcal{H}$ is stable under inclusion, i.e. if

$$A \in \mathcal{H} \text{ and } B \subset A \implies B \in \mathcal{H}.$$

An element of $\mathcal{H}$ is called a *hereditary set*.

Example 13.3.1. Let $\Gamma = (V; E)$ be a graph and let U_Γ be the set of subgraphs of Γ with no cycles. Consider the family $\mathcal{H}$ of all the subsets of U_Γ. Then $(U_\Gamma, \mathcal{H})$ forms a hereditary system, called the *graphic hereditary system*.

There is an *optimisation problem* associated to the hereditary system $(U, \mathcal{H})$, stated as follows. Let $f : U \to \mathbb{R}_+$ be a weight. The goal is to find a hereditary set A such that the weight $f(A) = \sum_{a \in A} f(a)$ is optimum (maximum or minimum).

For example, if E is the set of edges of a graph $\Gamma = (V; E)$ and $\mathcal{H}$ is the set of all subsets of E forming a forest, then determining a set of spanning trees of minimum weight of Γ is the optimisation problem associated to the hereditary system $(E, \mathcal{H})$. Kruskal's algorithm can thus be generalised to Algorithm 13.1.

To determine a set of trees of minimum weight spanning a graph $\Gamma = (V; E)$, we saw in Chapter 2 that the greedy algorithm – in this case Kruskal's algorithm – yields a suitable solution. For an arbitrary hereditary system, the greedy algorithm will always yield a solution, but not necessarily an optimal one; it will yield a hereditary set which does not necessarily have optimal weight.

Let us give an example. Recall that a matching in a graph $\Gamma = (V; E)$ is a set of edges having no common vertices; finding a pairing of maximum weight in a graph equipped with a weight on the edges means finding a pairing C such that the sum

```
Function Greedy(U: list; ℋ: list; T: table of elements): list
  list: L := null;
  begin
    order in T the elements of U by weight:
    T[U[1]] ⩽ T[U[2]] ⩽ ··· ⩽ T[U[m]];
    for every i from 0 to m do
      if L ∪ {U[i]} ∈ ℋ, then add U[i] to L;
    end
    return L;
  end
EndFunction
```

Algorithm 13.1. Greedy algorithm.

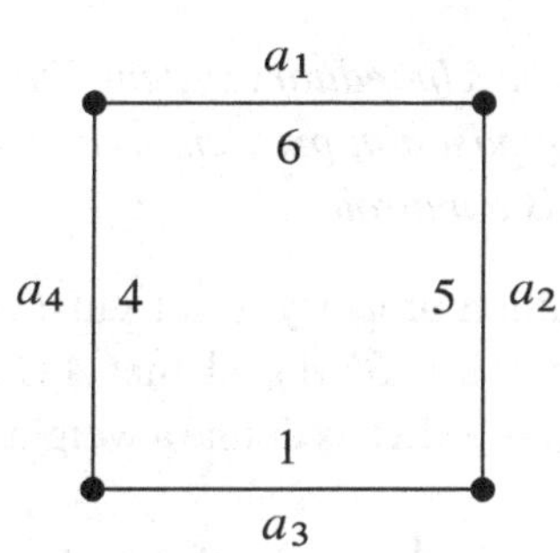

Figure 13.4. Searching for matchings with the greedy algorithm. The greedy algorithm chooses the edge a_1, which is of maximum weight 6, then the edge a_3 whose weight is 1. The total weight of this matching is thus equal to 7. But the two other edges of this cycle also form a matching, whose weight is 9: the greedy algorithm does not give an optimal solution in this case.

of the weights of the edges of C is maximum. The hereditary system associated to this problem is $(E, \mathcal{H})$, where $\mathcal{H}$ is the set of all matchings of Γ. The algorithm does not necessarily yield an optimal matching in this situation; see Figure 13.4 for an example.

The natural question is thus the following: under what conditions does the greedy algorithm give an optimal solution?

A hereditary system $(U, \mathcal{H})$ is a *matroid* if it satisfies the following property called *augmentation property*:

$$A, B \in \mathcal{H} \text{ and } |A| < |B| \implies \exists a \in B \backslash A \text{ such that } A \cup \{a\} \in \mathcal{H}.$$

Lemma 13.3.2. *The two following properties are equivalent:*

(i) *If $A, B \in \mathcal{H}$ and $|A| < |B|$, there exists $a \in B \setminus A$ such that $A \cup \{a\} \in \mathcal{H}$.*

(ii) *If $A, B \in \mathcal{H}$ with $|B| = |A| + 1$, there exists $a \in B \setminus A$ such that $A \cup \{a\} \in \mathcal{H}$.*

Proof. The fact that (i) implies (ii) is trivial. Let us show the inverse implication. Let $A, B \in \mathcal{H}$ and $|A| < |B|$. As $(U, \mathcal{H})$ is a hereditary system, for all $B' \subset B$ we have $B' \in \mathcal{H}$. We choose a $B' \subset B$ with $|B'| = |A| + 1$; thus there exists $a \in B' \setminus A \subset B \setminus A$ such that $A \cup \{a\} \in \mathcal{H}$. ∎

In the example of Figure 13.4, if we take $A = \{a_4\}$ and $B = \{a_1, a_3\}$, the augmentation property is not satisfied, and consequently the hereditary system associated to the maximum weight matching problem is not a matroid.

The notion of a matroid is related to the greedy algorithm via the following theorem.

Theorem 13.3.3. *Let $(U, \mathcal{H})$ be a hereditary system. Then the greedy algorithm gives an optimal solution to the optimisation problem associated to the hereditary system $(U, \mathcal{H})$ if and only if $(U, \mathcal{H})$ is a matroid.*

Proof. Suppose the augmentation property is not satisfied. Let $A, B \in \mathcal{H}$ and $|A| < |B|$ such that there does not exist $a \in B \setminus A$ such that $A \cup \{a\} \in \mathcal{H}$. By Lemma 13.3.2, we may assume that $|B| = |A| + 1$. Let us define a weight on U as follows: set $k = |A|$ and

$$f(a) = \begin{cases} k+2 & \text{if } a \in A, \\ k+1 & \text{if } a \in B \setminus A, \\ 0 & \text{otherwise.} \end{cases}$$

Observe that A is not an optimal solution (for the weight f) of the optimisation problem associated to the hereditary system $(U, \mathcal{H})$ since $f(A) = k(k+2) < (k+1)^2 \leq f(B)$.

The greedy algorithm first chooses all the elements of A since they are of maximum weight. But A is not optimal, so there exists $a \in U$ such that $f(A) < f(A \cup \{a\})$. But we have

- either $f(a) = 0$;
- or $a \in B \setminus A$.

In the first case, the weight of the solution is not increased, and in the second case, by hypothesis $A \cup \{a\} \notin \mathcal{H}$; thus the greedy algorithm does not give an optimal solution.

Conversely, suppose $(U, \mathcal{H})$ is a matroid. Let $A = \{a_1, a_2, \ldots, a_k\}$ be a hereditary set obtained by applying the greedy algorithm to f, and let $B = \{b_1, b_2, \ldots, b_l\}$ be an arbitrary hereditary set, where the elements of A and of B are ordered by decreasing weight. The augmentation property implies that $1 \leqslant l \leqslant k$, since otherwise there would exist a b_i which would have to be added to A by the greedy algorithm. Let $A_t = \{a_1, a_2, a_3, \dots, a_t\}$ and $B_t = \{b_1, b_2, b_3, \dots, b_t\}$ with $1 \leqslant t \leqslant l$. We will show by induction that $f(A_t) \geqslant f(B_t)$ for $1 \leqslant t \leqslant l$.

To start with, we have $f(A_1) \geqslant f(B_1)$ since the greedy algorithm first chooses an element of maximum weight.

Now assume that $f(A_t) \geqslant f(B_t)$ for all $1 \leqslant t < l$. Applying the augmentation property, there exists $b \in B_{t+1} \backslash A_t$ such that $A_t \cup \{b\} \in \mathcal{H}$. But then $f(a_{t+1}) \geqslant f(b)$, otherwise the greedy algorithm would have chosen b instead of a_{t+1}. We also have $f(b) \geqslant f(b_{t+1})$, since the elements of B are ordered by decreasing weight. Hence

$$
\begin{aligned}
f(A_{t+1}) &= f(A_t) + f(a_{t+1}) \\
&\geqslant f(A_t) + f(b_{t+1}) \\
&\geqslant f(B_t) + f(b_{t+1}) \\
&= f(B_{t+1}).
\end{aligned}
$$

Consequently, A is a hereditary set of maximum cardinality and maximum weight. ∎

Corollary 13.3.4. *All the hereditary sets of maximum weight have the same cardinality.*

Proof. Let A and B be two hereditary sets of maximum weight. As the weight is positive-valued, their cardinality is maximum. Suppose that $|A| < |B|$. By the augmentation property, there exists $b \in B \backslash A$ such that $A \cup \{b\} \in \mathcal{H}$, which contradicts the maximality of A. ∎

Corollary 13.3.5. *Let $\Gamma = (V; E)$ be a graph, and let $(U_\Gamma, \mathcal{H})$ be the associated graphic hereditary system; then $(U_\Gamma, \mathcal{H})$ is a matroid called the graphic matroid.*

Proof. In Chapter 2 we saw that Kruskal's algorithm gives a convenient solution. Thus, by applying Theorem 13.3.3 we obtain the result. ∎

Another matroid associated to a graph is the following. Let $\vec{\Gamma} = (V; \vec{E})$ be a digraph and let $\mathcal{H}$ denote the set of all subsets A of $\vec{E}$ such that no two arcs in A ever have the same terminal extremity; it is easy to check that $(\vec{E}, \mathcal{H})$ is a hereditary system.

Proposition 13.3.6. *The hereditary system $(\vec{E}, \mathcal{H})$ defined above is a matroid.*

Proof. A hereditary set of maximum weight can be constructed as follows: choose a vertex x whose entry degree satisfies $d^-(x) \neq 0$; the arc having this as a terminal vertex is then of maximum weight. It is easy to see that in the greedy algorithm then gives an optimal solution. ∎

Another interesting example of a matroid is the following. Let U be a set of n elements, and let $\mathcal{H}$ be the set of all subsets X of U such that $|X| \leqslant k$ for a fixed $1 \leqslant k \leqslant n$. Then $(U, \mathcal{H})$ is a matroid, called the *uniform matroid of degree* k. If $k = n$, this matroid is also called the *free matroid*.

Remark 13.3.7. The notion of a matroid was first introduced by Whitney [40] and Van der Waerden [34] in order to generalise the notion of linear independence of vectors. The generalisation of Kruskal's algorithm to matroids was given soon afterwards by D. Gale [14] and D. J. A. Welsh [37] in 1968. In this paragraph, we have preferred to give an algorithmic introduction to matroids starting directly with Kruskal's algorithm, a less theoretical approach.

Theorem 13.3.8. *Let U be a given finite set of vectors from V, and let $\mathcal{H}$ be the set of all subsets consisting of linearly independent vectors of U. Then $(U, \mathcal{H})$ is a matroid.*

Proof. It suffices to apply Lemma 13.3.2 (ii), which here is nothing other than Steinitz' exchange lemma on linearly independent systems in a vector space. ∎

13.4 Hypergraphs

The notion of a *hypergraph* is a generalisation of the notion of a graph; the term "hypergraph" was invented by C. Berge in the 1960s. In the context of hypergraphs, an edge no longer joins only one or two vertices, but an arbitrary number of vertices. Formally, a (finite) hypergraph H is simply a pair $(V; E)$ where $V = \{v_1, v_2, \dots, v_n\}$ is a non-empty set and $E = (e_i)_{i \in I}$ is a family of non-empty subsets of V. The elements of V are the *vertices*, and the elements of E are the *hyperedges*. A *simple* hypergraph is a hypergraph $H = (V; E)$ such that $e_i \subset e_j \Rightarrow i = j$. A hypergraph is *linear* if it is simple and if, for every pair of hyperedges e_i, e_j, $i \neq j$, we have $|e_i \cap e_j| \leqslant 1$.

13.4.1 Examples of hypergraphs

(1) At the athletic world championships, there are several events: the 100 meter race, the 200 meter race, the marathon, the shot put, the high jump and so forth. Suppose there are m events. Then we create a hypergraph as follows:

- the vertices are the athletes;
- the hyperedge e_i consists of the set of athletes entering the event i, for $1 \leqslant i \leqslant m$.

(2) Any matroid is a hypergraph.

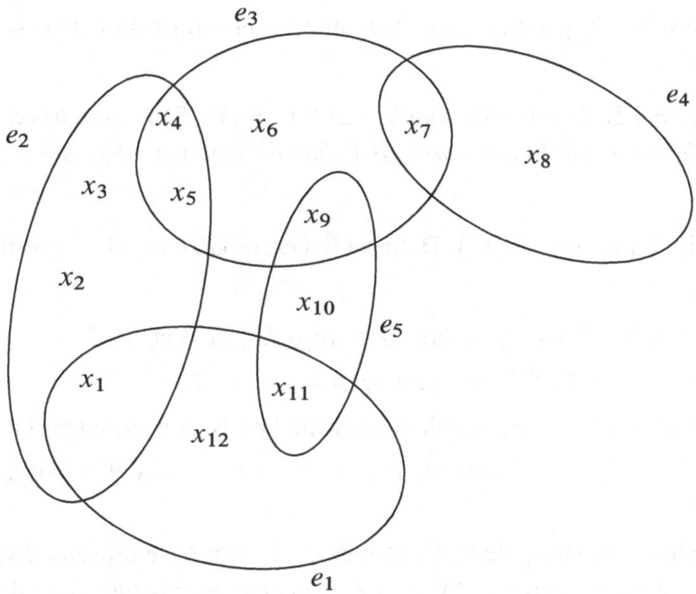

Figure 13.5. Representation of a hypergraph with 12 vertices and 5 hyperedges.

Let $H = (V; E)$ be a hypergraph with no isolated vertex, having n vertices and m hyperedges. The *incidence matrix* of H is the $n \times m$ matrix with coefficients 0 or 1 defined as follows:

$$A = (a_{i,j})_{(i,j)\in\{1,2,\ldots,n\}\times\{1,2,\ldots,m\}}$$

such that

$$\forall (i,j) \in [n] \times [m], \quad a_{i,j} = \begin{cases} 1 & \text{if } x_i \in e_j, \\ 0 & \text{otherwise.} \end{cases}$$

Observe that there is a one-to-one correspondence between hypergraphs and incidence matrices.

The transpose of the matrix A is the incidence matrix of a hypergraph (since H has no isolated vertex), which we denote by $H^* = (V^*; E^*)$. It is the *dual hypergraph*. In the dual hypergraph, the respective roles of vertices and hyperedges are exchanged. Note that the dual of an ordinary graph is a hypergraph which is often not a graph: for example, $K_{1,3}$.

Let $H = (V; E)$, $E = (e_i)_{i\in I}$, be a hypergraph with no isolated vertex. Just as for a graph, we can define the *line-graph* of H as the simple graph $(V_\ell; A_\ell)$ in which

- V_ℓ is the set I which indexes the hyperedges;
- $a = \{i, j\} \in A_\ell$ if $e_i \cap e_j \neq \emptyset$.

The following proposition gives the fundamental relation between graphs and hypergraphs.

Proposition 13.4.1. *Every finite simple graph* $\Gamma = (V; E)$ *is isomorphic to the line-graph of a hypergraph. Furthermore, if* Γ *has no isolated edge, the hypergraph is linear.*

Proof. Let $V = \{x_1, x_2, \dots, x_n\}$. Define a hypergraph $H = (W; A)$ as follows:

- $W = E$,
- A is the family $(A_i)_{1 \leqslant i \leqslant n}$ where $A_i = \{e \in E \mid x_i \in \varepsilon(e)\}$.

Let $\Gamma' = (V'; E')$ be the following line-graph:

- V' is the set $\{1, 2, \dots, n\}$ which indexes the family of hyperedges $(A_i)_{1 \leqslant i \leqslant n}$,
- $i, j \in V'$ are adjacent if and only if $A_i \cap A_j \neq \emptyset$, i.e. if and only if x_i and x_j are adjacent in Γ.

Then Γ' is clearly isomorphic to Γ. Moreover, H can have repeated hyperedges of cardinality 1: indeed when $e = \{x_i, x_j\} \in E$ is an isolated edge, then $A_i = A_j = e$ and H_ℓ is not simple. If Γ has no isolated edge, then $A_i \neq A_j$ for all $i \neq j$ and $A_i \cap A_j = \{i, j\}$ according to whether or not x_i and x_j are adjacent in Γ. This shows that H is linear. ■

Exercise 13.3. Let $\Gamma = K_{1,3}$.

(i) Show that Γ is not the line-graph of any graph.

(ii) Draw the hypergraph constructed in Proposition 13.4.1 of which Γ is the line-graph.

References

[1] N. Alon and J. H. Spencer, *The probabilistic method*. Second edn., Wiley Ser. Discrete Math. Optim., Wiley, New York, 2000

[2] C. Berge, *Graphs*. North-Holland Math. Libr. 6, North-Holland Publishing Co., Amsterdam, 1985

[3] N. Biggs, *Algebraic graph theory*. Cambridge Tracts in Math. 67, Cambridge University Press, London, 1974

[4] B. Bollobás, *Modern graph theory*. Grad. Texts in Math. 184, Springer, New York, 1998

[5] B. Bollobás, *Random graphs*. Second edn., Cambridge Stud. Adv. Math. 73, Cambridge University Press, Cambridge, 2001

[6] A. Bretto, A. Faisant, and F. Hennecart, *Éléments de théorie des graphes*. Collection IRIS, Springer, New York, 2012

[7] A. Bretto, A. Faisant, and F. Hennecart, *Éléments de théorie des graphes*. Édition revue et augmentée edn., Collection IRIS, Springer, New York, 2018

[8] T. Brylawski and J. Oxley, The Tutte polynomial and its applications. In *Matroid applications*, pp. 123–225, Encyclopedia Math. Appl. 40, Cambridge Univ. Press, Cambridge, 1992

[9] G. Chartrand and L. Lesniak, *Graphs & digraphs*. Fourth edn., Chapman & Hall/CRC, Boca Raton, FL, 2005

[10] R. Diestel, *Graph theory*. Third edn., Grad. Texts in Math. 173, Springer, Berlin, 2005

[11] P. G. Doyle and J. L. Snell, *Random walks and electric networks*. Carus Math. Monogr. 22, Mathematical Association of America, Washington, DC, 1984

[12] P. Erdős and A. Rényi, On random graphs. I. *Publ. Math. Debrecen* **6** (1959), 290–297

[13] L. Euler, Solutio problematis ad geometriam situs pertinentis. *Commentarii Academiae Scientiarum Imperialis Petropolitanae* **8** (1736), 128–140

[14] D. Gale, Optimal assignments in an ordered set: An application of matroid theory. *J. Combin. Theory* **4** (1968), 176–180

[15] M. R. Garey and D. S. Johnson, *Computers and intractability*. A Series of Books in the Mathematical Sciences, W. H. Freeman and Co., San Francisco, Calif., 1979

[16] E. N. Gilbert, Random graphs. *Ann. Math. Statist.* **30** (1959), 1141–1144

[17] E. N. Gilbert and H. O. Pollak, Steiner minimal trees. *SIAM J. Appl. Math.* **16** (1968), 1–29

[18] C. Godsil and G. Royle, *Algebraic graph theory*. Grad. Texts in Math. 207, Springer, New York, 2001

[19] J. L. Gross and T. W. Tucker, *Topological graph theory*. Wiley Ser. Discrete Math. Optim., Wiley, New York, 1987

[20] J. L. Gross and J. Yellen, *Graph theory and its applications*. Second edn., Discrete Math. Appl. (Boca Raton), Chapman & Hall/CRC, Boca Raton, FL, 2006

[21] F. Harary, *Graph theory*. Addison-Wesley Publishing Co., Reading, MA, 1969

[22] J. Köbler, U. Schöning, and J. Torán, *The graph isomorphism problem: its structural complexity*. Prog. Theor. Comput. Sci., Birkhäuser Boston, Inc., Boston, MA, 1993

[23] D. König, *Theorie der endlichen und unendlichen Graphen*. Kombinatorische Topologie der Streckenkomplexe; Mathematik und ihre Anwendung in Monographien und Lehrbüchern 16, Akademische Verlagsgesellschaft mbH, Leipzig, 1936

[24] D. König, *Theorie der endlichen und unendlichen Graphen*. Teubner-Arch. Math. 6, Teubner, Leipzig, 1986

[25] E. Kruja, J. Marks, A. Blair, and R. Waters, A short note on the history of graph drawing. In *Graph drawing (Vienna, 2001)*, pp. 272–286, Lecture Notes in Comput. Sci. 2265, Springer, Berlin, 2002

[26] J. Lauri and R. Scapellato, *Topics in graph automorphisms and reconstruction*. London Math. Soc. Stud. Texts 54, Cambridge University Press, Cambridge, 2003

[27] S. Lipschutz, *Theory and problems of general topology*. Schaum's Outline Series, McGraw-Hill, 1965

[28] B. Mohar and C. Thomassen, *Graphs on surfaces*. Johns Hopkins Stud. Math. Sci., Johns Hopkins University Press, Baltimore, MD, 2001

[29] M. R. Murty, Ramanujan graphs. *J. Ramanujan Math. Soc.* **18** (2003), no. 1, 33–52

[30] N. Robertson, D. Sanders, P. Seymour, and R. Thomas, The four-colour theorem. *J. Combin. Theory Ser. B* **70** (1997), no. 1, 2–44

[31] V. Rosta, Ramsey theory applications. *Electron. J. Combin.* **DS13** (2004), no. Dynamic Surveys, Paper No. DS13, 43

[32] W. T. Tutte, A contribution to the theory of chromatic polynomials. *Canad. J. Math.* **6** (1954), 80–91

[33] W. T. Tutte, *Graph theory*. Encyclopedia Math. Appl. 21, Cambridge University Press, Cambridge, 2001

[34] B. L. van der Waerden, *Moderne Algebra*. Springer, Berlin, 1940

[35] V. I. Voloshin, *Introduction to graph and hypergraph theory*. Nova Science Publishers, Inc., New York, 2009

[36] C. T. C. Wall, *A geometric introduction to topology*. Addison-Wesley Publishing Co., Reading, MA, 1972

[37] D. J. A. Welsh, Kruskal's theorem for matroids. *Proc. Cambridge Philos. Soc.* **64** (1968), 3–4

[38] D. B. West, *Introduction to graph theory*. Prentice Hall, Inc., Upper Saddle River, NJ, 1996

[39] H. Whitney, Congruent Graphs and the Connectivity of Graphs. *Amer. J. Math.* **54** (1932), no. 1, 150–168

[40] H. Whitney, On the Abstract Properties of Linear Dependence. *Amer. J. Math.* **57** (1935), no. 3, 509–533

[41] R. J. Wilson, *Introduction to graph theory*. Longman, Harlow, 1996

[42] R. J. Wilson and J. J. Watkins, *Graphs*. Wiley, New York, 1990

Index

Symbols used

$\gamma(\Gamma)$	genus of a graph Γ	185
$\gamma = \xi([0,1])$	simple curve	139
$\mathcal{H}(\mathbb{C}^n)$	set of Hermitian endomorphisms of $\mathbb{C}^n$	228
$h(\Gamma)$	isoperimetric number of a graph Γ	343
$h_w(\Gamma)$	Cheeger constant of a weighted graph Γ	440
$\mathrm{Hom}_G(G_1, G_2)$	set of morphisms of G-modules of G_1 to G_2	315
$\mathrm{Isom}_E\ (\Gamma_1,\Gamma_2)$	set of edge-isomorphisms from a graph Γ_1 to another graph Γ_2	301
$\mathrm{Im}(f)$	image of a linear map f or of a group morphism f	222
i and $t\colon \vec{E} \to V$	incidence functions of a digraph $(V;\vec{E})$	13
$\mathbb{1}_{\mathrm{P}}$	indicator function of a property P	55
$K^{m,n}$	set of matrices over a ring or a field K	223
K_n	complete graph of order n	9
$K_{1,n}$	star graph with $n+1$ vertices	33
$K_{n_1,n_2,\dots,n_k}$	complete k-partite graph	33
$\mathrm{Ker}(f)$	kernel of a linear map or of a group morphism f	222
$\kappa(\Gamma)$	vertex-connectedness of a graph Γ	109
$L(\Gamma)$	line-graph of a graph Γ	57
$L_2(n)$	lattice-graph	306
$\mathcal{L}(E,F)$	set of linear maps from E to F	222
$\mathfrak{L}(\Gamma)$	Laplacian of a multigraph Γ	333
$\lambda(\Gamma)$	edge-connectedness of a graph Γ	110
$\lambda^{\#}(\Gamma)$	second eigenvalue of a graph Γ	343
$\mathcal{M}_{m,n}(K)$	set of matrices over a ring or a field K	223
$\mu(\Gamma)$	girth of a graph Γ	11
$\mathbb{N}$	set of natural integers	2
$\mathcal{NP}$	non-deterministic polynomial complexity class	25
$\mathcal{NPC}$	non-deterministic polynomial complete complexity class	25
$\nu(\Gamma)$	nullity of a graph Γ	208
$\omega(\Gamma)$	maximal order of the cliques of a graph Γ	9
$\omega(f)$	total value of a flow f	123
$P(\Omega)$	set of subsets of Ω	390
$P_2(V)$	set of subsets of V with 1 or 2 elements	1
P_Γ	chromatic polynomial of a multigraph Γ	264
$\Phi(\Gamma)$	set of faces of a planar graph Γ	154
$\Phi(\Gamma, x)$	characteristic polynomial of a graph Γ	320
$\Phi(u,x)$	characteristic polynomial of an endomorphism u	227
$\mathcal{P}$	polynomial complexity class	25

$\mathring{P}$	interior of a polygonal line P	142
$P_k(\Omega)$	set of k-subsets of Ω	382
Q_k	k-cube or hypercube	35
$q(\Gamma)$	number of odd components of a graph Γ	277
$R(u, x)$	Rayleigh quotient for an Hermitian endomorphism u	229
$\mathbb{R}^k$	Euclidean space of dimension k	139
$\mathbb{R}_+$	set of nonnegative real numbers	40
$\mathcal{R} = (V; \vec{E}, s, p, c)$	network	121
$\mathrm{rank}_\Gamma(A)$	rank of $A \subset E(\Gamma)$	446
$\rho(\Gamma)$	rank of a graph Γ	208
$\mathrm{Sp}(\Gamma)$	spectrum of a graph Γ	320
$\mathrm{Sp}(u)$	spectrum of an endomorphism u	227
$\mathfrak{S}_V$	permutation group of a set V	297
$\mathfrak{S}_n$	permutations group of a finite set of cardinality n	297
$\sigma: V \sqcup E \to T$	drawing of a graph on a topological space T	139
$\sigma: \Gamma \to T$	embedding of a graph Γ in a topological space T	140
$T(n)$	triangular graph	306
$\theta(\Gamma)$	index of partition into cliques of a graph Γ	258
${}^tA$	transpose of a matrix A	223
$\tau: \Gamma \to \Gamma'$	planar isomorphism between two planar graphs Γ and Γ'	169
$\mathrm{tr}(A)$	trace of a square matrix A	328
u^*	adjoint endomorphism of u	227
$V(\Gamma)$	set of vertices of a graph Γ	2
$\mathrm{Vect}\{\mathcal{F}\}$	vector subspace generated by $\mathcal{F}$	324
$V(\vec{\Gamma})$	set of vertices of a digraph $\vec{\Gamma}$	12
$\chi'(\Gamma)$	chromatic index of a graph Γ	247
$\chi'_\ell(\Gamma)$	list chromatic index of a graph Γ	261
$\chi(\Gamma)$	chromatic number of a graph Γ	231
$\chi_\ell(\Gamma)$	list chromatic number of a graph Γ	261